Building Your Own House

Building Your Own House

Everything You Need to Know
About Home Construction
From Start to Finish

Part I: Exteriors
Part II: Interiors

by Robert Roskind

Ten Speed Press
BERKELEY TORONTO

Ten Speed Press
Box 7123
Berkeley, California 94707
www.tenspeed.com

Distributed in Australia by Simon & Schuster Australia, in Canada by Ten Speed Press Canada, in New Zealand by Southern Publishers Group, in South Africa by Real Books, in Southeast Asia by Berkeley Books, and in the United Kingdom and Europe by Airlift Books.

This material was previously published in *Building Your Own House: From Foundations to Framing* (Ten Speed Press, ISBN 0-89815-110-4, copyright © 1984), and *Building Your Own House II: Interiors* (Ten Speed Press, ISBN 0-89815-358-1, copyright © 1991).

Portions of the material in Part II were previously published in *The Do It Yourself Show Book of Home Improvements* (Addison-Wesley Publishing Company, Inc., © 1985). Reprinted by permission.

The author and publisher have made every effort to ensure that the information in this book fosters safety as well as efficiency. Because the specifics of each environment, as well as the design, materials, and uses of each building project vary greatly, the author and the publisher cannot be held responsible for injuries that might result from the use of this book.

Cover design by Catherine Jacobes
Book design by Hal Hershey

Part I
Illustrations by Bill Schaeffer, Marilyn Hill, Rik Olson, and Dan Corvello
Technical editing and additional writing (Doors and Trim, Stairs) by Steven George
Additional writing (Plumbing) by Peter Hemp
Developmental and production editing by Jessie Wood
Copyediting by Editcetera
Proofreading by Fuzzy Randall

Part II
Illustrations by Valerie Sharpe
Plans by Mark Robert Thomas and Kathleen Rousseau
Drawings on pages 13 and 54 by Jon Larson

Library of Congress Cataloging-in-Publication Data on file with publisher.

Printed in Canada

First printing this edition, 2000

1 2 3 4 5 6 7 8 9 10 — 05 04 03 02 01 00

To my wife, Julia, and my daughters,
Julie and Alicia, thanks for all the love
and all the laughs.

Contents

Acknowledgments

A book like this, with its numerous illustrations and its need for vast technical input, represents the merging of the energies of many people—more than I could possibly thank here.

Everyone who worked on this project put his or her heart into it. To all of those people, to the many I've met personally, and to those I've not yet had an opportunity to meet, I offer my heartfelt thanks. In particular:

To Valerie Sharp, for taking my scraps of paper and turning them into beautiful diagrams.

To Tom Cuthbertson, whose support and editing assistance helped keep me sane.

To Jon Larson, for the friendly illustrations.

To Kathy Rousseau and Mark Robert Thomas, for all the plan diagrams.

To Hal Hershey, the book designer, who smiled no matter what (and meant it).

To George and Phil, of Ten Speed Press, for all the patience as I fell farther and father behind my deadline.

To Maria Hirano, my editor, for her refinement of my mess.

To Kal Stein and Jim Phoenix, for the articles in the introduction.

To Blair Abee, for being my partner through so many beginnings and endings.

To Julia Holiman, for research support, pressing details, and hours of listening to me talk.

To Marnie Poirier, writer, compiler, organizer, and the person who kept expanding to hold it all.

To Steven George, who wrote the chapters on doors and windows and on stairs, and whose technical editing enhanced the whole book.

To Peter Hemp, who wrote the chapter on plumbing.

To Bill Shaeffer, whose beautiful illustrations speak for themselves.

To Marilyn Hill, illustrator, whose cooperative spirit and attention to detail make her a pleasure to work with.

To illustrator Rik Olson and consultant Dan Corvello, whose expertise added so much to the chapter on plumbing.

To Sal Glynn of Ten Speed Press for his direction and guidance.

To Jessie Wood, developmental and production editor, who continued to impose order on the chaos.

Building Your Own House

Part I: Exteriors

Contents

Introduction

OVER TWENTY-FOUR YEARS AGO, I began teaching people how to build their own homes. In 1979 I co-founded The Owner-Builder Center in Berkeley, California, the country's largest owner-builder school teaching people to do it themselves. In the beginning, I had some real reservations about teaching, and thereby encouraging, people to undertake a project as demanding as building their own home. Should people be encouraged? Was it a sane goal to attempt? Would graduates return in failure scolding us for ever having encouraged them? Although these questions haunted me, until we tried it, there was no way to know.

Since that time, hundreds of our graduates have built their own homes. Many of these people literally never held a tool before taking our courses. Instead of abuse from frustrated failures, we received praise from successful novices. My reservations placated, I have thrown myself into this work. I now know that people with little or no previous experience can successfully complete a major building or remodeling project.

Figure I-1. One story passive solar home. 1470 square feet, 3 bedrooms, 2 baths. Designer: David Norton, Habitat Construction, Nevada City, California.

Figure I-2. Victorian ranch house—a challenging project for a novice builder. 2670 square feet, 3 bedrooms, 3 baths. Designer: Richard Drace, Homestead Design, Nevada City, California.

Even with these successes, we have never tried to paint a rosy picture of the tasks involved. That would be irresponsible and do much harm. In all our courses and in this book, it has been our attempt to tell you of the frustrations as well as the joys; the ants as well as the picnic. The decision to build, contract or remodel your own home is a major one and should be well thought out. It ranks up there with moving, changing jobs, marriage and major investments. Unfortunately, many novice builders do not realize this until it is too late. Building a house for a novice cannot easily be compared to building a house with a team of 30 to 40 professional tradespeople (plumbers, electricians, framers, etc.).

Owner-built homes usually take from 1 to 3 years to complete. They require a stable marriage or partner relationship of the owner builder. An attitude of mutual support and understanding of the people involved is critical.

If you are planning to build, contract or remodel your own home, there are many factors that need to be wisely planned for the whole project to be a success. Buying land or a lot; choosing the site; planning water and waste systems, a source of power, and a driveway, road and bridge; obtaining permits; being aware of codes and inspections; getting insurance and financing; and determining whether you have what it takes to build your own house are all factors to consider before getting started. Since these issues have been dealt with in our book *Before You Build* (Ten Speed Press, 1981), they will not be covered in this book. Be sure you have given

ample thought to these factors before you begin to build.

There are three other areas, however, that are very important which novice builders can easily overlook. These are timing, design and estimating. They are the "soft" areas as opposed to the "hard" areas, such as building your foundation or roof. The rest of this book is about the "hard" areas, so let's look at these "soft" areas before beginning.

Timing

For years many banks have been reluctant to lend owner builders construction money. The reasons are several. The most obvious is that they have no way of evaluating the quality of the house you are going to build with their money.

Many bankers will also tell you that the rate of divorce is great with owner builders and this can endanger the entire project (and their money). For the most part I would have to agree with this statement. In Ken Kern's recent book *The Work Book: The Politics of Building Your Own Home* (Owner Builder Publications), he describes many case histories that have ended in broken partnerships. Fortunately, few of our graduates' marriages have ended in divorce. This is not, however, because they are more emotionally stable than others, but because we have always taken great pains to impress upon our students not to begin a project of this size unless the partnership is working well. The project can serve as a wedge if there are any cracks in the relationship.

Some partners enter this type of project with a "This is your project and don't expect me to help very much" attitude. One partner really wants to do it and the other is somewhat reluctant. Be careful of beginning the project in this case. Building or remodeling your own home is too large a project for it not to be continually spilling over into the entire family's life. I am not saying that one person cannot handle most of the responsibilities and take the lead. But that person will need the understanding and total support of a family who understands there will be sacrifices for them no matter what the preconstruction promises.

All this relates to timing. When, if ever, is the best time of your life to start such a project? If there are major demands in your life already (family, job, personal, financial, etc.), it may be best to put this off until they are handled. There may be no major stress areas in your life and you think it is time to begin, but maybe you are already operating at maximum efficiency. Add one thing and the whole house of cards falls. Be sure you have made enough extra room in your life for this type of project and that this is what you really want to do.

I remember a man in one of my earlier classes in Raleigh, North Carolina. For seven years he had hung wallpaper to save enough money to build a house for himself and his family. He was now ready to start. He was one of my most eager students, asking lots of questions and taking lots of notes. A year later, I passed him on a street in Chapel Hill, North Carolina, and asked him how his house was going. Much to my surprise, he told me he had decided not to build but to enter veterinarian school instead. His explanation made perfect sense. He had needed a challenge in his life (wallpapering only has so many challenges) and had decided to build his own house. After taking the class, he realized that he could build a house without great difficulty. He was then able to consider a greater challenge that he had been fearful of attempting: changing his profession and becoming a vet. His timing was perfect for becoming a vet, but not for building his own house. Ask yourself if this is absolutely what you are ready to do. What else could you accomplish with the same time, energy and money? Be honest with yourself.

Remember, we live in a society that is so rushed it has forgotten where it is going and why. A number of forces will come to bear on your decision, when to begin your home. You will think, the prices are going up, all the land is running out, if not now, then when, etc., etc. Stay evenly paced and know that when the time is right, you will know it. Wait for that knowing before beginning.

Design

Design is by far the most important single process of the entire project. Along with good workmanship and sound engineering, design is a keynote in the building of a successful house.

Figure 1-3. Two-story passive solar home. 1558 square feet, 3 bedrooms, 2½ baths. Designer: Lee Aaron Ward, Solar Solutions, Napa, California.

If it is designed poorly to begin with, no amount of expert craftsmanship can correct the mistakes that are written into the design. Before you ever start to build, consider this point well. A successful design is one that is the most supportive of the lives of you and your family, a place that feels like a loving home, a basis from which to grow and confidently confront the rest of the world. This is the basic function of a home. The farther away from that concept you get, the less successful your house will be. Most homes in America are usually built with one concept in mind—profit. If they are designed by an architect, they are often built to satisfy his artistic whim, a piece of sculpture to be placed in the woods, rather than a home for a family to live in. Frank Lloyd Wright once said that he could design a house that could cause a couple to have a divorce; to a certain extent that has been done in most modern American homes.

Let the design reflect you. Let this be your home. Imagine your ideal dwelling, just as you want it, and then build it. Unless your fantasy has always been a certain style home, let your mind run free, and find the house that is you. Be aware of your abilities, but do not be afraid to extend yourself. The object is not to extend your capabilities beyond yourself, but yourself beyond your capabilities. This is growth, this is moving one step further than where you are. Anything less is complacency or fear. Originality and character can be incorporated in any design, and still have that design remain simple.

Design must be seen as a dynamic, not static, process. It is not something that is drawn up in the beginning and then rigidly held to till the end. It is a process of becoming, a growth process. In the beginning, when you are working on design, do not accept your first draft. It is the one that comes at the beginning of your thought processes, and as you get deeper and deeper into them your design changes, as more and better ways become clear. After much thought and revision, form in your mind and on paper the design you plan to begin with. I say begin with because it is your basic plan, but should always remain open to change. As your construction begins, you will learn things about your house that you had no way of knowing about beforehand. This is one of the great advantages of the designer being the builder as well. The design needs to be flexible enough to incorporate this new information. By moving a window two feet on a wall, you get a better view of a rock formation you had not noticed before; by removing a wall, an upstairs storage room becomes a sleeping loft; by moving the stairs, you free more usable space. Ideas will flood you when you actually start to build and spend many hours of the day working and thinking about the house. Any good design should allow for changes to be incorporated. Always be looking to see how the design can be improved upon.

Design your home from the inside out. It is a good idea to spend time in other people's homes to get an idea of how the structure feels to live in. When you are designing, imagine

Figure I–4. "The Carlyle," a three-level home with 10-foot ceilings in the main living area. 1856 square feet, 3 bedrooms, 2½ baths. Designer: Cazwell, Inc., Chapel Hill, North Carolina.

Figure 1-5. Passive solar contemporary design with integrated solar greenhouse. 1400 square feet, 2 bedrooms, 2 baths. Designer: Mike Funderberh, Raleigh, North Carolina.

your inner living spaces and how you want them to be. Then tie all these parts into an integral whole with a pleasing exterior design. I am not suggesting that you ignore completely the aesthetic value of the exterior design, but only that you remember the primary function of the house is to live in, not to look at. Imagine in your mind each room. What kind of a feeling do you want in the room, where and how high do you want the windows placed, which rooms do you want adjacent to each other, which separate, what type of dividers and entrances between these rooms, what type of relationship and access to the outside, how high do you want the ceilings, etc. Do this for each room and the house as a whole. Once you have designed the entire house, live in it awhile in your mind. Work, play, entertain, cook and relax in that house; imagine all its functions. Then in your mind, go outside and see it on your site, and if all this pleases you, you have a good design to start with.

Design your total environment, not just the house. Remember that your living space extends past the walls of your house onto your land outside and the design should incorporate this concept. The relationship between your indoor and outdoor environment is a very sensitive one and should be developed with care and consideration. This should be done from the very beginning by blending your site with the land. This can often be a rather difficult task. It is possible to establish a union between your house and your land in such a way that the house seems to blend with the lot, almost grow from it, rather than having been imposed on it. Your land will

probably never be more beautiful than before you touch it; treat the land as gently and harmoniously as possible. The type of land, the terrain, the height of the surrounding trees and hills should all affect the design. Perhaps you might sit on your site for a few days and try to imagine your house growing up from the ground. Try to make the lines of the house merge with the lines of your land. If it is a very steep lot with high trees, the lines might need to be more vertical. If it is a slightly sloping or level lot, horizontal lines might be better.

Use your porches, verandas, entranceways, doors and windows as transitions between the indoors and the outdoors. Windows and especially doors opening to the outside or to a porch tend to bring the outdoors into your house and connect the two. Consider this as you design each room and the house as a whole. Porches are very inexpensive to build per square foot and serve as a very beautiful and usable living space, especially an open covered porch. Having an outdoor entrance into every living or sleeping area is also a good fire safety technique, which is of more importance in wooden structures that have fireplaces or are heated by wood, or are far away from the closest fire station.

Consider well the size of the rooms and size of the entire house in your design. Thoreau once said that people are rich in proportion to what they can do without. Something to consider when you are pondering what size your house should be. There are really few guidelines I can offer here, for only you can determine what size house and rooms you need. I can only suggest that you consider it well. The larger

Figure I-6. "The Bedford," a three-level home. 1875 square feet, 3 bedrooms, 2 baths. Designer: Cazwell, Inc., Chapel Hill, North Carolina.

the house, the greater the time and expense, both in construction and in upkeep. Yet, you still need enough space for all your indoor activities and so that you won't feel closed in. Everyone living in the house will need to have enough of their own private space in order to live together compatibly.

Often it is not how much space you have, but rather how the space is put together that is important. The total design and placement of rooms has a large effect on the volume of space enclosed. A well-designed 750-square-foot house can often feel larger than one twice that size which is poorly designed. Some people like all rooms opening to a large central living area, tying the whole house and the activities within together at all times. Others like the house more spread out and the rooms far away from each other, offering greater privacy and isolation of activities. Consider all these ideas well and take into account any additions or changes in the size or needs of the family. Children can wear thin during a long winter, and it is often a good idea to give them a playroom of their own at the far end of the house or build a separate playhouse for them.

You might want to design your house so that it can be built in stages or added on to as the need arises or as you acquire more time and money. Building a series of buildings, all tied together by walkways or courtyards, is another alternative. This further separates the activities and affords greater privacy and quiet. It also enables you to finish the most important structures first and move into them as you work on the others. Your work will improve with each building and

you can work out some of your mistakes on the smaller buildings before you begin with the main structures. The buildings can be joined in a whole indoor-outdoor living environment and afford much privacy and space. Just building your workshop or guest house or bedroom some distance from the main house can make a big difference.

The height of your ceilings and the number, size and placement of your windows and doors have a lot to do with your feeling of space. A relatively small room with the right number and location of windows and perhaps a skylight can seem much bigger than it actually is. Lofts are also effective where you want high or cathedral ceilings, yet want to use part of that space for a living area. Be careful of very large rooms with cathedral ceilings. They are often very hard to heat and they give some people the feeling of being so lost in the room's volume that it becomes hard to relax and feel intimate in the room. Again, it will be up to your personal tastes as to what feels right for you. There is a very subtle balance that needs to be achieved: large enough to feel spacious and unconfining, yet small enough to feel warm and comfortable.

A free-flowing, "open plan" type house is much healthier to live in than a series of boxes within a large box. Most houses built in this country are a series of tightly-enclosed small boxes arranged in a very orderly, linear fashion within another very orderly, linear box, grouped with other boxes in a very orderly manner on a very linear street. This is one reason we are all so crazy. The only person who ever thought that this was a "natural" way for man to live was probably the contractor, who makes more money by keeping costs down, since houses go up quickly in this manner. With labor costs being what they are, it would be hard to hire someone to build almost anything other than boxes, but sloping ceilings and off-square rooms are not out of the question. It is often not a matter of finances, but rather consciousness, that determines the design. A house whose living and activity spaces flow into one another with entranceways and movable panels rather than fixed walls and small doorways not only gives the occupant more of a sense of space, but makes the house more adaptable to the changes that future needs might demand. Allowing the house and its activities to flow together, rather than having distinct separation and divisions, will add continuity and order to your life in general.

Build with respect for the land. Perhaps for millions of years your land has been untouched by humans and it is beautiful and pure. For 30 to 60 years you will live on your land, so consider well what kind of impact your brief stay will have on that land. It is not really "your land," for we can never really own that land. We only have the privilege to use it as we will for a period of time. Before us and after us will come many other caretakers. It is possible for one man in his lifetime to destroy the aesthetic quality and usability of a piece of land for many generations to come. We are seeing this done all around us and for many of us, it is one of the many reasons we want to retreat into the shelter of our own home. Treat your land with love and respect, and appreciate the gift it is giving to you, with its beauty, its quiet, and its love. Design

and build with this in mind, and your land will reward you a hundredfold. Try to use as few non-decomposing materials as possible, plastic, fiberglass, asbestos, etc. Use instead materials that are only recently from the earth, and ready to return in a natural manner, such as wood, stone, etc. These items are usually much cheaper than manufactured items and have much more warmth and character than the mass-produced "dead" materials. If you have access to any used materials, make use of them. They are usually cheaper and often of higher quality than new materials, especially wood, since the wood will be drier than new green stock. But more important than this, in the recycling of used materials you will be helping the earth and all the people on it by not wasting more of our resources.

Let the function of your house determine its form. Be clear of your intent in building. Decide in which way exactly you want this house to serve you, and then design it and build it to do that for you. This is one of the hardest things to do sometimes — to determine how you want the house to function for you. Often our expectations and our fantasies overpower what it is that we really want. Often we follow a fantasy, whether it is in a design of a house or another situation in our life, only to discover the fantasy did not give the rewards we had hoped for and what we really wanted was something quite different. So try to determine what it is exactly that you want this house to do. If you want it to impress your friends and neighbors, build it that way. Maybe you want it to provide a warm, loving center for your family, one that will aid in

Figure I-7. Two-story passive solar home. 1680 square feet, 2 bedrooms, 1½ baths. Designer: David Wright, SEA group, Nevada City, California.

their communication and closeness, or perhaps just a small private place for you, or perhaps you and a mate. Maybe you want something that you can change over the years as you and your lifestyle change. Whatever that thing is for you, make it clear to yourself and then design and build with that intent in mind. Some may say that this plan is not always practical, efficient or functional — that people must use their intellects and build according to what their minds say is best, disregarding things of the heart and emotional centers. I think what is missing here is what is practical; what is most efficient for life on this planet, both for the individual and the species as a whole, is whatever aids a person to becoming a happy, secure, loving individual. *The most practical home for you is the one that feels most like you, the one that brings you the most peace of mind.* This is the ultimate criterion for what is reasonable and practical. If it feels right to you and it does not get in the way of someone else pursuing what feels right to them, *do it*. And if that thing is for you a house shaped like a bear wearing a petticoat and sitting full lotus on top of a replica of Charles de Gaulle's nose, then build it. *And* accept the responsibility for what that entails, both in building it and reselling it. This same idea applies not only to the house as a whole, but to each individual space or room also. Again, determine what you want that room to do for you. Do you want it to be open and expansive, with lots of space to serve the more social, extroverted areas of your nature, or do you want it warm and intimate, more like a cosy cave to serve the more private, introverted parts of your nature? If it is a

child's room, let it serve the child, at the age he is now, with adaptability as he grows. Determine its function, then determine its form.

Try to be realistic as to how much time, energy and money your particular design will take and see if there are some acceptable changes you could make if you feel the design is beyond your capacity. Needless to say, but here I am saying it, the more complex the design, the longer and harder it will be to build. Often a house can be pleasing and have character without being very complex. If you do decide to utilize a complex design, be aware of the extra time and energy required. The materials used to do custom work often take much longer to assemble than prefab components. They are usually individual boards and stones, rather than large preassembled products. They cost much less, but take more time. In this case, I feel it is definitely worth the extra time to be able to use "live" materials. It can make a difference in your life for as long as you live in your house.

Design the house so that the rooms have a relationship to each other that is determined by the activity in that room and the rooms close to it. Once you have determined what the function of a room is, locate it in the house in such a way that the activities in the rooms adjacent to it will not conflict, and if possible, design it so that adjacent activities will support each other. You might not want to locate your study next to the children's play area, but rather next to the courtyard or greenhouse which might be pleasing to you.

Figure I-8. Two-story passive solar design. 2200 square feet, 2 bedrooms, 3 baths. Designer: Richard Drace, Homestead Design, Grass Valley, California.

Figure I-9. One-and-a-half-story passive solar home with an integrated solar greenhouse. 1728 square feet, 3 bedrooms, 2 baths. Designer: Shannon T. St. John, Horizon Design, Anchorage, Alaska.

Estimating

The home which actually costs less than its estimate is so rare that it ascends into the realm of miracles. Houses go over their estimated costs so often, that doing so falls into the category of the inevitable. The consequences of housebuilding costs exceeding their planned amounts can be ruinous, and homeowners may face an additional several hundred dollars a month in loan or mortgage payments. Such a long-term burden can put an unhappy dent in your ability to realize other fantasies like vacations, a new car or boat, and travel, for twenty years or more. Before you take this risk, consider some things which may put you ahead in the art of estimating.

In estimating housebuilding costs, our fallibility usually falls in two areas. First, we try to incorporate more features in our designs than we can really afford. Second, adequate account is not made for unknowns in the job; the hidden costs.

For those of us that feel we are putting up the only house we will ever get to build, and one we may live in for many years, our desires to have everything we want will tempt us with peculiar potency. Cutting back on features we have dreamed of for years may appall us. Changes along the way that seem small may add up to a large increase in price. The tendency is to choose designs which are larger, more complex or more expensive than that which can actually be afforded.

It pays to remember the scale of your investment. In planning a house, we make decisions involving thousands or ten-thousands of dollars. Most of us lack familiarity with spending these amounts, and our perspectives can be lost and regained too late. Give yourself a healthy margin of estimating. Allow some decisions to be made later in the process; as construction moves along, final costs become more predictable. Then may be the time to decide on an extra wing, the types of furnishings and fixtures you will put in, or how the garage will be. Always try to remain conscious of how much money you can afford to spend and of how much house that money will buy.

Now, about those aspects of the art of estimation which make it tricky even for a pro. (The contracting business casualty rate is one of the highest, and it is poor estimating which often boots otherwise competent people out the door.) It must be said that very accurate estimates can be made on conventional or tract homes. Most of these homes are designed to use prefabricated components such as plywood, paneling, sheetrock and cabinets. These go up with a predictable speed and energy input. Also, since similar houses go up regularly, there are relatively few unknowns. Work can progress smoothly and costs may be calculated based upon square footage. Any knowledgeable person in the housebuilding industry in your area should be able to give you a reliable cost factor to multiply by square footage.

CUSTOM BUILDING, on the other hand, challenges any

Figure I-10. Passive solar two-story house. 1745 square feet, 3 bedrooms, 2 baths. Designer: Curt Burbick, Berkeley, California.

estimator. Many builders will operate only on a cost-plus basis. Any custom house is unique, a prototype. One like it has never been built before and the problems are unknown. These houses are very labor-intensive, which is also very expensive. I am not recommending that you forget custom building, for to my mind custom homes often possess more life and character than conventional ones. However, be aware of their added costs!

NON-HOUSE EXPENSES such as wells, roads, septic systems, water storage and filtering systems, and getting power to the land often overtip the cost of construction even when the basic estimate of the structure itself remains sound. Remember that there are often legal fees required, and custom building may necessitate costly structural engineering fees, as well.

INFLATION often leads to underestimation. Housebuilding costs for materials and labor may rise 5 to 10% just in the 6 months between the drawings and calculation of the estimate at the start of the work. Similar increases can occur during the construction period. Building materials can jump 10 to 30% overnight. This summer I saw cedar shakes go from $63 to $80 a square in a month, a 26% increase. Allow margins for these raises.

SITE AND LOCATION of a house can bear strongly on its cost. Costs rise with difficult or inaccessible sites. Foundations may need to be more complex on steep slopes, and construction is generally more difficult. Locations remote from hardware and building supply stores make the job more time consuming for the contractor, carpenters, and other workmen, and add to costs. Delays and added expense in the delivery of materials can frequently occur in remote areas. Also, power, roads and water in such instances can become prohibitively expensive.

INEFFECTIVE ORGANIZATION can be costly. The entire project requires a considerable amount of coordination between people, and a proper progression of processes. Poor organization results in less production. If contractors and workmen slip even 10 to 20% from maximum efficiency, it is liable to mean wasting thousands of your dollars. Try to insure that all the people involved work together well and are interested in providing maximum productivity for minimum time and effort.

Remember, housebuilding is a many-faceted undertaking. Estimating costs is a key process and can be a difficult one. You have a good headstart if you avoid some of the pitfalls I've been discussing. Watch for sudden or hidden costs. Remember there is a tendency in each of us at times to be somewhat too optimistic. We estimate on the low side, in an effort to convince ourselves that our dream house can be all that we want it to be. If you are doing that, remind yourself to take care and be cautious with your dollars. A successful project is one where you get the house that you really want, built for the price you can afford, built at a time when you are ready for the project!

Except for Figures I-4 and I-6, the drawings in this section are from *The Owner Builder Plan Book*, available from the Owner Builder Center, Berkeley, California.

1 Before You Build

IF YOU HAVE DONE all the preliminaries: if you have purchased the land and chosen a site; if you have made decisions on your water and waste systems and source of power; planned a driveway, access road and bridges; if you have dealt with permits, codes and inspections; and if you are sure about timing, design and estimating, you are ready to begin, well almost. In this chapter we will be providing additional information you will need before you build, such as when to hire or contract, insurance, purchasing, tips on lumber, work habits and attitudes, and how to use this book. Chapters 2, 3, and 5 are devoted to describing the different types of construction (2), tools (3) and types of foundations (5).

Included at the end of this chapter is a construction schedule showing the 60 steps necessary to build a house. The first 14 steps are in *Before You Build* (Ten Speed Press, 1981). Steps 15 through 33 are contained in this volume and Steps 34 to 60 are in the sequel to this volume.

In addition, there is a set of safety rules at the back which we have found essential to insure the safety of novice builders at the Center. At times they may seem excessive, but better 20% more cautious than 1% too lax. Both of these pages can be copied and placed at the work site for your convenience and safety.

When to Hire or Contract

You probably have chosen to build for a variety of reasons: you wanted to save money, you wanted to know what went into the construction of your home, you wanted to enjoy the feeling of accomplishment at having mastered the necessary skills. You may, however, want or need to hire professional contractors to help you with certain phases of construction. Though this can save time and money, hiring from a position of ignorance can present problems. Learn as much as you can about the building process, especially the phases for which you hire, so that you know what the job is worth and have a basis on which to judge the work done for you.

In order to decide when to hire a contractor, you will want to evaluate your own areas of expertise and where you lack confidence. Often, much of the anxiety of the entire project is directed toward a particular area, such as the foundation work or the electrical and plumbing systems. Sheetrock taping, tile work, cabinetry and finishing work are areas where you may feel you need the expertise of a professional. These are all areas of work that you can master if you have the time to educate yourself. What you need to do is sit down and plan the project, this time from the point of view of the actual knowledge and skills required to complete the job. Figure out where you need help and consult a professional.

You will want to begin this evaluation process at the design stage. You may wish to hire a designer or architect to help you shape your ideas into a workable plan. Perhaps you have drawn up a complete set of plans or know of a home you would like to duplicate. In this case, you may only need a draftsman. There is no law that says you cannot draft the plans yourself, but you may want someone with a designer's eye to check over your plans to see that you have not missed some basic flaw in the structure or a subtle problem, such as traffic flow. Money can be saved by getting advice from a professional at this stage.

The next step is to decide how much of the work you want to do yourself. For those owner builders who plan to use contractors for much of the work, there will be a fee, but contractors usually will not sign the plans. You will not be safeguarded the way you would be if you hired them to handle the entire job. Many contractors can be hired on a consulting basis. If you feel that you will need expertise at some point, try to figure out when: before you begin a particular phase, or after you have reached a point beyond which you can not proceed. Remember, many contractors will not want to finish work you have begun. They cannot be responsible for problems that arise from your mistakes, and it is difficult to troubleshoot when they don't really know what went into the work.

If you plan to hire a contractor to do a particular phase of the work, look around; talk to as many contractors as you can. There is a difference between a licensed professional and one that is unlicensed. A professional who is licensed can sign a legal contract and be taken to court if problems arise. One that is unlicensed is essentially your employee, and your only recourse is to fire them. The unlicensed tradesperson may be cheaper and just as skilled, knowledgeable and experienced, but you will have to cover them with workman's compensation insurance. Licensed people carry their own workman's compensation and build the cost into the bid. It all depends on your own willingness to carry the brunt of the responsibility.

Hiring a professional can be a learning experience in itself. Some contractors will be willing for you to work along with them as an assistant. Many, however, prefer to work completely on their own. The relationship between you and them can become confused if you hire them and yet you are their assistant. Who is the boss?

You may prefer to hire small contractors, even if they are not well known. A big contractor may have several projects underway at one time and their attention may be split. They may not be on your job all the time or even every day. Smaller contractors have less overhead, making the job cheaper. On the other hand, larger contractors are able to purchase materials at a better price because they are buying in greater bulk. You have to ask questions.

When you approach a contractor, make a list, not only of the tasks you want performed, but also of your fears and concerns about the job. Spell everything out, from when you expect the job to be completed to who will do the cleanup. Deal with as many of the details as you can. It is the gray area that will give rise to problems down the road.

There are a couple of other alternatives. You can hire individuals who hold down construction jobs to work for you on weekends or evenings. You can hire an unlicensed person who works for a licensed contractor. You could actually get the same worker you would get if you hired the contractor and probably for less money. You can go around to construction sites and watch the workers. If you see one particular person who seems to be working extra hard and/or well, you may be able to hire them during their off-hours. Problems that can arise here are that they may not be able to give you as much time as you need when you need it, or that they will suddenly be asked to work overtime for their main employer and you would be left out in the cold. If you really know what you are doing, you may just need to hire a go-fer, someone to hand you planks and nails.

As for inspecting the quality of your work, some people rely on the official inspectors. Be careful about this. Inspectors inspect for health and safety, not for quality. They may not be as regular about coming around to your job site as you might expect or need. If you are doing your own work and feel the need for someone to check what you have done, you can hire a contractor on a consulting basis. Don't take chances if you are in doubt, but don't think that you can't do it yourself. You are not alone out there. There are people who can and will help you.

Insurance[1]

It is easy for owner builders, being a bolder element of the population to begin with, to overlook the need for certain types of insurance while building. DON'T! Construction, by nature, is a dangerous trade. Building or contracting your own house takes a lot of optimism and trust. Healthy caution and insurance help fill out the package. Many lenders will require that you have certain types of coverage to safeguard their investments. Some types of coverage are optional, but definitely recommended.

Workman's compensation insurance covers people injured while working on your home. It not only pays for medical expenses, but for lost work and recuperation time as well. In many states it is mandatory and you cannot get your permit until you present proof that you have taken out a policy on the project. In any case, it is wise to have coverage for your workmen, yourself, your family and friends. One injury can cost a fortune in hospital and doctor bills and lost work. If you have workman's compensation for everyone working on the site, both paid and unpaid, you are protecting them in the event that something happens. If you do not have this insurance and something does happen, there is a good chance that you will be sued and lose, since it is negligence on your part not to have work people covered by workman's compensation.

Aside from protecting your workers, it is wise to cover the job site with liability in case anyone who is not a worker should get hurt. This is a blanket policy that covers everyone from delivery people to children who wander onto the job site. It is worth the money.

The type of property insurance you need is called a "course of construction" policy. Depending on what you need and how you and your agent write the policy, you can be covered for fire, theft and vandalism during the construction process. It is possible to have supplies and materials covered before they are actually attached to the structure. Your agent should be able to help you decide just exactly what kind of coverage you need.

If you are borrowing money to build the house yourself, much depends on you. Because of this, lenders may require life insurance. Should you die, they would then have enough money to hire people to finish the project.

You may want to cover yourself by both injury and life insurance. If something happens to you, money will be available to finish the project or help you if you are injured.

If you are remodeling your home, check with your insurance agent to see if you need to extend your current coverage. If you are working at a site that is removed from your home or other insured property, check with your agent to see if he can offer you the appropriate policies for the construction. Be sure that you talk to an agent that is knowledgeable in the field of construction insurance.

An investment in prevention can save time and money in the long run.

---

1. Based on *Before You Build*, by Robert Roskind (Ten Speed Press, 1981).

Purchasing

When you are planning major purchases for building, an organized approach pursued in a disciplined and professional manner can save both time and money. It is helpful to think of materials purchasing as involving four elements: budgeting, accounting, bidding, and accepting deliveries.

BUDGETING

1. Assume a total budget for your project, i.e., how much you have available to spend on the project.
2. Break the total budget into categories, e.g., lumber, sheetrock, nails, rebar, etc. This process must be done in conjunction with a "take-off" list from your building plans. The take-off tells you how much of each item will be needed. Take-offs can be obtained from architects, consultants, contractors or you can figure it out yourself. It is very important at this point to identify discrepancies between material needs and budget limitations; it is much easier to adjust your plans at this stage than later on.
3. The total of all budgets for each category must equal the total budget.
4. Budgets must include the *timing* of purchases. This is especially useful as it allows you to project your cash needs at each stage of building.

ACCOUNTING

1. Make a list of all categories of purchases.
2. Prepare a control sheet for each category. This is simply a piece of paper on which you will list all purchases within the category.
3. Code *all* checks and invoices for the category of purchase. There must be an invoice or receipt for all checks, except for those made out to "cash." "Cash" checks should still be coded for applicable categories. If your receipts and invoices and "cash" checks equal the total amount of checks written, you have a "balanced" system with no uncontrolled expenses.
4. Keep each category control sheet up to date and sub-total as often as possible. (It helps to sub-total with a different colored pencil. That way, you won't be adding in total figures and you won't have to re-add columns.) This process allows you to know when you are going over budget in a cost area and allows you to make adjustments as needed.

BIDDING

1. The first rule of getting bids is that you must get at least 3 competitive bids for each major area of purchase. This may not be worth it for small items, as there is a lot of effort involved, but it is certainly necessary for larger purchases where you want to be confident that you have gotten the best possible price.
2. Bids should be analyzed for: specifications, quantity, price, and delivery charges and dates. This is a screening stage for reviewing bids and it is here that you make sure you are comparing "oranges with oranges."
3. Make determination of factors other than initial costs, such as total costs; related costs; status of vendors; and vendor's rating and past history as to reliability, quality and performance. You should also give some consideration to the vendor's technical competence, maintenance of delivery dates, and overall stability.
4. Purchase materials at the lowest bid cost within the standards set for quality and service.

ACCEPTING DELIVERIES

1. Check materials for specification conformance and verify the invoice and/or shipping order. *Don't trust anyone!* The deliverer may be a great guy, but he did not pack the cartons or prepare the shipping tag. Examine the cartons carefully *in the driver's presence* for signs of damage, handling abuse or tampering. Check the contents against the shipping tag. Do it quickly, the driver is in a hurry. Ask the driver to note, date and sign for any evidence of damage or discrepancies in count. Be polite, but firm.
2. Make adjustments with the vendor for non-conformance to specifications, defective materials or rejects.

TIPS

1. Payments should always take advantage of "cash discounts" or "early payment discounts." If, for example, the terms of sale are 2/10 Net 30, this means that there is a *2%* discount if you pay within *10* days from the date of the invoice and that the whole amount is due in *30* days from the date of the invoice. A notation of "E.O.M." means

"end of month," so "10 E.O.M." means payment is due 10 days from the end of the month of the invoice. You should always take advantage of these discounts; they represent actual reductions in the price. A 2% reduction across the board in the price of materials for your project can have a large impact on the rest of your plans.

2. When dealing with salespersons, phone orders, shipments, public offices, etc., get their *name* and *write it down.* These people will be more responsible if you know their name. You can also save time and hassles by going back to the same person; they will already know your problems and needs.

3. When calling manufacturers or vendors long distance, ask them to call you back or if you can call them collect. Many companies have toll-free numbers for incoming calls or WATS lines for return calls. It pays to ask.

Supplies

Different owner builders have different degrees of knowledge about purchasing the supplies needed to build their home. Yet no matter what your level of expertise, you will find invaluable the advice of a reliable supplier you can trust.

It is, therefore, a good idea to get to know the people you will be dealing with. Go into the store a few times and make some small purchases to become acquainted with the counter people. Learn their names. They are more likely to be helpful and give discounts to someone they recognize.

There are advantages to dealing with a few, well-known suppliers. Convenience is one. Dealing with a few suppliers will make your paperwork simpler. Deliveries will run more smoothly. It will be easier to return faulty or extra materials or to order special materials the supplier does not ordinarily stock. In fact, this should be a very important factor to consider when you choose a supplier: pick one who gives you good service and one you trust. Even if the prices are a little higher, the comfort of dealing with people you like may make it worthwhile.

You will need to turn to a lumber yard, plumbing and electrical suppliers, and perhaps suppliers of tile, flooring, roofing, paint and appliances. All-purpose, do-it-yourself stores will offer you the convenience of location, counter knowledge and the display of their products. Large specialty houses, on the other hand, may offer more flexibility in prices and a greater depth of stock. Check to see if an all-purpose store has a "contractors' window" where you can get a better price for your larger orders.

When you are looking for your basic suppliers, follow a few sound principles. Make a complete list of the supplies you will need from that supplier. Make several copies and give them to a minimum of three suppliers to get fair price comparisons. Ask the suppliers for a total price, itemized piece by piece. When you get your lists back, be very careful. Do not look at the bottom lines only. Check each item carefully to make sure nothing has been omitted and that the prices are reasonable across the board. Sometimes savings are concentrated in a single category, e.g., rough lumber, sheetrock, hardware, windows, etc., while the other categories are more expensive than in other bids. Be careful to shop by category; you may lose your savings in the transportation and time involved in item-by-item shopping. Ask about refund policies, particularly if you are buying seconds or at a discount.

There are a lot of ways to save money when shopping for materials. One of them is the contractor's discount, which you may be eligible for since you are buying a large quantity of supplies to build a house. The amount of the discount you receive will depend initially upon your relationship with the supplier and how large an order you make. Attitude is important. Be honest. You do not need a fake contractor's license number, although it might be a good idea to have a business checking account (this will also make your record keeping easier) and a business card. When you go into the shop, lay your business checkbook on the counter. Let people know that you are serious, that you need to spend some money. If the person you talk to cannot give you the discount, ask to speak with someone who can—the manager or the owner. Be friendly and firm. Remember, you have nothing to lose and much to gain.

Consider setting up an account with the supplier if you are going to spend a lot of money there. Not only will it help them to see you as a credible customer, it will help you keep track of your purchases. Most suppliers will give you a 2% discount if you pay your bills by the tenth of the month. Check for discounts if you pay cash. Some suppliers will give up to a 10% discount for cash since it will save them paperwork and help them avoid bad accounts.

It may be possible to get together with another owner builder to buy supplies. Together you can make larger purchases and your discounts will be greater. Plus, between the two of you, you can cover more ground. You can seek out more bargains and share knowledge about prices and other important aspects of buying. A contractor might be willing to add your order to his, automatically giving you a discount. Sometimes it is possible to buy materials from a contractor who is doing installation work for you, like a roofer.

There are many ways to get good deals on supplies, but you need to be careful. The more you know about what you are buying, the better the odds you will actually be getting a bargain. Keep in mind there will be a break-even point between the money you save by looking around and the time you spend doing it. Often you can save about 15% of the total construction cost by hustling.

Find out where the price breaks are for materials. You may be able to save money by buying in quantity rather than buying just what you need. Nails, for example, usually come in 50-pound boxes. If you need 30 pounds, you may end up paying more if you buy 30 pounds loose than if you bought a box. You can always use, sell or give away the leftovers.

The wholesalers, distributors, and manufacturers of certain supplies may be willing to sell directly to you. They may have seconds or semi-damaged materials in their yards that you can use. It is possible to save enormous amounts this way, but you have to know what you are doing.

Another idea that might be fruitful, if you have the time and are making a considerable purchase, such as lumber, is to travel out of your immediate area to look for better prices. There are rural areas where the economy is depressed and goods are less expensive. You would either have to truck the supplies back yourself, which can be expensive if you do not have a truck, or pay freight and storage. Be sure to send out your supply list and compare quotes to those of local dealers.

A big area of potential savings is the use of recycled goods. Auctions are also a good source of materials. You can find notices of auctions in the papers. However, when you buy at an auction, make sure it is a good deal. Know what the materials are, what their quality is and what the going rate on the open market is. Sometimes people will actually bid above the market price. Even though you may not be able to buy all of the supplies you actually need (or you may have to buy more), you can still save money. Just be aware of what you are doing.

It is also a good idea to check the classified ads for special deals. You never know what might turn up. It is possible to find situations at job sites where a contractor might be willing to sell extra materials for a good price if he cannot use them on another job.

The name of the game is knowing what you need and where to find it. A friendly trusted supplier can be of great help to both the beginner and the more experienced owner builder.

PRICE VS. QUALITY

When building or remodeling your home, there is a strong desire to use the highest quality materials to produce the best results. Often we identify high quality with high prices. Yet with most materials, there is a certain point at which you are getting the best product for the money. You need to learn where the quality-vs.-price breaks are for different items.

For instance, foundation and framing materials are by-and-large standardized, i.e., they must conform to regulated or industrial standards that are universal. As long as they meet those requirements, you are not going to get a much better product by spending more. At the same time, you cannot replace your foundation or framing once the house has been finished, so you want to build it right the first time. It is the quality of the *work* that counts in this instance.

On the other hand, you will want to put money into quality materials when you get to the finishing stages, for this is what you will see for the rest of your stay in the house. So you may want to pay a little more. But be careful. There is still a quality line. For example, 3-×-4-foot windows may range from $30 for a cheap aluminum frame up to $140 for a double-paned, wood framed window. It may be that, after spending $90 on a window, the quality for the more expensive one does not increase as much as the price does.

Sometimes you may be able to substitute less expensive materials for the more traditional ones for certain jobs. You may, for example, find a good load of foundation grade redwood for a deck, rather than buying more expensive grades.

If you are working on a tight budget, you may be better off scaling down your plans rather than cutting down on the quality of materials you use. If you do not have the money initially, put up a good frame and do the finishing work as you can, and then replace inferior materials with those of higher quality later on. Or you can design your house so that you can build a portion of it now, move in, and add more as you can afford it.

The important thing is to learn to judge when the quality gained is worth the extra money.

SALVAGE

Looking for building materials at a price you can afford? If you take the time to learn the ins and outs of the salvage network in your area, you may be richly rewarded. With the evolving building codes, the constant changeover of owners and just changing needs, there is a lot of remodeling done. Old elements are often torn out to make way for new ones, and there are many occasions when new things are removed and the old elements restored.

Remodeling almost always begins with a certain amount of demolition. Depending on how skillfully this is done, some real prizes may be preserved for later use. Arched or stained glass windows, hardwood mantles, long lengths of redwood or other old-growth lumber, tile, ornate light fixtures, pedestal sinks and clawfoot tubs are often set aside in this way. These are examples of the "lucky finds" that make fleeting appearances at salvage yards before being taken to their new homes. The chance that you will discover something uniquely fitting and intrinsically valuable is what gives salvage shopping its spice.

Many contractors and owner builders search out salvaged materials because they are looking for a bargain. They know from experience that used items are generally less expensive than new versions of the same thing. One factor in the low price is the assumption that a lot of work will have to be done to bring the article to a usable condition. Sometimes prices are low because the item has cost little in money, time or skill, or because there is an oversupply with little buying pressure.

There are exceptions to this. Brick is an obvious one. The price of used brick in some areas ranges from the same to double the price of new brick. Lumber is another exception. Clean and square used lumber can cost 10% to 50% more than certain grades of new wood with the same nominal dimensions. Part of the higher price of used lumber may be traced to superior quality, since old-growth wood is frequently harder, denser and consequently stronger than new wood and it often has an extra $1/16$ to $1/8$ of an inch of material in each direction.

Another case in which used materials are more expensive is when much labor has been put into recovering or restoring the materials. The less the restoration, the lower the price. Finally, there is the possibility that the material is very scarce in relation to effective demand. In the case of used brick, this is probably the largest single element in the price. The seller knows that if *you* do not buy it, someone else will.

This is not to say that used materials are necessarily superior to new materials. Apart from any intrinsic quality the used item might have had when it was first made, anything old has been subjected to varying degrees of stress. This stress takes many forms, but its common result is to leave the item in a debilitated condition. Doors and windows, for example, are subject to dry rot in the joints and along the upper edges of any horizontal wood surface. Evidence of this malady includes shrinkage, cracking, separation of horizontal and vertical elements at the corners, metal reinforcing braces or plates, and softness or wastage of material. If in doubt, poke the suspicious area with a screwdriver or knife point. If the blade goes in easily, the wood is punky and unsuitable for any structural use. Wood that is merely dry can be strengthened by multiple coats of linseed oil which soaks into the wood, swelling and filling it out again.

To help you know what to look for in used materials, here is a list of the major categories and common problems that you may find.

WOOD: dry rot; powderpost beetled (tiny round holes); splits, cuts, other damage; knots

PIPE: rust deposits, plugs, bends and kinks; bent or crushed threads

DOORS: warping; broken glass; dog damage (deep grooves or bites taken out of edges); dry rot; broken panels; veneer separation of styles from rails, mullions from other elements, etc.

TOILETS, OTHER CHINA: cracks, chips, crazing (can make the item more valuable, as in old toilet bowls with a fine, lacey network of tiny cracks in the surface glaze)

CAST IRON TUBS, SINKS: chips, "rust stars," discoloration and stains, pitting, erosion of enamel by acidic water

WINDOWS: dry rot, cracked or broken glass, putty out, points missing

CABINETS: warped doors, missing hardware, veneer separations, dry rot

The building materials salvage network is an important resource for the owner builder. Learning what salvage has to offer can be enjoyable and educational due to the stimulating exposure to so many styles and potential uses. There is nothing quite like a sunny afternoon in a salvage yard to get the creative juices flowing!

DELIVERIES

In materials delivery, as in most aspects of building, organization pays off. It costs money to deliver, whether the cost is figured into the price of the materials, added on to the price or you pick it up. You do find some suppliers that will deliver for free if they have regular delivery routes that your site falls on, but generally you are going to pay to have things delivered.

Ask the supplier what their charge is. Most suppliers figure that it costs $25 to $40 per hour to deliver. If they say there is no charge, ask if they will give you a discount if you pick up the materials yourself. It may be worth your while, if you have a truck and the time. If you do not have a pickup, you may want to buy a cheap one for the duration of the job.

Ask the suppliers their delivery policy before you buy. Make sure that they will deliver near the job site, not on the street. If you have to carry the supplies all the way across your lot, you are going to waste valuable time and energy. If you are buying roofing material, make sure that they will deliver to the roof so you won't have to carry it up a ladder. Ask the suppliers what kind of truck they use. If they have a truck with a bed that will drop the whole load on the ground, then you and your help won't have to spend hours unloading the truck only to move all the material a second time to the job. If you are having a major order delivered, have them stratify the supplies so you won't be moving half the order off the materials you will use first. If your lot is on a hillside, arrange to have the goods delivered at the top. If you are getting small, frequent deliveries, have them planned out so that you don't end up receiving materials you are not ready to use and waiting for those you need.

When you are accepting a delivery, count it, inspect it, and double check it before it gets off the truck. If you are picking up supplies, count and inspect them before you load them. Once you have the load on the ground or in your truck, it will be harder to send anything back. People make mistakes. Check the price again to make sure it is the one you initially agreed to.

Be extra careful when accepting delivery or picking up bargains, since you may have a problem returning them. Inspect finishing and processed materials carefully. There is little room for error or waste with these products. If you are not sure about the goods, check with a professional. There is no guaranteed immunity from receiving bad materials. Prominent contractors who have been buying from reputable suppliers for years occasionally end up with something wrong. You are not going to offend anyone by checking everything carefully. You are more likely to surprise them if you do not.

You are going to need to pick up supplies as you go along. Running errands can run up the cost. Plan ahead to keep trips down to one a day. Consider installing a phone on the job site and let your fingers do the walking. A little organization can save a great deal of time and money.

STORAGE AND SECURITY

The most important aspect of storage is security. Most of your supplies will not suffer weather damage if they are covered properly. Wood and piping can survive, and while insulation and wiring should not get wet, you can find ways to keep it dry.

What you may run into trouble with is theft. No neighborhood is entirely safe. Know the area where you are building. If you have neighbors, meet them and make friends. Let them know what you are doing. Ask if they can help you keep an eye on your lot and your materials.

If possible and not too expensive, have minimum deliveries made. Stagger them as best you can so that your supplies do not pile up and present more of a temptation. Be extra careful not to leave your tools on the job site. If you have to buy your supplies in big lots, make other arrangements. If you are remodeling, use your garage for storage, making sure it is weatherproof and secure. If you have a neighbor near your site, ask about renting their garage, or consider renting space elsewhere, even if it's not near the job site. Another idea is to build a storage shelter that you could use later, or build your garage first to use for storage during construction. You can also build one section of the house first, make it tight and use it for storage while you are working on the rest of the house.

Present as little temptation as possible. The cost of prevention is worth avoiding the loss of tools and materials.

TIPS ON LUMBER

The quality of your workmanship will depend on five factors: 1) accuracy of your tools, 2) your skill, 3) the standards of quality you set towards your work, 4) your knowledge and 5) the quality of your materials. The tools will be discussed in Chapter 3. The skill should develop as you work, and you should make a *conscious* effort to improve it. The standard of quality, we discuss in the next section. The knowledge, I hope to impart to you in these volumes. Let us now take a closer look at the material.

Wood is a very pleasurable medium to work in. It is flexible, live and pleasant to the touch. It is beautiful and strong and has a certain integrity that man-made products often lack. If you are buying new lumber, try to find a lumber yard that has a high-quality *air- or kiln-dried lumber*. This is getting somewhat harder to locate and often there is an extra expense — 5% to 15% for dry lumber — but it is worth it. Most of the lumber today comes straight from the forest to the yard to the job site. The stud you set in the wall may have been part of a pine tree in Oregon two weeks ago. Wood is made up with the same main ingredients as commercial plastics — lignin and cellulose. Lignin is an adhesive which gives strength and rigidity to the wood; cellulose is nature's strong material. It is made up of long chains which run the length of the tree, which accounts for wood splitting along its grain. The moisture in the wood is not held within these molecules, so much as between them. When evaporation occurs, this moisture is lost and the molecules pull closer to each other, which is why wood shrinks much more along its width (laterally) rather than its length (longitudinally).

Wet lumber has many disadvantages. *Nails lose three-fourths their holding power when driven into wet lumber,* as the lumber shrinks away from the nail, reducing the friction between the wood and the nail. There is also a deterioration of moist wood due to the formation of iron oxides on a nail in the presence of wood acids. If dry lumber is not available or you cannot afford the added expense, you may offset the disadvantages of wet lumber by doing two things. First, use threaded nails throughout the house. The Virginia Polytechnic Institute found that *a house built with threaded nails and green lumber could withstand four times the racking load as a house built with green lumber and plain shank nails.* This is also an excellent reason to use threaded nails in earthquake areas. These are sometimes difficult to find but keep looking; they add only about $20 to the cost of the house. The second thing to do if you are using green lumber is to allow the house to season for 6 months to a year so that it will finish drying before you put on your finish material. So you could close the house in and wait awhile before doing finishing work. If you did this before the wood dried, the finishing boards would be pulled off as the wood shrunk.

Wet lumber has other disadvantages. It is much heavier and therefore harder to handle. This can be quite a problem when you are lifting and erecting your rafters. It is stickier and leaves a resin on the saw blade. (Keep some lacquer thinner on a rag near the saw to remove this resin.) Dry lumber holds paints and preservatives better and is more resistant to attack by fungi and insects. The main problem with wet lumber is that as it dries it shrinks, twists, and bows and thereby weakens your structure. This also throws off your finishing work.

Choosing your lumber

Most lumber comes stamped with the species, grade, moisture content and the producing mill. In each species of trees, there are different qualities in the wood. Usually in each area of the country the lumber yards take construction-grade lumber used for the framing from whatever local species of tree is most accessible. In the West, it is Douglas fir, ponderosa pine and spruce. In the East, it is white and yellow pine. Framing wood comes in 2-foot intervals and is easily available up to 16 feet. From 16 to 20 feet it is harder to find and is sometimes 10 to 15% more.

Above 20 feet it usually has to be specially cut. As you design the house, take this into account; order your lumber so that there will be as little waste as possible. Find a yard that has the wood you need at a good price and one that will let you pick your lumber. This is sometimes hard to do and the lumber company will want you to take the good with the bad. It is not uncommon for 20 to 30% of a stack of studs to be unusable. No matter what the lumber company says, there is no reason to have to buy anything that you cannot use. Sometimes there is a slight charge for picking your lumber. If they will not let you do it for free, pay the charge. Pick good straight boards that are not badly bowed, warped or twisted. At least one end should have a good straight cut. Their length is usually a few inches over the measurement to allow for an uneven cut end. No board will be perfect — it is not the nature of the material, but try to use judgment as to which can be used and which ones can't. Framing lumber is Number 2 or 3 grade, construction or better. Blocking and sheathing can be Number 4 grade.

Stacking lumber

If your lumber needs to be stored for more than a week, stack it in the following manner to allow air to flow throughout the stack, thereby helping it dry and keeping it from getting wet and rotting. Be sure it is supported at least every 4 feet and is level and that it is well weighted down so that the boards don't curl as they dry. Cover them well so that they will not get wet if it rains.

Sawhorses

Before starting your framing, build some sawhorses. Hardware stores sell metal brackets that 2×4s slip into or you can cut the boards yourself. Their height should be about ½ your height or whatever works for you. Be sure no nails are placed where the saw blade might hit them, as it will frequently cut into the top board. Drive a nail on each side to hang your tri square on.

Other fasteners

If you are not using threaded nails, be sure to use coated nails. These are coated with a coat of cement, increasing the holding power of the nail. Metal fasteners, now available at a very reasonable cost, are highly effective and have a much greater holding power than nails. These can often make the work go much quicker and can make certain framing situations much simpler as we will see in the floor-joist system. *In earthquake areas they are highly recommended because of their increased strength, especially in situations where the ground is moving laterally.*

Ordering and delivering the wood

If possible, try to order all your framing wood at one time, if you will be building constantly. If you are working only on weekends you might just want to order your floor joists, then the wall framing, and then the roofing lumber. This will prevent the boards being left at the site unprotected from thieves and the weather. When the lumber is brought to the site, decide on where and how it should be stacked to allow your work to go most efficiently. Be sure you have access to the lumber you will need in the order in which you will need it. If there is any fungus growing on the wood, be sure that there are proper air spaces between the different layers in the stack, even if it will be stacked only for a few days. Fungi can grow very fast and discolor the wood.

Some Thoughts on Work Habits and Attitudes

In ways that we may never realize, the attitude of the builder strongly affects the final outcome of the home. A house built with love and care will reflect it, and return the same to the dwellers. It is as simple as that. We have a chance, in building our own shelters, to do something few Western men or women do anymore. If we can keep this awareness throughout the job and find joy and satisfaction in each step of the work, then the work will become effortless and the house, and the experience of building it, nurturing. It is easy, with financial hassles, late scheduling problems, manpower shortages and all the other things that arise when building a house, to lose sight of this and be working only to finish. Before you begin each day's work, sit down and become aware of what it is you are doing, look at the ramifications it has in your life, realize the opportunities for growth and satisfaction creating your own environment is offering you, and when you feel that joy and excitement begin to rise (even on the 266th work day, after two weeks of rain, a broken Skil saw, and your third smashed thumb), begin your work.

It is a good idea to stop a little early each day and store and properly clean all tools, extension cords and materials. Clean up all the scraps and clean the entire work area. It is good to start each day in an orderly clean area and it can have an effect on your workmanship. A sloppy work area can mean sloppy work. Denail all boards as soon as the situation arises and keep your tools in an easily accessible box or hung on nails. This is especially important when a few people are using one set of tools. I have probably spent 10 hours work time over the last 10 years searching for my pencil, usually to find it stuck behind my ear. So have a set place for each tool; it keeps them from being stepped on, makes them accessible and gives the job a well-organized, professional feel. Learn to respect your tools and how to use them properly in the beginning and it will save you many cuts and perhaps a finger or two. Your power saw will become a very familiar tool to you after awhile, but do not allow this familiarity to lead to laxity. It has a sharp steel blade turning 6000 times a minute and is going to violently cut into anything that it touches, including you. When using a tool, *DON'T EVER LOSE TOUCH WITH WHAT YOU ARE DOING; YOU MAY LOSE A FINGER.*

How to Use This Book

This book was written for one purpose: to assist people who are building their first and probably last home. It is not meant to be a training manual for someone seeking to make house-building their profession (though I think it would be of great value to these people). If you were learning the skills to use for a lifetime as a profession, I would have taught some of these procedures differently. I would have stressed more production and speed-oriented techniques that are not really needed by someone building just one or two homes.

Also, know that this is definitely not the final word. No such thing exists in the trades. Ask 10 tradespeople for the right way to do something and you are likely to get 11

Photo 1-1. Safety: Never use a cutting tool so that it is being forced towards you. If it slips, a nice cut is possible. Remember too that well sharpened tools are much safer than dull ones.

that you understand how all the processes fit together. Since the housebuilding process is a matter of getting many different systems to blend together in time and space, you need to have an understanding of the whole house, all of the systems, before beginning. You should also read about electrical, plumbing, heating and finishing systems that are not mentioned in this book.

After you have finished your first reading, you can begin process by process. Before beginning the process, read the chapter all the way through in detail. Do not just start the process as you read along. Something mentioned late in the chapter might affect your entire approach. Underline or make notes of anything in the chapter that you think you will need to keep in mind during your construction. You may also want to check with your local builder to be sure that there is not some better way to do it in your area. Be sure that you are meeting all the local codes. This can be done by checking in your code handbook or by asking the inspector.

The *Most Common Mistakes* are mentioned to help you avoid them. Usually these mistakes are learned only after years at the job site and having made each one at least once. You can avoid all of these by understanding each mistake well and being sure that you avoid it. You still will make your share of mistakes, but this will help you in avoiding the larger ones.

The *Margin of Error* is there to let you know when you have done good work, work that does not affect the structural quality of the house and will not detract from it cosmetically. A lot of new builders go to great lengths to get everything perfect and spend way too much time doing the job. They can add months to the length of the project. Remember, you are not building a piano and good quality work is not perfect. Try to make each cut and measurement perfect, but use your margins of error and do not correct work that will not affect the integrity of the house either structurally or cosmetically. You just do not have the time to be that exact.

Photo 1-2. Safety: Surprisingly enough, one of the most common accidents is leaving a tool up high, on a ladder or a wall, only to remember, painfully, where you left it.

answers, and we are all sure we are right. On top of that, what might be right for one area of the country will not be for another. Be open to receiving different approaches than those laid out in this book and evaluating them as perhaps being better. Question others deeply as to why they think their approach will work better and if you agree, use it. Some of the best approaches have come from novices who had no preconceived ideas as to how something should be done, so don't discount your own common sense. On the other hand, many brilliant new ideas in construction often have a fatal flaw, so if you are trying a radically new approach of any impact, run it by a local professional first.

I think every owner builder should couple themselves with a local professional to act as their mentor and advisor during the construction process. A construction guru if you will. Pay them by the hour to advise you on key points. They can inspect the forms before the pour, inspect the framing, help with complicated roof framing, solve problems, correct mistakes, etc. The money will be well spent.

I advise reading the *entire book* before beginning construction. You do not have to read it in detail at first, but just so

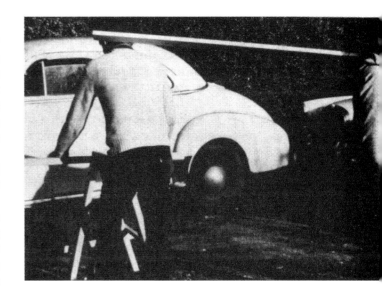

Photo 1-3. Safety: This may look comic, but hauling around long boards is dangerous, especially at a crowded site. Carry the boards low or call a general alert when passing through the site.

Photo 1-4. Safety: Distractions are a major cause of accidents. An accident will occur before you even realize you were getting lax.

The *What You'll Be Doing, What to Expect* section should help you conceptualize the process you are about to begin. It will also let you know what you can expect the procedure to be like for the first time. It might be frustrating, boring, easy, etc. Somehow, being told it will be frustrating makes the frustration easier to deal with when it happens.

The *What You'll Need* section will help you begin the construction procedure with all the needed tools, materials, permits and information. It will also tell you when you will need more than the information given in this book.

The *Reading the Plans* section should bridge the gap between the plans and the actual construction. It will show you how to get the needed information from the plans. On the plans only the information needed for that particular chapter has been shown in bold with unneeded information in the background.

The *Safety and Awareness* section is to remind you of the main things both about construction and safety that you need to keep in mind during the construction of that part of the process.

The *Worksheet* and *Daily Checklist* at the end of each chapter are fill-the-blank sections that allow you to take account and calculate the amount of materials needed. It's also useful in checking off tools and the number of workers you will need. The Daily Checklist can be photocopied and placed at the job site for your convenience.

GOOD LUCK! I hope all this works well for you and you have a safe and successful project.

Construction Site Safety

Since the construction site is unfamiliar ground for most of you, and since you will be concentrating on what you are learning and doing, the hazards around you are multiplied

considerably. It will be constantly necessary for you to remind yourself to maintain an active mental awareness of your immediate environment while working—beside you, behind you, above you, below you!

GENERAL

Be patient. If anger and/or frustration becomes a problem, stop, take a break and reassess the situation once you are calmer.

Working while under the influence of alcohol or other drugs is dangerous to yourself and those around you.

Be careful of sunstroke, sunburn and heat exhaustion.

POWER TOOLS

Power tools and many hand tools as well as construction operations are dangerous when being used by those lacking proper knowledge and experience. IF YOU HAVE ANY DOUBTS OR FEEL YOU NEED MORE INSTRUCTION: STOP! GET IT BEFORE MISTAKES ARE MADE.

Wear protective goggles *at all times* when power tools are in use.

Don't use power tools when standing in water—it can be a shocking experience.

If changing a blade or adjusting the cut, be certain the tool is unplugged.

Watch your cord placement when using the circular saw to be certain it is out of the way.

Set the circular saw blade to the appropriate depth.

There are specific safety rules for each power tool. Familiarize yourself with these rules prior to use.

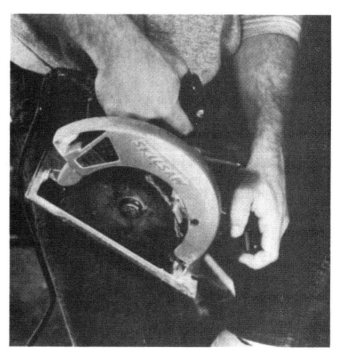

Photo 1-5. Safety: One of the most dangerous situations is working with a saw when the blade guard is tied back and having the moving blade brush against some part of your body. ALWAYS know where the blade is.

PROTECTIVE CLOTHING AND EQUIPMENT

On the job site, WEAR HEAVY SOLED WORK BOOTS (steel toes if available). Tennis shoes and other soft soled shoes are dangerous to the wearer. Use rubber soled shoes when working on the roof.

Wear gloves while handling lumber, metal or other materials capable of injuring your hands for extended periods. NON-POROUS GLOVES SHOULD BE WORN when handling or using solvents, preservatives or corrosives.

Safety glasses, goggles or face shields should be worn during any operation where there is an inherent probability (or even possibility) of eye injury from flying or falling particles, or from spraying or splashing of liquids. WEAR EYE PROTECTION WHEN USING POWER SAWS.

WEAR HARD HATS AT ALL TIMES WHEN WORKING UNDER OR AROUND OVERHEAD OPERATIONS OF ANY KIND.

Be sure long hair is tied back so it does not get caught in power tools.

SITE

Scrap lumber and debris should be kept reasonably cleared from all work surfaces, as well as normal foot traffic areas. Never leave lumber in a vertical position; stack horizontally in neat piles.

Within 6 feet of the building under construction, the ground should be kept reasonably free of irregularities and debris.

Place materials in stable stacks to prevent materials from falling, slipping or collapsing.

ANY BOARD OR OTHER MATERIAL THAT HAS NAILS PROTRUDING FROM IT SHOULD BE DE-NAILED OR THE NAILS BENT OVER FLAT in such a way to prevent punctures or other accidents—BEFORE THE PIECE IS DROPPED OR DISCARDED.

HAND TOOLS

When not using tools, keep them picked up and in a box. When using tools, keep track of them. Don't use tools which are dull or in poor condition—check the handles for looseness.

Select the appropriate tool for the job.

Hold edge tools (planes, chisels) so the cutting action is away from you. Beware, as edge tools will be quite sharp!

Don't throw or drop tools.

Don't leave tools on ladders, scaffolding or high areas—they are potential missiles. Avoid walking beneath ladders or scaffolding where people are working.

Don't carry sharp tools in your pockets.

LIFTING, CARRYING AND BENDING

Stand close to your load, bend your knees and lift from your knees, not your back. Reverse this technique to set an object down.

Photo 1-6. Safety: Cutting a board between two supports will often cause the boards to bind on the saw blade and result in the saw kicking back. Always cut with the board cantilevered over a support.

Do not twist or turn your body when lifting or carrying a heavy load.

Carry the load close to your body and at waist height.

Work at a level which avoids constant bending at the waist. When working at a low level, bend at your knees. Have a work bench or saw horses at a level which minimizes back strain.

If you anticipate an object being heavy, test it prior to lifting.

If an object is too heavy, get help!

When placing a heavy object down—do not drop it. Your fingers and your partner's fingers are at stake! If you drop lumber, you can crack it.

If you are over 5 feet high on a ladder, have someone hold the ladder.

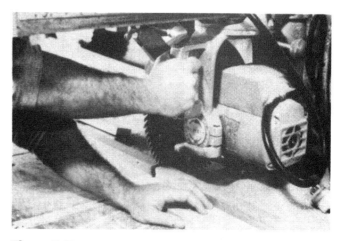

Photo 1-7. Safety: Never place your hand in the path of any saw blade, but especially not a radial arm saw, which is known to kick back.

Enjoy Your Work

Drudgery creates accidents. Hard work can be basically good and healthy, but don't get exhausted.

Be your own barometer as to when to rest or change to something different.

Take your time. Don't be afraid to just stare at the work a while before making your move.

Make it a *good* experience.

Don't fight your work. Arrange your immediate work area so that you're not reaching, straining, struggling. Move saw horses, pick up scraps under your feet, arrange the work for yourself.

THINK SAFETY

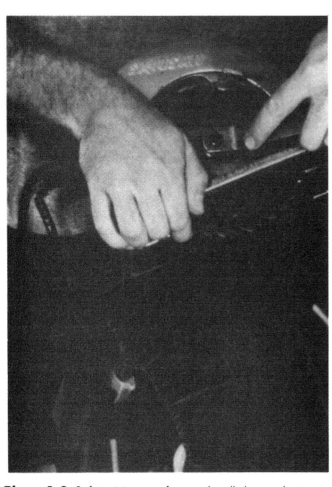

Photo 1-8. Safety: Many professionals still change their saw blades with the saw plugged in. Many professionals are also missing finger digits.

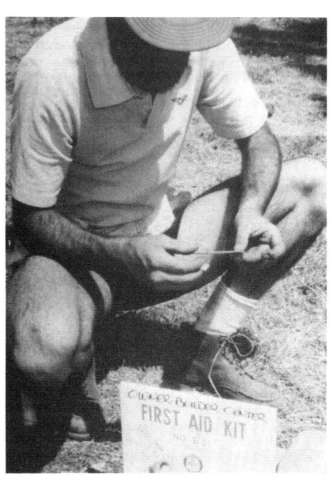

Photo 1-9. Safety: Keep a well-stocked first aid kit and booklet at the site at all times as well as emergency phone numbers.

Construction Schedule

by Charley Huddleston

THIS SCHEDULING GUIDE can be adapted to your project and circumstances. It follows a typical home-building project for the construction of a 1,500-square-foot one-story house with a perimeter foundation on a level site using standard wood-frame construction techniques. Note how certain tasks arrange themselves in clusters and that some activities must be successfully completed before you can proceed with subsequent jobs. This guide will also help you coordinate loan disbursements, ordering materials and inspections. You will need to schedule your work accordingly.

1. Submit copies of your house plans to your building department for the appropriate permits.

2. Deliver your plans to subcontractors for estimates and, if possible, firm bids.

3. Deliver plans to suppliers for quotes on materials. You're likely to need bids from subcontractors for the

following jobs: trenching, foundation, framing, plumbing, heating, sheet metal, electrical, wallboard installation and finish, roofing, finished flooring, tiling, painting, concrete walks and insulation. You can obtain quotes from suppliers for the remainder: cabinets, windows and doors, mirrors, countertops, hardware and fittings, appliances and a fireplace.

4. Apply for construction or mortgage loans at your bank. When contracting your own house construction, your application will probably include a personal financial statement, house plans (including elevation and lot drawing), a preliminary title report, copies of written bids and quotes and a lending institution Cost Breakdown sheet.

5. Contact and arrange with local utility companies for electrical, gas, water and telephone service.

6. Arrange for temporary facilities on the building site. These might include a chemical toilet, a temporary power pole, a construction shack, a telephone and living quarters.

7. Pick up your initial building permits.

8. Check with your bank for confirmation of your construction loan.

9. Complete loan arrangements with the bank and record your deed of trust.

10. Sign and return the contracts to your subcontractors. Set up tentative dates to start their services. Establish supplier orders and delivery schedules.

11. Get a surveyor to locate the lot boundaries.

12. Set the temporary power pole.

13. Have the electrical inspector inspect your power pole.

14. Have the utility company hook up power to the power pole.

15. Foundation contractor lays out the site for trenching, footings and foundation.

16. Now the foundation can be dug.

17. The foundation contractor installs the forms for the concrete.

18. Call for a foundation form inspection by the local building inspection department.

19. Order the framing lumber from the lumberyard.

20. The foundation contractor pours the concrete and later strips the forms.

21. The framing contractor lays the sill and joists.

22. The plumbing, electrical, and heating contractors do their underfloor work. The local building department makes an underfloor inspection (Uniform Building Code areas).

24. The framing contractor lays the subfloor.

25. If your bank uses a draw system to finance construction payments, you can call for your first disbursement, usually about 20%.

26. The framing contractor frames the walls.

27. The contractor frames the ceiling and roof.

28. Order the delivery of the windows and the exterior doors.

29. Carpenters install wall and roof sheathing, windows, siding and exterior doors.

30. With the house framed and the siding installed, you can install sheet metal and flashing.

31. The roofing contractor installs roofing.

32. At this stage your rough plumbing, electrical, heating, telephone pre-wiring, cable TV, etc., should be completed and ready for inspection.

33. Contact the local building department for a close-in inspection. You'll need this permit before covering any in-wall components of the electrical, plumbing and heating systems.

35. The insulation contractor installs insulation between wall studs and ceiling joists.

36. Call for second disbursement from the bank, usually about 30%.

37. The drywall contractor installs the wallboard.

38. At this stage you may also need to have a wallboard nailing inspection.

39. Order the delivery of interior doors and millwork trim.

40. Order and arrange for delivery of kitchen cabinets and bathroom vanities.

41. Carpenters can now install the garage door.

42. Arrange for the subcontractors to finish their plumbing, electrical and heating work.

43. Order appliances and electrical fixtures.

44. Call for a third disbursement from the bank, usually about 15%.

46. Various materials, finishes and household elements can now be ordered. These include mirrors, shower doors, tile, linoleum, carpeting, countertops and weatherstripping.

47. Painting contractor paints interior walls and trim, starts exterior.

48. Lay the resilient flooring.

49. Mechanical services are completed by the plumbing, electrical and heating contractors.

50. Install tile and countertops.

51. Install mirrors, shower doors, carpeting, weatherstripping and finish hardware.

52. Call for a final building inspection. This step allows you to establish electrical service.

53. Obtain a notice of completion and file with county recorder. The bank will make its own final inspection about this time. The lien period (usually about 60 days) begins, during which time your subcontractors still have lien rights on your construction.

54. Request your utility company to hook up electrical service.

55. You will receive another disbursement from the bank, usually about 15%.

56. Pay the final bills as they come due and take discounts on materials.

57. Have your subcontractors and suppliers sign lien releases as they are paid.

58. Receive the final disbursement from the bank, usually about 20%, as the lien period ends.

59. Move in.

60. Congratulations for hanging in there!

2 Types of Construction

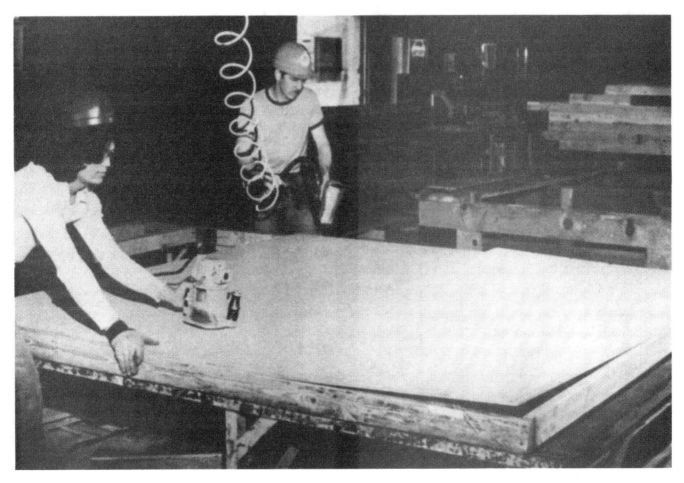

Photo 2-1. A modular wall section of a manufactured stud house being assembled in the factory with pneumatic tools.

MANY NEW BUILDERS do not realize that there are many different types of construction they can choose from. Most builders proceed building the standard stud-constructed house, never stopping to consider their other choices. This chapter will explain the most common types of construction methods used in the United States today. Your choice in the type of construction you will use should be well thought out. Your decision will impact your cost, your design, the feel of your home and your building process. There should be definite reasons for choosing one type of construction over another.

Along with your decision on the type of construction you will use, you also will need to decide whether to use a pre-fab kit home or build the entire house yourself at the site (a site-built home). This is a very important decision too; it can greatly impact your building experience. Let's discuss this building issue first.

The Site-Built Home

Until a few decades ago, almost all houses were site-built homes (except for those ordered from the Sears catalogue). Most owner builders have avoided factory-built homes, feeling that they could not get the design or quality they wanted from a kit. This is really no longer the case. You still may, however, choose to build the entire home yourself for several reasons. Needless to say, you do have more control over your home should you decide to build at the site. You can choose and inspect every piece of material and keep the quality of construction at whatever level you want. You are also able to make more design changes during the course of construction since the house is being cut and assembled as you go. Some types of designs may not be available in a kit home and you will need to build the entire house yourself.

Photo 2-2. A panelized wall section of a manufactured stud house being lifted into place by a crane. The entire house might be completed in a week or two.

Sometimes there is a certain feel you want to instill in the house, one of handcraftsmanship or maybe high tech, that kits do not offer. Finally, many people just want the satisfaction of knowing they built it themselves.

Manufactured Housing

(*Photos 2-1–2-3.*)

Until several years ago, I, like many other people, had a certain attitude about manufactured houses. Since I began as a custom builder, I felt manufactured houses limited people's creativity and design and often produced an inferior product. In the last few years, I have made a 180° turn. Most all of the existing manufactured housing companies, especially those that have been in business for some time, offer high-quality products and the design options are limitless. On top of that, it is often a less expensive way to create the same quality house you would build for yourself. Here are several of the advantages to manufactured housing.

One of the greatest advantages to the owner builder is that you are linking yourself with a support network, the manufacturer and your local dealer, who will assist you through all stages of construction — from getting your permit, to getting finances, to getting the house built, etc. This support system can be of great value to a novice builder.

Your construction time is greatly reduced since a lot of the work is done at the factory. And if you are working on a limited time budget, this can be very helpful to you. A lot of the work is done by the manufacturer; the amount of knowledge and information and work you will have to do is limited. Often, a manufactured home can go up in a third of the time of a job-site-built home. This reduced time can be translated into dollar savings since your construction loan, which is often a few percentage points over your mortgage loan, will be reduced. If you own another home and are paying a mortgage while you build, you will be able to move into your new home sooner, eliminating the cost of two mortgages.

Manufacturers are able to buy in large lots and can get superior-quality materials at a lower price. This saving is often passed on to the owner. Since factory labor is a lot less expensive than hired carpenters, the work done on the house is less costly. This is passed on to the buyer.

For the most part, a house using high-quality materials can often be had more cheaply from a manufacturer than one built by yourself at the job site.

Photo 2-3. A modular house being delivered to the site in several large sections.

There is one other added advantage. If you are building a home in a remote area where labor is expensive and unavailable, you can solve many of your problems by using a manufactured house.

A few problems with manufactured housing are that you may not always be able to get the design you want from every manufacturer, and occasionally there may be code problems in your local area with some manufactured housing. However, most of the larger manufactured housing companies have ICBO (International Congress of Building Officials) numbers, which will assist you in getting whatever permits you need to build the house in their area.

There are many different types of manufactured housing. Almost every style of home, including domes, logs, laminated timbers, stud, pole, etc., comes in a manufactured housing option. Some manufacturers simply provide the pieces for the weather-tight frame of the home. Sometimes the entire interior finishing materials are supplied as well. In some panelized packages, the entire wall section, including the interior finishing boards, the exterior walls, the plumbing and electrical fixtures, the frame, and insulation, are put together in panelized sections, shipped to the site and assembled (usually with the use of a crane). Sometimes the pieces are numbered, and you erect the entire thing like a giant erector set. Other modular homes have the entire house in two or three large sections with the carpets down, the appliances in, and the plumbing and electrical fixtures in, that they bring to the site and bolt together.

I highly recommend that you explore the manufactured housing alternative for any style house you are planning to build. Here are a few tips to consider before choosing whether to use this alternative and, if so, which manufacturer to choose.

1. Be sure the manufacturer's package will comply with the local codes.
2. Investigate the company. What is their credit rating? How long have they been around? Do they have a record with the Better Business Bureau? Have they ever been taken in front of a consumer affairs board? Have they ever been bankrupt?
3. Ask for references of other customers who have recently built and lived in one of their houses.
4. Ask to see all of the instructional materials and educational pamphlets.
5. Be clear as to what the package includes and what it does not include. Make sure you have worked this out in detail and know exactly what you are getting for your money and what you will still have to go out and purchase or subcontract. Often a lot of problems lie in this area. It is hard to compare one manufactured housing company to another unless you know they are both offering the same types of materials and services.
6. Be sure that you will be able to re-sell this style of home in your area should you ever move.

7. Investigate whether or not you will be able to make design changes or design the house exactly as you want it.

8. Find out what other assistance the company will provide. Will they assist you with financing? Will they assist you with permits, with site choosing, with site improvement, with construction, with finishing?

Standard Stud Construction

(Photos 2-4–2-8.)

Most of the houses built in the U.S. today are built using a standard stud construction. There was a time when almost all houses were either post and beam or log, but stud construction caught on with such a fury that, until recently, all others were almost forgotten. This did not come about because stud construction was, across the board, a better way to build, but it was quicker and easier, much as white bread replaced whole grain. The old construction method of post and beam makes use of large structural framing members, and therefore fewer timbers are needed, but construction usually required the skill of a highly trained joiner since a small number of joints carried the entire weight of the building. Because of this, each joint had to be exactly fitted. At the turn of the century there was a switch to stud construction; the work was quicker and took less skill. As cities grew, buildings needed to go higher in order to make better use of the expensive land. With the old post and beam method you would need longer and longer timbers in order to go higher. With stud construction, you simply stack one story on top of another, never needing a wall member over 8 feet long.

To build with stud construction, you first build a foundation, then on this you build a floor frame. The floor frame is built with 2-inch-thick construction-grade lumber (2×6, 2×8, 2×10 or 2×12) called floor joists. A floor joist is one of a series of parallel framing members that run underneath and support the floor. These are usually placed on 16- or 24-inch centers. After the joists are in place, you apply the subfloor (usually plywood) and then use this platform on which to build the rest of the house (hence the name platform building, as it is still occasionally referred to). The walls are built with studs, which are 2×4s or 2×6s that are about 8 feet long and again are placed on 16- and 24-inch centers. Where the structural integrity of the wall frame is violated by an opening for a door or window, a larger header spans across the top of the opening to carry the load to either side of the opening. Since the walls are carrying the roof load, these headers are essential. The roof frame is made up of roof rafters, which are also two-inch-thick stock that support the roof sheathing and roofing. These rafters rest on top of the walls, and the walls carry the load into the floor and onto the foundation and into the ground. If there is a second or third floor, you simply build your first-floor walls, place your second floor on top of

Photo 2-4. The floor joist system of a stud house. In this case, the joists are cantilevered as a design feature.

Photo 2-5. A stud frame house under construction. Note that the platform is built and walls are going up.

Photo 2-6. A stud house under construction.

Photo 2-7. A stud house using some post and beam in the front section. Note that large framing members are used for the wall, but the rafters, since they will not be exposed to the interior, are 2×10s.

Photo 2-8. A stud house with some post and beam construction.

Photo 2-9. A post and beam house built by students at Heartwood Owner Builder School in Washington, Massachusetts. Note that the horizontal beams and vertical posts shown constitute the entire structural framing system.

them and then build your second-floor walls. The third floor is added in the same way. So the roofing is supported by the rafters, which are supported by the walls, which are supported by the floor frame, which is supported by the foundation, which is supported finally by the ground.

There are many advantages to this type of building, especially for the novice builder. Everything in the construction world, at least as far as residential construction is concerned, is set up around this type of construction method. The materials and information are readily available and it will meet all local codes. In addition, since many different small pieces of wood make up the frame of the house, the load is spread over many joints and many pieces of wood. No one joint is essential and therefore a ¼-inch margin of error is allowable at the framing stages, taking the burden off the builder to make exact cuts. Finally, many of the building materials are provided in 4-×-8-foot sections: plywood, paneling, sheetrock, etc. These accommodate the 16- or 24-inch centers of the framing members. Heaters, ductwork, medicine cabinets, etc., are made to fit between the framing members.

Post and Beam Construction

(Photos 2-9, 2-10.)

Post and beam construction, which is sometimes called plank and beam construction, was used in the 18th and 19th centuries and was the dominant type of construction before stud construction was introduced in the United States.

This type of construction involves large timbers, usually 4 to 12 inches wide, that are placed both vertically and horizontally. The vertical timbers are called posts and these support horizontal timbers called beams. The load of the roof is carried onto the horizontal beams, then over to the vertical posts and onto the ground. The walls themselves do not carry the loads; they are used only to hang the siding, the doors and windows. The posts and beams carry the entire load. Therefore, fewer but larger pieces of framing materials are used than in the standard stud construction. Usually in this type of construction, the foundation and floor system are built as in conventional stud construction and then the posts

Photo 2-10. A metal bracket used to hold the post and beams together in earthquake areas.

Photo 2-11. Large timbers are bolted to poles in this pole house. Note how the house is resting on these timbers.

and beams are erected on the floor platform. Many times post and beam construction is used in certain areas of a house. For instance, highly exposed areas such as the bedroom or the livingroom might use a large post and beam type of construction, and the rest of the house may be stud. Four-inch-wide beams are used for the roof rafters and these are left exposed to the room below.

The greatest advantage of post and beam construction is the sense of solidity, character and integrity that the large exposed timbers lend to the structure itself. A home done with a stud construction, without the large exposed framing members, will feel quite different from the same house built with large exposed framing members. Where posts and beams, rather than the walls, carry the load of the roof, you can also have larger and more frequent openings for doors and windows, and large interior spaces. Post and beam houses can be designed to fit almost any style. Usually 4 to 5 people and perhaps even a crane may be needed during the frame-construction period.

The disadvantages in post and beam construction are few. In some areas, engineering may be required, but this is usually not too costly. It can also be difficult to find high-quality larger timbers. Since there are fewer framing members and each one carries a greater load, the cuts and your general work must be more exacting than in stud construction, where many small framing members carry the entire load.

In earthquake areas, as well as in hurricane and tornado areas, where the house stands a lot more stress and strain than in areas where these natural disasters do not occur,

houses have to be better built. Because of this factor, post and beam homes often have to be connected with large metal brackets. This is especially true in California. In this case, your work does not have to be as exacting since the joints are not made of wood but of metal brackets, bolted to the wood and framing members. Yet even this can have a very distinctive appearance which should be considered. As in all other types of construction, post and beam packages are available from a number of different manufacturers.

Pole Construction

(*Figure 2-1, Photos 2-11–2-20.*)

Pole construction involves vertical load-bearing poles that are embedded in the ground sometimes up to 6 or 8 feet deep. These poles continue up and support the floor frame, the second floor and the roof. The floors and the roof are attached by horizontal boards that are bolted and often notched into the vertical poles. This type of structure is becoming more and more popular. In the past, it has often been used for agricultural and farm use where the dirt floor between the poles was used for equipment and a place where livestock could move around within the building. Given some of its unique advantages, pole construction is now having somewhat of a renaissance.

Perhaps the greatest advantage of this type of construction is its practical use on a sloping lot. Often on sloping lots

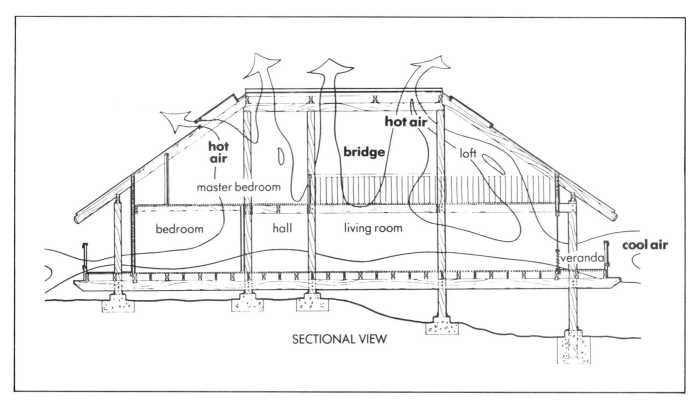

SECTIONAL VIEW

Figure 2-1. Since pole houses do not require as many interior bearing walls as standard construction, they are very well suited for natural ventilation in summer and transfer heat easily between rooms in passive solar homes. (Diagram courtesy of Pole House Kits of California, Irvine, California.)

Photo 2-12. A pole house in progress. Note how the walls are framed in typical stud fashion.

Photo 2-13. Poles for this pole house are kept out of the ground by fastening them to concrete piers. Note the small redwood sill block.

Photo 2-14. A pole house under construction. Note how few framing members are actually needed for the structural components of the house. The large poles serve as foundations and support the floors and roofs. Standard-size framing members will be added between the poles to support windows and doors and attach interior and exterior skins. The roof was added to protect the home early in construction. (Photo courtesy of Pole House Kits of California, Irvine, California.)

foundations can be very expensive; as the lot slopes, the foundation must become higher and higher to provide a level plane. With pole construction, however, longer poles are used at the lower end of the lot and shorter poles at the higher end to make adjustments for the foundation. This can save you many thousands of dollars in construction costs. A pole house also performs very well in hurricane, tornado and earthquake areas since the house is tied well into the ground with the long poles acting as the foundation.

Like post and beam construction, it uses some large framing members to give the house a sense of strength and solidity, but the exactness of post and beam work is not required with the round poles. The house can go up rather quickly and the roof can be put on before the walls are completed, thereby providing a protective work area as you finish the inside of the house.

Often it will take several people and perhaps even a crane to erect the large poles, but after this the work goes rather quickly. It is best to use some type of pressure-treated poles so the wood will not rot in the ground, or to build a system where the poles sit on top of a concrete pier foundation and are bolted to this foundation. This will also keep the wood out of contact with the ground.

Perhaps the greatest advantage of the pole house, aside from the strength and solidity of the large poles, is the openness it can allow. Load-bearing walls are not needed, as the poles carry the weight, and both large interior and exterior openings are possible. Also, large sweeping verandas are very simple to implement in your house design.

Though in the past most people have chosen to build their own pole house, nowadays there are a few kit companies available. If you are planning to build a pole house, it may be a wise decision to investigate one of these kit homes to see if there is a labor and money savings.

Photo 2-15. Pole houses are especially well suited for sloping sites, where the longer poles are used on the lower part of the slope. The savings over traditional foundation systems can be considerable. (Photo courtesy of Pole House Kits of California, Irvine, California.)

Photo 2-16. A finished pole house. Once again notice its application on steep slopes. (Photo courtesy of Pole House Kits of California, Irvine, California.)

Photo 2-17. Pole construction used for a commercial building. Note how well the poles work on a sloping site.

Photo 2-18. Pole houses are conducive to implementing wide-covered verandas as shown here. Note stringer attached to pole at upper right. (Photo courtesy of Pole House Kits of California, Irvine, California.)

Photo 2-19. Pole houses allow for open spaces and cathedral ceilings. The poles carry the load and can allow for large open interior areas without interior bearing walls. (Photo courtesy of Pole House Kits of California, Irvine, California.)

Log Homes

(*Photos 2-21–2-24.*)

Log homes were one of the first types of construction to come to the United States. All of us were familiar with the log homes built by the pioneers and our ancestors when they founded this country. Today there are still over 80,000 log homes built each year in the United States.

Log homes have a feel different from almost any other type of structure. Their massiveness and strength are apparent as soon as you walk in them or view them from the outside. A house built with a stud construction will feel entirely different from a house of the same design built from logs.

Log kit homes come in several different types. One type includes the walls only. This is usually made up of precut logs and does not include a roof system, doors, or windows. Another type of kit includes the logs for the walls, the roof

system, the doors, and windows, providing therefore a weather-tight shell. A third system includes the weather-tight shell, the floors, the interior walls, counters and stairs. And finally, the last system includes everything you need for the completed house.

Log homes are usually rather simple to erect and require less skill and exactness of construction than many other types of homes. It is really just like putting together a large lincoln-log set that we used when we were children. One great advantage of erecting a log home is that as the logs are erected, you are not only framing the house, but you are also insulating the house and finishing off the interior and exterior walls. As a bonus, log homes require little maintenance, usually just some wood preservative every few years.

Photo 2-20. Pole construction can also be combined with the beauty and strength of post and beam construction, as shown here. (Photo courtesy of Pole House Kits of California, Irvine, California.)

Photo 2-21. A hand hewn log building. This type of structure can be inexpensive to build and of greater strength and quality than typical stud construction.

Photo 2-22. A manufactured log house kit. Beautiful and solid.

Photo 2-23. The logs from a kit home stacked and ready for assembly.

Photo 2-24. Round log home built by owner from logs cut at the site.

There are many, many different log-home manufacturers and the design variations have multiplied considerably in the last few years. Log homes, being easy to assemble, usually require no more than 2 or 3 people and very novice skills.

Many people are concerned about the energy efficiency of a log home. Wood has an R factor of about 1.5 per inch, whereas good insulation can have an R factor of anywhere from 4 to 6.5. (R=Resistance to heat loss; the higher the R number, the greater the insulating property of the material.) Therefore, the R factor of a typical log wall is often no more than 8 to 10 while that of a stud home can be 11 to 14. This has discouraged some people from building a log home. You may want to consider building a laminated timber home, mentioned in the next section, which gives you the advantage of a solid-wall wooden home and at the same time provides the increased R factor of 23 or even higher. However, many people who live in log homes say that they are very comfortable in winter and summer and that the R factor of a true log home is not the only thing to be considered. The cellular nature of the solid wood acts as a much greater natural insulation by trapping air in the cells, making a log home more comfortable than the R factors would indicate.

Before choosing to build a standard log home, be sure that it fits all local codes. Usually this will not be a problem since log homes can exist anywhere. It should blend-in in your area. If you are building in the urban or suburban area, often times your design style will not fit in with your neighbor's.

Try to choose a dealer who can transport the heavy logs to your site without a tremendous increase in transportation costs. And, above all, be sure that you feel comfortable in the solidity and strength of a log home and that the tremendous amounts of exposed wood will go with your decorating scheme.

Laminated Timber Construction

(*Photos 2-25–2-33.*)

In the past few years, a new manufacturing process has been developed which gives the solidity, quality and strength of a solid log home or solid wood home, and at the same time gives a greater insulating factor than typical stud construction. The process involves laminating boards to both sides of styrofoam (or other types of insulating foam) core. Usually 2 inches of cedar are used on each side of a 2-inch-thick foam core. The R factor of this wall can be anywhere from 19 to 23 as compared with an R factor of 13 for a typical stud house.

These houses have the advantage over log homes in the sense that they provide a greater insulating property while still offering all the advantages of solid wood. They have an advantage over a hollow-wall stud home in that they are solid and add a much greater sense of stoutness and strength to the house and the hollow core of a stud home.

Usually they cost little or no more than a similar house of the same quality built in a stud or log construction. Another advantage of these houses is that, as you close in the house, you are finishing the exterior and interior walls, framing the home, and insulating it all in one process, much as you do with the stud home.

The sound transmission control, which represents the abil-

Photo 2-25. Mock-up of intersecting laminated walls. Note that the insulating foam core, sandwiched by cedar planks, offers a solid wall with good insulation. (Photos courtesy of Pre-Cut International, Woodinville, Washington.)

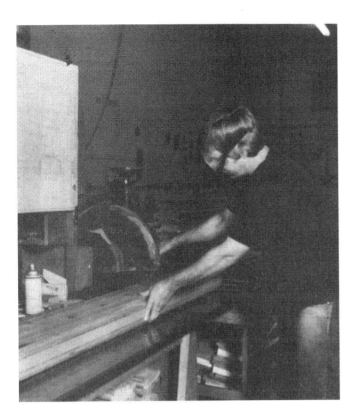

Photo 2-26. Cutting the laminated logs in the factory from a customer's design. Each house is cut according to the customer's own plans and then marked as to its location in the wall. (Photo courtesy of Pre-Cut International, Woodinville, Washington.)

Photo 2-27. Electrical wires are run through the bottom few courses of a laminated timber home. (Photo courtesy of Pre-Cut International, Woodinville, Washington.)

Photo 2-28. A laminated insulated timber home under construction. Note how no trimmers or headers are needed around the openings. (Photo courtesy of Pre-Cut International, Woodinville, Washington.)

Photo 2-29. A laminated timber house under construction. Logs are cut at the factory from the owner's plans. (Photo courtesy of Pre-Cut International, Woodinville, Washington.)

Photo 2-30. The interior of a laminated timber home. Interior walls look like walls covered with tongue and groove cedar planks. Interior walls can also be covered with traditional wall coverings. (Photo courtesy of Pre-Cut International, Woodinville, Washington.)

Photo 2-31. Finished solar home built from laminated insulated timbers. The exterior appears as a home sided with tongue and groove cedar boards. (Photo courtesy of Pre-Cut International, Woodinville, Washington.)

ity of the house to reduce the sound from within and without, is much greater because of the solidness of the home and the foam core. Its shear strength and wind-load factor is superior to that of the stud home. The wood also provides natural beauty to the home, and since wood breathes, it acts as a humidifier and dehumidifier. It is easier to maintain, requiring only a wood protection in the beginning and a light coating of a wood preservative every few years after that.

Energy-wise, it saves tremendous amounts of energy due to the increased R factor and the continuous insulating membrane provided by the solid home. Where a typical stud house uses 5,000 to 7,000 board feet of common-grade wet lumber, a laminated timber home uses 18,000 to 22,000 board feet of kiln-dried timber. This wood has minimal shrinkage and more strength.

The house can be designed according to your design plans, or it can be chosen from the manufacturers. You can design whatever style house you would like, anywhere from a southern mansion to a modern contemporary home, and it can be cut to fit this style. Should you choose to paint the outside of the home, it would only appear as if the house was sided with 1-×-8 siding boards put on horizontally. On the inside, the interior partition walls can be made out of sheetrock and even the interior surfaces of the exterior walls, which are usually made out of cedar, can be covered with sheetrock in order to make the house have a more traditional look. Both the exterior walls and interior walls can be left with the cedar exposed, giving a more rustic appearance.

Most of these houses are approved by the International Congress of Building Officials and will have no problems passing your local code.

In many ways, this is a superior building style. It gives you the solidity and strength of a solid timber home without the drawback of a log home, the lack of an insulating property. At the same time it allows you a tremendous amount of design flexibility and can appear to be a typical house built like any other house. It costs you the same price, but ends up to be a much higher-quality home.

Photo 2-32. Completed laminated log home.

Photo 2-33. A laminated timber home in traditional style. Exterior and interior look like T&G cedar siding. (Photo courtesy of Pre-Cut International, Woodinville, Washington.)

Photo 2-34. A dome shell under construction. A shell can be erected from a kit in a day or two. The triangular pieces are bolted together to form a pentagon which is then bolted in place to other assembled pentagons. (Photo courtesy of Cathedralite Domes, Medford, Oregon.)

Domes

(*Photos 2-34–2-39.*)

Domes are free-standing structures which need no load-bearing interior walls to carry the weight of the roof or the structure itself. They were developed in 1956 by Buckminster Fuller and Al Miller.

Domes have several advantages. For one thing, the structure encloses the maximum amount of living space with the least amount of exterior surface area for its volume. Also, domes are now accepted by most code offices throughout the United States and dome manufacturers usually have an ICBO (International Congress of Building Officials) number that allows them to pass all local codes. Presently there are over 200,000 domes in use world-wide and it is predicted that over one million domes will be erected during the 1980s.

There are two different types of dome kits: the panelized and the hub and strut systems. The panelized dome kit is made up of triangular panels, usually made of plywood and 2×4s or 2×6s, with struts on all three sides and exterior skins already in place. These plywood triangles are usually one of two sizes and are joined together in pentangles which are then lifted into place. Usually 3 to 5 people are needed

and scaffolding is erected on the dome floor to lift these pentangles into place. Often a riser wall, which is a vertical wall 3 to 5 feet tall, is built and the dome is erected on top of it. This gives the house some vertical spaces and also allows for a larger home. Openings for doors and windows are placed in different areas around the dome and skylights can be added by simply leaving out one of the triangles in the dome surface itself.

Another system, though not as commonly used, is the hub and strut system, developed in Arizona in the mid-1960s. In this type of system the frame of the dome or the skeleton is first erected similarly to that of a stud house. The frame is usually made up of 2×4s or 2×6s bolted together with metal hubs or plywood hubs. The external skin, which is usually plywood, is then applied just as the sheathing for a stud house would be applied over its frame.

Both kits usually contain only the shell with the interiors, the foundations and the finishing work being supplied by the owner.

Dome homes are some of the most energy-efficient homes

Photo 2-35. Domes can be used for commercial and industrial application as well as residential. (Photo courtesy of Cathedralite Domes, Medford, Oregon.)

Photo 2-36. A completed dome at the seashore. (Photo courtesy of Cathedralite Domes, Medford, Oregon.)

Photo 2-37. A completed dome built above a full-size first story with a slab foundation. (Photo courtesy of Cathedralite Domes, Medford, Oregon.)

Photo 2-38. The interior of a residential dome. Note how the shell can span the entire house without the use of interior bearing walls. (Photo courtesy of Cathedralite Domes, Medford, Oregon.)

Photo 2-39. The upstairs interior of a dome. Again note vaulted ceiling's interior partition walls. (Photo courtesy of Cathedralite Domes, Medford, Oregon.)

Photo 2-40. The floor system of a hexagon is a series of radiating 4-inch-thick floor beams. (Photo courtesy of Honeycomb Systems, Inc., San Juan Bautista, California.)

that you can build. Since the maximum amount of space is enclosed by the minimum amount of surface, the heat loss is cut to a minimum. Dome manufacturers claim that they will out-perform any other type of structure with a comparable amount of insulation. Domes provide natural air circulation, they have good acoustics, and they require less maintenance, since there are no corners for moisture or dry rot to build up in. Unlike many types of homes, no interior walls are needed to support the roof. Indeed, the entire dome can be one open room. Because of this you have a much greater design flexibility. Often the shell can be put up in a day or two and allow you to work inside with protection from the weather. Domes can also withstand a great amount of wind and snow and, because of their stable shape, they are very earthquake resistant.

The structure of a dome is entirely that of one large roof rather than walls and roofs as in many types of construction. Roofing, therefore, must be done properly to avoid leaks. Almost all manufacturers have developed good roofing techniques and this is no longer a problem, but be sure you do the roof properly. Since there are few vertical walls but many

sloping walls and angles, decorating can be a challenge in a dome home.

One problem with dome homes might be that financing may be difficult to obtain. However, more and more banks are getting used to seeing this type of construction, and if it is a problem with your local bank, with the help of your manufacturing company and perhaps some other local dome owners, you should be able to convince them to finance the home.

Before deciding to build a dome home, go and visit several different domes. From my experience in dealing with owner builders, I have found that most people who build a dome know almost instantly upon going in their first dome that this is the type of housing they want to build. They feel more comfortable with the angles and the sloping walls than they do with the true vertical and horizontal planes and the 90° angles that most other houses offer. In making this important decision, ask yourself these questions: How does the dome feel to you? Does it make you feel at peace? Do you feel more at rest in a dome, as opposed to a conventional type of structure?

Photo 2-41. The completed framing of a hexagonal home. (Photo courtesy of Honeycomb Systems, Inc., San Juan Bautista, California.)

Hexagonal Homes

(Photos 2-40–2-45.)

Another type of construction that is available in kits in the United States is hexagonal construction. As the name implies, this type of construction involves 6-sided buildings. This structure creates a rather unique living space with almost no right angles anywhere in the structure. Because of this and the fact that the structure needs no interior supporting walls, your design parameters and the feel of the house can be quite different from other types of construction.

The foundation system is built in a hexagonal shape. The floor system is built on top of this using beams that radiate in a hexagonal shape. A subfloor is applied and posts are erected around the exterior of the house. The posts will support the large roof beams. At the center of the home these roof beams are supported either by more vertical posts or by a "truss ring"—no vertical posts are needed in the center of the home. Walls constructed between these posts simply serve the purpose of attaching the siding, doors, windows, insulation and interior wall covering. Like the dome,

the hexagonal structure uses the strength of the triangle to give it stability and flexibility. It is a series of interactive triangles from which it derives great strength.

Hexagonal homes are available from several kit manufacturers or can be built by the owner from scratch. Since the pieces are all cut from patterns and the design is a simple one it is easy for the novice builder to build. Materials are easily available and it will usually meet all local codes. Since the weight of the roof is carried by the wall posts and not the walls themselves, windows and doors can be easily added and deleted. The buildings can be built up to three stories high or can be clustered together in a series of small hexagons. This allows you to add space as it is needed or can be afforded. Because of the vaulted ceilings and lack of right angles, the interior spaces can often feel larger than they really are. In many ways these homes reflect a certain change in lifestyle, one that is individual and yet strives for unity and order.

Photo 2-42. Hexagonal buildings have a good view from all sides. (Photo courtesy of Honeycomb Systems, Inc., San Juan Bautista, California.)

Photo 2-43. Hexagonal buildings can be built in clusters. This makes it easy to add on as the family grows or more funds become available. (Photo courtesy of Honeycomb Systems, Inc., San Juan Bautista, California.)

Photo 2-44. The interior of a hexagonal home. The large ceiling beams rest on the structural thrust ring, a feature which provides integral strength and flexibility to the hexagonal home. (Photo courtesy of Honeycomb Systems, Inc., San Juan Bautista, California.)

Photo 2-45. The interior of a hexagonal structure. Note the open spaces and the lack of right angles. (Photo courtesy of Honeycomb Systems, Inc., San Juan Bautista, California.)

Included in this chapter are pictures (Photos 2-46–2-51) of various styles of homes popular in different parts of the country.

Photo 2-46. Natural stone houses are rare, but permanent. These houses are labor-intensive, but beautiful.

Photo 2-47. Partial cut stone and post and beam structure.

Photo 2-48. Brick structures are common in some areas. Often stud frame houses are covered with a brick veneer.

Photo 2-49. Concrete block buildings are common in many areas. This one is in Florida where termites would pose a threat to wood frames.

Photo 2-50. A-frames are a quick, easy and inexpensive type of construction. I built this one in North Carolina in 2 months. It is 1200 square feet.

Photo 2-51. A house I built from a large wine barrel. Six inches of wine was left in it to create a "wine cellar."

3 Tools

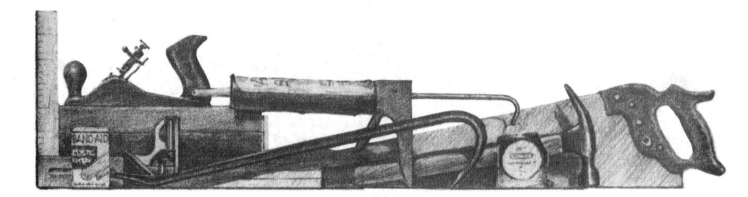

The following section on tools was written by Mel Berry. Mel has taught people to remodel their homes for many years and is a licensed cabinet maker.

THE FIRST CONSIDERATION, before starting to build portions of a house, should be to get good quality tools and understand how to use them. Frequently those who have not done a large construction project have gathered a collection of "used" tools made up of other people's cast-offs, slightly broken and usually dull. They have been satisfactory for putting a door back on its hinges, cutting firewood, replacing a stair tread or a piece of decking, but now the greater task of many hours in housebuilding demands a careful evaluation of tools available and their condition. The most highly skilled craftsperson would not be able to do satisfactory work in a reasonable period of time if his tool collection was not as carefully developed as his work habits and practices. The pleasures of building and creating as a craftsperson will change to frustration if dull screwdrivers tear out screw slots; dull chisels cause errors and even injury because of excessive pressure and force; dull hand saws leave one's arm aching and hand blistered; and dull power cutting tools smoke, burn and kick back.

Buying Tools

A builder should get the best tools he can afford. Choosing well-known brands is probably safer for a novice. In comparing both power and hand tools, consider balance, weight, comfort and durability. An electric drill with too small a handle feels "top heavy." A 32-ounce framing hammer for a large person will drive nails quickly, but would be too much for a smaller person. A chisel with an all-plastic handle will not take

as much pounding as a chisel with the steel of the blade extending all the way through the handle. In most cases the better tools will cost more, but it is a necessary expense and actually small, considering the size of the housebuilding project. Also, better tools frequently carry manufacturer's warranties and even lifetime guarantees. Once sharpened, good cutting tools keep a sharp edge. (Photo 3-1.)

Sharpening Tools

Most carpenters today have their saws sharpened by professional sharpening companies, but they sharpen their own plane irons, knives, screwdrivers and chisels. Regrinding plane irons, knives, and chisels is not usually needed unless the tool has hit a hard object during use and has chipped. If regrinding is necessary, it can be done by some saw sharpening shops or the carpenter can re-do the shape if he has his own bench grinder. For routine sharpening of tools, a craftsperson needs metal files, a combination bench stone and honing oil. Metal files are used for reshaping screwdrivers, hatchets, shovels, chainsaws and other tools. For some tools, filing is satisfactory sharpening. Other tools need honing on a bench stone.

A combination bench stone has a coarse side and a fine side. Use the same procedure for sharpening chisels and plane irons. With the bevel at the angle of the original grinding, move the tool in a figure 8 pattern over the surface of the stone until the edge is improved, usually about 30 to 40 strokes. Honing oil applied to the surface of the stone keeps the metal from filling the pores of the stone. Then turn the stone over and continue honing, using the fine side of the stone. The back or flat side of the tool is rarely ground or stoned, except for a few light strokes occasionally to remove

Photo 3-1. Avoid cheap tools. They can be dangerous when they fail.

the metal that curls over during sharpening, creating a burr. The figure 8 pattern on the stone is not essential, but helps to keep the oil distributed. If the stone gets "dirty," or clogged with sawdust, it must be cleaned with a solvent, such as kerosene. Other stones that can be used for fine honing include the natural Arkansas stones and Japanese water stones. After sharpening, a tool should be stored in a protective covering to keep its fine edge.

Basic Tools

These basic tools should be collected before attempting to build. They will all be needed frequently.

Combination stone	Nail sets
Metal files	Framing hammer
20- or 25-foot tape	Screwdrivers
Plumb bob	Phillips screwdrivers
Chalkline	Pliers
Pocket knife	Adjustable wrench
Utility knife	Block plane
Carpenter's pencil	4-in-1 file
Carpenter's square	Flat bar
Combination square	Nail claw
2-foot, 4-foot level	Twist bits
Crosscut saw	Spade bits
Rip saw	Circular saw
Coping saw	Sabre saw
Hacksaw	⅜ variable speed reversing
Wood chisels	belt sander
Cold chisels	Finishing sander
Diagonal cutting pliers	Heavy duty extension cord
Finishing hammer	Power drill
Speed Square	

Secondary Tools

Secondary tools have less frequent usage than the basic tools, but most of them should eventually be added to a builder's toolchest. Until then, they will have to be borrowed or rented when they are needed.

Bench grinder	Router and bits
50- or 100-foot tape	½-inch right angle drill
Builder's level or transit	Reciprocating saw
Sledge hammer	Power miter saw
Bit brace	Power plane
Hand drill	Com-a-long
Power auger bits	Rabbet plane
Forstner bits	Nail shooter
6-foot level	
Aviation snips	
Assorted files	
Ripping bar	
Countersink	

Specialized Tools

Specialized tools are infrequently needed tools, or an expensive purchase that can only be justified with a great deal of use on a single job or repeated professional usage.

Laser level	Pneumatic stapler and
Foundation stake puller	equipment
Hardwood flooring nailer	Radial arm saw
Shingle ripper	Table saw
Battery powered drill	Tool specialties from various
Pneumatic nailer and	trades
equipment	

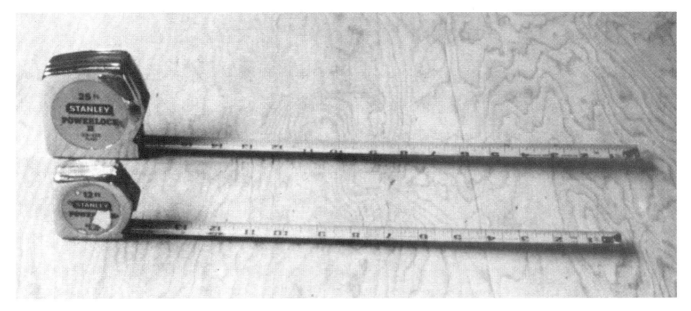

Photo 3-2. Metal tapes are an essential. I recommend a 20- or 25-foot tape with a 1-inch blade.

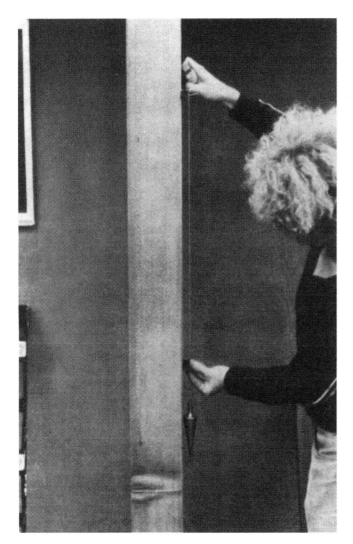

Photo 3-3. Here is an example of using a plumb bob to see if a post is plumb. Note that a block is used at the top and bottom of the string to see if it is an equidistance from the post. Plumb bobs are also used to locate points directly below other points.

Layout Tools

STEEL TAPES (Photo 3-2) have largely replaced the folding wood rules for most people. The usual lengths are 20 or 25 feet for general measurement, and 50 or 100 feet for larger layout jobs. The most desirable width for rough carpentry is 1 inch, since it allows the worker to extend the tape further before it "buckles."

A PLUMB BOB (Photo 3-3) is necessary to determine a reference of straight up and down or right angles to the earth. While any heavy object suspended on the end of a string will determine this reference by gravity, usually a craftsperson will purchase a machined plumb bob for greater accuracy. The heavier a plumb bob is, the less it will be affected by the wind while it is being used.

A UTILITY KNIFE (Photo 3-4) should be of high quality steel and fit comfortably in your pocket. Some craftspeople, however, prefer to have a larger knife and keep it in their tool pouch or on their belts. The choice is yours, of course, but a knife is necessary to cut cord, mark closely around hinges for mortises, sharpen pencils, etc. Most craftspeople will also have a utility knife for general use.

The CARPENTER'S PENCIL (Photo 3-5) is a flat pencil with large lead. It is easy to sharpen, draws broad lines that are easy to see and it won't roll away.

The T-BEVEL (Photo 3-6) or bevel square has an adjustable blade which can be set to any angle. It is usually used to transfer angles from the work spot to the cutting tools.

The COMBINATION SQUARE (Photos 3-7, 3-8) has several uses. Frequently it is used to mark a 90° line across a board to follow when cutting or when joining another board to it. It will also measure a 45° angle, check for square, and measure and transfer depths and distances.

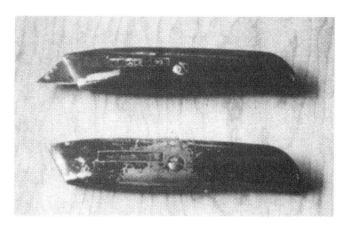

Photo 3-4. Utility knives are needed in many processes during construction, from cutting the layout strings to scoring the drywall. Be sure to get one that has a retractable blade; extra blades are stored in the body.

Photo 3-5. Carpenter's pencils are a must. They come in broad and narrow leads. Keep a supply on hand.

Photo 3-6. Using a sliding T-bevel to find an angle. This angle can then be transferred to the board you are cutting.

Photo 3-7. Using the combination square to mark a board at a 90° angle. This square can also be used to mark a 45° angle and to scribe the length of the board.

Photo 3-8. Using the combination square to mark a straight line down the length of the board. This method ensures that the line you are marking is an equidistance from the edge of the board even if the board is not straight.

Photo 3-9. Using a framing square to lay out a stair tread. Framing squares are used throughout the housebuilding process. Get a good one and be careful not to leave it lying around where it could get stepped on and bent. It is of no use once it is bent.

Photo 3-10. A chalkline is used to mark straight lines.

Photo 3-11. Using a 6-foot level to check the plumb of a post. Note the clips at the top and bottom of the level. These clips keep the level off the board which will have many irregularities and make it hard to determine true plumb. You will need a 2-foot and 4-foot level, and a 6-foot or 8-foot is advised.

Photo 3-12. A torpedo level or 1-foot level comes in handy. It's being used here to level a pier block.

Photo 3-13. Transits, or their cousins the builder's levels, are needed during the layout. You can rent these, but if you can afford one, they are very useful during many states of construction in checking for level and plumb.

The large CARPENTER'S SQUARE or FRAMING SQUARE (Photo 3-9) has many uses in addition to measuring squareness of work, as mentioned in other chapters of this book. Today most carpenters choose aluminum framing squares since they are lighter to handle. Some squares have one special scale measured in 12ths of an inch for reading scale drawings.

A CHALKLINE (Photo 3-10) is used to lay out straight lines for reference and assembly. It can be a length of line and a block of carpenter's chalk, or it can be a line inside a chalk box that applies powdered chalk as the string is drawn in and out.

Levelling Tools

A 4-FOOT LEVEL is used for spot checks of plumb and level. For more accurate measurements, a 6-FOOT LEVEL (Photo 3-11) is necessary. Some craftspeople extend the use of the

2-foot level by putting it on a longer straight edged board. A 1-FOOT LEVEL (Photo 3-12) can be used for levelling smaller surfaces.

A BUILDER'S LEVEL is not part of every carpenter's tool collection, but is essential for efficient layout of footings and foundation walls. The builder's level establishes a level line of sight; all points along that line of sight are the same as any other point.

The TRANSIT (Photo 3-13) has one more operating mode than the builder's level. It pivots up and down in a vertical plane and enables the operator to measure and transfer points in that plane. It can also be used to plumb work. These tools are expensive for one-time use and can be rented. (For an explanation on the use of a transit, see *Layout* chapter.)

A newer levelling device, the LASER LEVEL, projects a beam of coherent light onto surfaces for a reference indication. Laser levels, however, require electrical power, which may not be available in the initial layout of residential homes.

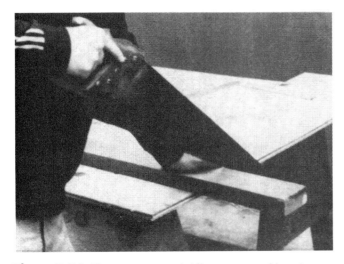

Photo 3-14. There are several different types of hand saws; their use depends on whether you are cutting across the grain (crosscut saw) or with the grain (rip saw). A finishing saw has teeth that are set closer together. You may want to buy a general combination saw that can cut in either direction.

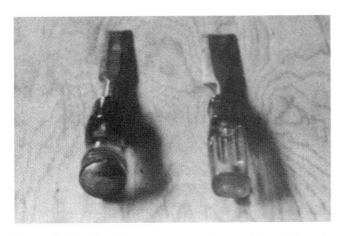

Photo 3-15. Chisels are a must. I advise a ¼-inch, ½-inch, ¾-inch, 1-inch and a 1½-inch. Get the kind where the metal shaft extends all the way to the butt, as shown on the left. The metal shaft on the right is embedded in plastic and will not last as long.

Photo 3-16. Several kinds of hammers are needed during house construction, i.e., a framing hammer, finishing hammer, drywall hammer, etc. Use ones with fiberglass or wood handles.

Cutting Tools

SAWS (Photo 3-14). Today's craftsperson cuts mostly with electric tools, but a crosscut saw, with about 8 knife-shaped teeth per inch, is used when power is not available or convenient and when the round-cutting pattern of a circular saw is not acceptable. When cutting stair stringers for example, the circular blade will not cut completely into the corner of the layout (see section on stairs). A sharp crosscut saw may also outperform the circular saw if lengthy extension cord layout is necessary. A blade cover should be purchased or fabricated to prevent damage to the teeth. Finish saws are usually slightly shorter and may have up to 12 teeth per inch.

A RIPSAW cuts with the grain of the wood and has chisel-shaped teeth to keep the saw from following the grain. Nearly all ripping cuts today are done with a circular whenever possible.

A COPING SAW or hand jigsaw is used especially to shape moldings to fit. Its fine blade allows intricate cuts over short distances.

A HACKSAW is necessary to cut metal stock and can be used for wooden moldings.

WOOD CHISELS (Photo 3-15) from ¼ inch to 2 inches in width are necessary to remove stock for hinges, shaping openings and patterns. High quality steel is necessary to keep a sharp edge longer. Strong handles that will take hammer blows are essential for a carpenter since mallets are rarely available. A cold chisel is necessary to cut through nails and undesired metal pieces.

Assembling Tools

A HAMMER (Photo 3-16) is probably one of the first tools associated with carpenters. Many carpenters have several hammers for different purposes and enjoy their unique features.

FRAMING HAMMERS are available in weights from 16 to 32 ounces with ripping and curved claws. Although few people find the heaviest weights comfortable, choose the heaviest hammer that can be used without tiring. The straight or ripping claw is useful for driving between pieces and as a lever. The curved claw gives more nail-pulling leverage. Framing hammers are also available with waffled faces to prevent slipping from the nail. FINISHING HAMMERS have smooth faces and range in weight from 7 to 16 ounces. Curved and rip claws are available. A general purpose hammer that could be used for both framing and finish work would have a smooth face and a 16-ounce head.

SCREWDRIVERS are necessary for both Phillips and slotted screws. Various sizes of each screwdriver are also necessary to fit all the screws used in construction. Standard screwdrivers are sized by the length of the blades and the size of the tips. Phillips screwdriver sizes refer to the size of the tips and are numbered from 0 to 4. A Number 2 Phillips screwdriver is the most common and will fit most Phillips screws, but other sizes will eventually be necessary to insure a proper fit on all screws.

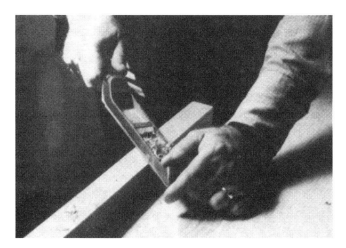

Photo 3-17. This tool works much like the plane.

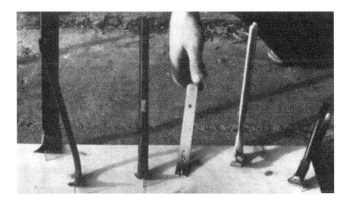

Photo 3-18. Above are various types of nail-pulling tools. Needless to say, it is highly recommended that all novice owner builders have a complete assortment as shown here. They will be well used. From left to right: large flat bar (wonder bar), small crowbar, cat's paw, small flat bar and a large and small set of nail clippers.

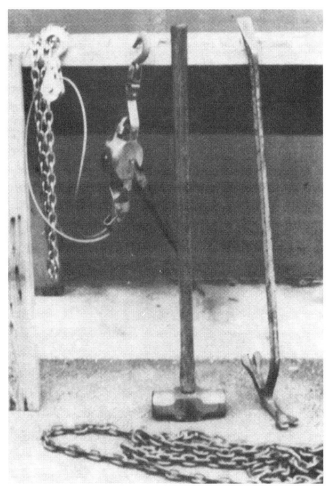

Photo 3-19. Shown here are some large wrecking and timber moving tools. It is a good idea to have the full assortment at the job site. From left to right: a hand winch (com-a-long), a sledge and a large crowbar.

Pliers and Wrenches

PLIERS are needed to hold steel pieces and fasteners during assembly.

For careful construction, nuts and bolts should have wrenches of appropriate size; however, a high quality AD-JUSTABLE WRENCH is very valuable in holding a variety of pieces.

Smoothing Tools

The BLOCK PLANE is necessary for jobs such as shaving small amounts from a door, window, or a finish piece of wood. While power planes are replacing the larger sizes of hand planes, the control necessary for small adjustments and shaping makes it important to always have a block plane handy. (Photo 3-17.)

Several files of various shapes, sizes and coarseness are desirable. The most universal file for wood is a 4-IN-1 FILE. This file has 4 different surfaces which are quickly available in one tool, and it is easily carried in a tool belt. One side has coarse and fine flat surfaces. The other side has coarse and fine curved surfaces.

Prying and Wrecking Tools

The RIPPING BAR, also called a wrecking bar or pry bar, is used to pull previously nailed pieces apart, withdraw nails and lever pieces into place.

A FLAT BAR is used to remove trim and molding, separate previously nailed framing pieces, and withdraw nails. It can be driven between pieces with a hammer and pried straight back or to the side to separate pieces.

The NAIL CLAW, also called a CAT'S PAW (Photo 3-18), is driven into the wood under the head of a nail to start to pry it out. Further prying is done with a hammer, flat bar or wrecking bar. Both novices and pros make mistakes, so have one of these in your tool box.

SLEDGE HAMMERS (Photo 3-19) are necessary to drive stakes, wreck or move framing.

Boring Tools

Most holes are drilled with ELECTRIC DRILLS today, and some drill motors are battery powered. Bit braces and hand drills are useful only when electricity is not available or convenient. The type of hole necessary determines the bit that is used. For holes up to ½ inch in diameter, machine or twist bits are frequently used. These bits are useful for metal or wood cutting. For larger holes, spade or power bore bits can be used. However, to cut a truly round hole, remove chips quickly, and drive easily, a power auger bit is used.

POWER AUGERS are used with large drill motors at lower speeds. When very accurate holes with crisp edges are needed, as for exposed dowels or plugs, a craftsperson can use FORSTNER BITS. When screw heads must be flush or below the surface, holes are first drilled with a countersink (Photo 3-20).

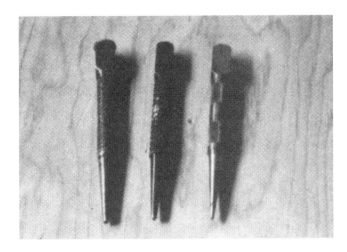

Photo 3-20. Nail sets or nail punches are available in different sized tips to countersink different sized nails.

Photo 3-21. Metal legs for saw horses are easy to use and more secure than wooden legs.

Power Tools

The CIRCULAR SAW (Photo 3-22–Photo 3-27) is usually one of the first power tools purchased. It is necessary to efficiently cut framing pieces. Since the saw will be used frequently it is important to choose a high quality tool that feels comfortable. The blade must always be sharp to prevent overloading the tool, damaging the work or causing a kickback toward the operator. To keep a sharp blade on the tool, some craftspeople choose carbide tipped blades which keep their keen edge longer. However with the rough service sometimes demanded of a blade, you may choose to have many less expensive steel blades to change frequently and have some at the saw shop for resharpening. Resharpening costs for steel blades are less.

Most Common Mistake

Setting a board *between* two supports to be cut. As the board droops during the cut it binds the blade and causes the saw to kick back.

Photo 3-22. Circular saws have come a long way from this original one.

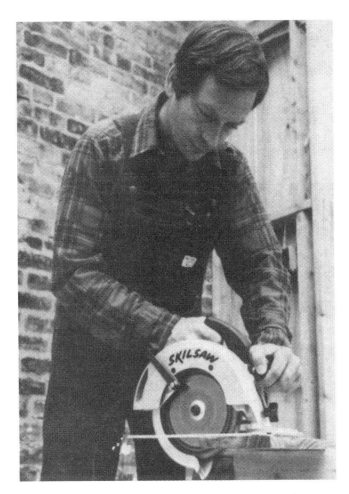

Photo 3-23. The power circular saw is perhaps the most commonly used power tool during the housebuilding process. Be sure to choose one that will last and that feels comfortable to you. The model above is referred to as a "sidewinder" as opposed to a "worm drive" model. Try them both and see which feels more comfortable.

Photo 3-24. With the proper masonry bit, power circular saws can be used to cut concrete block and brick.

Photo 3-25. A worm drive saw with blades. Note how dull blades are separated from the sharp blades on plywood boards. Have a good assortment of blades, both carbide tipped and standard.

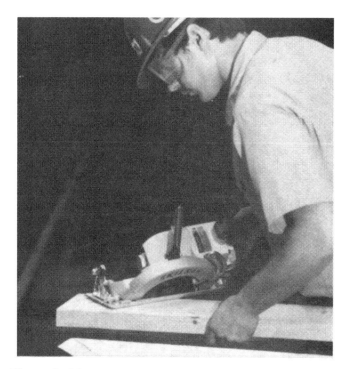

Photo 3-26. Most circular saws can be adjusted so that the saw is at an angle to the cutting platform. This allows you to bevel cut the board up to 45°.

Photo 3-27. This 5½-inch circular saw is good for cutting paneling, siding and trim. Not necessary, but useful if you can afford it.

The SABRE SAW (Photo 3-28) is a portable jigsaw used for cutting a variety of materials and shapes. Blades are available for cutting wood, plywood, metal and plastics. The tool is useful for making cutouts in the center of building materials, including sheets of plywood and laminated countertops. Because the sabre saw has a very small blade, it is also very useful in making curved cuts.

Helpful Hints and Most Common Mistakes

☐ Finished surfaces such as plastic laminates can be scratched by the base of the sabre saw. Covering the base of the saw with tape will prevent damage.

☐ Be careful not to pinch a finger between the blade chuck and the body of the tool while it is operating.

☐ Choose the appropriate saw blade to insure a fast and smooth cut.

Photo 3-28. A jigsaw is very useful at most building sites. It can cut curves and angles and fancy scrolls.

ELECTRIC DRILLS (Photo 3-29–Photo 3-31). The most universal drill for a carpenter is a ⅜-inch, variable speed, reversing drill. This tool allows bits up to ⅜ inch to be used for drilling. Occasional larger holes can be done with wide power bits with smaller shanks. For continued large hole boring and heavy-duty use, a more powerful ½-inch drill is more appropriate. As chuck capacities go up, drills are designed to turn more slowly to produce more torque with less likelihood of operator injury if the bit should jam.

It is also possible to use this ⅜-inch drill with a screwdriver tip (usually Phillips) in the chuck to make it a power screwdriver. The variable speed trigger plus forward and reverse features provide ample control.

Photo 3-29. ¼-inch drills are handy for smaller projects, but usually ⅜-inch or ½-inch drills are used for larger jobs.

Photo 3-30. A ⅜-inch, heavy-duty drill is a must for anyone planning to do their own plumbing and electricity.

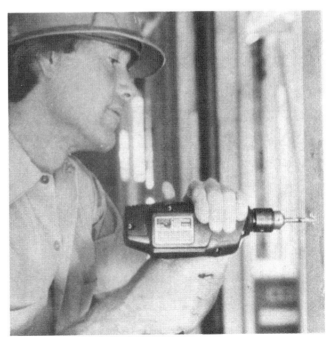

Photo 3-31. Cordless, rechargeable drills are handy during all stages of construction.

Photo 3-32. Small cordless screwdrivers are a new tool on the market. They come in very handy in many phases of building.

The BELT SANDER (Photos 3-33, 3-34) has a cloth-backed abrasive belt that travels between the two wheels of the tool. The base of the tool is flat steel and if the tool is held level, it will remove uneven and rough material, and create a surface ready for finish sanding. Practice and skillful handling are essential to the operation of this tool, so get a "feel" for the tool on a piece of scrap stock before sanding the finish piece.

Sanding should always be done by using progressively finer grits of abrasive belts and paper. Each finer grit removes the scratches of the previous grit until the desired finish is achieved.

Helpful Hints and Most Common Mistakes

☐ Belt sanders must be held level to achieve good results. Often it is necessary, when doing surfaces with pieces cut out of them, to imagine a completed surface and hold the belt sander to the imagined plane. Dipping the sander into the openings or applying more pressure to the corners will cause gouging of the work.

☐ Belt sanders must be kept moving from area to area or the belt will dig in.

☐ Sanding belts are usually directional and putting belts on backwards tears the seams. Some non-directional belts, however, allow additional sharp usage by reversing them.

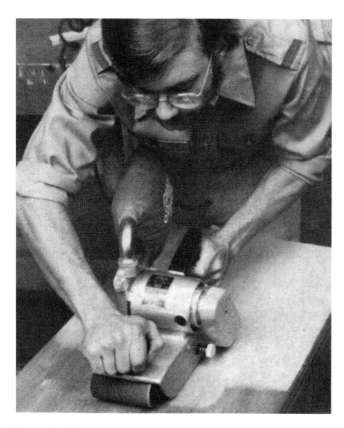

Photo 3-33. A heavy-duty belt sander like this is usually not needed in a typical housebuilding process. It is used for large sanding jobs where the weight of the machine is needed to make sanding easier.

Photo 3-34. The belt sander is used for larger sanding jobs. Wearing safety glasses as shown is recommended whenever you are using any power tool. If you wear glasses, use hardened glass or plastic.

The FINISHING SANDER or PAD SANDER (Photo 3-35) is useful in smoothing away tool marks and wood fillers. Some are available with orbital and "in-line" motions. Orbital action removes stock more quickly, while "in-line" action allows sanding with the grain of the wood for a smoother finish. Finishing sanders have a cushioned base of felt or rubber, and are designed for smoothing. To remove a lot of material and make a flat surface the belt sander is used.

Helpful Hints and Most Common Mistakes

☐ Different level wood joints sanded to flat with a finish sander will be smooth, but not flat and straight.

☐ Trim wood should be sanded with a grit of sandpaper as high as 220 to remove machine marks that will be evident when the stain is applied; but sanding with too fine a grit can burnish the wood cells closed and cause a blotchy stain effect. The highest sanding grits are used between coats of finish.

Additional Tools

ROUTERS (Photo 3-36) are basically edge-forming tools, but with additional guides they can be used to cut grooves, rabbets, dadoes and mortises for hinges. A collection of bits for routers can frequently exceed the cost of the router itself. Bits are available from high-speed steel or carbide-tipped steel. When pilot tips are part of the cutter, steel bits have a fixed pilot and carbide cutters usually have ball bearing pilots. The carbide cutters stay sharp longer, cut more cleanly, and the ball bearing pilots do not burn the wood. For infrequently used profiles, the steel bits allow a greater variety of cutters at a more economical price.

Helpful Hints and Most Common Mistakes

☐ The base of the router must be kept flat to the board during use to insure profile accuracy.

☐ Since the cutter revolves in a clockwise direction (as seen from the top) the router should be moved from right to left to move the cutter into the work. A full-depth cut from left to right will cause the cutter to "climb" the wood and start to self-feed; however, a shallow precut from left to right followed by a full depth cut from right to left will prevent splintering on difficult woods.

☐ Rapid movement can cause splintering of the wood. Slow movement with solid steel pilots will cause the cutter to burn the wood. Practice on a scrap of wood to develop the proper technique.

Photo 3-35. Finishing sanders are used for lightweight sanding such as furniture refinishing. Many people use them when sanding dry wall as well.

Photo 3-36. A router is an excellent finishing tool. It can round edges, make different types of grooves, finish formica, etc.

Photo 3-37. A heavy-duty extension cord with a four-way junction box is essential.

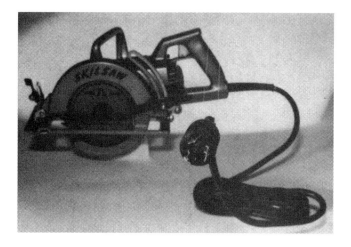

Photo 3-38. Many professionals use a locking plug as shown here. This male plug takes a corresponding female adapter. Once inserted and twisted, the plugs are locked together and will not pull apart. You may want to consider changing your regular plugs on your commonly used tools to this style.

HEAVY-DUTY EXTENSION CORD (Photos 3-37, 3-38). Along with the purchase of your first portable power tool should come the purchase of a heavy-duty (large wire diameter and heavily insulated) extension cord. When power must run a long distance to the tool, it loses effectiveness with small cords and can damage the tool. Nor will a small cord stand up to the wear and tear of use on a job site. Local safety requirements may dictate that the cord and tool be outfitted with twist-locking plugs, as shown, for secure connections.

A heavy-duty ½-INCH DRILL is essential for those who do their own plumbing and electrical work, and very desirable for boring long or frequent large holes for bolts and other fasteners. A right angle drill allows the operator to get in between framing members easily.

Helpful Hints and Most Common Mistakes

☐ Many bits used for drilling large holes have self-feeding pilot screws that assist in drawing the cutting edges into the wood. Care should be taken to keep face and hands clear of projecting handles. Frequently, injuries occur when a bit binds in a hole and the tool is wrenched from the operator's hands.

☐ Bit extensions and long bits are available to drill through several studs in one pass.

☐ When drilling deep holes, withdraw the drill bit occasionally to clear the chips.

The RECIPROCATING SAW (Photo 3-39) is similar to a sabre saw but much heavier in construction. It is the surgery tool of house construction. It is used to carve out holes for vents, cut openings for lights and other objects, notch studs for plumbing or electrical wires, and modify existing framing. Metal blades will cut nails, pipe, conduit and other pieces. It is an all-purpose saw, and its long-bladed snout allows the operator to reach into otherwise inaccessible areas.

Helpful Hints and Most Common Mistake

☐ Reciprocating saws bounce less and break fewer blades if the saw guide of the tool can be held tightly to the material.

POWER MITER SAWS (Photo 3-40) have largely replaced hand miter boxes in the woodworking shops and at construction sites. Nicknamed "chopsaws," they easily cut through material as thick as framing stock, bringing cabinet shop accuracy to the field, for clean miters on casings, handrails and all moldings. By setting a block to butt against, power miter boxes can be used to cut many pieces to the same length or angle without repeating measurements.

Helpful Hints and Most Common Mistake

☐ The power miter saw should not be overused. Straight cuts for framing that do not require this accuracy should be made with the circular saw at the spot they will be used. This saves time and handling effort.

A POWER PLANE quickly smoothes surfaces and trims materials to size. It is operated in the same manner as a hand plane, but since the rotating cutters smooth the surface, the tool moves more easily and the motion is not affected by knots or irregular grain.

Helpful Hints and Most Common Mistakes

☐ The blades are exposed on the bottom of a planer and subject to damage if they come in contact with a hard surface, while running or not.

☐ When starting cuts, pressure should be applied more to the front of the tool. When completing cuts, pressure should be applied more to the rear of the tool.

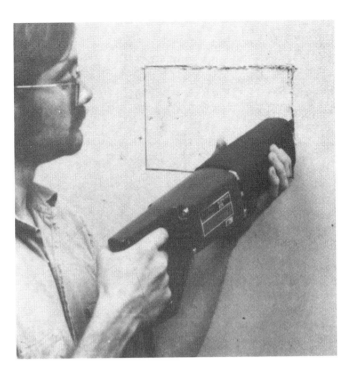

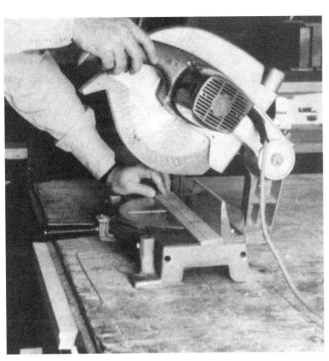

Photo 3-39. The reciprocating saw has a knife-like blade that protrudes from the front of the saw and cuts in a back-and-forth direction. When you need this tool, nothing else will do.

Photo 3-40. A power mitre saw (chopsaw) is a professional tool used for finishing work. You may want to rent one or buy one if you are building a large custom house.

Photo 3-41. Pneumatic tools, powered by compressed air, make many jobs quicker and easier. A professional tool, they can usually be rented at certain phases of construction.

Photo 3-42. A radial arm or contractor's saw is useful both during the framing and finishing stages. It is expensive and not necessary for a one-house project, but very useful if you can afford it.

PNEUMATIC TOOLS (Photo 3-41). Pneumatic (air driven) tools such as staplers and nailers are available to increase production. When a good hand rhythm is achieved, work goes quickly and smoothly. To operate pneumatic tools, the tool must be connected with an air hose to a compressor. This air hose "umbilical" is restrictive and of questionable value, especially considering the investment necessary for a single house.

Some contractors find pneumatic tools useful, however, for multiple units and large square footage jobs such as siding and roofing, and some framing sub-assemblies. The decision to use pneumatic tools or not depends on the job conditions, craftspeople, project design and total square footage. Pneumatic tools are enjoyable to use, less tiring, fast and effective, but not essential.

RADIAL ARM SAW (Photo 3-42). One of the first stationary power tools for many people is the radial arm saw. It is useful on the construction site for cutting beams, repetitious framing pieces, angles and compound angles. This saw is basically designed to draw the blade across (crosscutting) the wood. Ripping (cutting with the grain) in the long direction of a board is usually done with a portable circular saw or table saw.

Most Common Mistake
☐ Since the radial arm saw is a stationary tool the work must be brought to the tool. Making too many cuts with this saw, especially when work must be transported from another floor, is very time-consuming. Unless there is unusual thickness or angles, or a need for extreme accuracy is involved, the circular saw should be used for framing.

The TABLE SAW (Photo 3-43–Photo 3-46) excels at cutting sheet stock such as plywood, and cutting the long way of a board (ripping). It is more difficult to cut across the grain (crosscut) on a table saw since the whole board must be moved across the blade. For carpenters who will be doing their own cabinets and built-ins, a table saw will be very useful.

Most Common Mistake
☐ The table saw is outfitted with two guides to use in cutting stock, a miter gauge used at angles to the blade and a rip fence which is parallel to the blade. When cutting stock, choose the guide that will allow the longest edge of the stock to be against the guide. Do not have the stock touching both the miter gauge and the rip fence while cutting, or binding and kickback may occur.

Photo 3-43. Table saws are excellent for doing trim and finishing and cabinet work. They are not a required tool for the one-time owner builder, but are useful if you can afford them. The one above is a smaller homeowner's model. It should be all that is needed for a one-house project.

Photo 3-44. Large bench saws are not essential, but very useful if available.

Photo 3-45. Table saws can use dado blades to make grooves like these. Many other blades are also available.

Photo 3-46. Band saws are used mostly in finishing and sculpturing work. They are useful for cutting angles and curving pieces of wood. You can probably get by with the hand-held equivalent, the jig or sabre saw.

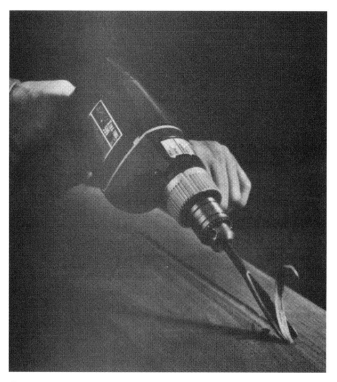

Photo 3-48. Shop vacuum cleaners are useful, especially during construction stages.

Photo 3-47. This "Xtra-tool" is several tools in one. It can serve as an electric screwdriver or drill, as well as a hammer-drill, or hammerchisel. It can drill into brick or concrete, chisel or gouge wood, and be used to remove paint or linoleum as well as chip concrete. A useful tool.

Specialized Tools

Specialized tools are not part of everyone's toolchest. They have specific purposes and are seldom needed, but essential when they are needed. You may want to buy, borrow or rent these tools. Also a tour of the local tool rental shop may find a tool you would like to have occasionally. A FOUNDATION STAKE PULLER saves time and money, for instance, but is a one-day need.

Other tools to consider are the com-a-long and hardwood floor nailer. A COM-A-LONG is a device with a cable and ratcheting handle used to pull framing sections into square and plumb. The FLOOR NAILER holds and aligns the nails along the edge of hardwood flooring. It drives a nail when struck with a hammer or mallet.

Additional hand tools with specialized purposes are the SHINGLE RIPPER for removing nails during roofing repairs, RABBET PLANE for shaping rabbet cuts in wood and a NAIL SHOOTER for driving nails in difficult spots.

Hundreds of tools also exist that are used for special trades only. Drywallers, carpet layers, tilesetters, roofers, plumbers, electricians and others have occasional need for more specialized tools. To become familiar with all the tools available, visit the trade supply distributors. While a one-time need will not justify a complete extension of your tool collection for every trade, some special tools may be helpful. (Photos 3-47, 3-48).

4 Layout

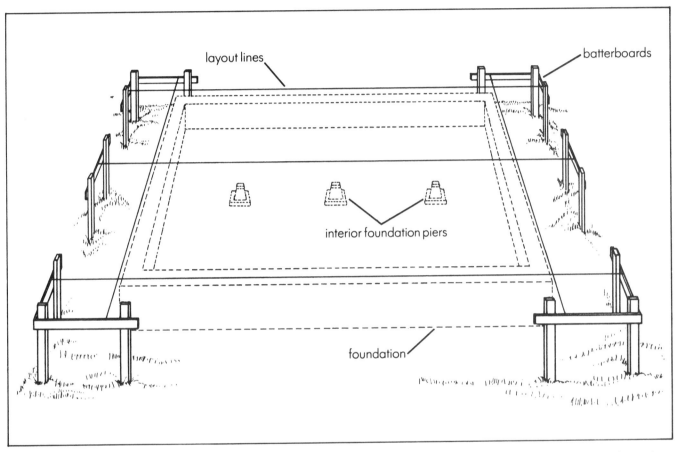

Figure 4-1. Example of a typical layout and its relationship to the foundation (dotted lines). Layout lines occur directly above the exterior edge of the foundation wall and over the interior foundation piers.

What You'll Be Doing, What to Expect

THE LAYOUT OR FOUNDATION LAYOUT, as it is sometimes called, is a process of suspending strings in the air over the building site. (See Figure 4-1.) These strings, called layout lines, indicate the exact edge of the future foundation walls as well as their exact corners. In addition to demarcating the perimeters and the corners of the foundation, the lines are all exactly level, thereby superimposing a level line on an otherwise irregular site. These lines are then used to build the foundation level. The structures that support these lines are called batterboards. Additional lines are run to indicate any interior foundation piers or walls. Lines are also used if you are planning foundation piers for the exterior walls instead of a continuous foundation. (See Photo 4-1.)

This process is one of the more mentally demanding processes in building, so be sure that you understand it well before beginning. Physically, it is a rather simple process, except for hammering the vertical supports into the ground, which can be difficult if the ground is hardpan. (In this case, some builders use round metal stakes as vertical uprights for the batterboards.) Mentally, the process is rather trying because it is essential that the layout be accurate as the foundation will be built from these lines. The process is one of trial and error and demands you keep working with it until the lines are level and accurate.

Since the batterboards and strings will eventually be taken down, there is a tendency not to do a good job or make the batterboards stable. Be careful of this tendency; the foundation will be off if the batterboards shift at all. Besides, you want to set healthy, high-quality building standards for yourself from the beginning.

Photo 4-1. A job site with the batterboards and layout lines in place and the foundation trenches dug.

Jargon

BATTERBOARDS. Temporary, wooden framework on which the layout lines or strings are attached.

BENCHMARK. A permanent mark on a stable object, such as a tree or the side of a building, to be used as a constant reference from which to compute elevations or levels.

BUILDER'S LEVEL. Telescope mounted on a tripod used in levelling and determining angles of wall.

CROSSMEMBER. The horizontal piece of the batterboards.

DUPLEX or FORM NAIL. A doubleheaded nail used in situations where the nails will be removed.

FOOTING. Concrete pad or support under the foundation wall or pier.

LAYOUT LINES. Strings suspended in the air, used to locate the perimeter and the corners of the foundation and as a level reference.

LEVEL. Perfectly horizontal.

PLUMB. Perfectly vertical.

PLUMB BOB. A heavy pointed weight on a string used to determine plumb or true vertical.

SETBACK. The regulated distance that a building is allowed to be built from the property line, easement line, road right of way, etc.

TRANSIT. Same as the builder's level but can also tilt up and down.

Purposes of the Layout

1) To locate the exact perimeters (outside edges) and corners of the foundation
2) To superimpose a level line on an otherwise sloping irregular site which will be used as a level reference during construction of the foundation
3) To assist in determining correct angles for intersecting walls
4) To serve as a guide for digging foundation trenches and erecting foundation form boards

Design Decisions

Before beginning your layout process, there is much to consider as to the exact location of the house. The subject of setting a home can fill a book on its own. Much has to be taken into account: solar, wind, privacy, view, setback requirements, parking, etc. Do not shortchange the setting process of the home; it is a crucial decision. Since I have discussed this subject thoroughly in our first book, *Before You Build* (Ten Speed Press, 1981), I will not go into it here. You must, however, be sure to consider all facets of the decision well before layout.

What You'll Need

TIME   24 to 32 person-hours per 1000 square feet of floor space on a relatively flat site, with a relatively simple house design.

PEOPLE   2 Minimum (1 skilled, 1 unskilled); 4 Optimum (2 skilled, 2 unskilled); 4 Maximum (1 skilled, 3 unskilled).

TOOLS   Levelling tool (builder's level, water or hydro level), hand or power saw, framing hammer, 3 metal tapes (one 100-foot tape), 4-foot level, knife, plumb bob, sledge hammer.
RENTAL ITEMS: Transit (optional)

MATERIALS   2-×-4 vertical supports, 1-×-4 horizontal cross members, short 2-×-2 pointed stakes, nylon string, 8d duplex nails, 8d standard nails.

INFORMATION   This chapter should give you everything you need for a typical house layout.

PEOPLE TO CONTACT   You may need to contact work people or suppliers you plan to use in the upcoming stages of construction.

INSPECTIONS NEEDED   Some subdivisions or local permit agencies may want to check your layout before you start to dig to be sure it does not encroach into setbacks or present other problems as to the siting of the house.

Reading the Plans

Your plans will tell you all you need to know about your foundation and thereby your layout. Your site plan will tell you the location of your house on your site. Be sure you have met any setback requirements as to frontyards and back-yards, there is enough distance between the house and the septic tank or well, and there is room for parking and the driveway. It is imperative to know exactly where your property lines are if you are building close to your borders. I know a couple who spent $1200 for a 2-×-100-foot piece of their neighbor's lot because they inadvertently built too close.

In addition to the site plan, the floor plans, foundation plans and structural plans will provide all the information you need for the layout.

Safety and Awareness

SAFETY: Be careful using the sledge, especially when someone is holding a vertical stake for you. Have them wear a hard hat and hold the stake a good distance below the top. Using gloves to hold the stake is also advised. Warn the person holding the stake not to look at the top in case there are flying splinters.

AWARENESS: Exactness is required throughout the process. Take your time. Be exact. Be careful not to upset the transit or batterboards once they are set. Be sure the batterboards are securely built. Remember it is a process of trial and error.

Overview of the Procedure

Step 1: Check the property lines

Step 2: Establish corner "A"

Step 3: Set up transit over corner "A"

Step 4: Locate approximate corner "B"

Step 5: Build batterboards behind corners "A" and "B"

Step 6: Run string directly over corners "A" and "B"

Step 7: Transfer corner "A" to string

Step 8: Measure along string "AB" to exact corner "B"

Step 9: Sight transit along string "AB"

Step 10: Rotate transit 90°

Step 11: Locate approximate corner "D"

Step 12: Build batterboards behind corner "D"

Step 13: Locate where string "AD" crosses batterboards at corner "D"

Step 14: Locate exact corner "D"

Step 15: Check intersecting lines using a 3-4-5 triangle

Step 16: Locate approximate corner "C"

Step 17: Build batterboards behind corner "C"

Step 18: Move transit to corner "B"

Step 19: Sight along wall "AB"

Step 20: Locate where string "BC" crosses batterboards at corner "C"

Step 21: Run string over corners "B" and "C"

Step 22: Measure along string "BC" to locate exact corner "C"

Step 23: Run string over corners "C" and "D"

Step 24: Check by measuring the diagonals

Step 25: Do layout for interior piers or walls

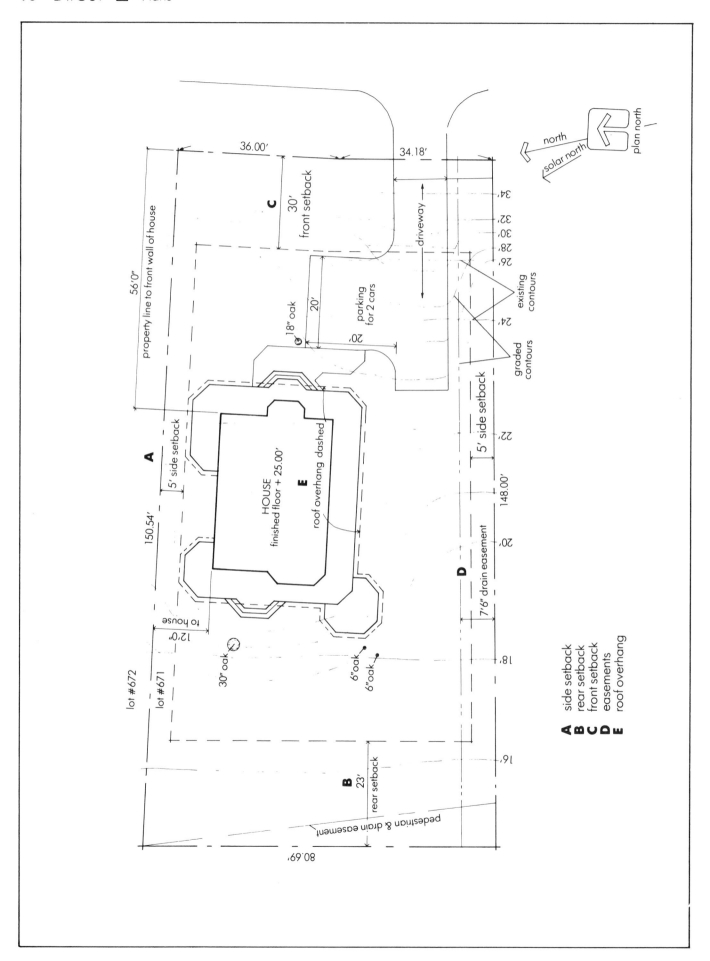

property line to front wall of house

56'0"

36.00'

34.18'

C 30' front setback

north

solar north

plan north

34'

32'

30'

28'

26'

driveway

existing contours

18" oak

20'

parking for 2 cars

20'

24'

graded contours

5' side setback

A

5' side setback

E

HOUSE
finished floor + 25.00'

roof overhang dashed

22'

150.54'

148.00'

D

7'6" drain easement

20'

12'0" to house

30" oak

18'

6" oak
6" oak

lot #672

lot #671

16'

B 23' rear setback

pedestrian & drain easement

80.69'

A side setback
B rear setback
C front setback
D easements
E roof overhang

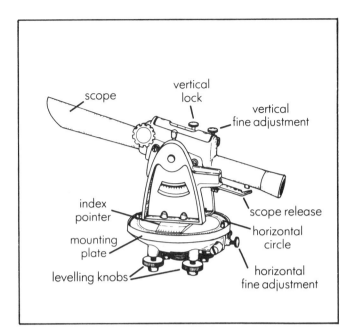

Figure 4-2. Transit.

Using the Transit

Before beginning the layout, it might be useful to become acquainted with the transit. This section has been added to show you how to use a transit and make you feel more comfortable with the idea of using one if you have never worked with one before.

Diagrams: Figures 4-2, 4-3.

Photographs: Photos 4-2, 4-3, 4-4.

The builder's level and the transit are very similar instruments, with the transit scope having the added capability of tilting. In our example, we will be using the transit and you will see how this ability to tilt comes in handy during the layout. If you are using a builder's level, you will have to make some adjustments at certain places where a tilting scope is needed; otherwise their uses are identical.

When using either instrument, first set the tripod on the ground. Adjust the sliding legs until the mounting plate is relatively level. Be sure that the pointed legs are securely embedded into the earth. If the instrument moves during the layout, it must be reset.

After the tripod is securely in the ground, attach the transit to it. You are now ready to adjust the transit so that the telescope is exactly level in any direction it is turned. Begin by turning the telescope so that it is directly over two of the adjusting screws. Now, turning the screws slowly and both at the same speed, either towards each other or away from each other, adjust them until the spirit level is exactly level and the feet of both adjustment screws are on the transit base. (See Photo 4-2.)

After the scope is adjusted over these first two screws, turn the scope 90° to the other two screws and adjust them in the same way. Turn the scope and adjust until both spirit

levels read exactly level in any direction the scope points. This could take some time, so be patient. Never force the adjusting screws as they strip easily. Never tighten them too much; you can warp the base plate.

Experiment with the transit. Looking through the scope you will note crosshairs. Look at an object and adjust the focus as if you were to drive a nail into that object exactly where the crosshairs meet. Swinging the scope to a new object, do the same. The two nails should be exactly level to each other. In this way, the transit is used to find exactly level points.

The transit can easily be used to find true right angles of intersecting walls. To use it, first twist the horizontal circle until it is at 0°, using the zero line of the index pointer as the mark. Turn the telescope and you will be able to see what

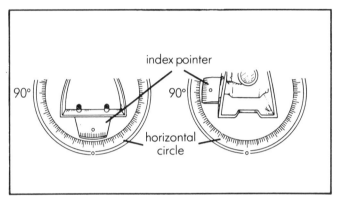

Figure 4-3. Rotating transit 90°, as viewed from above.

angle from the original line of sight you have turned by reading it on the horizontal circle, using the zero line of the index pointer as the mark. Locate a mark on an object using the scope and the crosshairs. Set the horizontal circle at zero and turn the scope until the zero lines of the index pointer read 90° on the horizontal circle. Look through the scope at another object and set a mark at the crosshairs. Lines coming to the transit from these points and intersecting at the transit would be at right angles. In this way, the transit can be used to find any angle size.

There are some other knobs and gizmos on the transit for fine turning and locking the instrument in place. Get acquainted with these before you begin the layout. If you are renting or borrowing a transit, check to see if it has been recently calibrated. If it has been badly taken care of, it will give inaccurate readings which can greatly affect the quality of your work. Be sure to treat the transit gently and with care at all times. It is a precision tool and can easily become misaligned if badly treated.

You might consider buying one of these tools, or at least having access to one, as they are useful for finding exact levels throughout the construction of the house.

Photo 4-2. Level a transit by adjusting 2 opposing levelling screws. Note bubble level on top of transit.

Photo 4-3. Making fine adjustment on the transit.

Step 1
Check the Property Lines

Before beginning the layout, make one final check of the house's location on the lot. Be careful to inspect any property line setbacks if you are building close to the borders of the property. If there are setback requirements by the inspection office or subdivision, are you within these? Do they refer to roof overhangs? Decks? Foundations? (Your house could look funny with all the eaves hacked off on one side.) Are you properly aligned for solar use? For view? For shade? Give it one final review. A total site evaluation should have been done before this. Some permit offices or subdivisions will want to check your layout lines before you begin to dig. See if this is so in your area.

Step 2
Establish Corner "A"

Margin of Error: Exactly where you want it.

Most Common Mistakes: Not starting at highest corner; not siting house properly.

Diagram: Figure 4-4.

Determine exactly where corner "A" will be. At this point drive a 2×2 stake securely into the ground with a nail in the center of the stake. This nail will mark the exact corner "A." It is usually best to choose the highest corner to begin with, so you will not run into the ground elsewhere in the layout. *You now have one definite corner, corner "A."*

Photo 4-4. The index pointer and horizontal circle are both set at zero before rotating transit 90°. Be very exact.

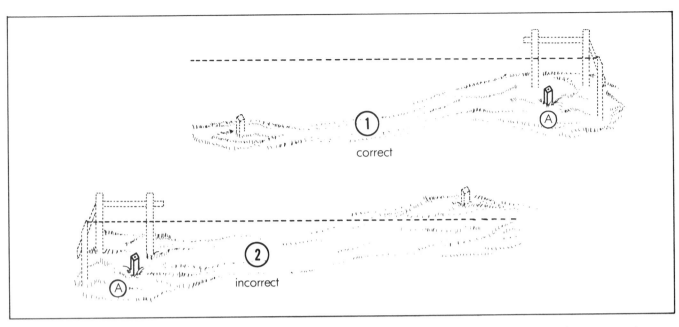

Figure 4-4. Example 1 is correct as layout lines will clear the ground at all points. Example 2 is incorrect as lines are too low at high corner.

Step 3
Set Up Transit Over Corner "A"

Margin of Error: Exactly over the point and exactly level.

Most Common Mistakes: Transit not level; transit not directly over nail in stake "A"; tripod legs not set securely into ground.

Photograph: Photo 4-5.

Set the transit up exactly over corner "A," aligning the point of the plumb bob that hangs from the center of the transit with the nail in stake "A." Using the method shown in the section called "Using the Transit" at the beginning of this chapter, level the transit and lock in place.

Step 4
Locate Approximate Corner "B"

Margin of Error: Approximate corner, but you are fixing wall "AB" permanently.

Most Common Mistake: Not fixing wall "AB" where you want it.

Diagram: Figure 4-5.

Determine which wall you want to lay out first (wall AB). Using a long steel tape and beginning at the nail in stake "A," measure out the distance along the wall line "AB" to your approximate corner "B." Drive a stake into the ground at this point and drive a nail into the top of the stake. This corner is approximate because you have no way as of yet to ascertain if you are holding the tape exactly level. This can throw your

Photo 4-5. Positioning a transit directly above corner "A." Usually you set up the transit and then locate corner "A" directly below.

measurement off. You are, however, fixing wall "AB" by setting this point. Exactly where along this wall corner "B" will fall is still undetermined. *You now have a definite corner "A," an approximate corner "B," and a definite wall "AB."*

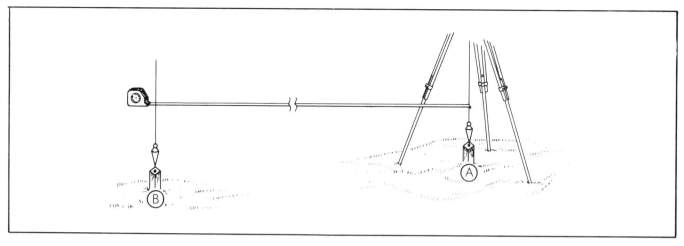

Figure 4-5. Locate the approximate corner "B" by measuring the distance of wall "AB" from point "A." Hold the tape as level as possible.

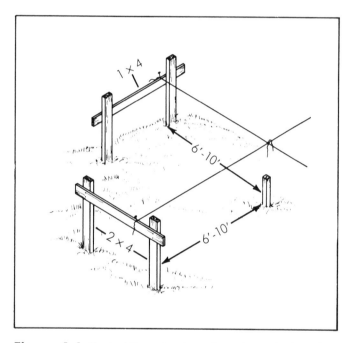

Figure 4-6. Typical 2-section batterboards used when digging with heavy equipment.

Step 5
Build Batterboards Behind "A" and "B"

Margin of Error: Horizontal cross members of the batterboards must be exactly level and all at the same level. The vertical stakes of the cross members do not have to be exactly plumb but they must be securely set.

Most Common Mistakes: Cross members not level with each other; batterboards too close to the corner; batterboards not securely built.

Diagrams: Figures 4-6, 4-7, 4-8, 4-9, 4-10.

Photographs: Photos 4-6, 4-7, 4-8.

The diagrams show different aspects of a typical batterboard. You will be building these behind all corners of the house using the same procedure as shown here. Note that these batterboards use 2×4s for the vertical pieces and 1×4s for the horizontal members. This is fairly typical. Two-section

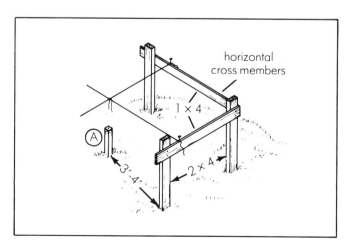

Figure 4-7. Typical 1-section batterboards used for hand digging or a careful backhoe operator.

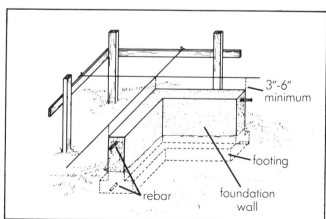

Figure 4-8. Note that layout lines are at least 3"-6" above foundation wall at its highest point. Forms and foundation will be built below these lines. Note lines directly above exterior edge of wall.

batterboards are often used to keep batterboards well behind foundation trenches, especially when using heavy equipment for digging.

The most essential part of building the batterboards is that all the horizontal cross members be exactly the same level. This is where the transit comes into play. First you must determine at what level you want the top of the cross members to be. Your layout strings will be running over the top of the horizontal cross members so the cross members will determine the level of the layout lines. If all of the cross members are at the same level, the layout strings will be level.

The layout strings should run at least 3 to 6 inches higher than the top of the proposed foundation wall; out of the way, but close enough so they can be used throughout construction of the foundation as a guide to check the level and location of the foundation forms. The foundation should be at least 18 inches above the ground at the highest corner. Check your local code on this. (I recommend increasing this minimum distance to 24 inches; it protects the wood of the house better and allows for a higher crawlspace in which to work beneath the house.) If your foundation has to be at least 18 inches high at the highest corner "A" and the layout lines run, say, 4 inches (minimum) above this, then the tops of the cross members have to be 22 inches above the ground at corner "A." Use an easily remembered number.

Photo 4-6. Using a transit to mark the location of horizontal cross members on vertical stakes at corner "A."

Photo 4-7. Marking the level points on vertical pieces of batterboards in order to locate level horizontal cross members. Batterboards are in place at corner "A."

Photo 4-8. Using a sledge to stabilize a vertical stake while nailing the horizontal cross member. Note how the top of the horizontal cross member is level with the mark on the vertical stake.

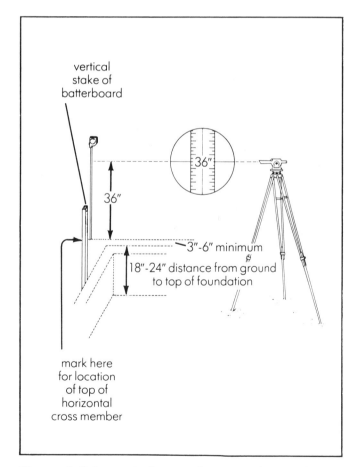

Figure 4-9. Locate the horizontal cross members on the vertical stakes. Be sure the top of the cross member is at least 3 to 6 inches above the foundation wall.

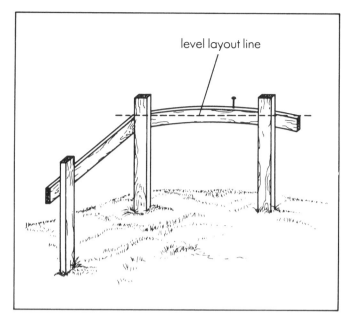

Figure 4-10. A bowed cross member throwing lines off level.

Using a tape or a graduated stick or rod, provided with the transit, determine how low below the crosshairs of the transit scope the tops of the horizontal cross members need to be in order for them to be a minimum of 3 to 6 inches above the foundation at corner "A." In our example, they will be exactly 36 inches below the crosshairs. Using this exact measurement, the top of the cross members at all batterboards can be located.

After you have driven in the vertical stakes of the batterboards in their proper locations and approximately plumb behind corners "A" and "B," use the transit to mark on these vertical stakes a point exactly 36 inches below the crosshairs. Mark each piece by having someone hold a tape or graduated stick exactly plumb until the crosshairs are at 36 inches. (You can tape your tape measure to a 4-foot level *or* mark the distance directly on a 4-foot level.) These points will fall at equidistant points below a level plane and will therefore all be level. By lining up the top of each cross member to these points, the tops of the cross members will also be level. Consequently, when the string is run over the tops of these cross members, the strings also will be level. Easy, isn't it?

Mark the vertical stakes, but be sure they are securely set into the ground. (If they ever move they will throw your entire layout off.) Mark all the vertical stakes.

Select a cross member and be sure that it is good straight stock, as any bow in the board will throw the lines off. Do this by looking down the piece to see if it is straight. Using your sledge to steady the vertical stakes, nail the cross members on so that their tops are on line with the marks. You will be repeating this process at each corner. Use double-headed (form) nails or regular nails not driven all the way in so you can take the batterboards apart after the foundation forms are finished.

Build these batterboards behind corners "A" and "B," using either 1-section or 2-section batterboards as described earlier.

Step 6
Run String Directly Over Corners "A" and "B"

Margin of Error: Exactly over the nail in each stake.

Most Common Mistakes: String not exactly over corners; using cotton string; throwing horizontal cross member off level when nailing.

Photographs: Photos 4-9, 4-10.

Diagram: Figure 4-11.

Using a plumb bob and nylon string, run a line that passes directly over the nails in each stake and tie the string taut by driving a nail at the top of the cross members at each corner. You can locate the exact point of the cross member at point "B" by sighting with the transit aligned exactly along wall "AB." (See "Step 9.") It is wise to support the cross member from underneath with a sledge placed between the cross member and your knee. This keeps the cross member from being thrown off level as you nail. An alternative to nailing a nail into the top of the cross member (which may throw the cross member off level) is to place the nail on the backside of the cross member as shown in Figure 4-11. This also allows you to make fine adjustments in the strings by placing knife notches in the top, backside of the board to move the string as needed. You now have a level line that is running exactly along the exterior edge of foundation wall "AB." This may be a good opportunity to let your children (if you have any) help. There is an old Russian adage: "Kids don't trample the wheat they sow."

Photo 4-9. Running out a string for the layout line that will pass directly above corners "A" and "B."

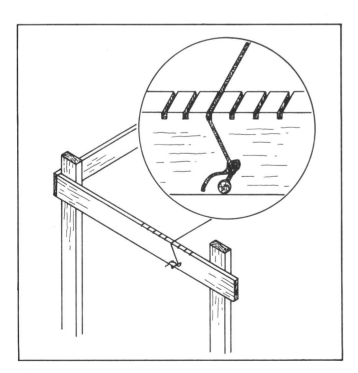

Figure 4-11. Use nails on the backside of the cross members and notches on the top to hold the string in its proper place.

Photo 4-10. A line passing directly above corners "C" and "B." Note line is touching the plumb bob line.

Step 7
Transfer Corner "A" to String

Margin of Error: Exact.

Most Common Mistake: Not transferring corner "A" *exactly* to string.

Photograph: Photo 4-11.

Using the plumb bob, transfer the exact corner "A" to the string by aligning the tip of the plumb bob directly over the center of the nail at stake "A." Tie a small piece of string at this point on string "AB." You have now transferred the exact corner "A" onto string "AB."

Step 8
Measure Along String "AB" to Exact Corner "B"

Margin of Error: Exact.

Most Common Mistake: Not measuring accurately.

Diagram: Figure 4-12.

You now have a level line along which to measure from point "A" to the exact corner "B" (which has been an approximate corner up until now). Simply use your long steel tape and have someone hold it at the piece of string on line "AB" that marks corner "A" and measure out the prescribed distance of

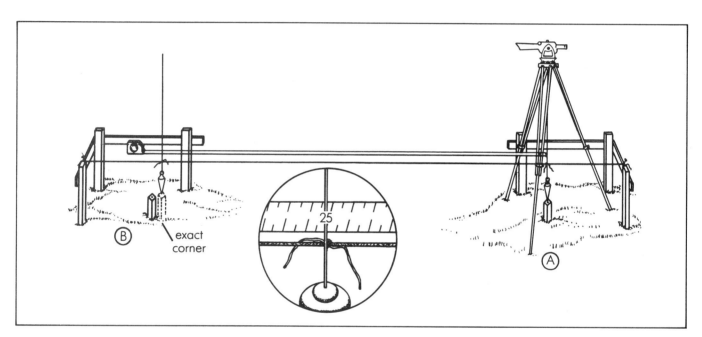

Figure 4-12. Measure the exact distance of wall "AB" and determine the exact corner "B." To hold the tape level, use the layout line for reference.

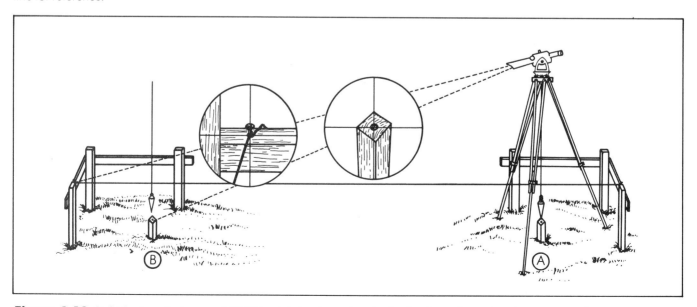

Figure 4-13. Sight the transit along wall "AB" by sighting the nail on the final stake or the nail on the cross member.

the wall (in our example 25 feet). Tie a string at that point. You may want to have someone support the tape midway to correct for any dip in the tape. This piece of string represents the exact location of corner "B." Reset stake "B," using your plumb bob, so that the center of the nail in the stake is directly below the piece of string. Check to be sure this final corner "B" is within your setback requirements. *You now have two definite corners, "A" and "B," and a definite wall "AB."*

Step 9
Sight Transit Along String "AB"

Margin of Error: Exact.

Most Common Mistake: Improper sighting.

Diagram: Figure 4-13.

Sight the transit along line "AB" and then rotate the transit 90° to locate wall "AD." At this time your transit is still directly over point "A." Before proceeding, check the transit. If it is not level, reset it. If you have to reset the transit, you will have to use new measurements—the transit has moved. The transit need not be exactly at the same level as the original sighting (36 inches above the string in our example). You need to know, however, how far above or below the original plane the new level is and adjust the measurement accordingly. If, for example, after releveling the transit, you are 34 or 37 inches above the previously set cross members, use this measurement for the remaining cross members.

Once the transit is level and is directly over corner "A," tilt the scope and sight the crosshairs either on the nail in the stake at corner "B" or on the nail holding string "AB" at the cross member behind corner "B." This means your transit is sighted directly along wall "AB." If you are using a builder's level which will not tilt, have someone hold a carpenter's level, exactly plumb, from one of these nails and sight on the side of it.

Step 10
Rotate Transit 90°

Margin of Error: Exact.

Most Common Mistake: Not rotating transit exactly 90°.

Photograph: Photo 4-12.

Diagram: Figure 4-14.

In order to find the intersecting wall "AD," which is at a right angle (90°) to line "AB," rotate the scope 90°. All points along the crosshairs in this new direction will be at a right angle from the original sighting on line "AB." Before rotating the scope, twist the horizontal circle until it is at 0°, using the center line of the index pointer as the mark. Be exact. Then turn the scope until you are reading 90° on the horizontal circle. (See "Using the Transit.") Lock the scope when this is done so it will not turn. You are now sighting at 90° from wall "AB."

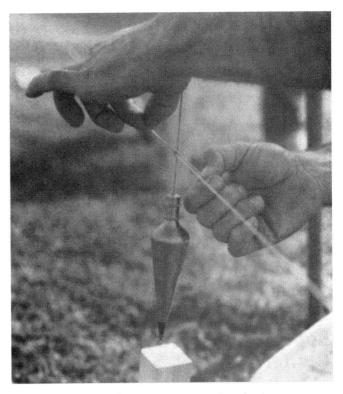

Photo 4-11. Transferring corner "A" directly above to string "AB." Use a small piece of string to mark "A" on line "AB."

Photo 4-12. Rotating a transit 90° from wall (line) "AB" to locate corner "D."

Step 11
Locate Approximate Corner "D"

Margin of Error: Approximate, within 1 or 2 feet.

Most Common Mistakes: None.

Diagram: Figure 4-15.

Run your steel tape out along wall "AD" until you can see it in the scope of the transit. Drive a stake at the measurement corresponding to the length of wall "AD." This stake is only approximate as you have no way of knowing if you were measuring along a level plane. *You now have two definite corners, "A" and "B," one definite wall, "AB," and a third approximate corner, "D."*

Step 12
Build Batterboards Behind Corner "D"

Margin of Error: Member must be exactly level with others.

Most Common Mistakes: Vertical stakes not securely set into ground; horizontal cross members not level with those of other batterboards.

As you did in "Step 5" for corners "A" and "B," build batterboards behind corner "D." Be sure that they are secure and can remain so throughout the construction of the foundation. Most important is that the tops of the horizontal cross members be at the exact same level as the tops of the cross members at "A" and "B." You will be using a point 36 inches below the crosshairs in our example unless you have changed the level of the transit and you need to compensate for this.

Step 13
Locate Where String "AD" Crosses the Batterboard at Corner "D"

Margin of Error: Exact.

Most Common Mistake: Sighting improperly.

Photograph: Photo 4-13.

Using the transit which has been turned at an exact 90° angle from wall "AB" (or a builder's level and a level-held plumb), sight the crosshairs on the horizontal cross member at corner "D." Drive a nail at this point. Run a taut string from this nail to a nail at the cross member at corner "A" so string "AD" crosses directly over corner "A." This string should be level with string "AB," at right angles to it, and passing directly over point "A" and over stake "D." The final location of stake

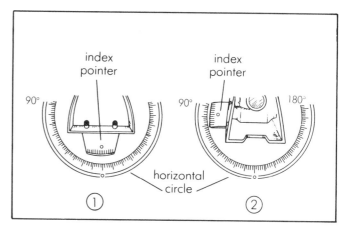

Figure 4-14. Rotate the transit 90° from wall "AB" (1) to sight along wall "AD" (2).

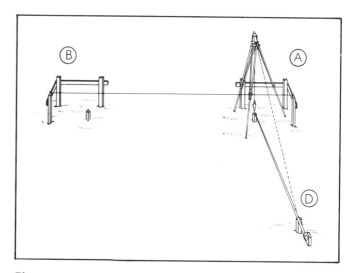

Figure 4-15. Locate the approximate corner "D" by rotating the transit 90° and measuring out the approximate length of wall "AD."

"D," however, is not yet set. *You now have two definite corners, "A" and "B," two definite wall lines, "AB" and "AD," and an approximate corner "D."*

Step 14
Locate Exact Corner "D"

Margin of Error: ⅛ inch.

Most Common Mistake: Improper measurement.

Photograph: Photo 4-14.

To find the exact corner "D," measure from where the two lines "AD" and "AB" intersect at corner "A." They should be intersecting directly over corner "A" or something is amiss. Measure along string "AD" the prescribed length of wall "AD" (in our example 25 feet) and tie a small piece of string there. Reset your stake so that it falls directly underneath this string. *You now have three definite corners, "A," "B" and "D," and two definite walls, "AB" and "AD."*

Photo 4-13. Using a transit at corner "A" to find the exact location of the string on the horizontal cross member at corner "D."

Photo 4-14. Using a tape to find the exact location of corner "D" along line "AD." Note the woman holding the tape in the middle to avoid slack in the tape. Mark "D" with the string on the layout line.

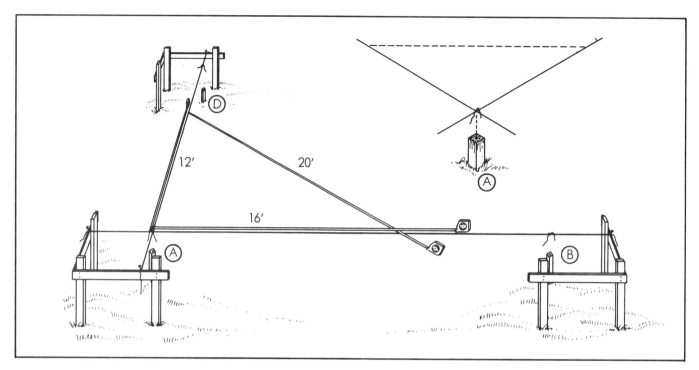

Figure 4-16. Using 3 tapes to check for the right triangle using 3-4-5 triangle (12-16-20). Be sure the strings cross *exactly* over corner "A."

Step 15
Check Intersecting Lines Using a 3-4-5 Triangle

Margin of Error: Within ¼ inch of true measurement when using a 12-16-20 triangle.

Most Common Mistake: Inexact measurements.

Diagram: Figure 4-16.

It is a good idea to check your walls to be sure the lines are definitely a right triangle. You should know this checking technique in case you do not have access to a transit or builder's level. It is a simple way of determining when lines are at a true right angle.

It is based on this mathematical principle: in a right-angle triangle the square of the hypotenuse (the side across from the right angle) is equal to the sum of the squares of the other two sides. This is also called a 3-4-5 triangle with 5 being the hypotenuse and 3 and 4 the two sides. Accordingly, 5 squared (25) = 3 squared (9) + 4 squared (16). This also applies if you multiply each number, 3, 4 and 5, by the same number.

To use this in checking your lines, measure out from corner "A" 3 feet (or a certain multiple thereof) in one direction, and 4 feet (or the same multiple thereof) in the other direction, and see if the distance between these two points is 5 feet (or the same multiple thereof). It is best to use the highest multiple the walls will allow, as this makes for greater accuracy. Builders usually use 12-16-20 feet. The diagonal should fall within ¼ inch to ½ inch of 20 feet, if the strings are at a true right angle.

Step 16
Locate Approximate Corner "C"

Margin of Error: Approximate, within a foot or two.

Most Common Mistakes: None.

Diagram: Figure 4-17.

Using two tapes, each at the measurement of the prescribed walls, locate an *approximate* corner "C." *You now have 3 definite corners, "A," "B" and "D," two definite walls, "AB" and "AD," and one approximate corner "C."*

Step 17
Build Batterboards Behind Corner "C"

Margin of Error: Cross members must be exactly level to other cross members.

Most Common Mistakes: Batterboards not secure; cross members not level.

Using the same method as explained in "Step 5," build batterboards behind corner "C." As before, be sure that they are well built, that the horizontal cross member is straight with no bows and that it is exactly level to the other cross members.

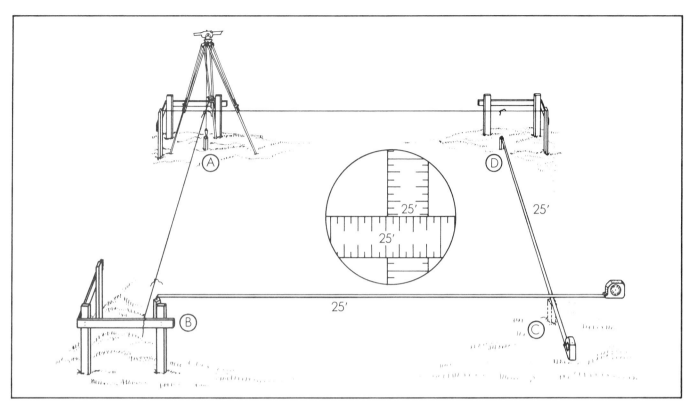

Figure 4-17. Using 2 tapes to locate the approximate corner "C."

Photo 4-15. All four batterboards are in place and the transit was moved to corner "B."

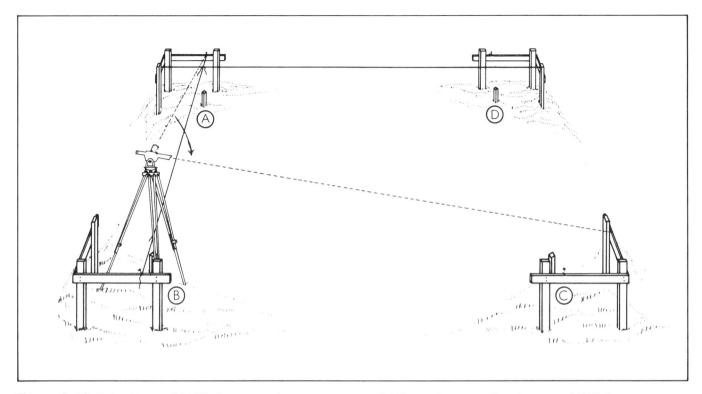

Figure 4-18. Sight along wall "AB" after moving the transit to corner "B." Rotate the transit 90° to locate wall "BC."

Step 18
Move Transit to Corner "B"

Margin of Error: Exactly over corner "B."

Most Common Mistakes: Transit not level; not compensating for new level.

Photograph: Photo 4-15.

Move the transit from corner "A" and set it up over corner "B." Adjust the transit *exactly* over corner "B" and level the transit. Use a previously set cross member to find how far above or below you are from your original level. Compensate for this in building your remaining batterboard. If your new level is now 32 inches above a previously set batterboard (distance from top of batterboard to center of crosshairs), use 32 inches to locate cross members set from this setting and forget the previous 36-inch measurement.

Step 19
Sight Along Wall "AB"

Margin of Error: Exact.

Most Common Mistake: Inaccurate sighting.

Diagram: Figure 4-18.

Now sight back to the nail holding the string at corner "A" so that the transit is again directly sighted to wall "AB." You are now looking for wall "BC," which is at a right angle to wall "AB," so simply rotate the transit 90°.

Step 20
Locate Where String "BC" Crosses Batterboards at Corner "C"

Margin of Error: Exact.

Most Common Mistake: Inaccurate sighting.

As you did in "Step 13," locate where string "BC" crosses the horizontal cross member. Drive a nail at this point.

Step 21
Run String Over Corners "B" and "C"

Margin of Error: Exact.

Most Common Mistake: Strings not running exactly over "B" and "C."

Diagram: Figure 4-19.

Run a string passing directly over corner "B" to the new nail at corner "C." This new string should represent the exact outside edge of the foundation wall "BC." Again check to see if it is running directly over corner "B." Use the 3-4-5 triangle to be sure you have a right triangle.

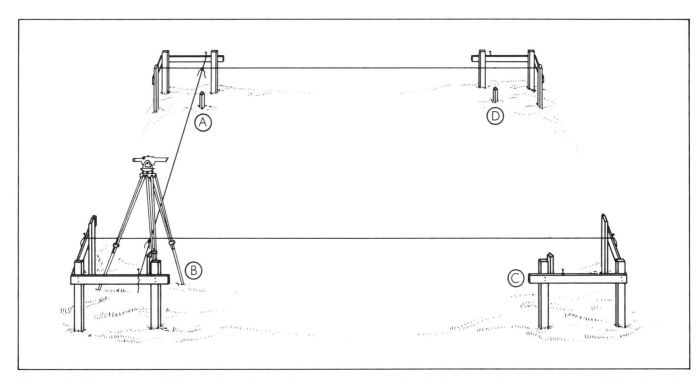

Figure 4-19. Run string "BC" from the nail set in the cross member behind "B," crossing directly over corner "C," to the nail behind corner "C."

Step 22
Measure Along String "BC" to Locate Exact Corner "C"

Margin of Error: Within ⅛ to ¼ inch of exact distance.

Most Common Mistake: Inaccurate measurement.

As you did in "Step 8" and "Step 14," measure along string "BC," from where it intersects line "AB" at point "B," the prescribed distance of wall "BC" to locate corner "C." Tie a string at this point and reset stake "C," driving a nail in the stake at the exact corner. *You now have four definite corners, "A," "B," "C" and "D," and three definite walls, "AB," "AC" and "BC."*

Step 23
Run String Over Corners "C" and "D"

Margin of Error: Exactly over these corners.

Most Common Mistakes: None.

Diagram: Figure 4-20.

Run a line, using your plumb bobs, directly over corners "C" and "D." When measured, this line should be the same as the distance prescribed for wall "CD." If not, something is off and you need to find out where.

Step 24
Check by Measuring the Diagonals

Margin of Error: ½ to ¾ inch over each 35 feet of diagonal.

Most Common Mistakes: Allowing too great a margin of error; improper measuring.

Diagram: Figure 4-21.

Photograph: Photo 4-16.

After all your lines are in place, you can check the rectangular or square sections of the building by measuring the diagonals. Ideally, the diagonals should equal each other. In practice, that seldom happens, nor does the foundation need to be that exact. If the diagonals are within ½ to ¾ inch over every 35 feet of diagonal, this will be sufficient.

If there is a greater discrepancy, do not be upset. Layout is a process of trial and error and it rarely works out the first time around, so do not expect it to. No matter how accurate you have been, it just never seems to work the first time. You now begin an investigation process of trying to determine where you are off.

Begin by measuring all the wall lengths to determine if they are the prescribed distance. If you are lucky, one may be off and the problem is quickly solved. Once you have checked the distance, use your 3-4-5 triangle or your transit or both to check to see if your corners are all 90°. Continue searching and adjusting until the diagonals work out within the allowable margin of error. Be patient, it may take time and get frustrating, but eventually you will make it work.

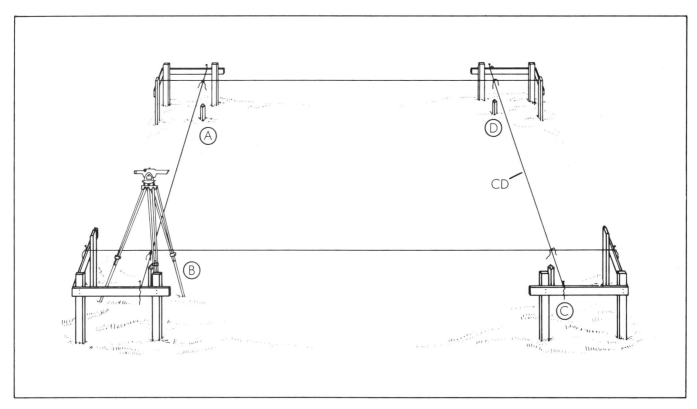

Figure 4-20. String "CD" is in place, crossing directly over corners "D" and "C."

Photo 4-16. Checking the diagonals to see if it is a true rectangle. Margin of error is ½ to ¾ inch over 35 feet of diagonal.

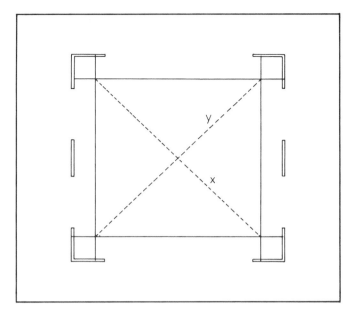

Figure 4-21. Measure the diagonals to check for accuracy of the layout (x=y).

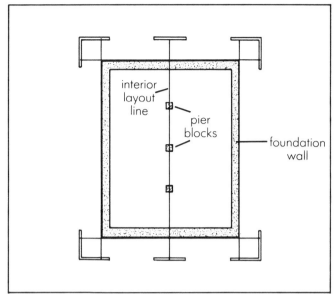

Figure 4-22. Single batterboards are in place to mark the pier blocks used for support girders.

Step 25
Do Layout for Interior Piers or Walls

Margin of Error: Within ¼ inch.

Most Common Mistake: Inaccurate layout.

Diagram: Figure 4-22.

Most houses will have some type of interior foundation as well as the foundation for the exterior walls. If the house is not very wide in some areas, 16 feet or less, intermediate piers and girder may not be required. Check your plans to see if any are needed.

If you have interior piers or full walls, you will need to set lines for them. (See Chapter 6, *Poured Concrete Foundations,* "Step 14B.") Actually, you can omit setting the lines at this stage. You can set them once the form boards are in place for the foundation by attaching them to the inside of the interior form boards. This will require digging the holes for the piers after the forms are in place. Whichever seems to work best for you is OK.

If you decide to dig the holes or trenches at this point, check your plans to see the location of the interior walls or trenches. Be sure you know what the measurement you are reading applies to, to the edges of the piers or wall or to the center of the piers or wall. THIS IS IMPORTANT. Measure that distance in on your perimeter lines and tie a piece of string at this point (in our example 10 feet). Build a single batterboard behind the lines at either wall. Be sure the cross members are level with each other. Run strings over the top of these cross members so they cross the pieces of string. If there is more than one set of interior piers or more than one wall, repeat the process for each one. Your layout is now complete.

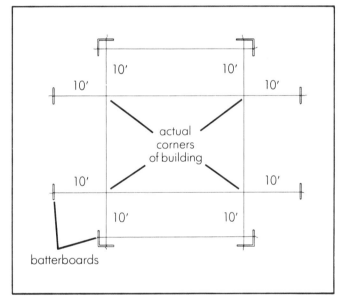

Figure 4-23. Setting strings and corners 10 feet out from the actual corners. Sometimes the batterboards are built further from the actual foundation to allow equipment to move without upsetting the batterboards.

Special Situations

Offset Hubs

Sometimes offset hubs are used instead of a standard layout. This is a process of setting the strings and corners a good distance, 10 feet or more, back from where the actual corners will be. Two-section batterboards are used instead of 1-section. You may want to consider using this process when you plan to use heavy equipment to dig your trenches in order to keep the batterboards a good distance from the equipment and digging. (See Figure 4-23.)

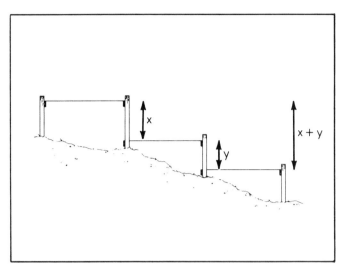

Figure 4-24. Stepped batterboards for steep slopes.

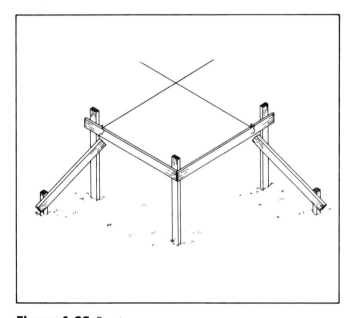

Figure 4-25. Bracing.

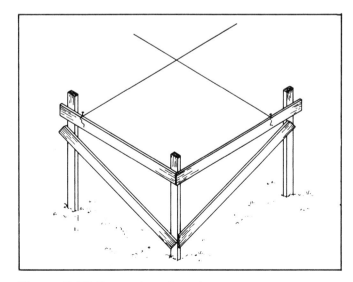

Figure 4-26. Bracing batterboards.

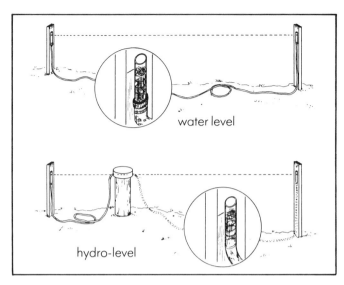

Figure 4-27. Less expensive hydro- and water levels that replace the levelling function of the transit.

Steep Slopes

Where steep slopes are involved, you may need to build several intermediate batterboards as batterboards are hard to stabilize if they are over 6 feet high. In this case the foundation will be stepped (see Chapter 6, *Poured Concrete Foundations,* "Step 2") and it is not essential that all the lines be at the same level, since the foundation will not all be at the same level (see *Knee Walls* chapter, "Purposes of the Knee Wall"). Since you can calculate the vertical distances between the strings, you can keep the different sections of foundation walls level. (See Figure 4-24.)

Braced Batterboards

If the batterboards are high (over 4 feet) you may need bracing as shown in Figures 4-25 and 4-26.

Water Level and Hydro Level

If you cannot afford a builder's level or transit and do not have access to one, you can buy a less expensive hydro level or water level that can serve the levelling functions of the transit or builder's level. These tools use water in clear tubes to level, on the principle that water always seeks its own level. In using these tools, be sure they are always in the shade or always in the sun when not in use. Changes in the temperature of the water can cause expansion and thereby change the water level. (See Figure 4-27.)

Digging Trenches or Excavating Before Final Layout

In some cases, where lots of excavating is involved, such as with basements, rough layout lines are set to show the backhoe operators where to dig. After the excavating is completed, exact layout lines are set up. In houses with standard foundations, some builders choose to set rough lines first, dig the trenches and then set up exact layout lines for building forms. This is done to keep the batterboards away from the digging.

Worksheet

ESTIMATED TIME

_____

_____

_____

_____

CREW

Skilled or Semi-Skilled

_____ Phone _____

_____ Phone _____

_____ Phone _____

Unskilled

_____ Phone _____

_____ Phone _____

_____ Phone _____

_____ Phone _____

_____ Phone _____

MATERIALS NEEDED

2×4s, 1×4s, 2-$\times$-2 stakes, nylon string, and nails.

ESTIMATING MATERIALS

2×4S:

_____ feet long: Quantity _____

1×4S:

_____ feet long: Quantity _____

2-$\times$-2 STAKES: Quantity _____

NAILS:

_____ lbs. of 8d duplex

_____ lbs. of 8d standard

STRING:

_____ feet of nylon string

COST COMPARISON

2×4S:

_____ Store:

Cost per foot _____ total $ _____

_____ Store:

Cost per foot _____ total $ _____

_____ Store:

Cost per foot _____ total $ _____

1×4S:

_____ Store:

Cost per foot _____ total $ _____

_____ Store:

Cost per foot _____ total $ _____

_____ Store:

Cost per foot _____ total $ _____

PEOPLE TO CONTACT

Concrete Supplier _____

Phone _____ Work Date _____

Concrete Pumper (if needed) _____

Phone _____ Work Date _____

If underfloor utilities need to be installed before floor framing:

Electrician _____

Phone _____ Work Date _____

Plumber _____

Phone _____ Work Date _____

Heating and Air
Conditioning _____

Phone _____ Work Date _____

Gas _____

Phone _____ Work Date _____

Daily Checklist

CREW:

_____ Phone _____

_____ Phone _____

_____ Phone _____

_____ Phone _____

_____ Phone _____

_____ Phone _____

TOOLS:

☐ builder's level

☐ water level or hydro level

☐ sledge hammer

☐ hand or power saw

☐ hammer

☐ three metal tapes (one should be 100-foot)

☐ 4-foot level

☐ knife

☐ plumb bob

RENTAL ITEMS:

☐ sledge

☐ transit

MATERIALS:

_____ ft. of 2×4s

_____ ft. of 1×4s

_____ no. of 2-×-2 stakes

_____ lbs. of 8d duplex nails

_____ lbs. of 8d standard nails

_____ ft. of nylon string

DESIGN REVIEW:

Be sure you have taken into consideration all the crucial factors in choosing an exact location before beginning layout.

INSPECTIONS NEEDED:

Inspector's name _____

Phone _____

Time and date of inspection _____

WEATHER FORECAST:

For week _____

For day _____

5 Types of Foundations

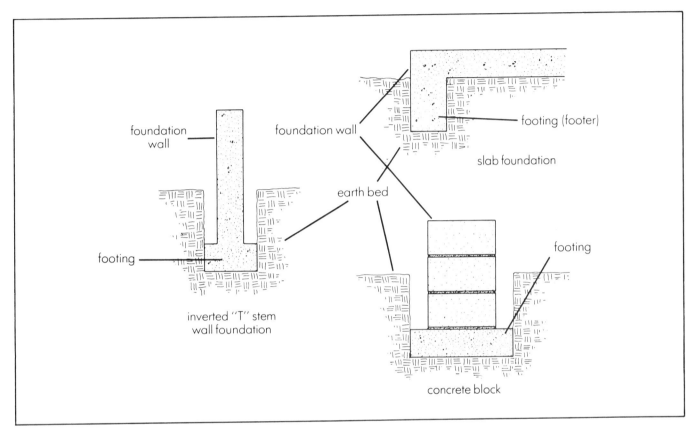

Figure 5-1. Parts of a foundation.

Introduction

BEFORE GETTING INTO the actual construction processes of various foundations, I thought it best to spend some time discussing what the different possibilities are. Often there are several options open to you, and a wise decision here can save you time, money and energy once construction begins. Other times, due to site, design or code specifications, no options are available and you will have to build one particular type of foundation. If the latter is the case or if you already have definitely decided what foundation you are going to build, you can skip over this section.

I have divided this section into three main areas: pier foundations, continuous-perimeter foundations (including basements) and slab foundations. Under each of these headings are examples with short descriptions of the construction details and a list of the advantages and disadvantages of each. Chapter 6, *Poured Concrete Foundations*, describes, in detail, how to build a concrete continuous-perimeter foundation.

Contracting Out the Foundation

Except for the most complex, hillside foundations, most foundation construction is well within the reach of the novice or first-time builder. It is not very difficult to understand, but it does require a greater accuracy than some wood construction; once it is poured, correction requires a lot of work. This factor often intimidates many owner builders, but if you know what you are doing and check everything out before the concrete is placed (poured), it need not present any real problems.

Some owner builders choose to contract out the foundation in order to save the time and energy involved in its construction. If it is a very complex or an extremely large foundation, I would recommend that you consider contracting it out. If your time is limited (or your energy) and you have allotted some money for labor, this may be the place to spend it. Building a foundation with a small, novice crew (like the lone owner builder) can take a long time and a lot of energy, but personally, I have always found a sense of satisfaction in preparing foundations for my houses.

Whether you do it or you contract it out, be sure that it is done properly: square, level and the proper dimensions. Inaccurate or sloppy work at this stage can cause problems throughout the building process and, if the work is poor quality, it can threaten the entire home.

Foundations

Almost all foundations (except slabs) have three main parts: the earth bed, the footing and the foundation wall. The earth bed is the final resting place for the weight of the house. The type and strength of the soil can have a bearing on the type of foundation chosen and its construction. (See Figure 5-1.)

The foundation footing spreads the weight of the house over a larger area of the earth so that a great load will not be concentrated on a small area.

The foundation wall lifts the house above the moist ground and provides a level building plane. It also serves to carry the load of the house down to the footings and from there to the earth bed. The main purposes of a foundation are:

1. To anchor the house to the ground;
2. To provide a level building plane on an irregular sloping site;
3. To carry the weight of the house down to the soil;
4. To keep the vulnerable wood away from the moisture and insects that might attack it.

Before we get into the different types of foundations, consider the following factors as you determine the type of foundation that you plan to use.

Factors to Consider in Choosing a Foundation

The Soil

The type and strength of the soil directly below the foundation must be considered. Weak or unstable soils may require larger footings or even engineered foundations. Some soils are so weak that a slab may be required to spread the weight over the greatest possible area. Know your soil, especially if there is a history of soil problems in your area, before you begin.

Codes

Codes can be very restrictive when it comes to foundations. In areas covered by the Uniform Building Codes, all houses must have continuous-perimeter foundations; many types of pier foundations are not permitted unless engineered. Some areas require reinforcing bars (rebar) in the foundations, others do not. Codes regulate the size and thickness of the footings and foundation walls, the depth of the footing, and other construction factors in foundations. Before beginning, be sure that you are clear on what the code requirements are in your area.

Slope of the Site

Steeply sloping sites may require high foundation walls at their lower ends. Unless you are using piers or a pole constructed house, this can be very costly. You may want to consider a half basement on steep lots. This would affect the type of foundation to be used. Long foundation walls running across a slope can create big drainage problems. Drain tile is recommended.

Weather

In colder climates, you may want to use a solid perimeter foundation instead of an open pier foundation to keep the area directly under the house warm and protected from cold winds. If you are using a pier foundation in cold climates, it is recommended that you insulate the floor. In areas having heavy rainfall, slabs may not be a wise decision, due to possible flooding in low areas.

Seismic Areas

In the event of an earthquake, the foundation will receive the brunt of the quake and therefore needs to be built much stronger than foundations built in non-earthquake areas. It is often recommended that these foundations be continuous foundations so that the entire foundation will move as one unit should an earthquake hit. In earthquake areas, local code offices should be consulted for requirements not found in the national codes.

Passive Solar Systems

Very often concrete foundations are used for heat storage in passive solar systems. This is especially true of slab foundations. Continuous-perimeter walls can also be effectively used as storage areas (if they are within an enclosed solar greenhouse, for instance). Consider this as you choose a type of foundation.

Local Foundation Problems

Many areas of the country have specific problems that often affect the type and design of foundations used. In many areas of Maine, for instance, bedrock is encountered a few feet below the ground and the footings need to be anchored to the rock by inserting steel stakes into holes drilled into the rock. Areas of Florida have sand to a great depth and long poles need to be used. Check with local builders or the code office about any problems in your area.

Cost Factors

Your decision as to the type of foundation you plan to use can make the difference of hundreds (and in some cases thousands) of dollars. Research well the difference in costs before you make your decision; otherwise you are making it with incomplete data, and that can be dangerous. Pier foundations are much less expensive and easier and quicker to build than continuous-perimeter foundations. If allowed by code, the pier foundation can serve just as well as the continuous foundation, which is greatly overbuilt to begin with. The floor can be insulated to help with the energy efficiency; and it allows

Photo 5-1. Natural stone piers (dry laid). This foundation is not allowed in code-enforced areas.

the crawlspace greater ventilation and access, which to me is a great advantage.

Slab foundations serve as not only the foundation but the first floor as well. This saves you the cost of having to build a floor frame and subfloor. The savings can be considerable in time and labor as well as money. Their great drawback is that they keep the house close to the moist ground.

Aesthetics

In choosing the type of foundation, consider the impact the foundation will have on the looks of the home. This is especially important if there are to be any high foundation walls. Houses built close to the ground on slabs will appear different than those built on foundations that raise them above the ground a certain distance.

Pier Foundations

Natural Stone

This type of foundation was the most common before the advent of concrete. Many older homes and barns are still standing, level and secure, on natural stone piers. Few new homes are built this way, but if no codes exist it can be considered in some areas. It is simple, inexpensive and, to some, aesthetically pleasing. I built a home a few years ago in North Carolina using this type of foundation and as yet there have been no problems. It can be laid with mortar or dried-laid stone (no mortar). In earthquake or code-enforced areas it is not allowed. (See Photo 5-1.)

Advantages
Inexpensive
Aesthetically appealing
Easy to build

Disadvantages
Not allowed by many codes
May affect resale value
Not advised in earthquake, hurricane or tornado areas
Large, flat stones are hard to find

Wood Columns

Though more common than stone piers, these are seldom seen except in rural buildings. If they can be approved by the code office, you may want to consider using these. Using any type of wood for a foundation is an "iffy" business. Even the new pressure-treated woods are only guaranteed for 20

Photo 5-2. Wood posts used as foundation piers. Use a rot-resistant wood.

years (some manufacturers swear they will last 50, but this is still not a long time for a foundation). (See Photo 5-2, Figure 5-2.)

Advantages
Inexpensive
Easy to build
Can use on steep sites by using longer poles on lower areas

Disadvantages
May not be allowed by codes
Lower resale value
Wood is prone to rot

Brick

Brick piers offer the permanency of stone with the low cost and ease of construction of pier foundations. (See Figure 5-3.) Brick piers are often seen on older and sometimes newer homes in the East, especially the Southeast. (See Photo 5-3.) In earthquake areas they are prohibited, since they are likely to fall apart in a good shake. You may want to consider them if the code will allow.

Advantages
Easy to build
Inexpensive
Attractive

Disadvantages
May not be allowed in some areas
May affect resale value
Not advised in earthquake, hurricane or tornado areas

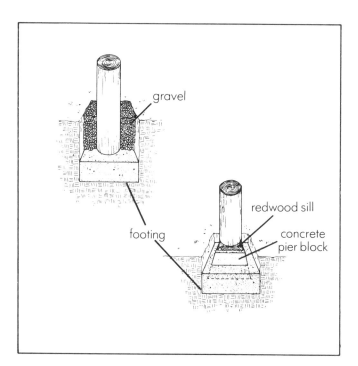

Figure 5-2. Wood piers showing footings. Note how the use of a pier block on the right keeps the piers out of contact with the moist ground.

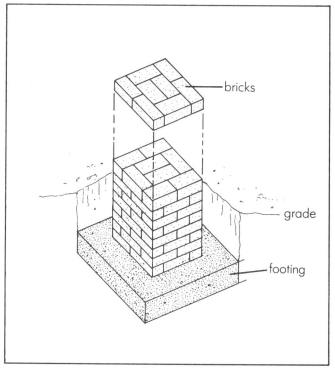

Figure 5-3. Brick pier.

Photo 5-3. Brick piers used for foundation.

Concrete Block

Piers of concrete block are still quite prevalent in the building trade. (See Photo 5-4.) They are used for foundations constructed totally from piers as well as piers for girders between continuous-perimeter wall foundations. Pier holes need to be dug accurately so that all the piers will be the same height when completed. In earthquake prone areas, they are not advised for the perimeter foundation but can be used for interior piers. In this case it is advised (and often required) to run reinforcement bars through the cells of the block and then fill the cells with concrete (grout).

Advantages
Easy to build
Inexpensive
Permanent
Allowed by many codes

Disadvantages
Not allowed in some code areas
May affect resale value

Photo 5-4. Concrete blocks used for foundation piers.

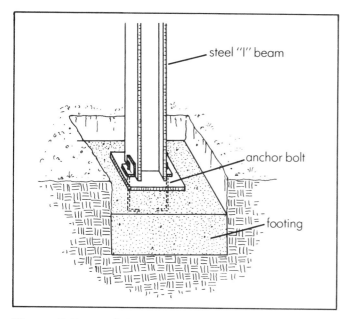

Figure 5-4. A steel I-beam used as a pier.

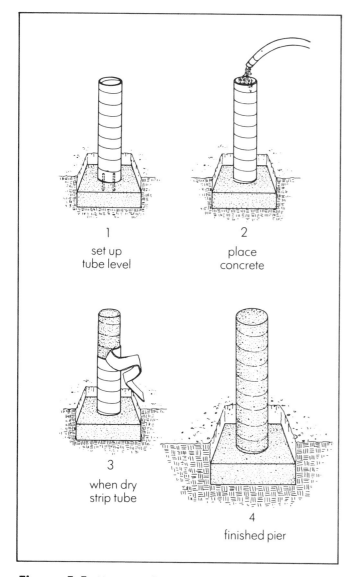

Figure 5-5. Using cardboard tubes as forms for concrete piers.

Steel Columns

Using steel columns for the entire foundation is rarely seen in single-family residence construction. (See Figure 5-4.) Some of the cliff-hanging homes in the California area use these systems, which must be professionally engineered. Steel columns are, however, commonly used as supports for wooden and steel girders in basement areas.

Advantages
Easy to build
Permanent

Disadvantages
Unattractive
Expensive
Not allowed by many codes

Concrete Columns

Concrete piers are also commonly used in construction. A common way of building concrete piers is with cardboard tubes. (See Figure 5-5.) These are cardboard cylinders that come in different diameters and heights. The fresh concrete is poured directly into the tube and then the cardboard is peeled away when the concrete is hard. (See Photo 5-5.) Forms can be built to whatever shapes and sizes you want. I once saw a home that was built entirely on concrete piers shaped like toadstools.

Advantages
Permanent
Easy to build
Inexpensive
Can be built to any shape and size

Disadvantages
May not be allowed by some codes
May affect resale value
Can be unattractive

Pole Construction

Pole construction is basically a different type of construction procedure than those shown so far. (See Figure 5-6.) Large poles are sunk into the ground for the foundation. The floor frames and roof are attached by the use of horizontal girders or stringers that are bolted to the poles. The weight of the building is totally carried by the poles and not by the wall, as is the case with standard stud construction. Studs are placed between the poles to hold the windows, doors, siding and insulation and interior trim.

In the past, pole construction was used mostly in agricultural buildings, where the earth was left for the floor so the tractors and livestock could move around. More recently it has become popular for one-family dwellings and larger buildings as well. It lends itself well to hillside lots, where longer poles can be used in the lower areas. Often the large poles are exposed, giving the house a greater sense of strength than found in stud construction. You may seriously want to consider this type of construction. Check the bibliography for books available on the subject.

NOTE: One major disadvantage of pole construction in the past was the need to sink wooden poles into the ground for the foundation. These would rot. You can modify the process slightly to prevent this, as shown in Figure 5-6.

Advantages
Inexpensive
Good for steep sites
Allows for large wall openings (up to 14 feet)
Large poles give house a sense of strength
Allowed by many codes
Earthquake resistant

Disadvantages
Trim and finish work around poles may be difficult
Foundation is wood (unless you use an alternate method)
Some engineering may be required

Photo 5-5. Poured concrete pier.

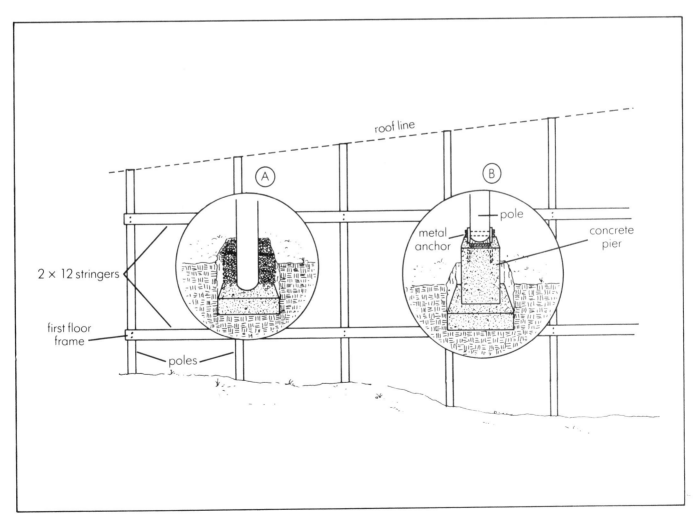

Figure 5-6. Pole construction. "A" shows poles sunk into the ground. "B" shows recommended way that keeps pole out of contact with the ground.

Photo 5-6. Natural stone continuous-wall foundation under mill built in the 1700s. Foundation is still level and intact though no mortar was used.

Continuous-Perimeter Foundations

Natural Stone Continuous Wall

As were their stone pier counterparts, natural stone continuous foundations are common to houses built before the advent of concrete. Many old houses used a natural stone, either pier or continuous. (See Photo 5-6.) These stones were either dry laid, simply stacking the stones using no mortar, or they were laid using mortar. Large, flat rocks, called baserocks, were used for the footings. Many are still standing in excellent shape. Though seldom used in this form in new construction, in the proper conditions and if no codes exist, they can work well and add the beauty of natural stone to your home.

Advantages
Inexpensive
Attractive

Disadvantages
Not allowed by most codes
Not advised in earthquake, hurricane or tornado areas
Large flat stones can be hard to find
May affect resale value
Time consuming to build

Natural Stone Using Forms and Mortar

An adaptation of the natural laid stone is sometimes used whereby forms are built, stones are placed inside the forms and then the spaces are filled with mortar or concrete. (See Figure 5-7 and Photo 5-7.) Sometimes only an interior form is used. It is simpler than laid stone and the rocks do not need to be as flat, since the concrete or mortar will bind them together. Often this system will not pass code, but it is used in rural non-code areas. It offers the permanency and beauty of natural stone without requiring as much effort or as many flat stones as does the laid-stone method.

Advantages
Permanent
Attractive
Inexpensive

Disadvantages
May not be allowed by codes
Not advised in earthquake, hurricane or tornado areas
May affect resale value
Somewhat difficult to build

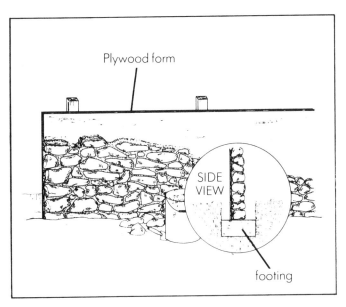

Figure 5-7. A laid stone foundation. The plywood form will be removed after the mortar has set.

Photo 5-7. Using form (plywood) and natural stone for continuous-wall foundation.

Continuous Brick

Brick foundations are sometimes used in new homes, especially in the eastern states. They can be either solid brick or a brick veneer covering a 4- or 8-inch concrete block foundation. (See Figure 5-8.) These foundations are aesthetically appealing, permanent and allowed by many codes. Bricklaying, for the beginner, can be both frustrating and time consuming. You may want to consider contracting a brick foundation, unless you are ready to accept the challenge of mastering a new skill such as bricklaying.

All-weather Wood

The all-weather wood foundation is new to the building industry. (See Figure 5-9.) It was developed mostly for contractors in the colder regions of the country so they could continue to build year-round, even when the weather was too cold to pour concrete. Since its introduction into the industry, it has been pushed and praised (often by the plywood associations). It is built with no concrete, gravel serves as the footing, and it uses pressure-treated lumber and plywood. This wood is guaranteed for only 20 years, though manufacturers claim it should last longer. Basically, it costs about the same as a poured concrete foundation, but its lifetime is not nearly as long. I would not ever recommend using any type of wood for a foundation.

Advantages
Can be built in freezing weather
Easy to insulate
Easy to build for novice builder

Disadvantages
Limited lifetime (20 to 50 years)
May affect resale value
May not be allowed by some codes

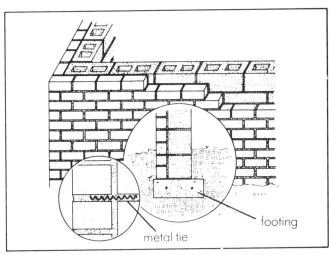

Figure 5-8. A concrete block foundation with brick veneer. Metal tie holds brick to concrete block.

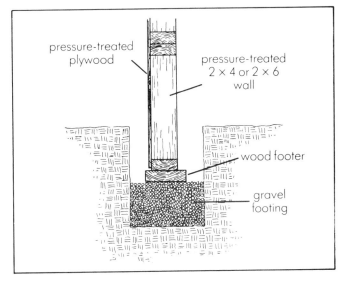

Figure 5-9. An all-weather wood foundation.

Photo 5-8. Concrete block foundation in progress.

Concrete Block

Concrete block foundations comprise approximately 38% of residential foundations. (See Photo 5-8 and Figure 5-10.) They are cost-effective and within the abilities of a novice builder. Their cost is equal to their poured concrete counterpart, though they can take longer to build if you are slow at laying blocks. In areas that are inaccessible to a concrete truck, they may be the best way to go.

Advantages
Within the abilities of a novice
Permanent
Accepted building practice
Cost-effective

Disadvantages
Work can go slowly for novice builder
More costly than pier foundation

Concrete Block (Earthquake Areas)

Using concrete blocks in earthquake areas is a good bit more difficult than in other areas. Should the foundation receive a good shake, the unfilled and unreinforced blocks could easily come loose from the mortar and the foundation fall apart. Because of this steps have to be taken to keep the blocks together in the event of an earthquake. Rebar (reinforcing bar made of steel rods) is run both vertically and horizontally through the cells of the blocks. (See Figure 5-11 and Photo 5-9.) Special blocks, called bond beam blocks, are laid every few courses to allow the rebar to run horizontally. Once the blocks are laid, grout (concrete with pea gravel) is poured in some or all of the cells. (See Photo 5-10.) This holds the foundation together. In essence, the blocks act as a form for the concrete.

Advantages
Permanent
Cost-effective
Can be built in remote areas

Disadvantages
Slow and difficult to build

Photo 5-9. Horizontal rebar being placed in a concrete block foundation in an earthquake area. Note the slots left in the blocks to receive the horizontal rebar.

Photo 5-10. Filling cells of concrete blocks with grout (concrete with pea gravel).

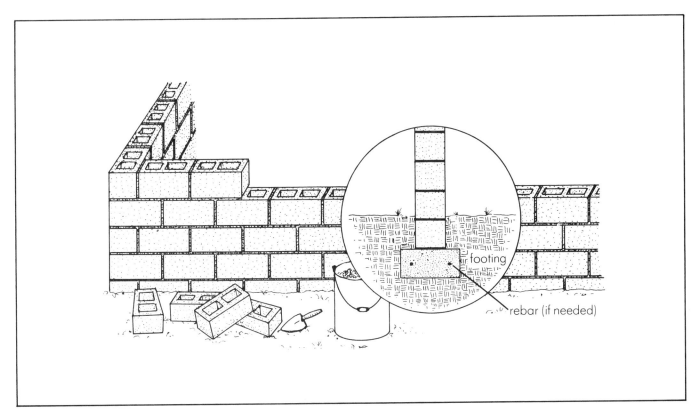

Figure 5-10. A concrete block foundation.

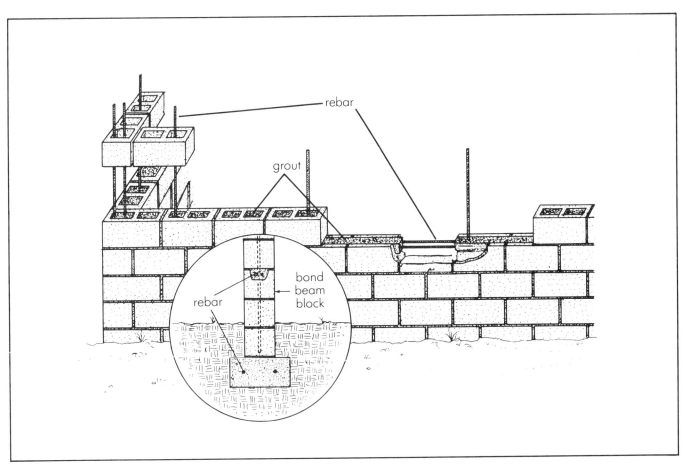

Figure 5-11. A concrete block foundation for earthquake areas. Note that the rebar (reinforcing bar) is running both vertically and horizontally and that all the cells are filled with concrete (grout).

Photo 5-11. Concrete block basement walls. Walls are waterproofed (black areas) where they are below grade. Note the concrete block piers to support the floor frame system.

Concrete Block (Basement)

Basically, this foundation is similar in construction to the concrete block perimeter foundation except that it is higher, contains more doors and windows, is below grade, and must be well waterproofed. Since the walls are earth, retaining some areas requires rebar and filled cells. The floors are often slabs poured between the walls. (See Photo 5-11.) This type of foundation offers a large challenge to the novice builder and should not be undertaken without careful thought and a complete understanding of what is involved.

Advantages
Cost-effective
Permanent

Disadvantages
Can be difficult and slow for the novice builder
Ground water can cause leakage

Poured Concrete

Poured concrete foundations are by far the most common in residential construction (58%). Their construction involves digging footing trenches, building wooden forms, placing the concrete and then stripping the forms. (See Photo 5-12.) Though it often appears intimidating, it is well within the

scope of the novice builder. If a perimeter foundation is desired or required, either a poured concrete or concrete block is suggested. For owner builders it is recommended that the floor joists be used as the foundation forms to help cut costs. Construction of this type of foundation will be discussed in the chapter to follow.

Advantages
Permanent
Good resale value
Within the scope of the novice builder
Typical code foundation

Disadvantages
More expensive than comparable pier foundations
Takes considerable time and energy
Demands considerable accuracy

Poured Concrete (Basement)

Basically the poured concrete basement wall is similar to the poured concrete stem wall foundation. There are a few differences: the walls are higher, they must be waterproofed, there is a slab floor between the walls, there are more openings for doors and windows and they will be below grade. They are perhaps the most difficult of all foundations to build.

Photo 5-12. Poured concrete basement. Note that the forms are removed as the foundation progresses.

Photo 5-13. A grade beam foundation being poured. The round columns will be joined together with a 12×12 grade beam of concrete and rebar.

Photo 5-14. After the foundation is poured, girders are set on the piers. The joists will rest on top of these or be hung from them. These foundations can be very costly if you need to go very deep into the soil.

Photo 5-15. Slab preparation. Forms are in place and trenches are dug for the footing for both exterior and interior load-bearing walls. Rebar is in and pipes are wrapped where they pass through the concrete slab.

Photo 5-16. Footings are being poured on first pass. Note that the underbed is ready; and the wire mesh, and wooden screed board, to be used in levelling the slab, are in place. A boom pumper is being used.

Photo 5-17. Screed boards and a long float are being used to smooth and level the top of the slab.

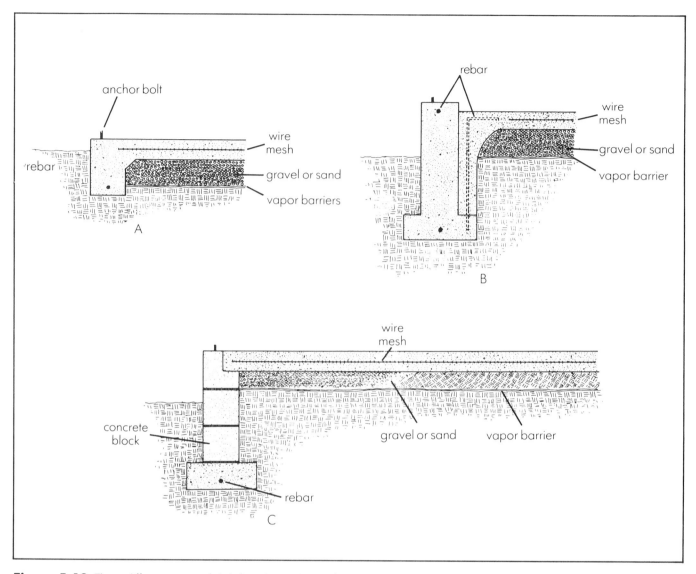

Figure 5-12. Three different types of slab foundation. Vapor barriers are sometimes placed on top of the sand or gravel (check local practices).

They involve high forms with tremendous pressure against them from the concrete and accurate slab work as well. It is advised that you consider contracting these types of foundations to a professional (a good one, as they can run into problems). As with all basement foundations, great care must be given to waterproofing and also insulation. Drainage is very important with this type of foundation, so determine how local builders install their drain tiles.

Advantages
Permanent
Adds to the living space
Good resale value
Commonly accepted foundation

Disadvantages
Requires buying or renting tall forms
Very difficult and time consuming for the novice builder
Expensive
Must be properly waterproofed

Grade Beam

The grade beam foundation (Photos 5-13, 5-14) is a rather new addition to the family of foundations. It is often used on the West coast, where there is a chance of earthquakes, to tie a foundation together so that it will move as a unit. Ordinarily, on a steep site, a poured concrete stem wall foundation that is stepped is used. This being expensive and time consuming to erect on a steep lot, a grade beam foundation is often used in its place.

A grade beam foundation is made up of individual round piers connected by concrete "beams" that run below grade and tie the entire foundation together. The beams are made by trenching into the ground along the grade of the slope, placing rebar into the trench and then pouring the concrete into the pier forms and trenches at the same time. This foundation is easier and cheaper than a comparable poured concrete wall foundation, but may require some engineering.

Slab Foundations

Slab foundations are becoming increasingly more popular. (See Figure 5-12.) They help in cutting costs, since no floor frame or subflooring system is required. The slab foundation serves as the floor frame and subfloor. They are also being used more and more as part of direct-gain passive solar storage systems. (See Photos 5-15–5-17.) They are relatively easy to construct once you know what you are doing and are well within the reach of most novice builders. Their two greatest drawbacks are that they make for a hard, unyielding floor and they bring the house very close to the ground. This is not a problem to most builders, who give only a few years' warranty. You, however, may want to take this into consideration for the house you are building for yourself.

Once a wooden house is placed near the ground, the earth, in a very natural and positive way, begins to attack it, causing it to decay and turn to organic topsoil to be able to support and grow yet more trees to make more lumber for yet more wooden houses. The earth uses fungus, insects, moisture, vapor and water to decompose the house. It is a beautiful system, but not one that you may particularly want working on your home. Because of this, be sure you have a good drainage and gutter system if you plan to use a slab foundation.

Advantages
Cost-effective
Lends itself well to passive solar heat storage
Time efficient

Disadvantages
Can crack on unstable soil
Brings house close to ground
Floors are hard and unyielding
Can be problematic if not built correctly

6 Poured Concrete Foundations

Photo 6-1. The house after the concrete has been "placed" and the forms removed.

What You'll Be Doing, What to Expect

THIS CHAPTER WILL discuss the construction of a typical poured concrete stem wall foundation that is stepped (Photo 6-1, Figure 6-1). This type of foundation is common in many areas of the country. We will be using the earth as the forms for the footing, suspending the foundation forms over the earth and then placing the rebar, wedge ties, and anchor bolts inside the forms. The size and location of the rebar change depending on your locale and whether you are in an earthquake area or not. Check your local building practices before beginning, to see if there are any special requirements due to local codes or conditions. More than any other part of the house, foundations vary from area to area.

It is important that this part of construction be very accurate, not only because you are working with concrete, which is difficult to change in case you make a mistake, but also because all the rest of your work, the floors, the walls, the roof, etc., will be built on this. If the foundations are not square at the corners, or if the dimensions are off, or if the tops of the foundation walls are not level at this stage, it can cause tremendous headaches down the line. At the end of this chapter there is a checklist of things to inspect before the concrete is placed. Go over this carefully. If possible, have a professional inspect your work before you pour. Be accurate and careful, it will pay later.

Jargon

ANCHOR BOLTS. Bolts embedded in the concrete foundation used to hold the sill plates to the foundation wall.

CLEAT. A small piece of wood used for reinforcing.

CONCRETE PUMPER. A machine used to pump concrete through a hose.

CROWBAR or PRY BAR. A metal bar used for prying.

DRAIN TILE. Perforated pipes placed near the footing of the foundation and surrounded by gravel used to drain ground water away from the foundation, basement or crawlspace.

FLOAT. A wooden plaster or concrete working tool.

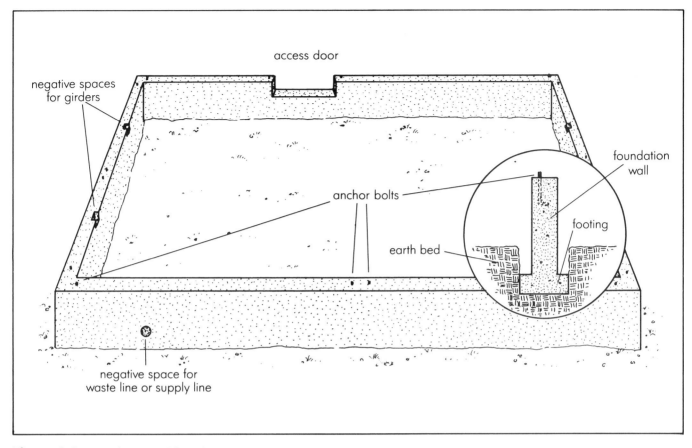

Figure 6-1. Typical stem wall foundation.

FOOTING. Concrete slab at the base or bottom of the foundation wall, used to spread the load over a greater area.

FORM TIES. Metal straps or bars that hold the forms together until the concrete has set.

FOUNDATION. The supporting portion of a building below the first-floor construction.

FOUNDATION FORMS. Wooden, plywood or metal panels that hold the concrete in place until it has set.

FOUNDATION STAKES. Wooden stakes that hold the forms suspended over the ground until the concrete has set.

FROSTLINE. The depth below the surface to which the soil freezes and thaws.

KICKERS. Wooden stakes that hold the forms upright until the concrete has set.

NEGATIVE SPACES. Holes or spaces left in the concrete foundation to be used as passages for plumbing and gas pipes, electrical wires, girder pockets, etc.

PIER BLOCKS. Small concrete foundation blocks, usually used to support wooden posts.

REBAR (REINFORCING BAR). Steel bars that are set in the concrete in such a manner that the two materials act together to resist force.

REBAR CUTTER. A tool used to bend and cut reinforcing bar.

SILL PLATE. A board bolted to the top of the foundation wall, on which the floor joists are attached (see *Sill Plates*, Chapter 7).

SCREED. To form concrete to the desired shape or level by using a metal or wooden tool (float).

Purposes of the Foundation

1) Raise the house above the moist earth
2) Provide a level building plane
3) Anchor the house to the ground
4) Carry the weight of the house to the earth bed

Design Decisions

You will need to determine at this point the manner in which the joists are to be joined to the girders. You need to know this in order to determine the exact heights and locations of the negative spaces. There are several different ways this is done. Each way has advantages and disadvantages. You need to decide at this time which system you are going to use. (Figure 6-2.)

The advantage in System 1 is that the metal fasteners being used help the floor frame hold together in an earthquake. The joists are also on line for the plywood subfloor installation (see *Subflooring*, Chapter 10). Its only real disadvantage is that there is less room in the crawlspace than in the systems where the joists are on top of the girders.

System 2 still allows your joists to be on line and also gives you more room in the crawlspace. System 3 eliminates the need to cut each joist exactly, but adjustments must be made in order to provide nailing surfaces (end joists) for splices in the plywood. System 4 is an older system originating before metal hangers were around. It is not recommended as it

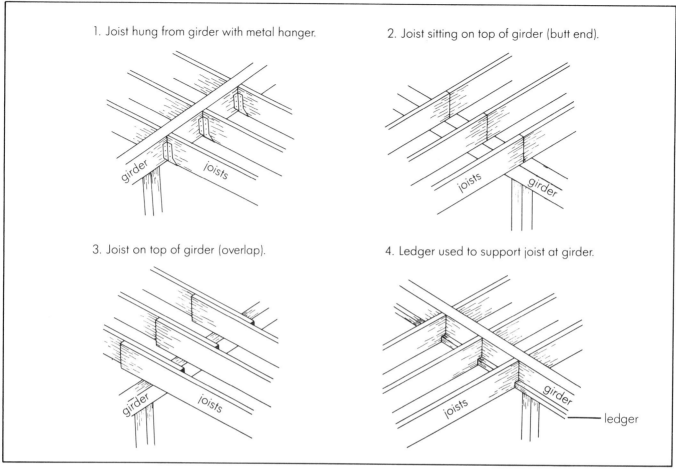

1. Joist hung from girder with metal hanger.

2. Joist sitting on top of girder (butt end).

3. Joist on top of girder (overlap).

4. Ledger used to support joist at girder.

Figure 6-2. Different ways of attaching the joists to the girders. Each affects the level of the girder and thereby the placement of the girder pockets in the foundation wall.

takes too much time to notch each board and the notches weaken the joists. (For more on notching, see end of *Floor Framing,* Chapter 9: "Special Situations—Notching.") Just make your decision before you pour the foundation. (See photos in Chapter 9, *Floor Framing,* "Step 7.")

Review the decision you made for the locations of access doors and other negative spaces. Be sure the final height of the foundation wall is certain. You are finalizing the location of the house, so give it one last thought. You will need to determine the locations of the access door for the foundation, the entrance of any utilities as well as the exit of the sewer lines in order to leave negative spaces for these (see "Step 6," this chapter). See Chapter 5, *Types of Foundations,* to be sure you have chosen the proper foundation.

What You'll Need
Steps 1 to 10

TIME   TO DIG TRENCHES: According to backhoe operator's estimate. TO BUILD FORMS, SET REBAR, TIES AND

ANCHOR BOLTS: 50 to 70 person-hours per 120 linear feet of foundation wall averaging 3 feet in height or less.

PEOPLE   2 Minimum (1 skilled, 1 unskilled); 4 Optimum (2 skilled, 2 unskilled); 6 Maximum (2 skilled, 4 unskilled).

TOOLS   Sledge hammer, framing hammer, levelling device: transit, builder's level, hydro level or water level, steel tape, saw horses, combination square, safety goggles, nail-pulling tools, nail belt, large crayon, pencil, carpenter's level, extension cords, plug junction boxes, saw blades.

RENTAL ITEMS: Transit, rebar bender and cutter, electric shovel (optional), backhoe.

MATERIALS   You are better off purchasing the concrete from a concrete company than mixing it yourself. The form boards can be rented, but most owner builders use their joist stock as forms and clean them and use them as floor joists. Anchor bolts are ½×10 inches (in earthquake areas ⅝×18 inches is recommended) and there are several varieties of form ties that can be used. We will be showing wedge ties in our example.

ESTIMATING MATERIALS NEEDED

REBAR: Rebar is rather simple to order. Just measure the length of the walls and how many levels of rebar there are in each wall. Add 20% to this figure for bends and waste. Baling wire is used to tie the bars together for bends and waste.

ANCHOR BOLTS: You need one anchor bolt for each 6 linear feet of foundation wall (3 in earthquake areas for added protection). In addition, there needs to be an anchor bolt within 12 inches of the end of any sill plate stock.

FORMS: If you are using floor joists, calculate the length and height of the concrete forms to see if the joist stock will be enough for the forms. If not, you can use some of your rafter stock. Have an additional 100 feet or so of 2×4s and 2×6s for cleats and bottom form boards (more, if you are building high forms over 3 feet). Remember, you don't want to cut your floor joists when using them for forms.

STAKES and KICKERS: You will need 2 vertical stakes for every 3 to 4 linear feet of form (one for exterior form, one for interior). Also you will need 2 kickers for every 4 to 6 linear feet of wall. Kickers must be long enough to extend from the top of forms, at an angle, into the ground 10 to 12 inches; usually, 1×4 precut stakes can be bought and used.

Also rented metal stakes can be used for kickers and stakes (see "Step 3A: Installing the Vertical Stakes").

WEDGE TIES: Wedge ties usually occur every 2-4 feet along the wall and between each course of form boards.

INFORMATION
This chapter is sufficient for building a standard poured concrete foundation. Many of you may have unique situations that are not covered here. You may need to consult a professional or a good book (of which there are few). Many problems you may need to solve at the job site using your common sense. Be sure your solutions will work, pass code and are the simplest solutions you can think of.

PEOPLE TO CONTACT
Concrete supplier, concrete pumper. If underfloor utilities are to be placed before the subfloor, you can give any subs you plan to hire notice at this time: plumber, electrician, gas, heating and air conditioning installers, and solar panel installer.

INSPECTIONS NEEDED
After the forms are completed and before you pour, you will need to get your rebar and forms inspected.

Reading the Plans

Most plans will provide the details about your foundations and footings. Since this is probably the most important section of the house and since a poorly designed or poorly constructed foundation can cause total house failure, the permit department will probably want detailed foundation specifications.

Plans should include the following:

A) Width of foundation wall
B) Width of footing
C) Height of foundation wall
D) Height of footing
E) Size and placement of reinforcing bar (rebar)
F) Depth of footing trench
G) Location and size of any pier blocks
H) Location of grade, both interior and exterior
I) Location of any drain tile
J) Size and locations of anchor bolts
K) Waterproofing (if any)
L) Any special details
M) Locations of access door and vents

If your plans are not complete, be sure you gather the proper data before beginning. A mistake in this area can be very difficult to correct. (See Figure 6-3.)

Safety and Awareness

SAFETY: Form building is not a particularly dangerous part of the housebuilding process. As always, be careful with the power equipment. Be careful moving the long lengths of rebar around other people. Be careful with the sledge.

AWARENESS: What you are seeking in form construction is accuracy and stability. The tops of the forms need to be high enough above the bottom of the trench to provide the needed footing thickness. The trenches need to be clean, on stable soil and deep enough to place the entire footing beneath the frostline. The forms need to be square and the proper dimension within the allowed margin of error. The rebar and anchor bolts need to be accurately and securely set. All negative spaces must be provided. The forms and dams must be stable so as not to break or fall during the pour.

Overview of the Procedure

Part One: Digging the Footing Trenches and Building the Forms

Step 1: Transfer the batterboard lines to the ground
Step 2: Dig the footing trench
Step 3: Build the forms
Step 3A: Install the exterior vertical stakes
Step 3B: Nail on the exterior form boards
Step 3C: Plumb the exterior forms
Step 3D: Locate and place the interior form stakes
Step 3E: Nail on the interior form boards
Step 4: Place the ties
Step 5: Place the rebar
Step 6: Provide for negative spaces and step dams
Step 7: Set the anchor bolts
Step 8: Oil the bottoms of the forms and stakes
Step 9: Clean out the trench
Step 10: Call for code inspection

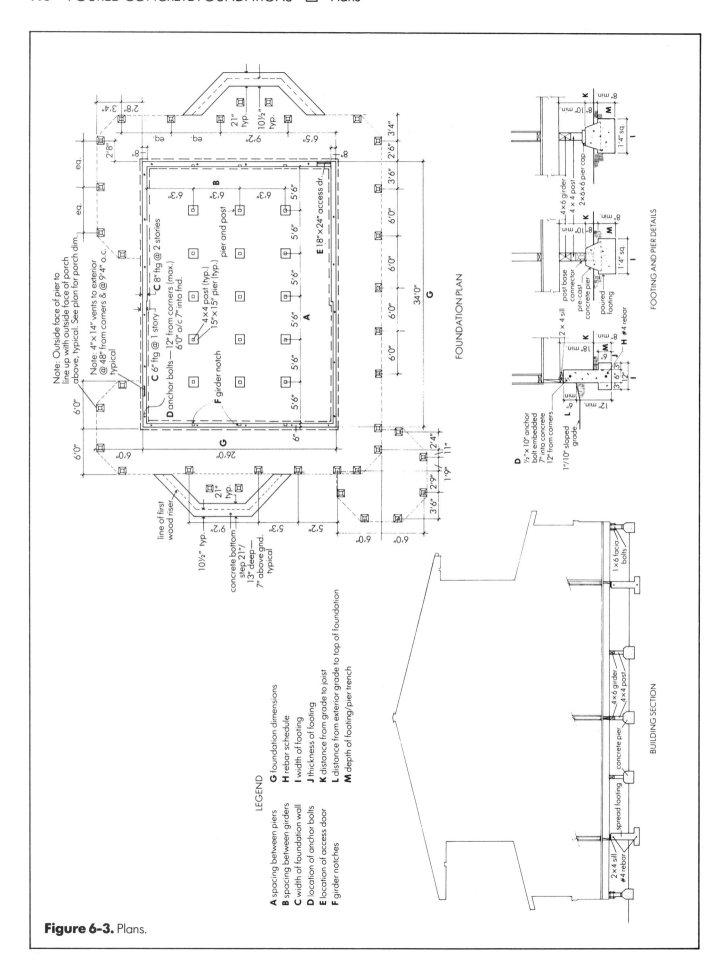

Figure 6-3. Plans.

Photo 6-2. Chalking a line on the ground as a guide for the backhoe operator. Here the line represents the outside edge of the ditch and the footing.

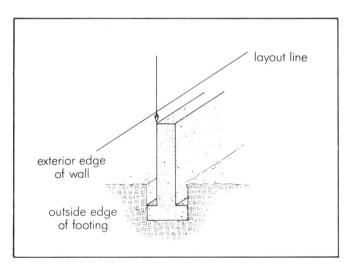

Figure 6-4. Relationship between foundation wall and layout line.

Step 1
Transfer the Batterboard Lines to the Ground

Margin of Error: ½ to ¾ inch.

Most Common Mistake: Forgetting which part of the foundation wall aligns with the layout line.

Diagrams: Figures 6-4, 6-5.

Photograph: Photo 6-2.

At this point your layout lines are up and should be checked for squareness and dimension accuracy one last time. Their accuracy will determine the accuracy of the foundation wall. Once you have the lines where you want them, you need to "transfer" these lines to the ground in order to know where to dig your foundation trenches. Remember, these lines usually represent the exterior edge of the foundation wall, not the interior of the wall or the middle. Also note that the exterior edge of the footing falls to the outside of this line.

To transfer these lines to the ground, simply use a plumb bob to locate each corner of the foundation on the ground, draw a taut string between the corners and mark along the string with lime or powdered chalk. This line will now be your guide for digging your footing trench.

If you want, lines can be run to represent the exact sides of the footing trench.

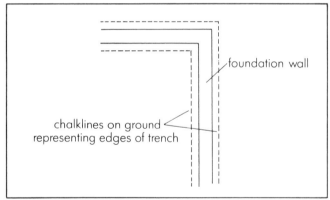

Figure 6-5. Birdseye view of trench lines marked on the ground.

Step 2
Dig the Footing Trench

Margin of Error: 1 to 1½ inch off alignment with mark.

Most Common Mistake: Not digging deep enough or into stable soil.

Diagrams: Figures 6-6, 6-7, 6-8, 6-9.

Photographs: Photos 6-3, 6-4.

At this point you know where to dig the trench; the next question is how deep. To answer this you must know the frostline in your area. The frostline is the depth to which the moisture in the earth thaws and freezes. This can vary from several feet in the cold states to no frostline in warmer areas where there is no freezing (in these areas a nominal 6-inch frostline is often used). In most areas, the code (as well as good building practices) demands that the *entire* footing be below the frostline. This is needed so that as the earth thaws and freezes, heaves and rises, it will not affect the footing or foundation. The ground below the footing must be stable also and the soil in the area of the frostline is not. So if your frostline is 12 inches and the thickness of the footing is 6 inches, the trench must be at least 18 inches deep to keep the entire footing below the frostline. If you plan to step the footing or foundation (see "Steep Slopes" in Chapter 4, *Layout*, p. 94), to avoid digging deep trenches and pouring lots of concrete, be sure you are still at the proper depth to keep the footing below the frostline. The soil at the bottom of the trench should remain undisturbed and stable or the foundation could sink. Never throw any fill dirt into the bottom of the trench.

It is best to hire a backhoe contractor to dig your trenches. If this is not possible, it can be done by hand. If that is the case, renting an electric shovel is advised. These are jackhammer-like tools that use a small electric motor and are very easy to dig with. If you are hiring a backhoe operator, be sure he or she understands where you want the trench, where the steps are to be (if any) and how deep to go. Also be sure they are capable of good work or you will have to do a lot of hand digging to adjust for mistakes and perhaps place a lot of wasted concrete. While they are at the site digging the footing trenches, see if there is other work you need them to do, such as driveways, parking pads, septic systems, water, gas or power trenches, stump removal, etc. Be sure all dirt is placed 4 to 6 feet to the outside of the foundation; you want to keep a clearance between the ground and the under parts of the house. The dirt should also be out of the way during the pour.

At this time you can dig any pier holes that are needed. By referring to the plans, run strings over any lines of pier blocks and mark on the ground the outline of the pier footings. Remember, these footings need to be entirely under the frostline. Refer to "Special Situations: Post and Beam" in Chapter 11, *Wall Framing*, p. 250, to see if any piers are needed under load-bearing posts. Cover these holes with plywood so you do not step in them while working. One of our students sprained an ankle that way.

After the trenches and holes are dug, set the batterboard lines back up (they will need to be removed during the digging) by simply rehooking the lines to the nails on the batterboards and test the accuracy of the trench and pier holes. If they are within the margin of error allowed, you are ready to build the forms. Before beginning, clean out the trenches and holes of any debris that could become embedded in the concrete footing and decay, leaving an air pocket. If your soil crumbles easily, you can bevel the sides of your trenches to keep dirt from falling in.

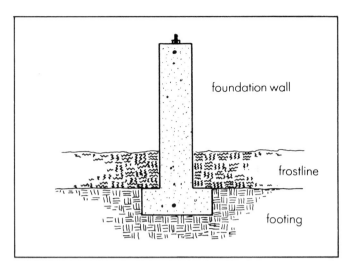

Figure 6-6. Note that the entire footing is below the frostline.

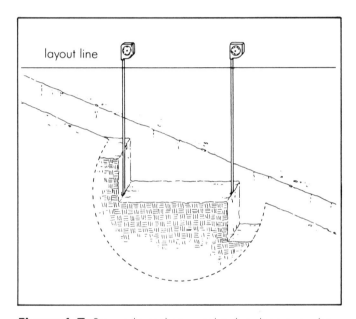

Figure 6-7. Stepped trenches must be dug deep enough to keep the footing below the frostline at all points. Tape determines level of step by measuring from a level line (layout line).

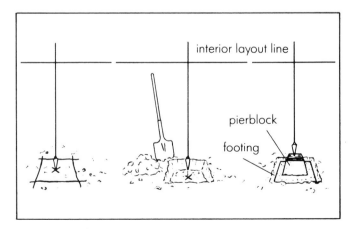

Figure 6-8. Locating, digging and placing pier footings and blocks.

Photo 6-3. Digging the trenches with a small backhoe. These can often be rented.

Photo 6-4. Digging the trenches with an electric shovel. This is a common rental tool.

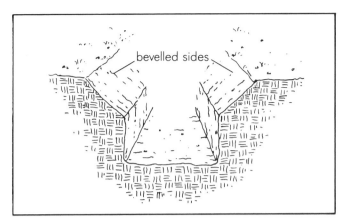

Figure 6-9. Bevel edges of trench in loose soil to keep walls from collapsing.

Step 3
Build the Forms

Diagram: Figure 6-10.

Photograph: Photo 6-5.

Note that the vertical stakes keep the forms suspended over the footing in such a manner that the top of the footing is at the same height as the bottom of the lowest form board. Because of this, the width of the bottom form board must be chosen to allow the footing to come out the proper thickness. The concrete will ooze under the form boards to the trench walls, thus forming the footing. The diagonal kickers keep the forms plumb and properly aligned; the ties and wedges hold the forms together under the pressure of the fresh concrete.

Step 3A
Install the Exterior Vertical Stakes

Margin of Error: ¼ inch off alignment, ¼ inch off plumb.

Most Common Mistakes: Using flat metal stakes; not setting stakes back from layout lines the thickness of a form board.

Diagram: Figure 6-11.

Photograph: Photo 6-6.

After the layout lines have been set up and rechecked for their accuracy, you can begin to install the exterior vertical stakes. You can use wooden stakes (either 2×4s or 1×4s) or metal stakes (either flat or round). The metal stakes can be rented if you do not need them for too long; stake pullers are made for the round stakes which make stake pulling very easy. The flat metal stakes are much harder to pull out of the footing concrete and therefore are not recommended. The 2×4s work pretty well if the ground is not hardpan; if it is, switch to the metal stakes.

The stakes are usually 3 to 4 feet high and are spaced 3 to

4 feet apart. They are set back from the layout line (the outside edge of the foundation) the thickness of 1 form board (in our example 1½ inches).

Locate the 2 corner stakes, run a string between them and place the remaining stakes along the string. Do not push the string out of line with the stakes.

Photo 6-5. The house with trenches dug and forms beginning to be installed. Note: the batterboards and layout lines are used for reference in building the forms.

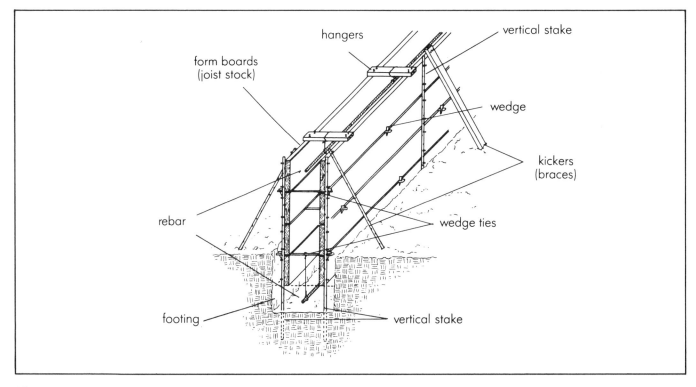

Figure 6-10. Cross-section of typical stem wall form.

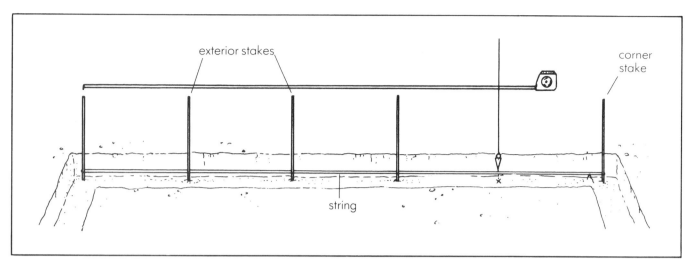

Figure 6-11. Locating corner stakes and running string between these to locate other stakes.

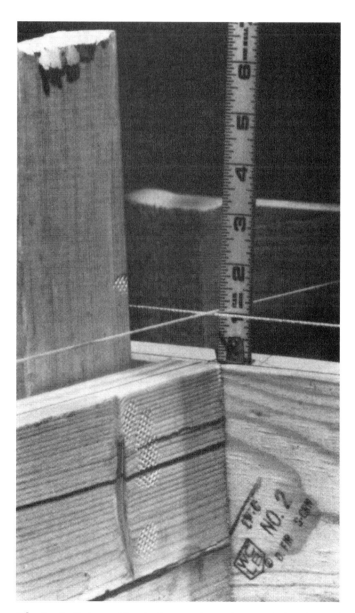

Photo 6-6. Layout line being used as reference to level top of form.

Photo 6-7. Dam being used to step foundation wall. Note how forms (floor joists) are not cut but extend past dam.

Step 3B
Nail on the Exterior Form Boards

Margin of Error: ⅛ to ¼ inch off level.

Most Common Mistakes: Not using straight stock for top boards; not leaving space for form ties; not aligning inside edge of form board with layout line.

Diagrams: Figures 6-12, 6-13.

Photograph: Photo 6-7.

Now it is time to start nailing on the exterior form boards. Usually these are the same boards you plan to use for the floor joists.

Use duplex (double-headed form nails) to nail through the stakes into the form boards so the nails can be easily pulled out when the forms are to be stripped. The metal stakes are predrilled so that you will be able to drive the nails through them. Hammer the nails through the stake while holding the sledge on the interior side of the form board for support. Take care not to step on the edge of the trench.

As you put this first top form board in place, take great care that it is exactly level. Since the concrete wall will be poured to the top of this board, the level of the foundation wall will depend on the level of this board. You can use a long 8-foot level, a builder's level or transit, or use your layout line as a level reference and check the level every 4 feet or so along the top of the board. *Important: Be sure the top board is a good straight board so that the top of the forms will be level.* Remember that the width of the bottom form board will need to be chosen so that the footing comes out the proper thickness. (Sometimes form boards become off-level or out of line, after they have been accurately erected, due to the wet boards warping as they dry or the stakes sinking into the ground.)

Do not line up any splices over the form boards in suc-

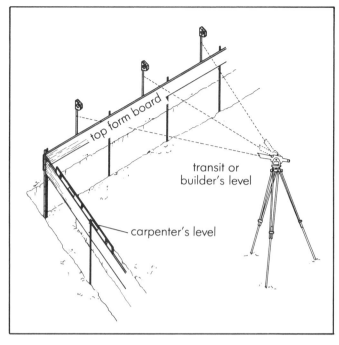

Figure 6-12. Using a builder's level or transit to check level of top form board. A carpenter's level can also be used, but it is not as accurate.

cessive courses, but rather stagger them so as not to create a hinge effect. Nail cleats over the splices. If there are steps in the foundation to reduce the use of concrete, these boards do not have to be cut at the steps. These full-length boards may be needed later. The boards can simply overlap and the dams can be placed between the 2 boards. (See Photo 6-7.)

A space must also be provided between each layer of form boards so that the form ties can be placed in. Use the shank of a nail to gap the forms between each layer of boards. Complete all the outside form boards, using the spacer, and then go on to the next part of Step 3.

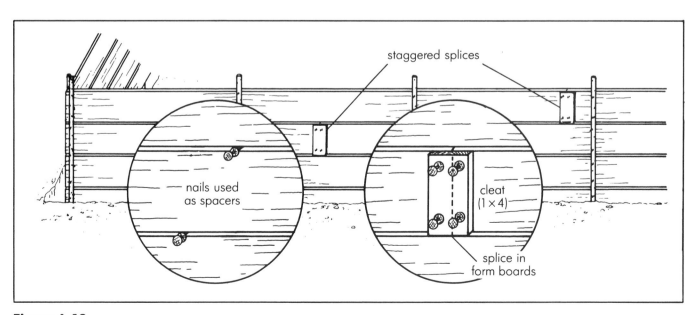

Figure 6-13.

Step 3C
Plumb the Exterior Forms

Margin of Error: ¼ to ½ inch off true plumb; ⅛ to ⅜ inch off alignment with layout lines.

Most Common Mistake: Forms off plumb.

Diagram: Figure 6-14.

Photographs: Photos 6-8, 6-9.

If your stakes were properly placed, the forms should be pretty well lined up directly underneath the layout lines. The *interior* edge of the exterior forms is directly beneath the layout line. Drive your kickers into the ground (6 to 12 inches) at an angle and begin driving 2 nails in the kicker where it intersects the top of the form. Then, using a plumb bob or a level, align the *interior* edge of the form directly underneath the layout lines and drive in the nails the rest of the way. Place these kickers every 4 to 6 feet. It is usually best to cut their tops off flush with the tops of the forms to make levelling the concrete easier. Flat or round metal stakes can also be used for kickers. Check the forms with your level to see if, after their inside edges are aligned directly below the layout line, they are also plumb. The form can be within ¼ to ½ inch plumb — but if it is off by more than this you will need to remove the stakes and start again. The most important thing is that the inside edge of this form be directly below your layout line. Remember, this line marks the outside edge of your foundation wall.

Important: After all the exterior forms are erected, aligned and plumbed, check all your dimensions and check for squareness by checking the diagonals. All measurements should be

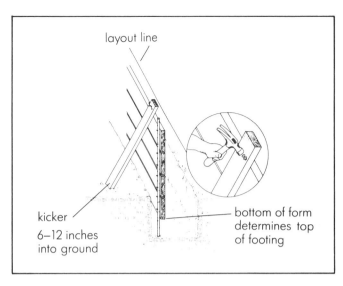

Figure 6-14. Be sure the form is directly below layout line before nailing kicker to top of form.

within ¼ to ½ inch. Your diagonals can be off as much as 1 inch every 50 feet of diagonal. Also place some extra bracing at the corners, as these are often weak areas that burst during the pour.

Photo 6-8. Using plumb bob to align inside edge of form directly below layout line. A level could also be used.

Photo 6-9. Cutting kickers.

Step 3D
Locate and Place the Interior Form Stakes

Margin of Error: ¼ inch off exact placement.

Most Common Mistake: Stakes not set in proper place.

Photograph: Photo 6-10.

Locating the interior vertical stakes is rather simple. You can use a spacer block as shown, locating an interior stake across from each exterior stake. This assures you that after the forms have been completely constructed, the open space between them will be the exact thickness of the foundation wall. You could also locate the corner stakes, as you did the exterior stakes, run a line between them, and locate the rest of the stakes along the line. If the size of the foundation wall changes, such as in switching from a one-story section to a two-story section (from a 6-inch to an 8-inch foundation wall), simply change the spacer to change the location of the interior stakes. The exterior stakes will always be on line, as the outside edge of the foundation wall is always straight.

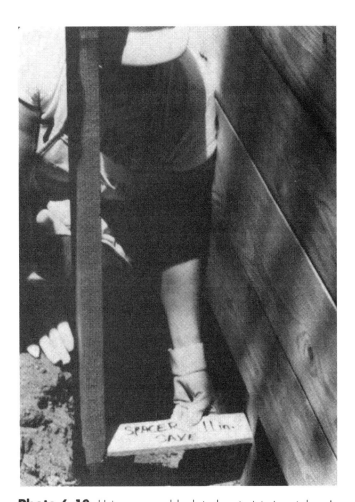

Photo 6-10. Using spacer block to locate interior stakes. In this example an 8-inch foundation wall is being built with 1½ inch thick form boards.

Step 3E
Nail on the Interior Form Boards

Margin of Error: ⅛ to ¼ inch off level.

Most Common Mistakes: Not using straight stock for top boards; not leaving space for form ties; not aligning inside edge of form board with layout line.

Diagrams: Figures 6-15, 6-16.

Photograph: Photo 6-11.

Starting from the top and using a form nail as a spacer between each course of boards, nail on the interior form boards. Again, be sure to use good straight pieces for the top boards so the forms will be level. With these boards, be sure that they are not only level, but also at the same height as the exterior form boards. Remember, the width of the bottom board needs to be chosen so that the footing is the proper depth, give or take 1 to 1½ inches. As with the exterior forms, provide extra bracing at the inside corners. After you have completed all the interior forms, you are ready to place your wedge ties and rebar. The wedge ties will automatically set the interior forms the exact distance away from the exterior forms. If the exterior forms were aligned and plumb, then the wedges will pull the interior forms into alignment. No kickers are set for the interior forms until after the wedge ties are in place.

Photo 6-11. Top interior form being installed.

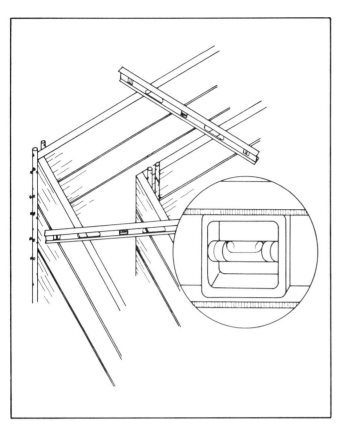

Figure 6-15. Using level to set interior form to the same height as the exterior form boards.

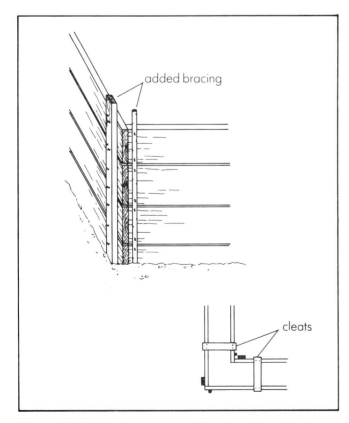

Figure 6-16. Bracing at corners. Top view shows bracing at inside and outside corners, and cleats.

Step 4
Place the Ties

Most Common Mistakes: Insufficient ties; forgetting to put in wedges; wedge ties in upside down.

Diagram: Figure 6-17.

There are several different varieties of form ties. The ones we will be showing are wedge ties and they are recommended. They not only hold the 2 sets of forms together until the concrete sets, but also automatically set the proper spacing between the interior and exterior forms so the foundation wall will be the proper width. Ties are available in different sizes for different wall widths. Six-inch and 8-inch ties are the most common, though larger and smaller ones are available. (*Note:* Sometimes in high walls, the ties are put in *as* the rebar is placed. See "Step 5.")

After the concrete has hardened, the ties will be stuck in the concrete forever, only the little wedges are recyclable. (Also note that many ties are concaved. Always point the open side up so that it can fill with concrete; otherwise you may leave a hole in the concrete.) Ties go between each course of boards and are usually staggered between courses and spaced every 32 inches. On higher walls there will be a great pressure on the bottom of the forms due to the heavy load of concrete. In these cases the spacing between the ties is reduced to handle the extra load. If you are not sure of the spacing for your wall, ask someone who knows. A busted form is a mess.

After you have placed the ties, hammer in their wedges till the form boards are tight against the prongs. Do not forget to put in the wedges on *both* sides of the ties. The flat side of the tie goes against the form, *not* the angled side.

After all the tie wedges are hammered in, our interior forms are aligned, spaced and plumb. Now nail on kickers for the interior forms directly across from the exterior kickers.

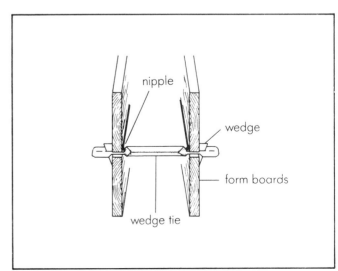

Figure 6-17. Wedge ties. Wedges are placed with straight, not angled, sides against the forms.

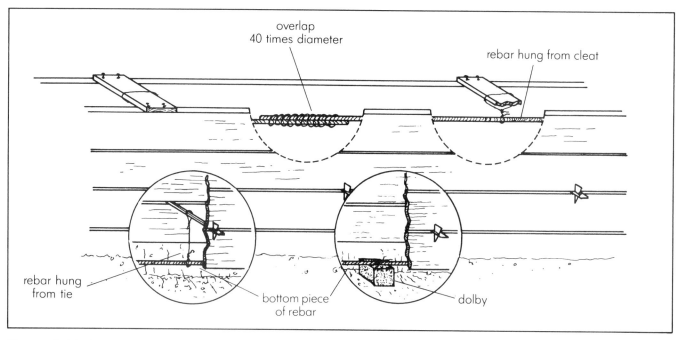

Figure 6-18. Supporting rebar inside of forms.

Step 5
Place the Rebar

Most Common Mistakes: Improper placing of rebar; rebar not well supported; rebar touching form or ground.

Diagram: Figure 6-18.

Photographs: Photos 6-12, 6-13.

Rebar (reinforcing bar) is used to strengthen the concrete by giving it greater bending strength. It helps the pour resist cracking and holds it together in the event of an earthquake or other movement. Most poured concrete foundations will require some rebar. Its location is shown in the rebar schedule of the plans (see "Plan Reading" at the beginning of this

chapter). In earthquake areas there can be quite a bit of rebar, as much as 1 bar every 2 feet vertically and horizontally, and sometimes more. Here are a few things to note about rebar:

☐ The number rebar always refers to the number of eighths of an inch diameter; i.e., #4 rebar equals 4/8 inch or ½ inch.

☐ When joining 2 lengths of rebar together, overlap the pieces 40 times their diameter; i.e., to overlap ½ inch rebar, the formula is ½ inch × 40 = 20 inches.

☐ No rebar can touch any form or dirt; it will be exposed after the pour and may cause the rebar to rust into the wall. No rebar can be within 3 inches of any earth or 1½ inches of any air.

☐ When oiling the forms (explained later in this chapter), be careful not to get any oil on the rebar, as oily metal will not adhere well to the concrete.

☐ Rebar is bent to extend around the corners 18 to 24 inches and also to join interior solid foundation walls with exterior walls.

You are now ready to place the rebar. I highly advise renting a rebar cutter that cuts and bends rebar; it will make the process much easier. The rebar can be supported in various ways as shown in the diagram. Note that in some cases the rebar is placed at the same time as the ties. If more than 2 levels of rebar are to be used or if you foresee any problems in threading the rebar through the ties, then I advise placing the rebar as you set the ties. Insert the lower level of rebar and then the ties, then rebar, then ties, till finished. Tie all rebar together wherever it overlaps with baling wire.

Photo 6-12. Rebar, wire and ties.

Photo 6-13. Using a rebar cutter and bender (a common rental tool).

Step 6
Provide for Negative Spaces and Step Dams

Margin of Error: ¼ inch off proper size and placement.

Most Common Mistake: Leaving out needed negative spaces.

Diagrams: Figures 6-19, 6-20, 6-21, 6-22.

After the rebar is all in place, build all needed dams at steps and provide any needed negative spaces. Negative spaces (holes or indentations left in the concrete) may be needed for any of the following:

☐ Pipes, sewer, underground electricity, etc.

☐ Girder into wall

☐ Access doors or vent space

Often your plans will not show the locations of all these and you will have to make job site decisions. Refer to *Floor Framing*, Chapter 9, "Step 7," for a discussion on how joist-girder connections may affect a foundation's negative spaces. (Negative spaces for girders should be ¼ inch wider and deeper than the girders.)

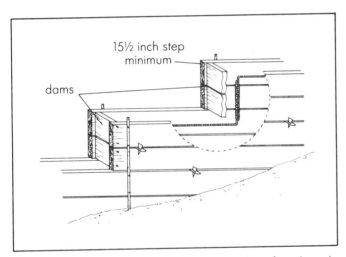

Figure 6-19. Dams and rebars at steps. Note: form boards can be left uncut. First step must be at least 15½ inches to avoid having any cripple studs less than 14 inches high.

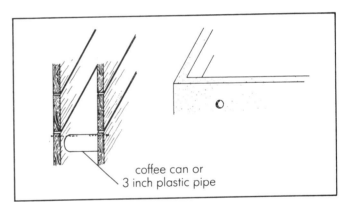

Figure 6-21. Plug being used for negative space for utility or water. Coat with newspaper so plug will not stick to concrete.

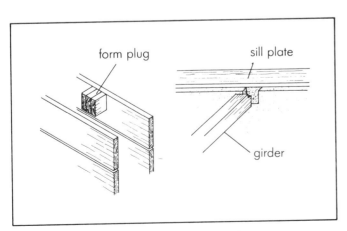

Figure 6-20. Negative space for girder when joists sit on top of girders.

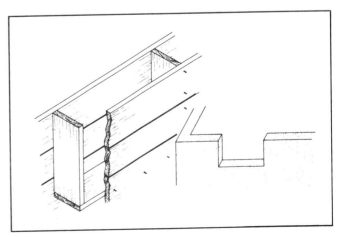

Figure 6-22. Form for access door.

Step 7
Set the Anchor Bolts

Most Common Mistakes: Anchor bolts set too deep; no anchor bolt within 12 inches of end of sill board.

Photograph: Photo 6-14.

Before the pour, all anchor bolts must be set. A jig (a 2×4 with a hole drilled in it to support the anchor bolt) can be used, or the bolts can be set in the fresh concrete and will support themselves until the concrete hardens. The anchor bolts are used to fasten the sill plates (2×6s that lie flat on the foundation) to the concrete foundation. Here are some things to know about anchor bolts:

□ They are usually ½ inch in diameter and 10 inches long; in earthquake areas a ⅝ inch diameter and 18-inch length are recommended.

□ Code usually requires them every 6 feet; in earthquake areas every 3 to 4 feet is recommended.

□ Code requires that at least 7 inches of the bolt be embedded in the pour.

Set the bolts in so they will be in the center of the sill plate. This may not always be the same as the center of the foundation wall. You could have an 8-inch wall and a 6-inch sill plate. The sill will be set flush to the outside edge of the foundation wall. For the bolts to be in the center of the 6-inch sill they will need to be 3 inches in from the edge, not in the center (4 inches from edge) of the wall. Do not keep the bolts exactly on line, but offset them about ¾ inch so as not to split the sill plate along the grain.

There must be a bolt within 12 inches of the end of any sill plate, so be sure you know where 2 pieces of sill stock will butt together along the wall. Also be sure there is a bolt within 12 inches of all corners (end of sills). Every piece of sill must have at least 2 bolts.

Step 8
Oil the Bottoms of the Forms and Stakes

Most Common Mistake: Getting oil on rebar.

Diagram: Figure 6-23.

After everything is ready for the pour, it is usually best to oil certain pieces so they will not stick to the concrete when it is time to remove them. I advise spraying or wiping old motor oil on the vertical stakes where they will be embedded in the footings and on the bottom of the lowest form boards, as they have a tendency to get embedded in the concrete as well. Some builders spray the entire interior of the forms, but I have never really found that necessary. You will need to scrape clean the form boards after they are removed and before they are cut. In very dry areas, you may also want to wet with water the inside of the form boards and the dirt so they do not draw the water from the concrete too quickly.

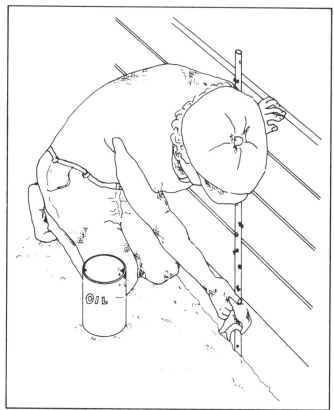

Figure 6-23. Oiling bottom of vertical stakes where they are embedded into concrete footing.

Photo 6-14. Anchor bolt jig to suspend bolt in form.

Step 9
Cleaning Out the Trench

Before the pour, be sure that the trench is clear of any debris. Any organic debris, such as limbs, mud clods, leaves, etc., will decay and leave a hole in the concrete. Be careful not to brush against any poison oak roots as you do this: an entire class of our students once got poison oak that way. Cut a piece of the root and smell it; poison oak roots smell foul.

Step 10
Calling for Code Inspection

Before the pour, most inspection bureaus will want to check your rebar, the depth of the footing trenches and the quality of the final earth bed. They will not, however, be inspecting for the quality and completeness of the forms. The following checklist will assist you in being sure you are ready for the pour. If possible, I advise having a professional inspect before you pour. It is very hard to un-pour concrete.

Checklist Before the Pour

☐ Are all forms stout and properly braced?

☐ Are all dams in place?

☐ Are all ties and their wedges in?

☐ Is the first step at least 17 inches high so no cripple wall stud will be less than 14 inches?

☐ Are vertical stakes and bottom of forms oiled?

☐ Is all rebar properly overlapped, wired together, bent around corners and the proper distance from the form or earth?

☐ Are all negative spaces provided for: girder pockets, electricity, sewer, access door, water, etc.?

☐ Is there an anchor bolt every 6 feet (3 feet in earthquake areas) and within 12 inches of the end of any sill board?

☐ Are anchor bolts the proper height above the foundation wall (2½ to 3 inches)?

☐ Are all forms level, square and the proper dimensions?

☐ Are extra ties and bracing provided at high walls (over 4 feet)?

☐ Do any bridges need to be built over the forms so you will not be stepping on them?

☐ Are pier holes dug?

After the forms have been built and everything checked for the pour, you are now ready to place the concrete. This is one of the most exciting and tiring days of the entire project. Be ready for it: psychically, mentally and physically. Be sure you are well prepared and rehearsed before the concrete trucks arrive. They create a lot of frantic energy at the site with all the noise they make. They also allow you only so much time; after that there is a charge of $30 to $70 per hour or any fraction thereof. Stay calm during the pour and move carefully, even if it means paying for overtime; it is better to do that than make a mistake, cause an accident or spend a generally frantic day.

About Concrete Companies, Pumpers and Booms

Unless there is no concrete delivery possible, I highly recommend against mixing your own concrete; it is not worth the effort and in the end can cost more than having it delivered. In most areas concrete companies can deliver to the site. They mix the concrete at their yards and keep it ready till it is delivered to your site. They may help you figure out how much you will need, or you will have to calculate your needs in cubic yards. Don't forget to figure for your pier footings. They deliver in a timed progression of trucks until your order is filled.

If you are new at this, you may ask them to spread the deliveries out, giving you a longer time between trucks than usual. Also they may have 5½- or 6½-bag mixes. This designates the number of bags of concrete in each cubic yard (concrete is always measured in cubic yards). Unless recommended otherwise, I would go with the cheaper 5½-bag mix. You will have to pay for all the concrete you order, even if you ordered too much. You can have some step pads or well house slabs ready in case there is an excess of concrete.

If the concrete trucks can get uphill and within a certain distance (about 25 feet) of all parts of the pour, no boom or pumper will be needed; the trucks can use their chutes. If only a small portion of the pour is outside of the reach of the trucks' chutes, you can finish it off with wheelbarrows. However, if your site does not allow for this, a boom pumper or standard pumper should be called in. These machines pump the concrete, both vertically and horizontally, and can place concrete a great distance from the concrete truck. If you plan to use one of these pumpers, let the concrete company know which type. The pumpers require a special concrete mix that is more liquid than most since it must pass through the hoses.

The boom truck is more expensive to hire than the standard hose pump, but it may actually be cheaper in the long run. A standard pumper needs a 6½-bag mix, which is more expensive than the boom truck's 5½-bag mix; so, after a certain number of yards of concrete, the boom is cheaper. Also the boom is many times easier to handle. The extremely heavy hose filled with fresh concrete is held by a tall, hydraulically controlled boom which is controlled by the driver. One other person at the spout can control the pour of the concrete. The standard pumper takes one person at the spout and 4 to 7 more to move the heavy hose around. My recommendation, therefore, is to use the boom when possible. Also, be sure everyone who handles the concrete wears thick rubber gloves to protect their skin.

Worksheet
Steps 1 through 10

ESTIMATED TIME

_____

_____

_____

_____

CREW:

_____ Phone _____

_____ Phone _____

_____ Phone _____

_____ Phone _____

_____ Phone _____

_____ Phone _____

MATERIALS NEEDED

Rebar, baling wire, anchor bolts, floor joists, stakes, kickers, wedge ties.

ESTIMATING MATERIALS

_____ feet of REBAR (#4)

_____ coils of BALING WIRE

_____ ANCHOR BOLTS

$2 \times$ _____ FLOOR JOIST _____ feet long = Quantity

_____

Additional material for forms _____

_____ precut stakes or _____ feet of 1×4 =

No. _____ stakes No. _____ kickers

_____ linear ft. of forms ÷ _____ ft. of spacing

between = No. _____ wedges

COST COMPARISON

REBAR:

_____ Store:

$ _____ per M = $ _____ total

_____ Store:

$ _____ per M = $ _____ total

_____ Store:

$ _____ per M = $ _____ total

FORMS:

$ _____ per M = $ _____ total

_____ Store:

$ _____ per M = $ _____ total

_____ Store:

$ _____ per M = $ _____ total

PEOPLE TO CONTACT

Concrete Supplier _____

Phone _____ Work Date _____

Concrete Pumper (if needed) _____

Phone _____ Work Date _____

Plumber _____

Phone _____ Work Date _____

Electrician _____

Phone _____ Work Date _____

Heating and Air
Conditioning _____

Phone _____ Work Date _____

Solar Panel Installer _____

Phone _____ Work Date _____

Gas _____

Phone _____ Work Date _____

Daily Checklist
Steps 1 through 10

CREW:

_____ Phone_____

_____ Phone_____

_____ Phone_____

_____ Phone_____

_____ Phone_____

_____ Phone_____

_____ Phone_____

TOOLS:

☐ sledge hammer

☐ framing hammer

☐ levelling device

☐ transit

☐ builder's level

☐ hydro level

☐ water level

☐ steel tape

☐ saw horses

☐ combination square

☐ safety goggles

☐ nail-pulling tools

☐ nail belt

☐ large crayon

☐ pencil

☐ carpenter's level

☐ extension cords

☐ plug junction boxes

☐ saw blades

RENTAL ITEMS:

☐ transit

☐ backhoe

☐ electric shovel

☐ rebar bender and cutter

DESIGN REVIEW: Review your decision for the location of access doors and other negative spaces. Also be sure the final height of the foundation wall is certain.

MATERIALS:

_____ feet of rebar

_____ quantity of anchor bolts

_____ coils of baling wire

_____ quantity of floor joists

_____ quantity of stakes

_____ quantity of kickers

_____ lbs. of duplex nails

_____ negative space materials

_____ lumber for cleats, miscellaneous needs

_____ no. of wedges

INSPECTIONS NEEDED:

Inspector _____

Phone _____

Time and Date of Inspection _____

WEATHER FORECAST:

For Week _____

For Day _____

What You'll Need
Steps 11 to 17

TIME TO POUR, SCREED, SET ANCHOR BOLTS, PULL STAKES AND SET FORMS: 24 to 48 person-hours per 120 feet of foundation wall 3 feet high, with standard pumper. 10 to 20 person-hours per 130 feet of foundation wall 3 feet high, with boom pumper. (It will take much longer if you are mixing your own concrete.)

PEOPLE This varies, depending on different factors about the pour. You need to carefully evaluate how many people will be needed and be sure that you have enough. Too few people can result in added expenses for delaying the concrete truck or, worse yet, allowing areas of the foundation to harden too soon.

If you are using chutes or a pumper, fewer people are needed than if you are using the pumper with the long hoses that lay on the ground. WITH CHUTES OR BOOM: 2 people to manage chute or boom. WITH HOSE PUMPER: 4 to 6 people to manage the hose.

In addition to this you will need: 1 person to run the vibrator (if you are using one), 1 to 2 people to screed the concrete, 1 person to set the anchor bolts and keep the concrete from trapping the form boards.

TOOLS Screeds, floats, vibrator (if needed), framing hammer, tape, nails, shovels, rubber gloves, water hose, power saw, brush or insecticide spray can for oiling and watering, oil, sledge hammer, wrench for anchor bolt nuts, level, torpedo level, levelling device (transit, builder's level or water level), wheelbarrow, crow and pry bars, wire brush to clean anchor bolt threads, stake puller, pencil, nail belt, safety goggles, extension cords, plug junction boxes, saw blades.

RENTAL ITEM: Transit.

MATERIALS The materials you will need are concrete, anchor bolts and pier blocks (if needed).

Contact concrete and pumper companies to determine size of aggregate and number of bags of cement in the mix.

In our example, we will be placing precast concrete pier blocks, with redwood sills embedded in the top of the pier. You can also build your own forms and pour these yourself, though this is more time consuming. On these blocks we will place 4-×-4 posts to support the girders. (See Figure 6-24.)

Another possibility is to use concrete block piers to support the girders. This method is often used on the East Coast. In this case you must adjust the depth of your footing hole so that the top of the piers will come out at the proper height. Each hole must be individually adjusted because you are working with 8-inch intervals for the blocks (4-inch blocks can be used at the top). Don't forget the small sill plate when figuring the height. A termite guard may be needed, a thin piece of metal on the pier, to prevent termites from tunneling into the wooden girder through the hollow cell of the block. (See Figure 6-25.)

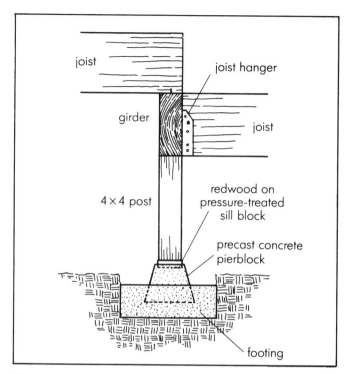

Figure 6-24. Precast concrete pier block and two ways of attaching joist to girder.

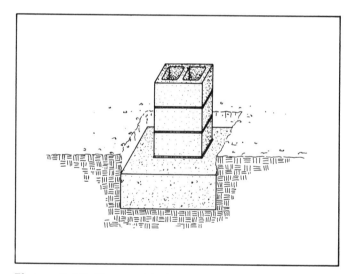

Figure 6-25. Concrete block pier.

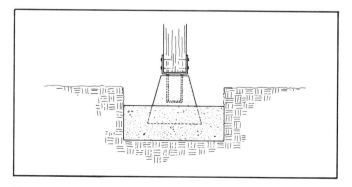

Figure 6-26. Precast concrete pier block with metal fastener.

Pier blocks with metal fasteners are sometimes used. They provide a greater bond between the pier and the wooden post, but are also somewhat more difficult to work with. You need to be more exact when you set them; the post must fit directly between the metal prongs. (See Figure 6-26.)

ESTIMATING MATERIALS NEEDED

CONCRETE: Concrete for foundation walls is always rather hard to estimate. Many concrete companies will do it for you if you give them the details of the project. The problem is, if you buy too little, you will need to call for a partial load, which will cost almost as much as a full load. If you order too much, you will have to pay for it; concrete companies will not take it back. It is best to order a little extra and have forms ready for something else, such as a door stoop, deck piers, etc., where any extra can be used.

To estimate, you will need to measure the widths and heights of the forms and trenches, and then calculate how much concrete measured in cubic yards they will hold. You can only get so exact in this. Then figure out how much the pier footings will use and add these together and order the concrete.

INFORMATION For the most part, the information in this section of the chapter is all you will need to successfully place your concrete, set the anchor bolts and strip the forms. If there is any section that is not clear to you, be sure that you talk to a professional or get clarity in some manner before you pour. Afterwards, it is too late.

INSPECTIONS NEEDED In most areas you will need an inspection before you pour, sometimes both from the government agency and any subdivision regulatory agency.

Do not pour if heavy rain is possible before the concrete has set. Hard rain will wash the cement from the stone aggregate in the concrete. A light rain is OK. (You can cover the freshly poured concrete in rain, but it is generally difficult and dangerous to do a pour in the rain). If there is a chance of freezing weather, wait if you can; the water in the concrete must not freeze before it dries. If you cannot wait, ask the concrete company to add an anti-freeze to the mix. In hot, dry weather, either keep the pour damp for a day or ask about having a drying retardant added to the mix.

Reading the Plans

See "Reading the Plans" in the beginning of this chapter.

Safety and Awareness

SAFETY: Unlike form building, the pour is rather dangerous. There are a lot of people, activity, noise and a need to finish by a certain time. Who could write a better script for an accident? Be sure everyone understands what they are doing, that one person is in charge and that people consciously seek to remain calm and move slowly in an otherwise anxiety-prone setting. The people staffing the concrete hoses or chutes need to be very careful.

AWARENESS: You are seeking to bring the level of the concrete to the top of the form and no higher. Be sure all your dams and negative space inserts hold up under the pressure of the falling concrete. Be sure the rebar remains in its proper location, and that the anchor bolts are accurately set high enough and their threads kept clean of concrete. Be sure the forms are not upset during the pour by knocking them, standing on them, letting the falling concrete pour against them, etc. Make sure the pour ends up as you planned it.

Overview of the Procedure

PART TWO: THE POUR

Step 11: Place the concrete — first pass
Step 12: Place the concrete — second pass
Step 13: Screed the concrete
Step 14: Place the interior pier blocks (if any)
Step 14A: Dig the holes
Step 14B: Reset the layout lines

Step 14C: Pour the footings
Step 14D: Place the interior piers
Step 15: Pull the vertical stakes
Step 16: Strip the forms
Step 17: Keep the foundation damp

Photo 6-15. Concrete truck dumping concrete into standard pumper.

Step 11
Place the Concrete—First Pass

Most Common Mistake: Letting concrete beat against form and push form off.

Diagrams: Figures 6-27, 6-28, 6-29.

Photographs: Photos 6-15, 6-16, 6-17, 6-18, 6-19, 6-20.

After you have decided whether to use the chutes, boom pumper or standard pumper, check to be sure you have all the help, time and tools you will need. Go over the duties with your crew and be sure that one person will take charge and be in control during the pour. Lack of leadership at this point can cause problems. Many builders lightly spray the interior of the forms and the footing trenches with water before the pour. This is especially advised in very dry climates, as the dry wood and earth can draw the moisture out of the concrete too soon, causing it to be too brittle.

Next, begin to place the concrete (place, not pour, is the trade word, just in case you want to sound like a pro). Usually it is best for the first pass to complete the footing and go 6 inches into the forms. Return for the second and final passes *before* the first has set up. This process of 2 or 3 passes takes some of the pressure off the fresh concrete and prevents it from oozing around the bottom of the forms and embedding the form boards. On high walls, 3 passes are recommended. Remember, avoid pouring fresh concrete on top of set concrete. This will set up a "cold joint" which will be a weak spot. If you cannot finish the pour in one process, rough up the edges and place pieces of rebar in the first part of the pour before it dries.

As you pour the concrete be sure that the fresh concrete does not beat against the side of the forms. As much as possible pour straight down. A piece of plywood can be used for a splash. The pressure of the concrete beating against the sides can push the forms off plumb.

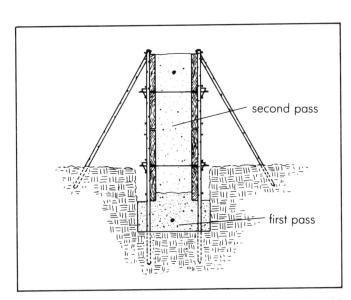

Figure 6-27. Placing the concrete in two passes. Avoid "cold joints."

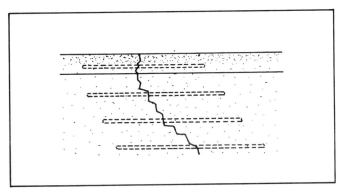

Figure 6-28. Using rebar on irregular joint when connecting a concrete pour to one which has already hardened.

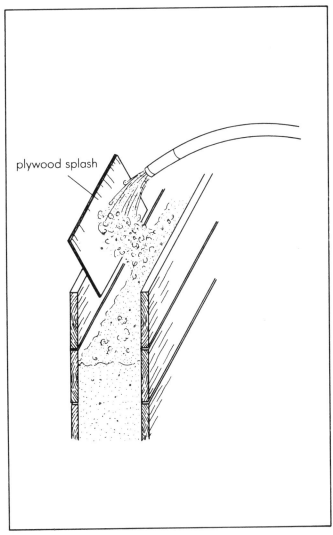

Figure 6-29. Plywood splash being used to prevent concrete from battering form boards.

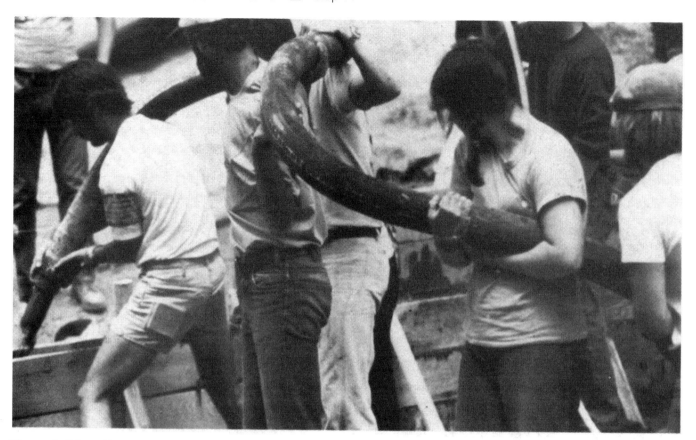

Photo 6-16. Holding heavy hose of standard pumper. The hose is hard to hold as it has a surging action.

Photo 6-17. The boom pumper with suspended hose.

Photo 6-18. One person directing hose of boom pumper.

Photo 6-19. Concrete truck with chutes. This is easiest and cheapest if the chutes can reach all areas of the pour.

Photo 6-20. Pouring concrete straight down.

Photo 6-21. Using vibrator to settle concrete.

Step 12
Place the Concrete—Second Pass

Margin of Error: No concrete over top of form boards after screeding; ⅛ inch lower.

Most Common Mistake: Concrete poured too high or too low.

Photograph: Photo 6-21.

After your first pass, begin the second pass. At this stage you may want to use an electric concrete vibrator. This assures that the concrete is completely settled and there are no holes or any honeycombing effect. One person follows directly behind whomever is pouring the concrete and sticks the vibrator completely in and then quickly pulls it back out. Do not leave the vibrator in too long or it will cause the concrete to push out around the bottoms of the forms. Should any concrete creep up outside the bottom form boards and embed them, push the concrete away while it is still fresh or it will entrap the bottom form board. If you are not using a vibrator, you can use a piece of rebar and plunge it up and down, and beat on the sides of the forms with a small hand sledge to be sure that the concrete has settled properly.

As you pour, bring the concrete to the top of the form and ½ inch higher. Another person should be working behind you screeding the concrete and being sure the anchor bolts and negative spaces are in properly.

Note: Sometimes, as the concrete is compacted by the vibrator, you may have to add more to bring the concrete to the top of the form again.

Photo 6-22. Using 1 × 4 as screed board. 1 × 4 is being moved in short strokes perpendicular to the forms.

Step 13
Screed the Concrete

Margin of Error: Flush with top of forms.

Most Common Mistake: Concrete too high.

Photograph: Photo 6-22.

A common screed is a 1×4 or 2×4 used on edge and moved back and forth across the width of the foundation wall to level out the pour. Small hand concrete floats can be used also. Be sure this is done well and especially that there are no high spots that will later have to be chiseled down. Check that all anchor bolts are in, that they are plumb and that no concrete is in their threads. Screed around anchor bolts as you go along. If you don't have pier blocks to place, your next step will be to pull the vertical stakes that are sticking into the concrete footings. We will discuss that in "Step 15." "Step 14" will discuss the placement of pier blocks for interior girders.

Step 14
Place the Interior Pier Blocks (If Any)

Pier blocks are used to support any girders required in the floor frame. Girders are used for additional support of floor joists spanning from foundation wall to foundation wall. This step will be broken into sub-steps A through D.

Step 14A
Dig the Holes

Most Common Mistake: Pier holes off line.

Be sure that the holes are on line, the proper dimension and depth (deep enough for the entire footing to be below the frostline) and in good stable soil. Before pouring any concrete, be sure the hole is free of any debris.

Step 14B
Reset the Layout Lines

Margin of Error: ¼ inch.

Most Common Mistake: Inaccurate lines.

Diagram: Figure 6-30.

Photograph: Photo 6-23.

If the holes are dug and ready, reset the layout lines for the interior pier blocks. These lines can be running on either side of the blocks or in the center of the block; it does not matter as long as you know what part of the pier it designates and you place the pier accordingly. An easier way is to run new lines connected to the inside of the interior form boards. Keep these lines close to the ground and this will make locating your blocks much easier. Tie a little piece of string on the line to mark the exact center of the block. Hang a plumb bob from that string to locate where the center of the pier block should be placed.

Photo 6-23. Using plumb bob and layout line to locate piers.

Step 14C
Pour the Footings

Diagram: Figure 6-31.

If possible, pour the footings before the concrete trucks leave, rather than having to mix your own concrete or having to call them back with a partial load. Pour the footing the proper thickness and screed it on the top so that it is relatively level. The only time this footing pour must be very level is when you are using concrete blocks. If this is the case, it is best to set up level lines so that you can measure down from them to be sure the top of the footing is where you want it to be. A water level, transit or builder's level will work well here.

Step 14D
Place the Interior Piers

Margin of Error: ⅛ inch off level; ¼ to ½ inch off proper location.

Most Common Mistakes: Pier block off level, or off line; piers not sunk deep enough.

Photograph: Photo 6-24.

While the footing concrete is still wet, press a pier block into it until about half of the block is embedded in the fresh concrete. Then using a small torpedo level, level the top of the block in both directions and diagonally. Continue using this method until all pier blocks have been placed.

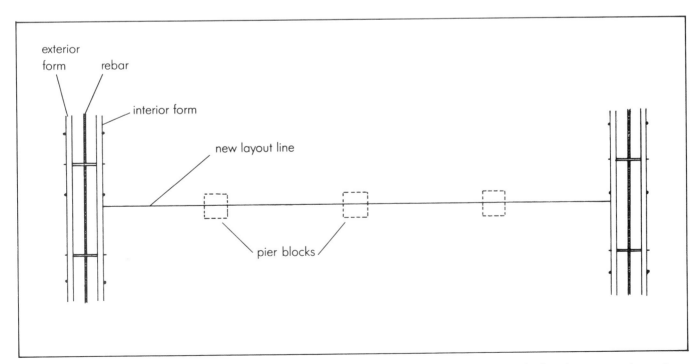

Figure 6-30. Attaching new pier layout line to inside of interior form board (top view). Note that the line demarcates the center of the piers in this example.

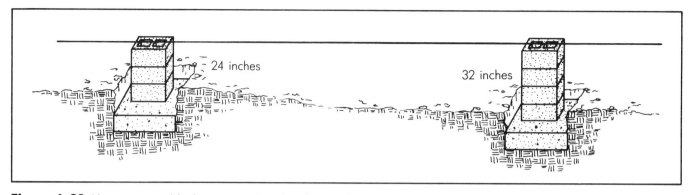

Figure 6-31. Using concrete blocks as piers. Note that the top of the footing must be located an even multiple of 8 inches below the bottom of the girder.

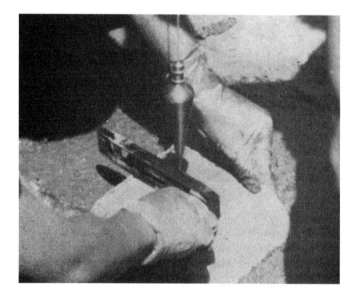

Photo 6-24. Using torpedo level to level pier block.

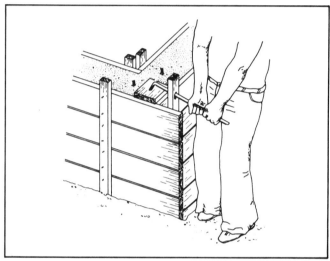

Figure 6-32. Pulling out a wooden vertical stake. Note the hole drilled for crowbar leverage.

Photo 6-25. Foundation immediately after the pour. Note: pier blocks in place.

Step 15
Pull the Vertical Stakes

Most Common Mistake: Allowing the concrete to become too hard.

Diagram: Figure 6-32.

Photograph: Photo 6-25.

Before the concrete has entirely set up, you need to pull out all the vertical stakes that are stuck into the concrete footing. Don't pull these too early, as they are still serving to hold the forms suspended over the footing trench, but if you wait too long you will not be able to pull the stakes loose from the concrete footings. It is hard to say exactly how long after the pour to wait before pulling the vertical stakes. It depends on shading conditions, humidity, temperature, dryness of the ground, mix of the concrete, etc. We have had to pull stakes one hour after the pour on a hot, dry California day and could pull them the next day on a wet, humid North Carolina day. Use your judgement. Usually, pull the stakes when you can press hard into the concrete with a hard object and make an indentation, but you cannot really penetrate the concrete. Test the stakes; when you feel you can wait no longer, start to pull. Sometimes you will have to come back after dinner or work at night under the lights to pull them at the right time, but don't let them get stuck. We still have $100 worth of metal stakes sticking in a foundation in the Sierra foothills. We waited too long. If you are using the round metal stakes, you can rent a special stake puller for these.

Figure 6-33. Stripping form using hammer (or crow or pry bar).

Step 16
Strip the Forms

Most Common Mistake: Straining your back.

Diagram: Figure 6-33.

Photograph: Photo 6-26.

Usually the forms can be stripped the next day. This is a rather simple process unless some of the bottom form boards have gotten embedded in the concrete. Simply remove all the wedges from the ties, and remove all cleats, stakes and kickers. Then, using a large crow or dry bar or a pick-ax, start to peel the boards off. Clean these boards by scraping them with a shovel and stack them neatly to be used as floor joists.

Step 17
Keeping the Foundation Damp

Though many builders ignore this step, it is usually advisable to keep the pour damp for 1 to 3 days. This allows the concrete to cure gradually and makes for a stronger pour. Often, if it cures too quickly, it can be very brittle. Sometimes chemicals (pozolith) are added to the concrete to slow the process. There are a number of ways of keeping the concrete damp. If possible, you can spray it with water every few hours or drape damp burlap bags over it. Some builders just strip the interior forms, leaving the exterior forms on for a few days; the concrete will stay damp under the forms. You are probably creating no great threat to the house if you skip this process, but, if possible, it is better to include it.

Special Situations

Foundation Drain
(Figures 6-34, 6-35, Photo 6-27)

Many areas of the country either require or recommend foundation drains. These drain pipes keep ground water from flowing into the crawlspace or basement or from undermining the footing. Consult with your local builders or inspection office to see if they are needed in your area.

Basically, they are perforated plastic pipes, about 3 to 4 inches in diameter, that are placed at the same level as or directly on top of the footing. Gravel is placed above and below them to keep dirt from clogging the holes. Tar paper is often placed above the gravel before the dirt is backfilled to further keep the sill from filling the holes.

If at all possible, place the pipes above the footings. If you place them at the same level, you will have to build a wider footing trench and form on one side of the footing wall. Check to be sure which is recommended before proceeding. If you are using the type of pipe that is only perforated for half the diameter of the pipe, point the holes down towards the ground, not up!

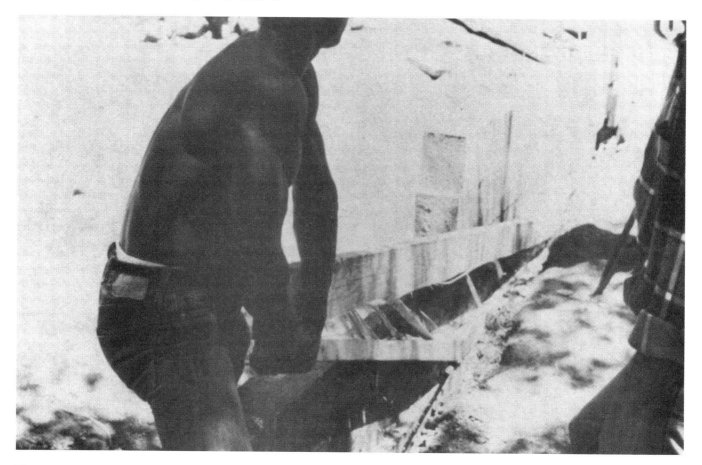

Photo 6-26. Prying bottom form board loose that has gotten embedded in footing concrete.

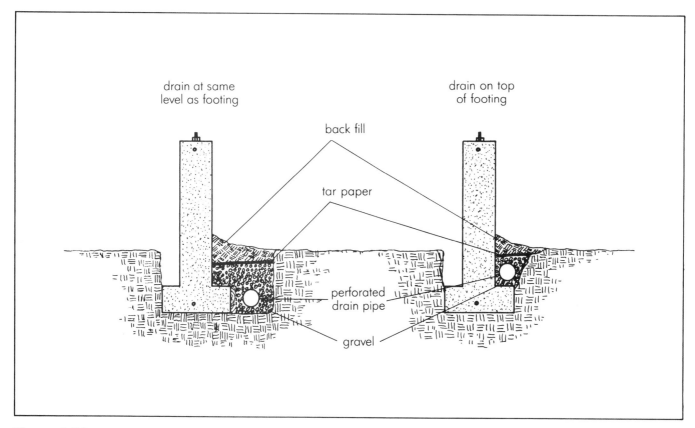

Figure 6-34. Two ways of installing drain pipe or tile at foundation footings.

Photo 6-27. Drain pipe going in.

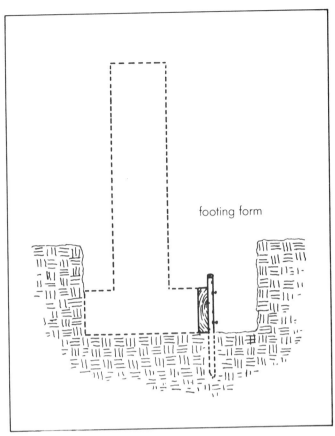

Figure 6-35. Footing form used in larger footing trench when placing drain pipe at same level as footing.

footing form

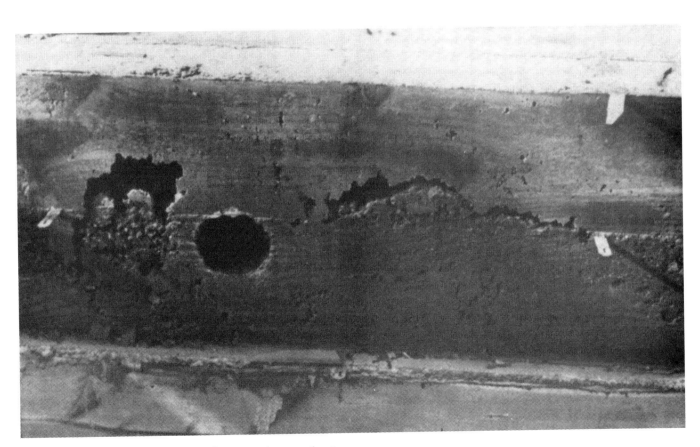

Photo 6-28. Air pockets left by insufficient packing or vibrating.

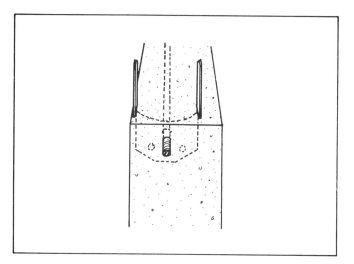

Figure 6-36. Sill anchor (not recommended).

Special Situations

Sill Anchors
(Figure 6-36)

Some builders use sill anchors instead of anchor bolts. It is tempting, since they make the application of the sill plates quicker and easier; you do not have to drill any holes. I would recommend against these, however, as they do not have the ability to pull any irregularities out of the sill plate board stock as the permanently set anchor bolts do.

Wet Sills

See wet sills in Chapter 7, *Sill Plates*, p. 157.

Air Pockets
(Photo 6-28)

Sometimes, due to insufficient vibrating, air pockets will be visible after the forms are removed. Unless this is severe, it will not affect the quality of the foundation. Simply take some mortar and fill in the spaces.

Photo 6-29. Building forms on a steep slope.

Building Forms by a High Bank
(Photo 6-29)

Sometimes if you are building the forms in front of a steep bank you may not be able to get a hammer between the bank and the outside of the form boards. In this case, the forms are built on the ground first and then lifted into place, levelled and plumbed.

Worksheet
Steps 11 through 17

ESTIMATED TIME

_____

_____

_____

_____

CREW

Skilled Leader

_____

_____

Unskilled

_____

_____

_____

_____

_____

For Hose Pumper (Additional)

_____

_____

MATERIALS NEEDED

Concrete, anchor bolts, pier blocks (if needed).

ESTIMATING MATERIALS

Number of cubic yards in foundation walls

and wall footings _____

Number of cubic yards for pier footings _____

Yards of concrete _____

Number of anchor bolts _____

Number of pier blocks _____

COST COMPARISON

_____ Pumper Company:

$ _____ per hour; $ _____ per day

_____ Pumper Company:

$ _____ per hour; $ _____ per day

_____ Pumper Company:

$ _____ per hour; $ _____ per day

_____ Concrete Company:

$ _____ per yard; $ _____ per hr. overtime for truck

_____ Concrete Company:

$ _____ per yard; $ _____ per hr. overtime for truck

_____ Concrete Company:

$ _____ per yard; $ _____ per hr. overtime for truck

Daily Checklist
Steps 11 through 17

CREW:

_____ Phone _____

_____ Phone _____

_____ Phone _____

_____ Phone _____

_____ Phone _____

_____ Phone _____

_____ Phone _____

TOOLS:

☐ stake puller   ☐ sledge hammer

☐ vibrator   ☐ wrench

☐ floats   ☐ torpedo level

☐ brush   ☐ level

☐ spray can   ☐ levelling device

☐ oil   ☐ transit

☐ sledge   ☐ builder's level

☐ wheelbarrow   ☐ water level

☐ rubber gloves   ☐ crow and pry bars

☐ screeds   ☐ wire brush

☐ framing hammer   ☐ pencil

☐ tape   ☐ nail belt

☐ nails   ☐ safety goggles

☐ shovels   ☐ extension cords

☐ rubber gloves   ☐ plug junction boxes

☐ water hose   ☐ saw blades

☐ power saw

RENTAL ITEMS:

☐ transit

MATERIALS:

_____ concrete

_____ anchor bolts

_____ pier blocks

INSPECTIONS NEEDED:

Government Inspector _____

Phone _____

Time and Date of Inspection _____

Subdivision Inspector _____

Phone _____

Time and Date of Inspection _____

WEATHER FORECAST:

For Week _____

For Day _____

7 Sill Plates

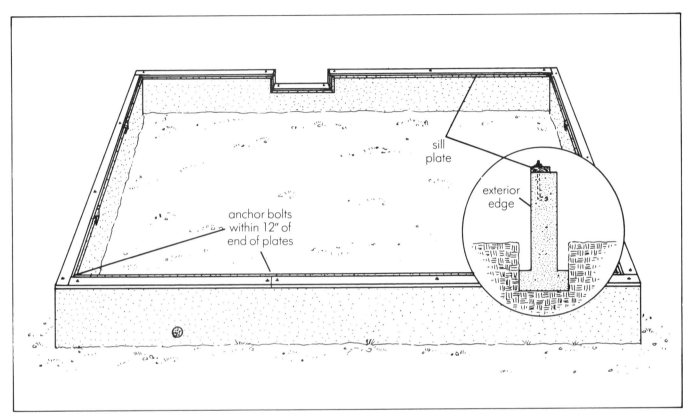

Figure 7-1. An overview and a closeup of sill plates. Note that the sill plate is always flush with the outside (exterior) edge of the foundation wall.

What You'll Be Doing, What to Expect

SILL PLATE INSTALLATION is easy and enjoyable. There is little physical strain or effort. Your emphasis is on accuracy and correcting discrepancies in the foundation wall. It is just a matter of attaching the sills (see Figure 7-1) on the foundation wall so that they are square, the proper dimensions and, most important, level.

As in many areas of construction, there are several approaches to sill plate installation. I have chosen the one I feel is best for the novice builder. It involves a process of checking the squareness of the foundation before installing the sills. Although it is a little more time-consuming than other approaches, it assures you that the sill plates will be true and square even if the foundation is not. This is very important because the floor frame and the rest of the house will be built on these sills, and if they are off square it can cause problems further down the line. (See Photo 7-1.)

Jargon

ANCHOR BOLT. Bolt embedded in the concrete foundation that holds the sill plate (and thereby the house) to the foundation.

PRESSURE-TREATED OR TREATED LUMBER. Wood that has been thoroughly penetrated with a decay and termite preventative chemical. The chemical is inserted under pressure and replaces the natural moisture in the wood.

SHIM. A thin wedge of wood, metal, or a piece of plywood used for levelling or plumbing wood members.

SILL PLATE. Horizontal framing member that rests on top of the foundation wall and is bolted to the foundation. Usually a 2×6, pressure-treated or redwood, lying flat.

Photo 7-1. A sill plate installed on top of a foundation wall.

Purposes of Sill Plates

1) To attach the wooden house to the concrete foundation
2) To level the top of the foundation wall
3) To keep the framing lumber out of contact with the concrete, where it could rot

Design Decisions   None.

What You'll Need

TIME   10 to 20 person-hours per 120 linear feet of foundation wall.

PEOPLE   1 Minimum (skilled); 2 Optimum (1 skilled, 1 unskilled); 4 Maximum (2 teams of 2).

TOOLS   Framing hammer, power saw, level, carpenter's level, chalkline, string, 100-foot tape, 25-foot tape, combination square, wrench, pencil, nail belt, safety goggles, drill, bits, extension cords, plug junction boxes, saw blades, shims, chisels, charges and nails for nail gun (if needed).

RENTAL ITEMS: Transit, concrete nail gun (if needed).

MATERIALS   Usually only two types of materials are used for sill plates: foundation grade redwood and pressure-treated stock. Most codes will require one of these two. (Occasionally codes allow an untreated softwood to be used.) Often the stock sold for sill plates is very low quality; the sill is not in a structural position and cannot fail. Your greater concern is rotting, which is why a rot-resistant material is required. Be careful in buying your stock; badly bowed boards simply will not work. Knots in the boards are OK.

Pressure-treated lumber is usually a little cheaper than the redwood. If the pressure treatment has thoroughly penetrated the thickness of the stock, it should work as well as the foundation grade redwood. Also check to be sure it is as straight as the redwood. Often it is not. Of late, the foundation grade redwood, which is fairly inexpensive, has been such straight stock that we have been using it for deck planks.

ESTIMATING MATERIALS NEEDED

Since there must be an anchor bolt within 12 inches of the end of any piece or splice of sill stock, the length of the pieces needs to be coordinated with the placement of the anchor bolts. See *Poured Concrete Foundations,* "Step 7."

INFORMATION   This chapter should get you through without any other needed info.

PEOPLE TO CONTACT   None.

INSPECTIONS NEEDED   None.

Reading the Plans   None.

Safety and Awareness

SAFETY: This process of construction is rather safe, but as always never lose respect for power tools. If you are using the concrete nail gun (in order to apply vertical plates at steps in the foundation), be sure you understand its operation. Always use safety goggles and turn your head away when you fire. This is one of the most dangerous construction tools I know.

AWARENESS: When applying the sill plates, you are seeking exactness in level, squareness and dimension. Do not assume the foundations are true in any of these. Work out any foundation discrepancies with the sill plates.

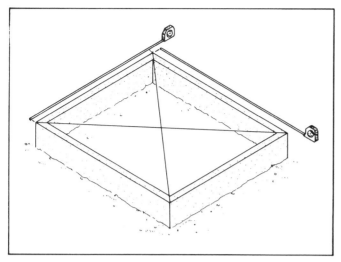

Figure 7-2. Checking the foundation wall for squareness and the proper dimensions before installing the sill plates.

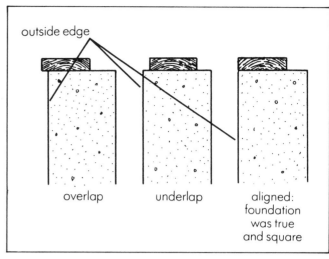

Figure 7-3. Three possibilities of sill plate alignment with the outside edge of the foundation wall.

Overview of the Procedure

Step 1: Check the squareness of the foundation wall
Step 2: Select the stock
Step 3: Measure the sill plates
Step 4: Establish inside corners of sill plates
Step 5: Snap chalklines for inside edges of sill plates
Step 6: Mark and drill anchor bolt holes
Step 7: Install and level sill plates

Step 1
Check the Squareness of the Foundation Wall

Margin of Error: ½ inch off square or off dimension.

Most Common Mistake: Allowing too great a margin of error.

Diagrams: Figures 7-2, 7-3.

Before installing the sill plates, check the squareness and dimensions of the foundation wall. Refer back to the plans to be sure you have the required dimensions. No matter how accurate you are, there will always be some error; it is the nature of the tools and materials you are using. Installing the sill plates precisely will allow you to make up for most minor discrepancies created at the foundation stage.

During this step, determine any errors in the foundation wall. This will give you a sense of how much adjustment will be needed in the sill plate stage. Even if the foundation wall is not exactly true and square, install the sill plates so that *they* are. This may mean that the outside edge of the sill plates will not align exactly with the outside edge of the foundation wall.

If the difference between the outside edge of the foundation wall and the outside edge of the sill plate is too great (over 1 to 1½ inches), only allow about a 1-inch difference and build the sill plates out of square. You may be able to adjust a little more in the floor framing stage.

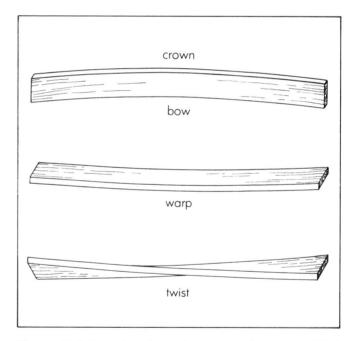

Figure 7-4. Examples of irregularities in lumber. Be careful of sill stock that is badly bowed.

Step 2
Select the Stock

Margin of Error: Bows — ½ to ¾ inches out of true.

Most Common Mistake: Bowed stock.

Diagram: Figure 7-4.

Photograph: Photo 7-2.

It is important that the stock be good straight pieces. All pieces will be somewhat bowed; it is the nature of wood. A small bow can be pulled or pushed out. Do not use badly bowed pieces. Warps in the boards are OK; they will flatten out once the weight of the house is placed on them.

Photo 7-2. Using the foundation wall as a straight edge to test for straightness of the sill plate. A ½ to ¾-inch bow over a 12 to 14-foot board is allowable.

Step 3
Measure the Sill Plates

Margin of Error: ¼ inch off measurement.

Most Common Mistake: Inaccurate measurements.

Photograph: Photo 7-3.

After you have selected your stock, it is time to measure and cut. This is relatively simple and straightforward. The only thing to remember is that there must be an anchor bolt within 12 inches of the end of any sill plate and sill plates cannot butt together over any opening in the foundation wall, such as a vent opening or an access door.

Step 4
Establish Inside Corners of Sill Plates

Margin of Error: Exact.

Most Common Mistake: Not being exact.

Diagram: Figure 7-5.

After you have figured how far off the foundation wall is, mark these discrepancies on the top of the foundation wall indicating whether the true outside corner is farther in or farther out from the actual one.

Now using the exact outside corners as references, mark the corresponding inside corners of the sills on the foundation wall. To do this, you must know the exact width of the sill stock. In our example we are using 2×6 sills, so we will use 5½ inches as the exact width measurement. Do this at all corners of the foundation. Be sure you are marking for the inside, not the outside corners of the sill plates.

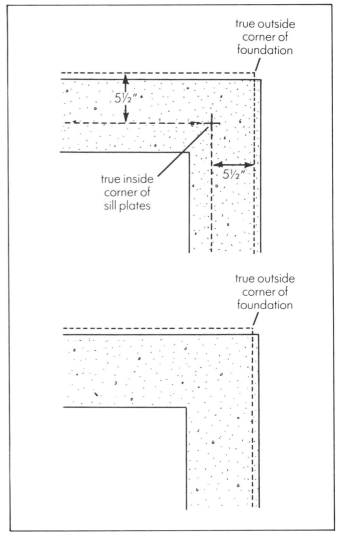

Figure 7-5. Locating the true inside and outside corners of the foundation wall.

Photo 7-3. Measuring for sill plates. Remember that anchor bolts are needed 12 inches from the end of the wall and 12 inches from where the sill ends come together along the wall.

Step 5
Snap Chalklines for Inside Edges of Sill Plates

Margin of Error: Exact.

Most Common Mistake: None.

Diagram: Figure 7-6.

After you have determined the exact inside corners of the sill plates, snap chalklines between these points. You will then have an outline on the top of the foundation wall of the exact inside edges of the sill plates. Once this is finished, check the dimensions and squareness of these lines to see if they are accurate.

Step 6
Mark and Drill Anchor Bolt Holes

Margin of Error: ¼ inch of exact location of bolts.

Most Common Mistake: Not drilling holes larger than anchor bolts.

Diagrams: Figures 7-7, 7-8.

Photographs: Photos 7-4, 7-5.

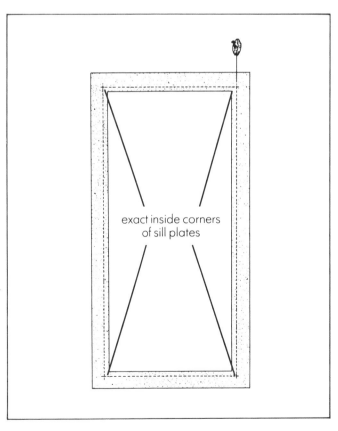

exact inside corners of sill plates

Figure 7-6. The dotted lines (chalklines) represent the inside edges of the sill plates.

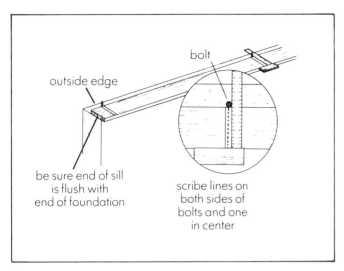

Figure 7-7. Marking the sill plates for anchor bolt holes.

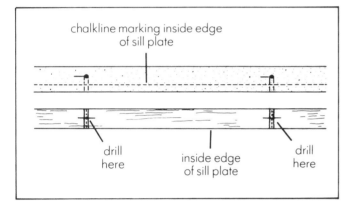

Figure 7-8. Locating the anchor bolt holes on the sill plates.

Photo 7-4. Marking an anchor bolt location on the sill plate.

Now place the sill pieces you have cut on top of the foundation wall. They should be placed on the inside of the bolts, aligning the ends into position. Press the sill plates against the inside of the anchor bolts and scribe lines on the sill, marking both sides of each bolt.

Draw a third line on the sill plate exactly between these first two lines. This new line, in the middle of the first two, marks where the center of the anchor bolt will be. Now you must determine where on this centerline to drill the hole. To do this, measure from the center of the anchor bolt to the line that you snapped marking the inside edge of the sill plate. Then transfer this measurement along the scribed bolt centerline from the inside edge of the sill plate.

After locating the marks, drill the anchor bolt holes. It is wise to use a drill bit that is ⅛ to ¼ inch larger in diameter than the anchor bolt itself to allow for final adjustment. Be sure that you drill a straight hole. You are now ready to install the sill plates.

Photo 7-5. Drilling holes for the anchor bolts. These holes should be ⅛ to ¼ inch larger in diameter than the bolts.

Step 7
Install and Level Sill Plates

Margin of Error: ⅛ to ¼ inch off level. ⅛ to ¼ inch off proper alignment with chalklines.

Most Common Mistake: Sill plates off level.

Diagram: Figure 7-9.

Photographs: Photos 7-6, 7-7, 7-8, 7-9.

Now simply place the sill plates over the bolts and check to see if the inside edges of the sills are aligned with the chalklines on top of the foundation wall. Also check to see if the inside corners are aligned with their marks. If the board is at all bowed (which most probably it will be), you may need to push or pull the sill while tightening the bolt.

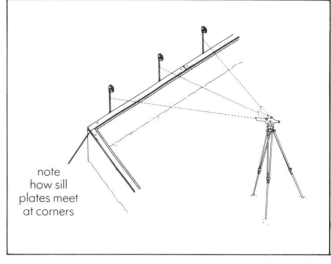

note how sill plates meet at corners

Figure 7-9. Using a transit to check the level of the sill plates.

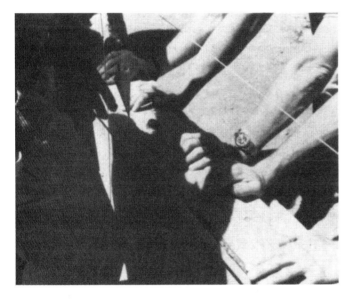

Photo 7-6. Pushing the crown out of a sill plate. Note that a plumb bob is being used to align the outside edge of the sill directly below the layout line.

Photo 7-7. Using levels to level the sill plates.

Photo 7-8. Wooden shims are being used to level the sill plate where the foundation wall was poured off level. Wooden shims may not be allowed by all codes.

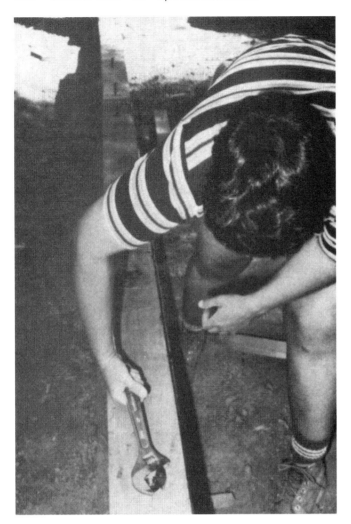

Photo 7-9. Sill plates are finally being secured after all the levelling and alignment adjustments have been made. Note the string along the edge to test for straightness.

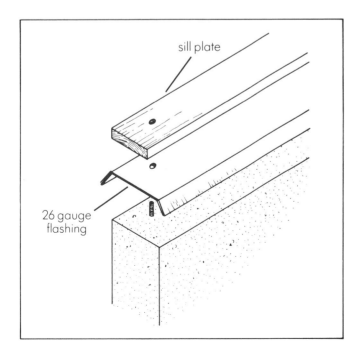

Figure 7-10. Termite guard.

Before you tighten the nuts permanently, partially tighten them and check the sill to see if it is level. This is a very important process. The entire house will be resting on these sills and if they are not level, it will throw off the floors and walls. You can check the level by placing a 4-foot, or preferably an 8-foot, level on top of the sill. A more exacting process, and one that I would recommend, is using a transit or builder's level and shooting the level every 3 or 4 feet along the sill.

If the sill is too low in areas, it can be shimmed. If you plan to use wood shims, I would suggest plywood shims as opposed to the standard cedar shims. Plywood shims will not compress under the weight. You can also pack the low areas with mortar; in some counties this may be required by code.

If the sill plate is too high, the high areas of the concrete will need to be chipped away. If a knee wall (cripple wall) is to go above the high area, you can install the sill as is and adjust the heights of the cripple studs to make the top of the knee wall level. (See next chapter, *Knee Walls.*)

After the adjustments are made to level the sill plates, tighten the bolts using a large washer underneath each bolt. Using your chalkline, be sure the plates are properly aligned as you tighten the bolts.

You are now finished with this stage; unless the foundation was aligned badly or the wood stock was of poor quality, it should have gone relatively smoothly and prepared you for the somewhat more difficult stage of knee walls and floor framing. You may want to check the square and dimensions of the finished sill plates to see how accurate you have been. Keep your standards high and at the same time use your margins of error to keep the work flowing at a good pace.

Special Situations

Termite Guards
Diagram: Figure 7-10.

In areas where there is a high risk of ground termite infestation, a termite guard is advised. It is especially suggested in areas where you are using concrete block foundations because the termites can tunnel through the hollow cells of the blocks. It also stops them from building mud tunnels on the inside of the foundation walls and finding their way into the wood. When in doubt, use it; it's cheap and quick, and may save a lot of problems down the way. It also keeps the wooden sill plate out of contact with the concrete, which can draw the moisture out of the wood.

Sill Insulation (Sill Sealer)
Photograph: Photo 7-10.

In colder climates or in houses where the underfloor areas are to be used as part of the heating ventilation system, or for storage or as a basement, a sill sealer or sill insulation is recommended. This prevents air from flowing between the sill plate and the foundation wall. Sill insulation can be bought in 6-inch wide by 1-inch thick fiberglass rolls, similar to stan-

Photo 7-10. Sill insulation is being applied before the sill plates are installed.

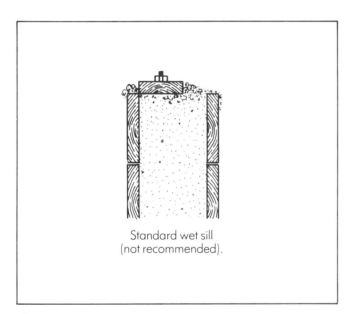

Figure 7-11. Two types of wet sills.

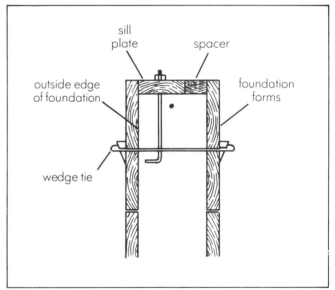

Figure 7-12. An alternate wet sill.

dard insulation. It is applied between the sill and the foundation wall or between the sill and the termite guard (if you are using a termite guard).

Wet Sills
Diagrams: Figures 7-11, 7-12.

Some builders use a "wet sill." This is basically a sill plate that is installed into the fresh concrete and levelled before the concrete dries. The anchor bolts are inserted into the sill and then the sill and bolts are set into the freshly poured concrete. This has a serious drawback. Since the anchor bolts are not strongly anchored into hardened concrete as in the standard procedure, you cannot use them to pull out any bow or warp in the sill plate. It is a quicker way of applying the sill, and is therefore popular with many builders, but it is not advised for high quality construction.

An alternate process of using a wet sill may work to pull out any bows or warps. As shown in Figure 7-12, the foundation forms are built 1½ inches higher than the actual pour. The sill plate and anchor bolts are set into forms and the concrete is poured. Spacers are used if the foundation wall is wider than the sill plates. The straight foundation forms can then be used to pull out any warps or bows.

Photo 7-11. Sill anchors are being nailed to the sills. Use these only if the sill stock is very straight.

Photo 7-12. Using a concrete nailing gun to secure vertical sill plates. Be careful, the gun can be dangerous.

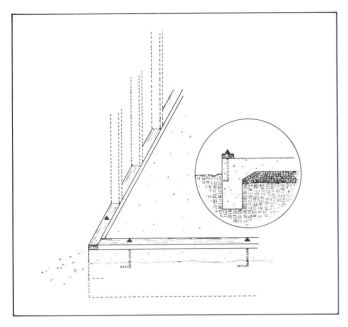

Figure 7-13. A sill plate on a slab foundation.

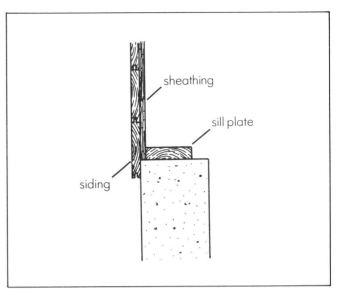

Figure 7-14. A sill plate kicked back to allow the siding to hide the sheathing when viewed from underneath (not recommended).

Sill Anchors
Photograph: Photo 7-11.

Sill anchors are sometimes used instead of anchor bolts. These are easier and quicker to put on because neither drilling nor measuring for anchor bolt holes is required. They are not advised either, because they do not assist in pulling out any bows or warps as do the anchor bolts.

Sill Plates on Concrete Slabs
Diagram: Figure 7-13.

Sill plates are installed in basically the same way on slabs as on concrete walls or concrete block foundations. Find the exact inside corners, pop lines and apply the sills. Usually these sills, which are often pressure treated or redwood, are 2×4s (unless you are building a 2×6 wall) and serve as the bottom (sole) plate of the wall. Often the entire wall is built using the predrilled sill as the bottom plate, and the wall is lifted and bolted to the slab.

Sill Plates Around Openings and at Steps
Photograph: Photo 7-12.

At steps and around the vertical walls of openings, sills are often required. Use the same material as the standard sill. A concrete nailing gun can be used to secure these to the foundation wall. Be very careful when using the concrete gun; it is dangerous.

Offset Sill Plates
Diagram: Figure 7-14.

When using a sheathing material below the siding, some builders kick the sill back towards the interior the thickness of the sheathing. The siding then overlaps the gap between the sill and the foundation wall. The only advantage here is that it hides the exposed end of the sheathing from view if you are looking from underneath.

Worksheet

ESTIMATED TIME

_____

_____

_____

_____

CREW

Skilled

_____ Phone _____

_____ Phone _____

_____ Phone _____

Unskilled

_____ Phone _____

_____ Phone _____

_____ Phone _____

_____ Phone _____

_____ Phone _____

COST COMPARISON

_____ Store:

Cost per M foundation grade redwood $ _____

Cost per M pressure-treated stock $ _____

_____ Store:

Cost per M foundation grade redwood $ _____

Cost per M pressure-treated stock $ _____

_____ Store:

Cost per M foundation grade redwood $ _____

Cost per M pressure-treated stock $ _____

MATERIALS NEEDED

Lumber for sill plates.

ESTIMATING MATERIALS

$2 \times$ _____ SILL STOCK:

_____ pieces _____ feet long

_____ pieces _____ feet long

Daily Checklist

CREW:

_____ Phone _____

_____ Phone_____

_____ Phone _____

_____ Phone_____

_____ Phone_____

_____ Phone _____

MATERIALS:

_____ sill stock

_____ insulation (if needed)

_____ termite guard (if needed)

WEATHER FORECAST:

For week _____

For day _____

TOOLS:

☐ framing hammer

☐ power saw

☐ level

☐ carpenter's level

☐ chalkline

☐ string

☐ 100-foot tape

☐ 25-foot tape

☐ combination square

☐ wrench

☐ pencil

☐ nail belt

☐ safety goggles

☐ drill

☐ bits

☐ extension cords

☐ plug junction boxes

☐ saw blades

☐ shims

☐ chisels

RENTAL ITEMS:

☐ transit

☐ concrete nail gun

8 Knee Walls

Photo 8-1. Knee walls partially erected.

KNEE (CRIPPLE) WALLS are wooden extensions of concrete foundations. They are used on sloping sites to avoid the time and expense of building high concrete foundation walls. The foundation walls are stepped and then extended with wooden foundation walls. Not all of you will need to use these, especially if you are building on relatively level ground or if you are using a concrete or concrete block basement. If that is the case, this chapter can be skipped over, unless you care to read it for your own general information. If you do have a sloping site that requires a continuous concrete perimeter foundation, using knee walls at the lower areas is highly recommended.

What You'll Be Doing, What to Expect

If you do have a sloping site that requires a continuous concrete perimeter foundation, the use of knee walls at the lower areas is highly recommended. (See Figure 8-1, Photo 8-1.)

Since building knee walls is very similar to building standard stud walls, I advise reading the *Wall Framing* chapter before beginning knee wall construction. Although this part of construction is not difficult (you are working close to the ground with short walls), it is very important that the walls are built very level. Knee walls are an extension of the foun-

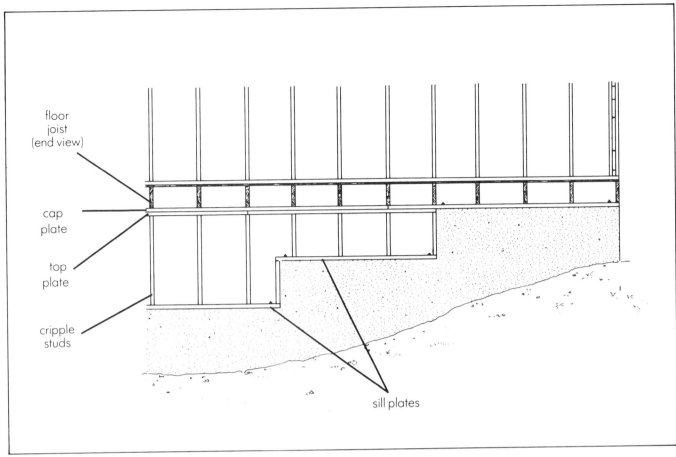

Figure 8-1. A typical knee wall. Note how the upper sill plate is at the same level as the cap plate of the knee wall.

dation on which the house will be built. Go slowly—be accurate.

In the event of an earthquake, knee walls can fail. Check with your local codes and take this fact into consideration, even if knee walls are allowed. At minimum, be sure to use a plywood sheathing skin.

Jargon

BOTTOM PLATE (SOLE PLATE). Horizontal base member of the wall partition.

FACE NAIL. To nail perpendicular to the finished face.

KNEE WALL OR CRIPPLE WALL. Wood wall extension of the concrete foundation; most often used in stepped foundations on sloping sites.

LAYOUT. Method of marking the location of a series of framing members on a sill, band, plate, girder or ridge.

ON CENTER. A term to indicate the spacing of framing members by using the measurement from the center of one member to the center of the succeeding one.

PLUMB. Exactly perpendicular or vertical.

STUDS. One of a series of vertical wood structural members in a wall or partition.

TOENAIL. Driving a nail at an angle.

TOP PLATE. Horizontal top member placed on the wall and supported by the studs.

Purposes of the Knee Wall

1) To extend the concrete foundation wall with a wooden wall (thereby avoiding the use of the more expensive concrete).
2) To create a level building surface on a sloping site.

Design Decisions

Consider carefully the location of the access door(s) for your crawlspace. The plumbing code often requires that the access door be within 15 feet of any drain cleanout plug. Check this out with your local authority. If it is difficult to place the door near the cleanout plug(s), the plug can poke through the house to the outside or a trap door can be built through the floor from the inside.

You must also consider if you are ever going to use this crawlspace for anything else: a half basement, storage, a full basement, etc. See if this affects any of your decisions. You can, for instance, add some rough window or door openings at this stage and cover over them until you build the basement. If you plan to use the crawlspace for storage, you may want to enlarge the access doors for wheelbarrows, lawnmowers, etc.

What You'll Need

TIME 8 to 16 person-hours per 30 linear feet of wall 3 feet high.

PEOPLE 1 Minimum (skilled); 3 Optimum (1 skilled, 2 unskilled); 6 Maximum (2 teams of three).

TOOLS Framing hammer, nail belt, tape, pencil, combination square, power saw, carpenter's level, builder's level (optional), extension cords, plug junction box, saw blades, nail-pulling tools.

RENTAL ITEMS: Transit, com-a-long.

MATERIALS The wood used is common framing grade lumber, either 2×4s or 2×6s. The only exception is the bottom plate of the knee wall, which is the sill plate. It will be either foundation grade redwood or pressure-treated lumber (see *Sill Plates*). All nails nailed into the sill plate to hold the bottoms of the knee wall studs must be hot-dipped galvanized, non-rusting nails. The top and cap plates are also framing grade lumber.

ESTIMATING MATERIALS NEEDED

Estimating the materials will take a little calculating, since many different length boards will be needed for the varying stud heights. Try to calculate what length board will create the least wastage.

INFORMATION For a better understanding of this procedure, it is essential that you go over the *Wall Framing* chapter, especially for door and window openings. Check your local codes for any mandatory specifications on knee walls in your area. Other than this, you should be able to make it through without problems.

PEOPLE TO CONTACT None.

INSPECTIONS NEEDED None.

Reading the Plans

If your plans are well detailed (and few really are), all the information you will need to know about the knee (cripple) walls will be in the plans. This will include: where they are to be used, the size, length and spacing of the studs, location of crawlspace vents, size and location of access doors, etc. Often the plan will not provide all this information, especially if the plans have not been designed for a specific site. In this case, you will have to provide the knee walls wherever you decided to step the foundation walls. The depth and location of these steps is what determines the location and height of the knee walls.

Be sure you check with your local permit office as to the required size and spacing of the studs. If the knee wall is over 4 feet high, it is considered a separate story; you may be required to use larger (2×6s) studs or closer stud spacings. The code also says that no knee wall stud can be less than 14 inches in height. This must be taken into consideration as you step your foundation.

Don't forget to use a double, not single, top plate on your knee walls. I once used a single plate on a house we built in California and the inspector made me jack the 2200-square-foot house up, cut 1½ inches out of the top of the studs and insert another plate. It was 2 days' work for 5 people. We found out later we could have corrected the problem by simply nailing a 2×6 on the side of the studs.

Safety and Awareness

SAFETY: No special considerations other than usual care, especially with power tools.

AWARENESS: Mainly you want to be sure that the wall is level, as this is an extension of the foundation walls on which the entire house will be built. Also be sure the walls are plumb.

Overview of the Procedure

Step 1: Lay out the studs
Step 2: Determine the heights of the studs
Step 3: Toenail the studs to the sills
Step 4: Face nail first top plate to studs
Step 5: Nail second top plate (cap plate) to first top plate
Step 6: Plumb corners and walls, and brace
Step 7: Check knee walls

Step 1
Lay Out the Studs

Margin of Error: ¼ inch during the layout.

Most Common Mistake: Placing stud on wrong side of layout mark.

Diagrams: Figures 8-2, 8-3, 8-4, 8-5.

Photograph: Photo 8-2.

At this point your sill plates should be level and fastened to the foundation wall. Before beginning with the knee walls, check to be sure the sills are level and in their proper place, as described in the *Sill Plates* chapter. If the sill plates are level and accurate, you are now ready to do the layout for the location of the cripple studs on the sill plate.

As in all standard construction, the framing members will be on 12-, 16- or 24-inch centers so that 4-×-8-foot panels of drywall, plywood and paneling can be used and they will always break on the center of a framing member. In all layouts, the first stud is set 15¼ inches on center rather than 16 inches, so that the edge of the first panel will fall at the edge of the framing member rather than at the center. After this first layout mark, use a 16-inch layout for the rest.

After you have completed the layout on the sill plates, choose a good, straight piece of stock that will be your top plate for the knee wall. Place it against the marked sill and transfer these marks to the top plate. While you are doing this, be sure that the top plate is properly aligned with the end of the wall and that any break in the top plate falls in the center of a stud.

After the top plate is marked, you are ready to cut the cripple studs. Corners and doors are done as shown in the *Wall Framing* chapter.

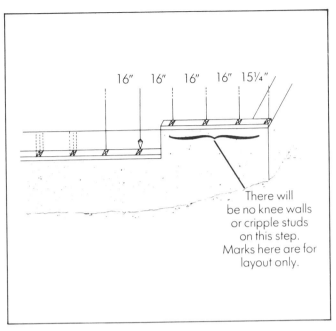

Figure 8-2. Laying out studs on the knee walls.

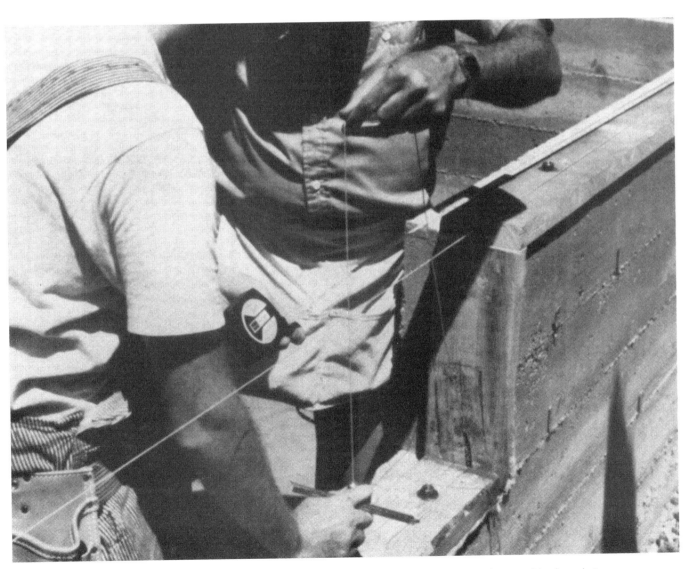

Photo 8-2. Using a level string line and plumb bob to locate the layout marks for studs on each step of the foundation.

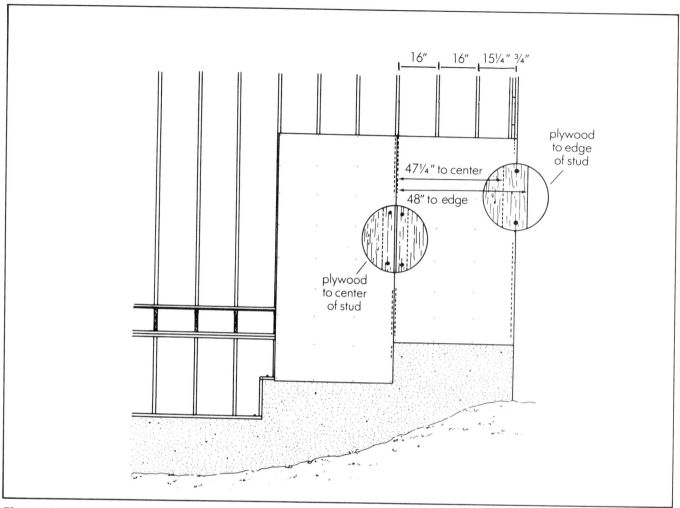

Figure 8-3. Dimensions are from center to center of studs. First sheet of plywood comes to edge, not center, of first stud. This accounts for the "lost" ¾ inch.

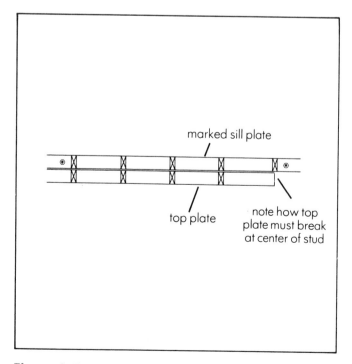

Figure 8-4. Transferring layout marks from the sill plate to the top plate of the knee wall.

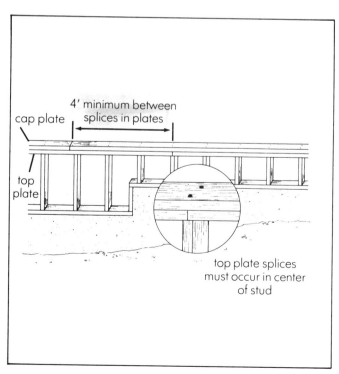

Figure 8-5. A finished knee wall showing plate splices.

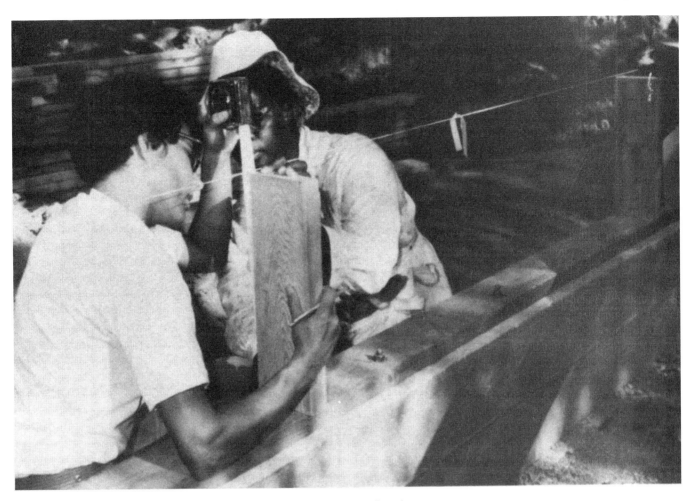

Photo 8-3. Using a level string to determine the height of the knee wall studs.

Step 2
Determine the Heights of the Studs

Margin of Error: Within ⅛ inch of proper height.

Most Common Mistakes: Not figuring for second top plate (cap plate); not allowing for irregularities in level of foundation wall.

Diagrams: Figures 8-6, 8-7, 8-8.

Photograph: Photo 8-3.

Be very accurate in determining and cutting the heights of the knee wall studs. The level of the top of the foundation wall will be determined at this stage. If the concrete wall is off level at any spot and the discrepancy has not been properly adjusted with the sill plate, you can make adjustments now by adjusting the height of each knee wall stud.

If the concrete foundation wall is perfectly level or if all discrepancies have been adjusted by shimming the sill plates, all the studs can be cut the same height in any one section (or step) of the knee wall. When you determine the stud heights, remember that the *second* top plate (the cap plate), not the first, will be at the same level as the sill plates at the highest

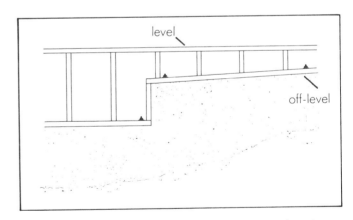

Figure 8-6. Knee wall studs are cut at varying lengths to correct an off-level foundation wall.

part of the foundation wall (where no knee walls are to be built).

To determine these stud heights, run a string, level with the bottom edge of the top plate, and measure from there to sill plates, as in Figure 8-7. A transit or builder's level can also be used.

After you have determined the height and number of studs, proceed to cut. Note that in some areas the minimum height allowed for knee wall studs is 14 inches.

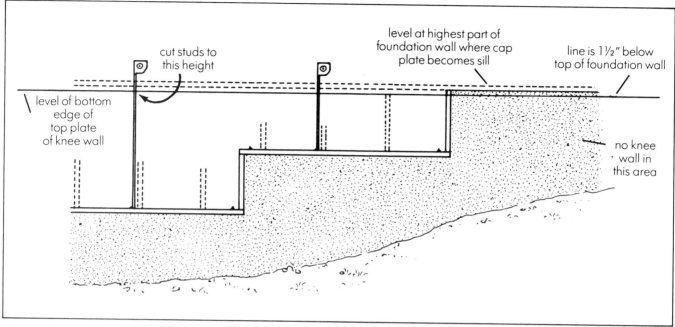

Figure 8-7. Determining the heights of the cripple studs.

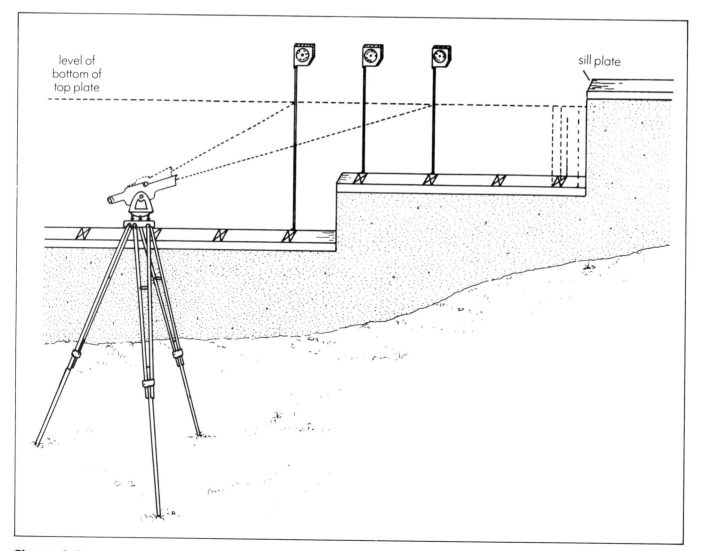

Figure 8-8. Using a transit to determine the heights of the cripple studs.

Step 3
Toenail the Studs to the Sills

Margin of Error: ⅛ inch on line with mark.

Most Common Mistake: Nailing to wrong side of layout mark.

Diagram: Figure 8-9.

Photograph: Photo 8-4.

Now toenail the studs to the sill plates at the proper marks as shown. Use 8d hot-dipped galvanized nails, two on each side of the stud. Be sure the stud is on the proper side of the mark. The "X"s make this easy. You may want to start the nails in the studs before you place the studs on the sill. You can also use a block of wood, cut to the exact size of the space in between the studs (use 14½ for 16-inch layouts), to help hold the stud in place as you nail. Be careful to have fairly straight studs, as bowed studs will create irregular surfaces for future siding application.

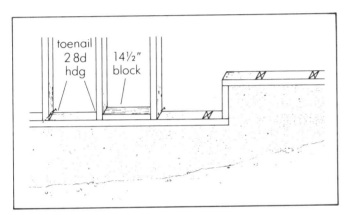

Figure 8-9. Nailing studs to the sill.

Photo 8-4. Knee wall studs nailed to sill plates.

Photo 8-5. Checking the level of the knee wall with a transit after the first top plate has been nailed to the studs.

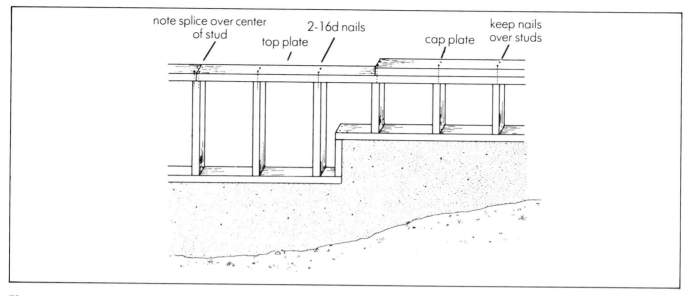

Figure 8-10. Nailing the top and cap plates to the cripple studs.

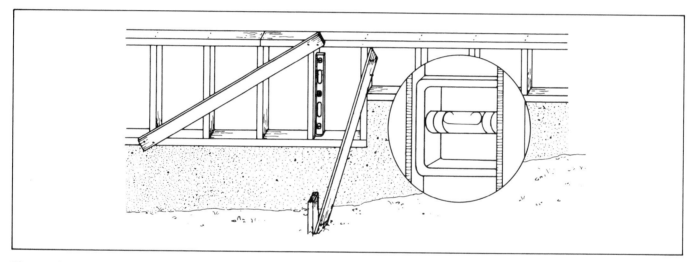

Figure 8-11. Bracing the knee wall plumb in all directions.

Step 4
Face Nail First Top Plate to Studs

Margin of Error: ⅛ inch on line with layout mark.

Most Common Mistakes: Using crooked plate stock; not nailing on layout mark.

Photograph: Photo 8-5.

Now nail the top plate over the studs. Using two 16d standard nails, face nail through the top plate into each stud. Be sure that splices in the top plate break in the center of a stud. The top plate should be of good straight stock or the wall will be crooked.

At this point, you should check the level of the wall. You can use a level or a transit to do this. If the level is off, determine why, and make the proper corrections. *It is important that the top of the wall be level within ⅛ to ¼ inch.*

Step 5
Nail Second Top Plate (Cap Plate) to First Top Plate

Margin of Error: Aligned with first top plate within ⅛ inch.

Most Common Mistakes: Using crooked stock; not nailing both plates flush to each other.

Diagram: Figure 8-10.

As you nail on the second top plate, also called the cap plate, be sure again you are using good straight stock, or the top of the foundation knee wall will be crooked. It is advisable that you nail the second top plate on with two 16d CC sinkers, directly over each stud. Keeping the nails over each stud will keep them out of the way of drill bits should you later run your plumbing or electricity through the top plates. Any splice in the cap plates must be at least 4 feet away from any splice in the top plate.

Figure 8-12. Two ways of checking the completed knee wall for level.

Step 6
Plumb Corners and Walls, and Brace

Margin of Error: Within ⅛ to ¼ inch off plumb.

Most Common Mistake: Off plumb.

Diagram: Figure 8-11.

Using a transit, level, or plumb bob, plumb the corners, which in turn should automatically plumb most of the studs. If the corners are plumb but some of the studs are not, brace till they are plumb. After the wall is plumbed, temporarily brace it by nailing boards diagonally across the studs to hold the wall in place as the floors and walls are being built.

Step 7
Check Knee Walls

Margin of Error: ⅛ to ¼ inch of level.

Most Common Mistake: Off level.

Diagram: Figure 8-12.

Before going on to the next step of floor framing, check to be sure that the knee wall is the proper dimension, square, level and plumb. This is very important as the floor frame will be built on top of the knee wall and it needs to be accurate. Take your time here and get it correct. It will save you hours down the line trying to adjust for sloppy work.

Worksheet

ESTIMATED TIME

_____

_____

_____

_____

CREW

Skilled

_____ Phone _____

_____ Phone _____

_____ Phone _____

Unskilled

_____ Phone _____

_____ Phone _____

_____ Phone _____

_____ Phone _____

_____ Phone _____

MATERIALS NEEDED

2×4s or 2×6s, lumber for plates, nails (8d hot-dipped galvanized and 16d CC sinkers).

ESTIMATING MATERIALS

STUDS:

2 × _____ BOARD _____ feet long: Quantity _____

2 × _____ BOARD _____ feet long: Quantity _____

2 × _____ BOARD _____ feet long: Quantity _____

PLATES:

2 × _____ BOARD _____ feet long: Quantity _____

2 × _____ BOARD _____ feet long: Quantity _____

NAILS:

_____ lbs. 8d galvanized

_____ lbs. 16d CC sinkers

COST COMPARISON

_____ Store:

$ _____ per M = $ _____ total

_____ Store:

$ _____ per M = $ _____ total

_____ Store:

$ _____ per M = $ _____ total

Daily Checklist

CREW:

_____Phone _____

_____Phone _____

_____Phone _____

_____Phone _____

_____Phone _____

_____Phone _____

TOOLS:

☐ framing hammer

☐ nail belt

☐ tape

☐ pencil

☐ combination square

☐ power saw

☐ carpenter's level

☐ builder's level (optional)

☐ extension cords

☐ plug junction box

☐ saw blades

☐ nail-pulling tools

RENTAL ITEMS:

☐ transit (optional)

☐ com-a-long

MATERIALS:

_____ feet of framing stock

_____ lbs. of 16d CC sinkers

_____ lbs. of 8d hot-dipped galvanized nails (for toenailing into sill plate)

DESIGN REVIEW: The location of the access door for the crawlspace will be finalized at this stage. Decide how you will use the space, whether to add windows, how big the doors need to be, etc. Be sure you're in compliance with plumbing codes concerning drain clean-out plugs.

WEATHER FORECAST:

For week _____

For day _____

9 *Floor Framing*

Photo 9-1. Floor joists are completed here and the subflooring is being applied.

What You'll Be Doing, What to Expect

FLOOR FRAMING (Photo 9-1) is the simplest of all framing processes and serves as an excellent introduction to carpentry. It is also a very exciting time of building, because the work goes very quickly and the house starts to take shape. Except for the joists becoming off level or out of line (discussed later), there are few mistakes. Don't be deceived by this stage of construction, however; almost no other stage will progress quite this rapidly.

This chapter will deal with the construction of a typical floor frame in a house that has a continuous concrete wall foundation (Figure 9-1) around the exterior of the home and a pier foundation (Figure 9-2) to support the girders on the

interior of the home. By using this example, you will learn to construct a floor frame for both continuous wall foundations and pier foundations. Some houses may have all continuous wall foundations, both exterior and interior, others may use all pier foundations.

The example given here is fairly straightforward: a typical joist system, using girders and piers; an opening; and a load-bearing wall running parallel to the joist system. Your floor frame may not be that simple, so I have included other details of floor framing toward the end of the chapter.

Keep the high standards of quality you have set for yourself as you frame. (See Figure 9-3.) Make good tight-fitting cuts—it will increase the strength of your floor frame and help you practice good carpentry techniques. Remember, however, that a house is not a piece of custom furniture; make the demands on your work reasonable. Enjoy yourself!

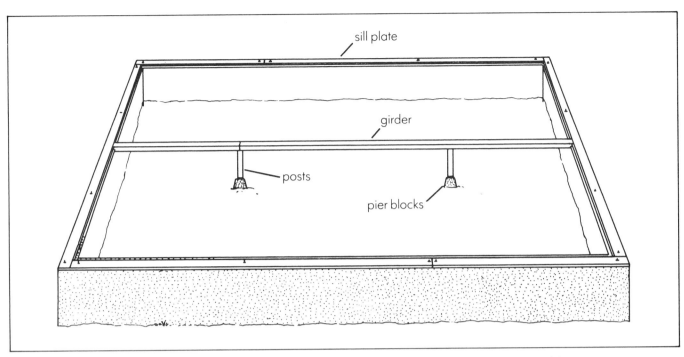

Figure 9-1. A concrete continuous-wall foundation with a girder supported by posts to reduce the span of the joists.

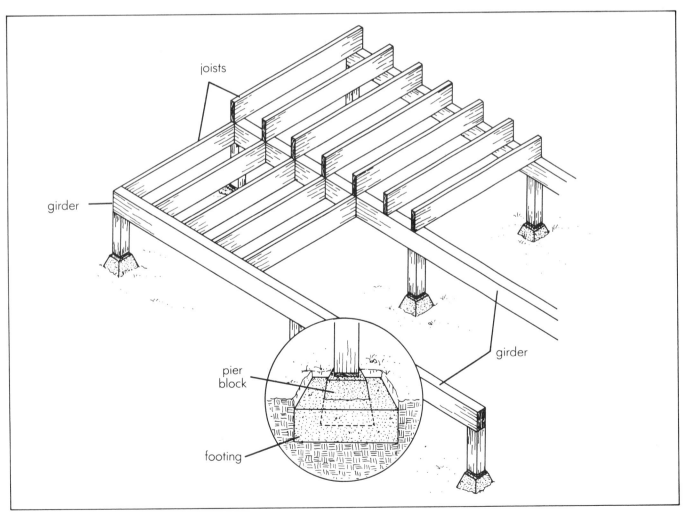

Figure 9-2. A pier foundation using girders to support the joists. The joists can either sit on top of the girders or be hung from them as shown.

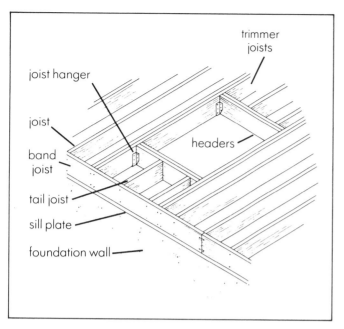

Figure 9-3. Typical pieces of a floor frame.

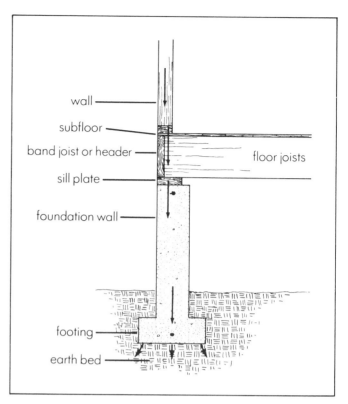

Figure 9-4. Transfer of weight through the floor frame to the earth bed.

Jargon

BLOCKING OR BRIDGING. Metal straps, diagonally nailed 1×4s, or pieces of joist stock nailed between joists used to stiffen and add rigidity to the floor frame and decrease lateral movement.

CANTILEVER. A structural member that is supported at one end only and supports a projected load.

GIRDER. A large beam used to support a portion of the joist system, which serves to decrease the span of the joists. They are usually nominal 4- or 6-inch-thick stock or built up from 2 pieces of 3-inch-wide stock and are spaced from 4 to 14 feet apart.

JOIST. A parallel floor framing member used to support the floor. Usually it is a nominal 2-inch-thick board (2×6, 2×8, 2×10, 2×12, 2×14) on 12-, 16- or 24-inch centers.

JOIST HANGER. A metal fastener used to fasten the joists to the girders (they are required in earthquake areas).

JOIST HEADER. A piece of stock, the same size as the joists, that runs perpendicular to, and at the same level as, the joists, at the ends of the joists. It serves to keep the joists plumb and erect.

JOIST LAYOUT. Marking the locations of the joists on the sill plate, header or girder.

LOLLY COLUMN. A round metal support post, often used to support girders in the basement.

O.C.–ON CENTER. The distance from the center of one piece of framing member to the center of the adjacent member.

POST OR COLUMN. A vertical support used to support the girders or beams. They are made of concrete blocks, wood or metal.

R.O.–ROUGH OPENING. The opening in the floor, wall or roof frame for a window, door, stairway, chimney, etc.

SCAB. A scrap piece of wood used to join other pieces of wood together.

T & G–TONGUE AND GROOVE. A joint made by cutting a groove in one board and mating tongue in the other.

TAIL JOIST. A relatively short joist used around openings or girders supported by a header at one end and a foundation wall or girder at the other.

THREADED NAIL. An annular ringed or barbed nail used for greater holding power.

TRIMMERS. Double or single joists around openings that run parallel with the common joists and along the edges of the openings.

Purposes of the Floor Frame

1) Transfers the weight of people, furniture, materials, etc., from the subfloor, to the floor framing, to the foundation wall, to the footing and finally to the earth bed. (See Figure 9-4.)

2) With the subfloor, it serves as a building platform for the walls of the house (hence the name, platform building).

Design Decisions

In floor framing, perhaps more than in any other framing process in the house, there is a large variety of choices. Aside from the standard joist system, there are two other floor framing systems currently in use: the floor truss system and the girder system. Both of these are briefly discussed at the end of the chapter under "Special Situations."

Even in conventional joist systems there are choices to be made. In one floor system you could, for instance, choose between 2×12s on 12-inch centers, or 2×10s on 16-inch centers or 2×8s on 12-inch centers. Or you could add an extra girder, thereby decreasing the span of the joists, and switch to yet another size and spacing of the floor joists. Whoever drew up your plans may or may not have had a good reason for choosing the system they did. You may want to rethink the decision and, if possible, discuss it with a professional.

I always like to be sure that the floors are a little over built. It's nice to be able to jump on a floor and have the house feel solid. When the kids run through the living room you don't want the cups to rattle in the kitchen cabinets. To build a solid floor, add an extra girder or increase the required size of the joist or decrease the required spacing. It costs a little more, but it makes for a stouter house.

What You'll Need

TIME 32 to 60 person-hours per 1,000 square feet of frame.

PEOPLE 2 Minimum (1 skilled, 1 unskilled); 3 Optimum (1 skilled, 2 unskilled); 5 Maximum (2 skilled, 3 unskilled).

TOOLS Framing hammers, combination square, hand saw, power saw, saw blades, saw horses, pencil, steel tape, nail belt, sledge hammer, safety goggles, nail-pulling tools, transit or hydro level, chalkline, 4-foot level, extension cords, junction box, framing square.

MATERIALS Almost all framing is done with Number 2 or better framing grade material of the local evergreen (Douglas fir, southern yellow pine, spruce, ponderosa pine, etc.). Lumber is graded according to the number and size of the knots as well as the general quality of the stock. Number 1 grade stock is used in structural situations where there is a tremendous stress factor, which is usually indicated on the plans by an architect or engineer. In non-structural situations a Number 3 or utility grade stock is sometimes used. Unless you indicate otherwise, the suppliers will always deliver the Number 2 or better framing stock when you order nominal 2-inch material. (Number 2 grade lumber or better indicates 60% is Number 2 grade and 40% is Number 1 grade.)

KILN DRIED OR WET LUMBER

Wood is made up of the same main ingredients as commercial plastics: lignin and cellulose. Lignin is an adhesive which gives strength and rigidity to the wood; cellulose is nature's own strong material. It is made up of long chains running the length of the tree, which accounts for wood splitting along its grain. The moisture in the wood is not held within these molecules so much as between them. When evaporation occurs, this moisture is lost and the molecules pull closer to each other, which is why wood shrinks much more along its width (laterally) than its length (longitudinally).

In choosing your floor frame material, you must make the major decision to use either wet or kiln dried materials. Wet lumber may well be in the forest a few weeks before you get it and will go through its drying process after it has been installed in your house. This causes certain problems. Kiln dried stock, on the other hand, has been dried in a heated kiln and a high percentage of the moisture content has been driven out of the material, both from within the cells and between the cells. The wood is therefore preshrunk before it is installed. Until the last ten or fifteen years only kiln dried materials were used. When we switched to using mostly wet framing lumber, the quality of houses went down.

Wet lumber has two advantages:

1) it is cheaper by 20 to 30% than kiln dried, and

2) it does not split as readily.

Kiln dried, on the other hand, has many advantages:

1) Because it is dry and preshrunk *before* it is installed, it will not shrink and twist after installation. Because of this your trim work will not be pulled off, your doors and windows are less likely to stick, and your nails will hold much better, thereby making a stronger house. Your finishing work will also be easier, since the boards have not warped and twisted.

2) Dry lumber is more resistant to fungus, mold and insects.

3) Dry lumber is lighter and easier to carry and handle.

4) Wet lumber, with a higher sap content, gums up saw blades more quickly.

5) Kiln dried stock, though graded as framing grade, is often higher quality than the same grade wet lumber.

6) Nails lose three-fourths their holding power when driven into wet lumber; as the lumber shrinks away from the nail, the friction between the wood and the nail is reduced. There is also a deterioration of moist wood due to the formation of iron oxides on a nail in the presence of wood acids.

If dry lumber is not available or you cannot afford the added expense, you may offset the disadvantages of wet lumber by doing two things.

First, use threaded nails throughout the house. The Virginia Polytechnic Institute found that a house built with threaded nails and green lumber could withstand four times the load as a house built with green lumber and plain shank nails. This is also an excellent reason to use threaded nails in earthquake areas. These are sometimes difficult to find but keep looking; they add only about $40 to the cost of the house.

The second thing you can do if you are using green lumber is to close the house in and allow it to season for 6 months to a year so that it will dry before you put on the finishing material. For many owner builders, it's about a year between these two stages anyway. If you did the finishing work before the wood dried, the finishing boards would be pulled off as the wood shrank.

Though it costs a little more ($400 to $600 per 2000-

square-foot home), I advise using kiln dried lumber for the entire frame.

How to check for moisture content. To determine the moisture content of your lumber, pick 6 random pieces. Cut out a 1-inch slice across the entire width and at least 1 foot from the end. Trim these samples so that they measure 6 inches in width and place them in a warm dry place for 48 hours. The 6-inch dimension is then measured to determine the shrinkage. If it shrinks over ¼ inch in the 6-inch dimension, it is not satisfactory for framing. If it shrinks over ⅛ inch, it is not satisfactory for interior trim or finish.

ESTIMATING MATERIALS NEEDED

JOIST STOCK: Figure joist needs by calculating on center measurements and the length the joist needs to be to span the supports. Include band joist and material for blocking, if necessary.

GIRDER STOCK: Figure how many pieces are needed and lengths so girders break over centers of posts.

POSTS: Order enough length of post stock to span piers to girders.

NAILS: 15 pounds of 16d CC sinkers per 1500-square-foot house.

METAL FASTENERS AND FASTENER NAILS: Figure the number needed.

INFORMATION This chapter deals with the most common floor frame situations. If your plans call for any unusual situations, contact a professional for further information. Some situations can be worked out by novices using common sense. Just be sure that your solution will work, pass codes and is the easiest.

All lumber used as joists, girders, posts and other load-bearing pieces, and all plywood used as subflooring, must be graded and show the grading mark. Often recycled or rough cut lumber with no marks will be allowed if inspectors think it is high quality. Other times they will require a lumber grader, structural engineer or architect to grade it. Check first. Blocking can be ungraded stock.

Check the local codes for specifications on joist and girder spans, and the sizes of framing materials, required in the course of the framing process in your area.

PEOPLE TO CONTACT If underfloor utilities need to be inspected before the subfloor is placed and you do not plan to do the work yourself, contact subcontractors: electrician, plumber, solar panel installer, and gas, heating and air conditioning services.

INSPECTIONS NEEDED None.

Reading the Plans

Most plans (Figure 9-5) will tell you everything you need to know about the floor framing system. All of the following should be included:

A) Type (species) of wood; e.g., df (Douglas fir), syp (southern yellow pine), etc.
B) Size, location and spacing of the: posts or columns, joists, girders or beams
C) Direction of both the joists and girders
D) Manner in which the joists connect to the girders, see "Step 7"
E) Location of any load-bearing interior walls that run parallel to the joists, see "Step 6"
F) Location of any toilet drains, see "Step 6"
G) Size (rough opening) and location of any openings in the floor for stairs, crawlspace access, chimneys, etc.

Aside from the above, the following specialized information may also be included:

H) Any cantilevering situations, see "Special Situations"
I) Any changes in floor levels
J) Any special structural situations needed to carry certain loads
K) Any special metal fasteners needed in earthquake areas, see "Step 4"
L) Any special details needed to connect floor frame into a slab floor
M) Location, size and spacing of any large beams to be used as exposed timbers (in the second or third floor frame)
N) Type of blocking or bridging, see "Step 8"
O) Clearance from soil to girders and joists

Safety and Awareness

SAFETY: As always, be careful when using the power tools; also, be careful walking on the joists—be sure they are stable; many accidents happen when people fall through the joists.

AWARENESS: Be sure your layout is accurate and the joist placed to the proper side of the layout mark. Be sure joists are installed crown up and their tops are flush with the top of the band and the girder. Use straight girders.

Overview of the Procedure

Step 1: Check the level, squareness and dimensions of the sill plates
Step 2: Erect the posts
Step 3: Erect the girders
Step 4: Erect the band
Step 5: Lay out the joists on the band and the girders
Step 6: Layout for special situations
Step 7: Install the joists
Step 8: Install the bridging or blocking
Step 9: Check the system

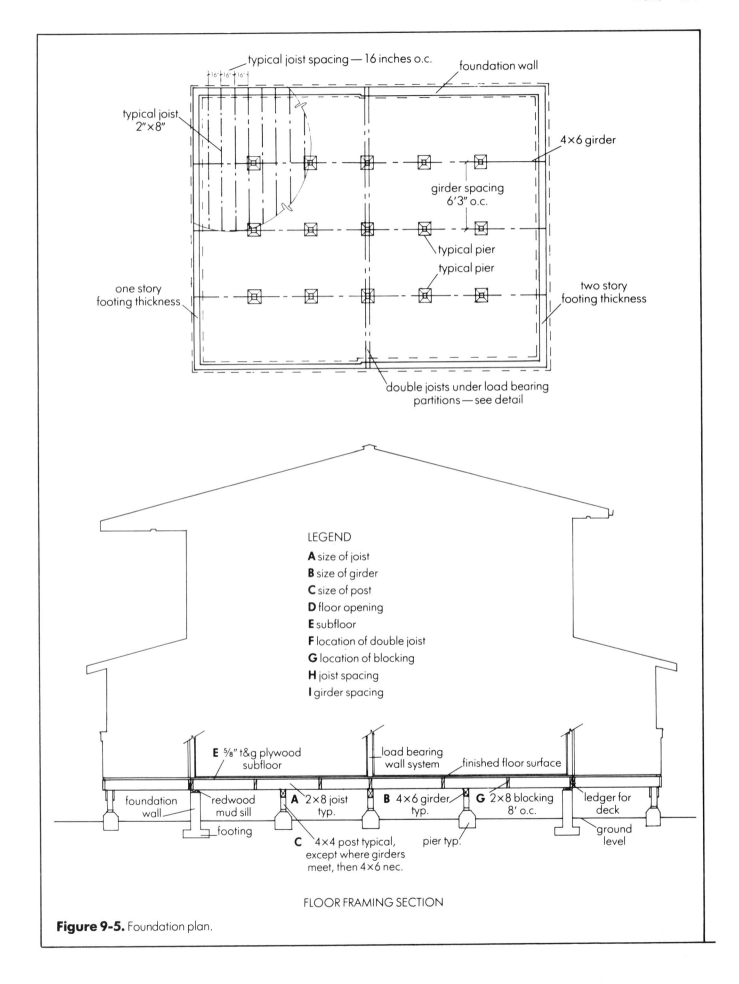

typical joist spacing — 16 inches o.c.

foundation wall

typical joist
2″×8″

4×6 girder

girder spacing
6′3″ o.c.

typical pier

typical pier

one story
footing thickness

two story
footing thickness

double joists under load bearing
partitions — see detail

LEGEND

A size of joist
B size of girder
C size of post
D floor opening
E subfloor
F location of double joist
G location of blocking
H joist spacing
I girder spacing

E ⅝″ t&g plywood
subfloor

load bearing
wall system

finished floor surface

foundation
wall

redwood
mud sill

A 2×8 joist
typ.

B 4×6 girder
typ.

G 2×8 blocking
8′ o.c.

ledger for
deck

footing

C 4×4 post typical,
except where girders
meet, then 4×6 nec.

pier typ.

ground
level

FLOOR FRAMING SECTION

Figure 9-5. Foundation plan.

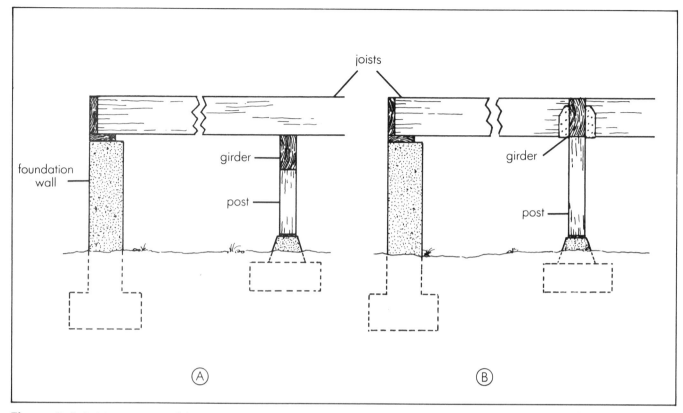

Figure 9-6. (A) Joists on top of the girder (the top of the sill is level with the top of the girder). (B) Joists hung from the girder (the top of the sill is flush with the bottom of the girder).

Step 1
Check the Level, Squareness and Dimensions of the Sill Plates

Margin of Error: For level—⅛ inch (shim or notch to correct). For dimensions—¼ inch (adjust floor frame). For squareness—¼ inch (adjust floor frame).

Most Common Mistake: Allowing too great a margin of error.

Diagram: Figure 9-6.

Before beginning the floor frame, use a transit (a water or hydro level will work) to check the heights of your sill plates. (See *Sill Plate* chapter, "Step 7.") The height of the joists and girders is determined by the level of the top of the sill plates. The joists will rest on the top of the sill plates. The top of the girders will be level with either the top of the sills or the top of the joists, depending on whether the joists sit on top of the girders or are hung from the girders. Because of this, it is wise to check the sill plate levels every 6 to 8 feet to see how you are doing.

Mark any areas that are off level. The joists and girders in these areas can be shimmed or notched to bring the entire floor frame to level. If the sill plates and foundation walls are off dimension or not square, you can build the floor frame true and square, even though it will underlap or overlap the sill plates, to correct this error. Be sure you check it carefully; it will make many of the procedures down the road much easier.

NOTE: You may want to backfill around the foundation at this time to make working on the floor frame easier.

Step 2
Erect the Posts

Margin of Error: In straight line—¼ inch. Tops of posts level—⅛ inch.

Most Common Mistakes: Post tops not level; posts not centered under girder.

Diagrams: Figures 9-7, 9-8, 9-9, 9-10, 9-11.

Photographs: Photos 9-2, 9-3.

There are four possible posts to use: concrete block, brick, metal or wood (either solid or built-up). The wood posts are the most common and easiest to use, except where there is a basement, in which case round, metal lolly columns are the most common. Bricks and concrete blocks are more difficult to use because their level cannot be adjusted as easily as that of wooden posts, which can be cut, and metal posts, which screw out to any height. When using concrete blocks or bricks as posts, the height of the footings must be accurately set so that when using 4-inch-thick bricks or 8-inch-tall blocks, their tops can come out at level heights. See *Poured Concrete Foundation,* "Step 14C." In our example we will focus on wooden posts; they are the simplest and the quickest.

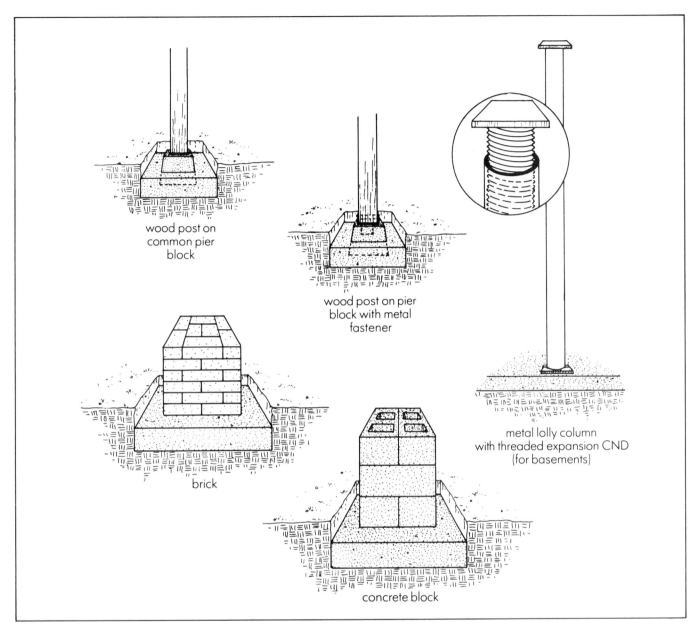

Figure 9-7. Common foundation posts are being used to support girders.

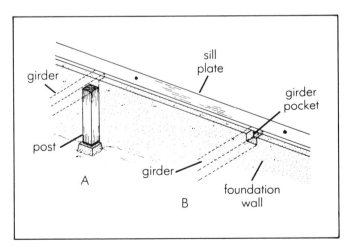

Figure 9-8. (A) Girder supported by a post at the foundation wall. (B) Girder supported by a girder pocket provided in the foundation wall.

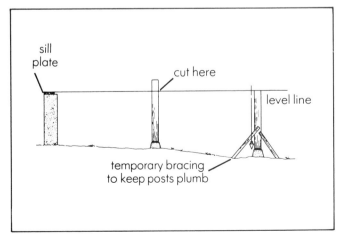

Figure 9-9. A string is being run at the top of the posts, the same height as the bottom of the girder. (In this example, the joist will be hung from girders with metal joist hangers.)

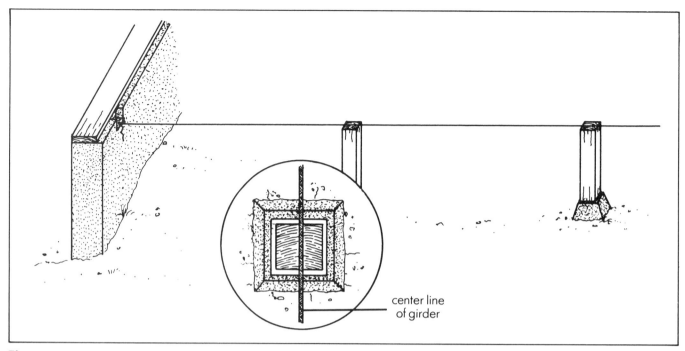

Figure 9-10. Using a string to align the posts with the center of girder. (In this example, joists will sit on top of the girder.)

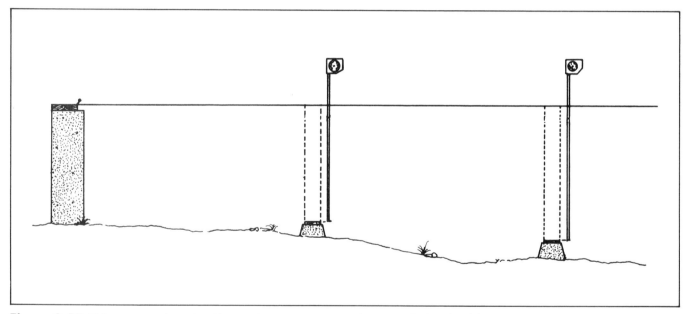

Figure 9-11. Using a properly set level line and a tape measure to determine the height of the posts.

Usually 4-×-4 or 4-×-6 posts are used. All posts must be as wide as the girder and a 4×6 is often required where girder stock is joined over a post. As described earlier, the pier blocks for the wooden piers should already be set in their concrete footings and properly aligned. Proper negative spaces should have been left to receive the girders both in the continuous foundation wall and any cripple or knee walls. (See *Poured Concrete Foundations,* "Step 14" and "Step 6.") If a negative space is missing, you will need to set another post up against the foundation wall to support the end of the girder. If all the pier blocks are properly placed and all the

negative spaces are there, check to be sure the surfaces where the girders are to be set (top of posts and girder pockets) are perfectly flat so the girders will not be resting on any high spots.

Remember: the height of the posts will depend on the manner in which the joists are connected to the girder—they can be either on top of the girder or hung from the girder.

You are now ready to cut the posts. The best way to do this is to use a transit or hydro level and locate and place 2 nails on each foundation wall, where the girders will rest, at the level of the bottom of the girder (which is the level of the

Photo 9-2. Posts are being nailed to metal fasteners embedded in the concrete pier blocks. Metal post caps are also being used here to attach girders to posts (recommended in earthquake areas).

top of the posts). Run a taut line between these 2 nails. Next, temporarily set the posts on their pier blocks. Be sure they are plumb and that the top of the pier block is clean of all debris. Mark the posts where they cross the string. Remove the posts and cut them at that mark; nail them to the pier block or metal fastener; and brace them plumb. They should now all be cut at exactly the same height, allowing the girder to sit level on them.

When installing the posts, it is more important that they sit directly underneath the girders than that they all be directly in the center of the pier block. As you install them, use a straight string to line them up directly underneath the girder that will be placed on top. Remember: where two pieces of girder stock meet over a post use a 4-×-6 post. Check girder stock lengths at this time to determine where to place these 4×6s.

Another way you can cut the posts is to measure down from the level string (set by your transit) to the pier block, record the exact dimension and cut the posts the corresponding height. This is somewhat quicker and can be just as accurate if you take your time. A water or hydro level can also be used. No matter which method you choose, be sure that you make a good flat cut at either end of the pier, providing a good bearing surface. Some builders also seal each end of the post with a wood preservative. It couldn't hurt!

Photo 9-3. A notched girder being inserted into a girder pocket of a foundation wall. Note the use of metal flashing to keep the wooden girder from touching concrete.

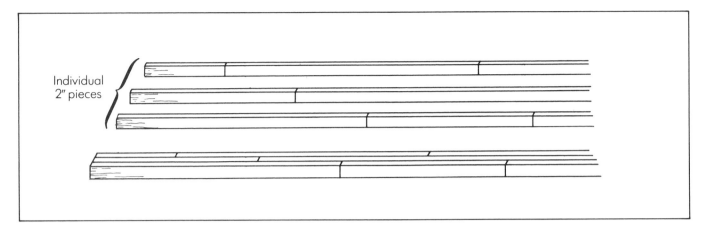

Figure 9-12. A girder built from individual 2-inch pieces of stock nailed together. Note how the splices are staggered.

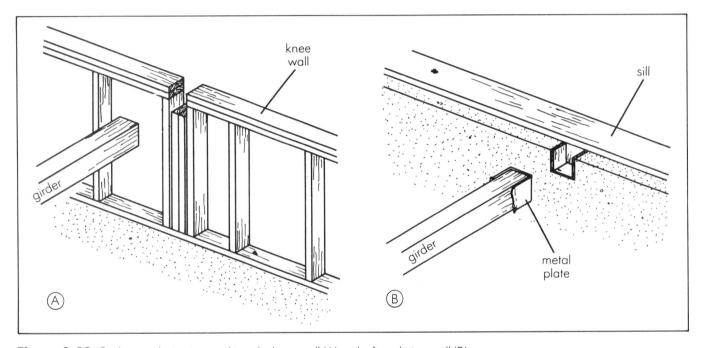

Figure 9-13. Girders are being inserted into the knee wall (A) or the foundation wall (B).

Step 3
Erect the Girders

Margin of Error: Off level ¼ inch.

Most Common Mistakes: Using badly bowed stock; not using a 4-×-6 post under splice in girder.

Diagrams: Figures 9-12, 9-13, 9-14, 9-15, 9-16, 9-17, 9-18.

Photographs: Photos 9-4, 9-5, 9-6.

As with posts, there are several ways you can go in your choice of girders: solid timbers, built-up wooden girders, steel beams or glulams. Each has its advantages and disadvantages. Metal beams will be able to span the greatest distances and may be the best choice when few or no piers or interior supports are called for, as in a basement. Also they can span greater distances with smaller beam sizes, thereby creating greater headroom in the basement or crawlspace. A 6-inch-high steel beam may support the same load as an 8-inch or 10-inch-high wooden beam. Being metal, they are more expensive than wood and harder to work with. They are seldom used unless the design calls for it.

Wooden beams are most common and the cheapest and easiest to use. Often solid pieces are badly bowed and can create a rise in the floor unless the crowns are pulled down. If you can find good, straight large timber girder stock (4×6, 4×8, 6×6, etc.), use it; you will save time by not having to build up the girders. Check the price of manufactured glulam beams—beams created from individual 2-inch-thick material glued and laminated together. You may want to consider using them for girder stock; they are often very straight with few crowns or warps.

On the East coast where there are fewer large trees, and therefore fewer large timbers, girders are usually built up from nominal 2-inch stock, nailed together well to act as one

Photo 9-4. Metal I-beams and metal columns in the basement are used as girders to support the first-floor joists.

Photo 9-5. Built-up girders on concrete block piers. Note the sill plate and metal termite guard on top of the piers. Note also that the splice in the girder occurs over the piers.

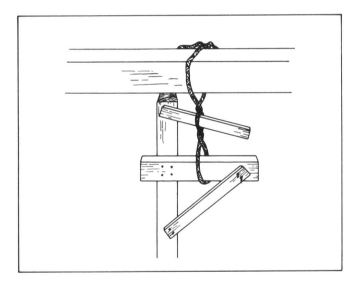

Figure 9-14. Using a rope twister to draw the crowned girder down to the post. Twisting the board shortens the rope and draws down the girder.

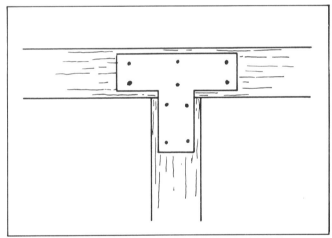

Figure 9-15. A metal girder post fastener. These are used in earthquake areas and to keep crowned girders down on posts. These are also used to join splices where 2 pieces of solid girder stock meet.

piece. Because of the opposing grain patterns created when using different pieces of stock, the built-up girder is actually stronger than the one large, solid piece of girder stock. You can often work the bows and crowns out of the built-up girders by nailing boards together in such a way that their bows point in opposite directions.

After you have decided what type of girder you plan to use, build or cut them in such a way that any splices in the solid girder (and if required, joints in a built-up girder) occur over a post or column. With built-up girders nail the individual boards together according to code requirements.

Where the girders meet the concrete wall foundation, you

should shim on both sides as well as underneath the girder. (The negative space left should be larger and deeper than the girder.) This helps to avoid rot. Cedar shims are OK on the sides, but underneath the girder, where a great amount of weight will be bearing, you should use plywood or even metal shims. In some cases, metal plates which extend 2 inches on either side of the girder are inserted in the fresh concrete. NOTE: In concrete block foundations, fill the cells of the blocks directly below the girders with concrete for greater bearing strength. Where the girder intersects the wooden knee wall, a special pocket is built as shown in the diagram.

If the tops of the posts are all level to each other, and the

Photo 9-6. A house during the floor frame stages. Here the posts and girders are erected and the joists are about to be installed.

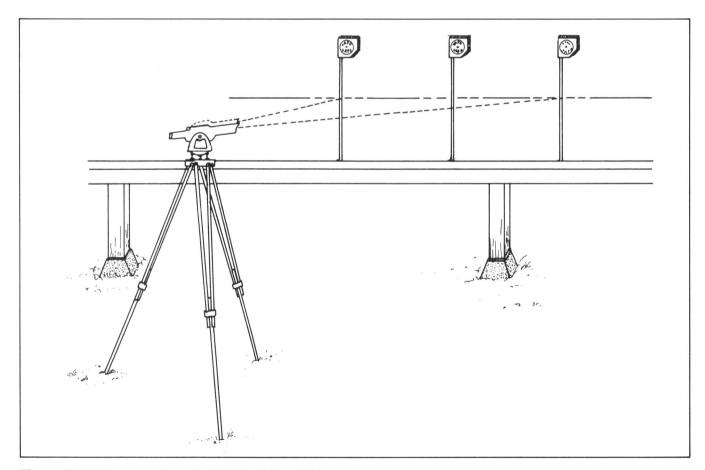

Figure 9-16. Using a transit to check the level of the girder.

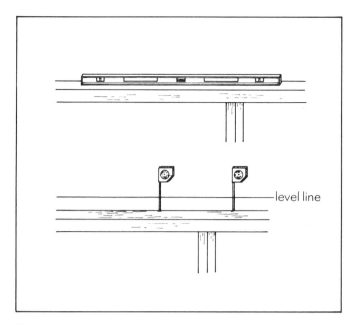

Figure 9-17. Using a level or level line and measuring tape to determine the level of the girder.

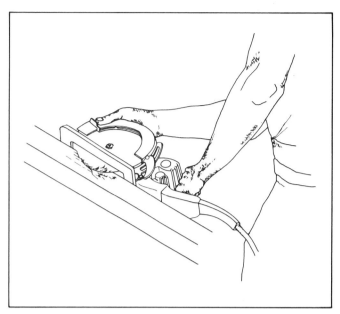

Figure 9-18. Using a power saw to remove the crown from the girder.

girder is not resting on the posts but a little above them, your girder is bowed. Rather than shim it, thereby trapping the bow, you should work on pushing the crown down so that the girder rests on the top of the post. This can be done by having someone stand on it while you nail it to a metal or plywood plate that fastens the post to the girder. If this does not work you can use a rope and crank it down as shown in the diagram. If you are still unable to bring down the crown, do not shim, but allow the girder to float above the post in the hope that in time the crown will settle out. Check it at later stages of building.

After the girder is in place, it is a good idea to check along the length of the girder to be sure that it is level. If it is not level, it will throw all the joists off level, and thereby that entire floor area off. You can check the level by using your

transit or using nails at each end of the girder with a string attached at level points of the nail. Then note the distance from the string to the top of the girder. An 8-foot level, placed directly on top of the girder, can also be used.

If there is a bad bow in the girder, you can either choose not to use it at all or to cut out the high section of the crown. You can also notch the girders or joists where they rest on the girder. No matter what you choose to do, be sure that your girders are true and level or that the girder or joist is notched or shimmed for the high and low spots so that *the tops of the floor joists are level.*

Where individual pieces of solid girder stock come together, they should meet over a 4-×-6 post and be joined by 24-inch-long scabs on both sides (2 inches thick and the same width as the girders) or by metal plates.

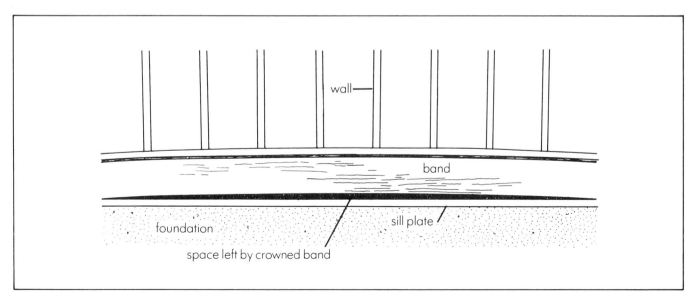

Figure 9-19. How a bowed band joist can push up a wall.

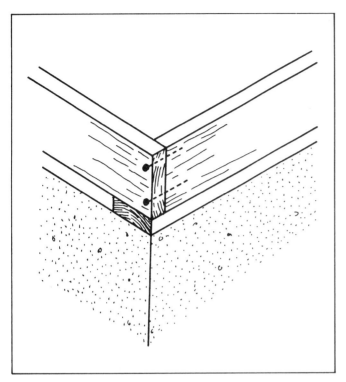

Figure 9-20. The outside edge of the band is flush with the outside edge of the sill plate.

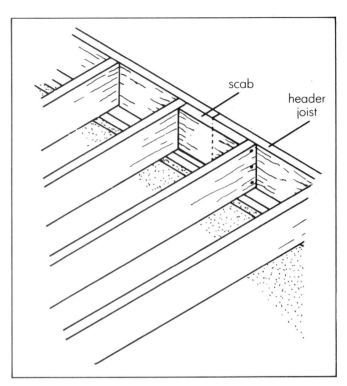

Figure 9-21. A scab is being used to join a splice in the header joist.

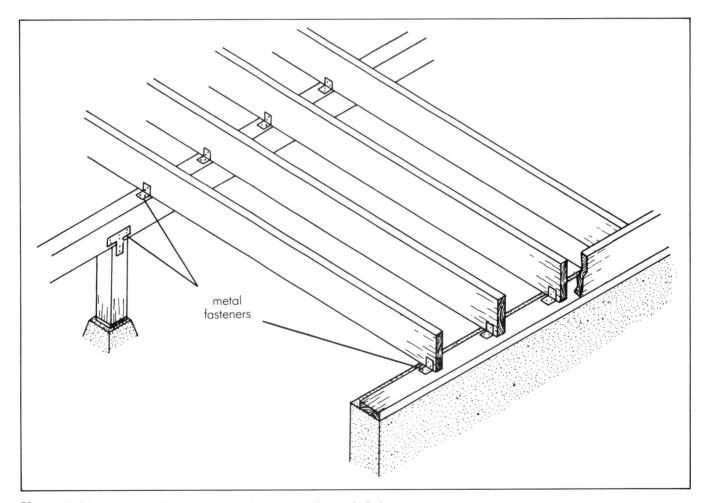

Figure 9-22. Using metal fasteners to attach joists to girders and sill plates.

Photo 9-7. A band being installed flush with the outside of the sill plate.

Step 4
Erect the Band

Margin of Error: Off level ¼ inch. Flush with sill — ⅛ inch, unless adjustment is needed to make frame square or proper dimension.

Most Common Mistakes: Using bowed stock; building band off square or with improper dimensions.

Diagrams: Figures 9-19, 9-20, 9-21, 9-22, 9-23.

Photograph: Photo 9-7.

After the girders are all in place, erect the band. The band is made up of the pieces running perpendicular to the joists (called the header joists) and the first and last joists. The band encircles the joist system and is made from the same size stock as the joist material. Choose good straight pieces of stock so that they do not rise above the sill plates. They need to sit flat on the sill so as not to push up the wall. Nail these pieces into the sill plates with 10d nails, every 16 inches. Unless you need to make adjustment for mistakes in the sills or foundation, the outside edge of the band should be flush with the outside edge of the sill plates. Splices that occur in the header joists can either meet in the center of a joist or be joined by a scab.

Where the first and last joists meet the header joists, face nail these together. Check your diagonals and your mea-

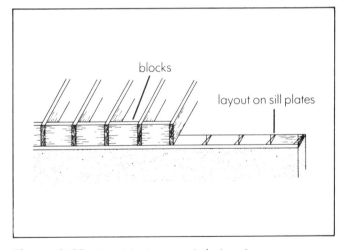

Figure 9-23. Using blocks instead of a band.

surements after the band is completed to see how true your floor system is. You may also want to test the band joist with a string to see if it is straight.

In some areas, where there are earthquakes or hurricanes or tornadoes, metal fasteners are used to further attach the band to the sills. Some builders do not use headers; they use blocks at the end of the joists instead. I like the header method better, but either will pass code. When using the blocks, the joist layout is done on the sill plates rather than the headers.

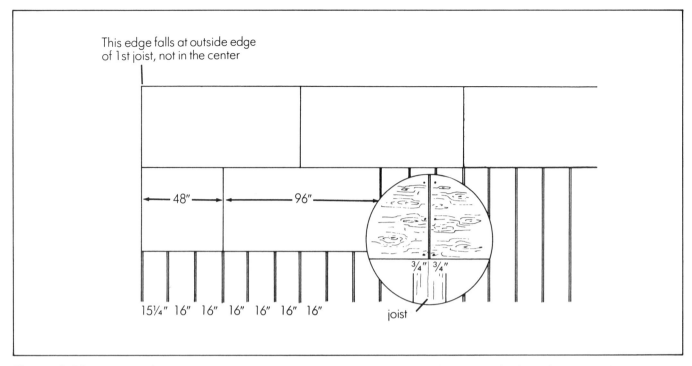

This edge falls at outside edge
of 1st joist, not in the center

48" 96"

15¼" 16" 16" 16" 16" 16" 16"

¾" ¾"

joist

Figure 9-24. How 16-inch on center joists accommodate 48-inch or 96-inch panels. The bubble shows how 2 panels meet over a joist. The distance from the outside edge of the 1st joist to the center of the 4th joist is 48 inches, and to the center of the 7th joist, 96 inches.

Step 5
Lay Out the Joists on the Band and the Girders

Margin of Error: Exact.

Most Common Mistakes: Not starting with 15¼ inches (or 11¼ inches or 23¼ inches) layout; placing "X" on wrong side of line; using framing square to do layout.

Diagrams: Figures 9-24, 9-25, 9-26, 9-27.

You are now ready to do your first layout. You will do similar layouts for wall framing and roof framing. Laying out framing members is simply marking where the individual members of a series will be placed and nailed. In this case we will be marking the location of the joists where they are nailed to the girder and the band.

Before beginning the layout, study your plans well. You may want to build a plan table to support your plans. Note the direction the joists will run and how they will be attached at the girder. Also note the location of any load-bearing partitions running parallel to the floor joists, openings, load-bearing columns or posts, or toilet drains. These will be dealt with in the next step. The most common spacing for floor joists is 16 inches on center but 12-inch and 24-inch centers are also used. All these distances are used to accommodate the 4-×-8-foot plywood subfloor panels and provide a nailing surface where 2 panels meet. In our example, we will use the 16-inch O.C. (on center) measurement.

The band and girders are marked with straight lines made with a combination square and an "X" marked on one side of

the line to show the position of the actual joist. These lines will mark the edge of the joists, not the center. Use a thick-leaded carpenter's pencil.

It does not really matter on which end of the building you begin your layout, unless there is already a foundation cripple or knee wall in place. In that case, begin the floor frame layout at the same end where you began your cripple wall layout. Your wall and roof rafter layout will also begin from this end. This enables all your cripple studs, floor joists, wall studs and rafters to line up directly over one another (assuming they have the same on center distance). This provides greater strength, but more important, allows open straight channels through which to run any heating or cooling duct-work or large plumbing waste pipes.

To begin the layout, hook the tape at the end of the header joist and measure out 15¼ inches (23¼ inches for 24-inch O.C. layouts, 11¼ inches for 12-inch O.C. layouts). Drive a nail at this point and draw a line on the band at this spot. Hook your tape on this nail and mark the header every 16 inches from this point. Your "X"s (joists) will fall to the right (see diagram) of these marks.

The reason 15¼ inches, not 16 inches, is used as your first measurement is so that your first piece of plywood will come to the outside edge of your first joist, not to the center of it. All pieces, except the first and last, need to fall on the center of the joists to provide a nailing surface for the adjoining piece of plywood. The first and last pieces do not have adjoining pieces of plywood on one side. By reducing your first measurement by ¾ inch, half the thickness of a joist (1½ inches), you shift the first piece of plywood from the center of the first joist to the outside edge of it.

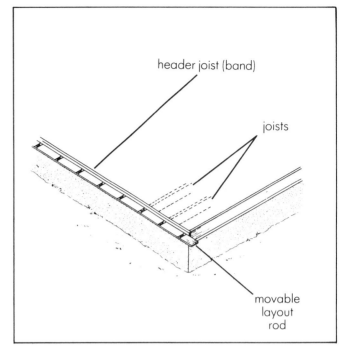

Figure 9-25. Laying out joists on the band.

Figure 9-26. Using a layout rod to mark joist locations on the band.

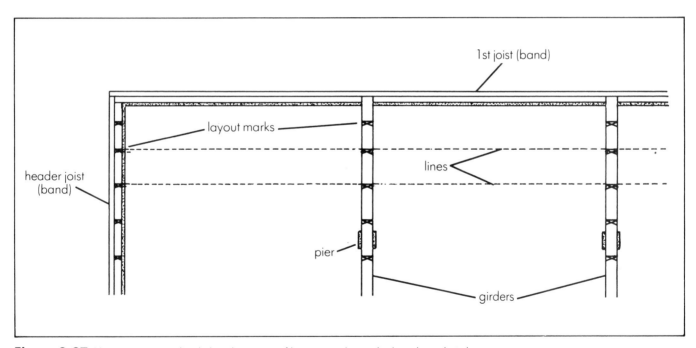

Figure 9-27. Using strings to check the alignment of layout marks on the bands and girders.

Some builders mark a layout rod with the proper measurements and use that at the band and girders. Other builders lay out the bands before they are nailed to the sill plates. Continue laying out the band at 16-inch intervals down the entire length. The distance between the last two joists may not be exactly 16 inches. This is OK, as long as it is not greater than 16 inches O.C. Be sure you always make your "X"s on the proper side of the line. It is a common mistake, even among professionals, to place the "X" on the wrong side of the line.

After you have laid out the first band, repeat this process on the girders and the band on the opposite side. Once I have marked the band and girders, I usually run a straight chalkline from band to band to see if all the marks line up with each other.

After you have laid out all the common 16-inch O.C. joists, you are now ready to mark for any load-bearing partitions, openings in the floor, toilet and bathtub drains, large ductwork or other special situations.

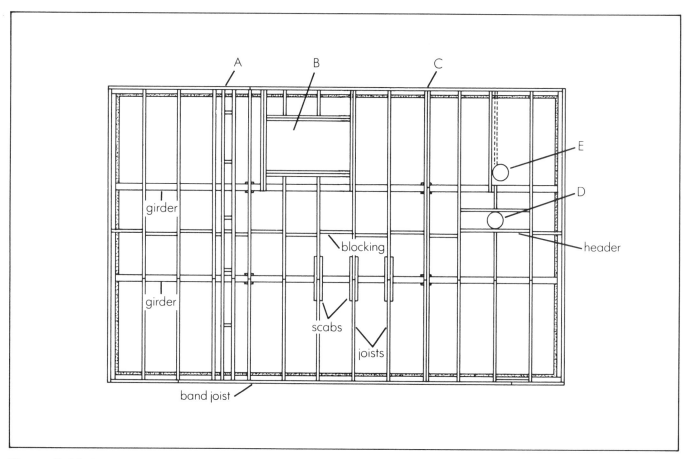

Figure 9-28. Some common floor frame situations: (A) Spaced double joists for a load-bearing interior wall directly above. The spacing allows for pipes and wires to run into the wall. (B) Typical floor opening. (C) Double joists (no spacing) for an interior load-bearing wall above. (D) Bathtub, shower or toilet drain. Note how the joist is broken by headers. (E) Bathtub, shower or toilet drain. Note how the joist has been moved.

Step 6
Layout for Special Situations

Margin of Error: Within ¼ inch of proper placement.

Most Common Mistake: Forgetting to allow for special situations.

Diagrams: Figures 9-28, 9-29, 9-30, 9-31, 9-32, 9-33.

Photograph: Photo 9-8.

Refer back to your plans and see if there are any details in the plan that may need special attention during layout. I will deal with a few of the common situations here and a few not so common ones, such as cantilevering, toward the end of the chapter.

Openings in the Floor Frame

Sometimes an opening is called for in the floor frame. Usually these are for chimneys, stairs, and access doors into the crawlspace. The plans should give you the location in the floor frame and the size of the rough opening. As with the window and door openings in the wall frame, the size of the opening will be listed as the "rough opening" or R.O. This represents the dimensions from the inside of one trimmer to

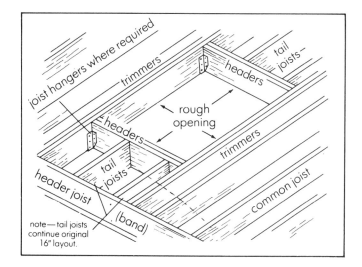

Figure 9-29. Typical opening in floor.

the inside of the other and from the inside edge of one header to the inside edge of the other. A trimmer is a double or single joist at the edge of an opening that runs parallel with the common joists.

The width of the opening will determine how many common joists must be cut. This will then determine how much

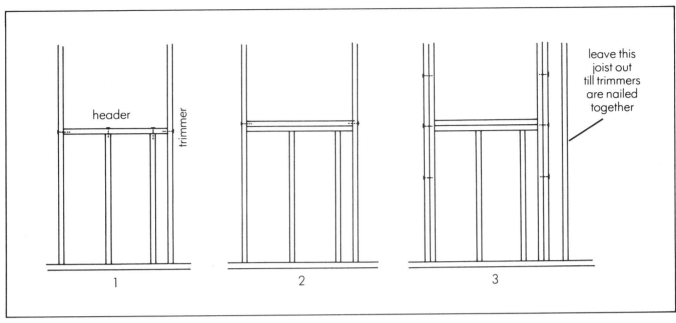

Figure 9-30. The sequence of nailing around an opening when using doubled headers and trimmers.

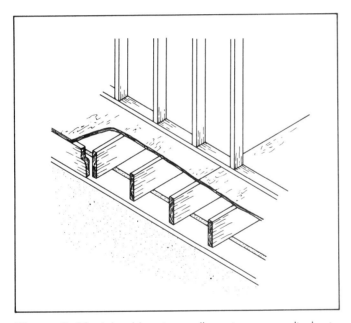

Figure 9-31. A load-bearing wall running perpendicular to the joist system. No special framing is needed.

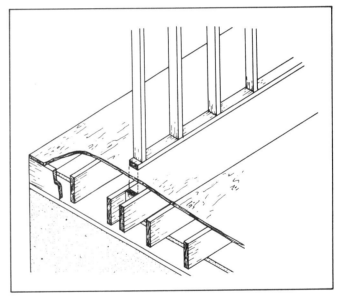

Figure 9-32. A load-bearing wall running parallel to the joist system. Note the double joists separated by blocks to allow pipes and wires to pass into the wall.

weight will be carried by the headers and trimmers. The more common joists that are cut, the more weight they will carry. Usually the trimmers and headers are doubled when the opening is over 4 feet wide. If the opening is over 6 feet wide, special metal hangers are needed to attach the headers to the trimmers. If the tail joists are over 12 feet long, joist hangers are used to attach them to the headers. Note that the tail joists are placed along the original 16-inch O.C. marks to assure a continuous layout that will provide the center of a joist at each splice in the plywood subfloor.

The trimmers and headers are nailed together in a certain sequence so that there is never a need to nail through a doubled piece of stock. Note the diagram for the sequence.

Load-Bearing Interior Walls

Load-bearing walls are interior stud walls that carry a load. Non-load-bearing partition walls only section off a space; there is no load or weight placed on them. Typical loads on walls are from second-floor joists, or rafters or roof truss systems. When these walls are running perpendicular to the floor joists, nothing special is needed, because their weight is distributed evenly over several floor framing members which, in turn, carry the weight over to the foundation. If, however, these walls are running parallel to the floor joists and are falling between the joists, the weight being carried by the wall can be great enough to cause the plywood subfloor to bend and even fail. Because of this, codes require that a

Photo 9-8. An opening in the second-floor floor frame for stairs. Note the double headers, double trimmers and metal hangers supporting the headers and tail joists.

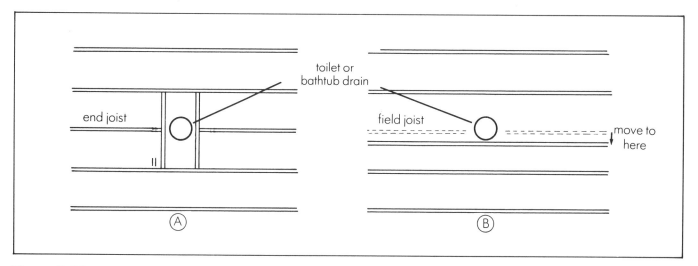

Figure 9-33. Providing for a bathtub or toilet drains. (A) Box for end joists. (B) Field joists can be moved a few inches.

double joist be placed below all load-bearing walls that run parallel to the joists. Often these joists are separated by 2-×-4 blocks placed every 4 feet. This allows any ductwork, plumbing or electrical wires to pass into the wall from below. If a 6-inch-diameter drain pipe is located in the wall, it will be made of 2×6s, not 2×4s, and the joists can be separated by 2-×-6 blocks.

Sometimes one of the wall-supporting joists needs to occur at the same spot as a common joist, in which case a joist is saved. Often they fall between the common joists, and the common joists will be located only a few inches away from the double joists. If this is the case, you may leave out the close common joists if: 1) the distance between joists does not exceed code regulations, and 2) the joist being eliminated is not an "end joist," a joist that is needed to support the splice between two pieces of plywood subfloor, but a "field joist," one where no splice occurs. Measurements can determine this, since the joist at every 4-foot interval will be an end joist.

Toilet and Bathtub Drains

Toilet and bathtub drains occur on the floors within rooms, whereas other plumbing drains are located within the wall cavities. Because of this, their locations have to be anticipated during the floor framing stage so that there is not a joist directly below the drain. If a joist is scheduled to fall directly below the drain, you can either box around it or move the joists a little bit to avoid the drain. If the joist is one that occurs at the splice of two pieces of plywood subflooring (end joist), it cannot be moved. You will either have to box around it or relocate the drain.

Large Ductwork

Sometimes ductwork for furnaces or air conditioning or fan-assisted passive solar circulation systems is too large to fit between the standard joist spacing. In these cases, openings have to be provided to allow for the ductwork.

Step 7
Install the Joists

Margin of Error: Height of crown—½ inch. Proper location—¼ inch.

Most Common Mistakes: Not turning crown up; placing joist on wrong side of line.

Diagrams: Figures 9-34, 9-35.

Photographs: Photos 9-9, 9-10, 9-11, 9-12, 9-13.

There are different ways to join the joists to the girders. Four ways were discussed in the "Design Decisions" section of Chapter 6, *Poured Concrete Foundations*. You should have already decided which method you will use and the proper negative spaces for girder pockets should have been allowed for when pouring the foundation.

The layout having been completed, you now measure and cut the joists. If everything is true and straight, you can just measure at a few spots and cut all your joists; if it isn't, and it rarely is, it may be best to measure for each joist, record the measurements and then cut all the joists.

When installing the joists it is very important to check each joist to locate the crown of the bow. Almost every joist will have one no matter how slight. Always point this crown UP as it will settle down during construction. If there are any large knots on the edge of the boards, place them on the top; it is stronger than having them on the bottom.

Try to cut the boards to fit as tightly as possible. Even though the common margin of error in framing is ¼ inch, the tighter the floor frame fits, the stronger and more rigid it will be. Nail the joists to the header joists with 3 or 4 16d nails. These nails, like all framing nails, should be no closer to each other than half their length and no closer to the edge than a

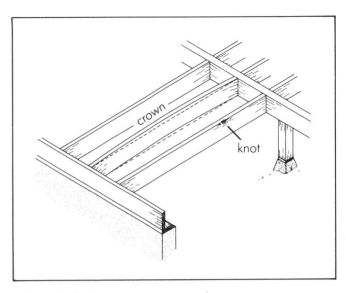

Figure 9-34. Point crown up. Note: knot is up.

quarter of their length. When nailing at the header joists, be sure joists are installed plumb. If you are using joist hangers, special short nails are available to attach the hanger to the joists.

As you nail the joists in place, be sure that they are being nailed on the proper side of the layout line. Placing them on the wrong side is a common mistake in building. It is a good idea to run your tape out every few joists to be sure there is a center of a joist at every 4-foot interval. When nailing the joists to the header joists and girders, see that they are flush along their top edges. If they are low, shim using cedar shims; if they are too high, notch the joists so they will sit lower.

Continue down the frame until all the joists are installed and openings are completed.

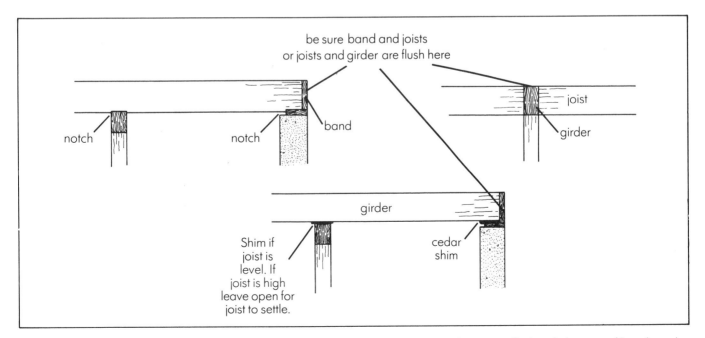

Figure 9-35. Use notches and shims to be sure the girders are level and the tops of joists are flush with the tops of bands and girders.

Photo 9-9. Joists are on top of the girders and joined by scabs. Note a scab is used where 2 band joists are spliced.

Photo 9-10. Joists are hung from the girders with metal joist hangers.

Photo 9-11. Eyeing the joist to locate the crown. Point the crown up.

Photo 9-12. The joists are being nailed to the band. Be sure that the joists and band are flush.

Photo 9-13. Floor framing in progress.

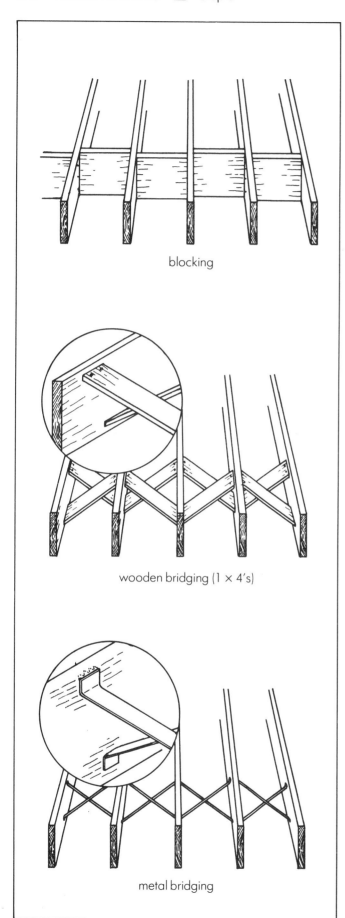

blocking

wooden bridging (1 × 4's)

metal bridging

Figure 9-36. Three types of blocking or bridging.

Step 8
Install Bridging or Blocking

Margin of Error: Solid blocking—exact. Bridging—¼ inch.

Most Common Mistake: Cutting blocking too long or too short.

Diagrams: Figures 9-36, 9-37, 9-38, 9-39.

Photographs: Photos 9-14, 9-15, 9-16.

Most floor frames and most codes require blocking or bridging to be installed in the frame. Blocking or bridging is the placing of metal straps, diagonally nailed 1×4s or pieces of joist stock between the joists to provide greater rigidity and strength to the floor, to prevent the joists from twisting and to transfer the weight from one joist to another.

The metal strap method is rather new and seems quick and easy to apply, but I do not know if it adds as much rigidity and strength as the other systems do. The solid blocking method, using pieces of joist stock, has been the most prevalent in recent years, almost totally replacing the older system of bridging with 1×4s diagonally nailed between the joists. It has replaced it probably because it is quicker, not particularly because it is better. After years of using the solid blocking system, I am leaning more and more towards the 1×4 bridging system. It offers several advantages that the solid blocking system does not.

1) Solid blocking tends to keep bows in place rather than allowing them to settle. Diagonally placed wooden bridging is nailed only at the top in the beginning and late in construction it is nailed along the bottom, giving the bows in the joists time to settle down.

Photo 9-14. Measuring at the end of joists for blocking that will be in the middle. This ensures that the joists will be parallel and straight.

Photo 9-15. Blocking is staggered to allow for face nailing through the joists into the blocks. Here joists are supported (hung) by girders at both ends.

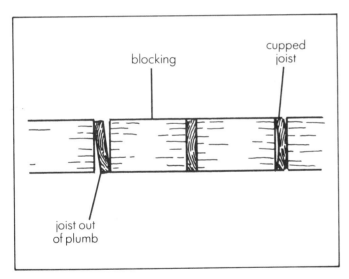

Figure 9-37. Blocking throwing the floor frame out of line because of cupped or out of plumb joists.

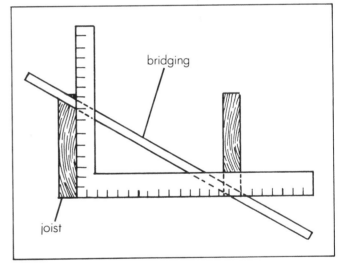

Figure 9-38. Use a framing square to determine the length and angle to cut bridging pieces.

2) When using the solid blocking, if you cut the blocks too long or too short, they will push or pull the joists system out of line. Even if they are cut perfectly, or bought from the lumber yard precut, a cup or twist in the joists will not allow the block to sit flat against the joists, and it will throw things off.

3) The diagonal bridging is easier to run pipes and wires through.

In any of the systems, blocking or bridging is usually spaced every 8 feet or in the center of the span, e.g., a 14-foot span would have blocking 7 feet in. Measure the proper distance from each end of the floor frame, pop a line across the top of the joists with a chalkline, and begin installing either the blocking or the bridging. (Some builders install the solid blocking as they place each joist; this avoids having to nail in the small areas between the joists.)

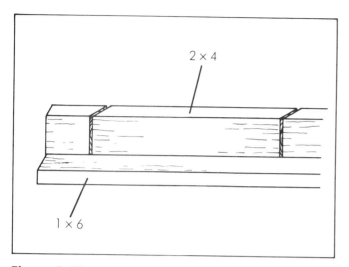

Figure 9-39. Jig to cut bridging.

Photo 9-16. The bridging (1×4s) is in place. Only the top is nailed; the bottom is nailed after the house is completed, to allow the crowns in the joists to settle.

Solid Blocking

When using solid blocking, stagger the blocks on opposite sides of the center chalkline so they can be face nailed on both sides (see photo). Be sure the tops are flush with the tops of the joists. In the case of second- and third-floor frames, where sheetrock ceilings will be applied to the bottom of the joists, be sure the blocks are flush with the bottom of the joists as well. When measuring the space between the joists for the size of block to be placed, measure at the end of the joists where the joists are nailed to the band rather than where the blocking is to be installed; there may be a warp in the joist that could throw the measurement off. Be sure the joists are straight and in line as you block.

Bridging

When bridging, be sure that the 2 pieces do not touch or they will cause the floor to squeak when it is walked on. The top of bridging should be about ⅜ inch *below* the top of the joists and the bottom of the bridging should be the same distance *above* the bottom of the joists. The top of the bridging is always directly on line with the top of the adjacent piece of bridging located in the adjoining space. Nail only the top of the bridging; the bottom will be nailed later when the joists settle.

You may want to nail a 1-×-6 board across the tops of the joists as you nail the bridging. This will keep the joists in their proper place until the subflooring is installed. The bridging, unlike the solid blocking, will not hold them in place.

When cutting bridging, you can figure out the length and the angles of the cuts by using a framing square. It is worthwhile to build a small cutting jig to enable you to cut many pieces at once.

Step 9
Check the System

Margin of Error: ¼ inch.

After the floor frame is complete, check for the following before going on to the subfloor:

☐ Dimensions

☐ Squareness (check diagonals)

☐ All joists parallel and on proper mark

☐ All crowns up

☐ Joists flush with girders and headers

☐ Proper rough openings/squareness

☐ Center of joist every 4 feet

☐ All joists properly nailed

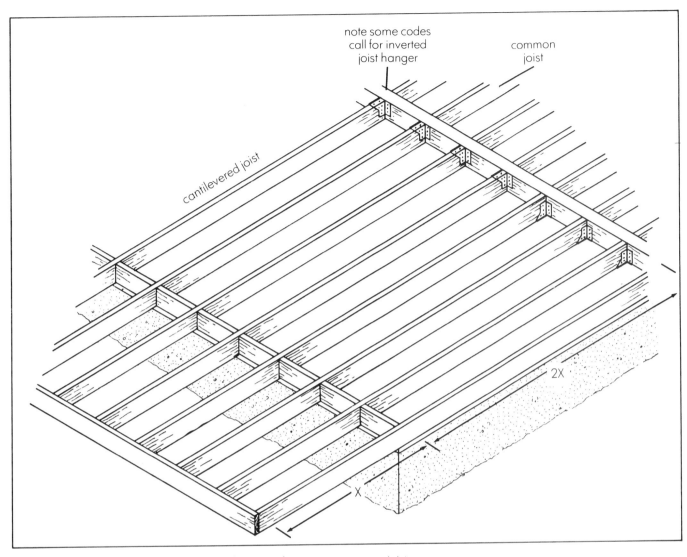

note some codes
call for inverted
joist hanger

common
joist

cantilevered joist

2X

X

Figure 9-40. Cantilevered joists running the same direction as common joists.

Special Situations

Cantilevering

Diagrams: Figures 9-40, 9-41, 9-42.

Cantilevering is used as a design effect both for porches and decks as well as for projections from the building itself. They overhang the foundation wall and are suspended without any vertical support posts. Where part of a room is being suspended, you may have to get this okayed by the building code office; some building codes read that there must be a foundation system (usually a solid wall foundation) under all exterior walls.

If you plan to cantilever a deck, remember you will be exposing your floor joist system to the weather. This exposure could cause rot which, in time, could travel into the house and cause structural damage to the joist system. You can switch to a rot-resistant wood for these joists (redwood, cedar, pressure treated), which will be more expensive than common joist material. Since joist lumber is also weaker, it will have to be either larger or more closely spaced than the

other joists. If you plan to expose your common framing lumber, at least treat the lumber to help prevent rot. You may want to consider avoiding cantilevering altogether and just build a standard deck with vertical supports.

Should you decide to use a cantilevered system, it will make a difference whether the cantilevered joists are running parallel or perpendicular to the common joist system. If they are parallel, you simply run the joists out past the foundation wall. If your cantilevered joists are to run perpendicular to the common joists, you have to double up a common joist and tie the cantilevered joists into this double joist. There is always a 1 to 2 relationship between the length of the total joist and the distance it can be cantilevered past the foundation wall (it can be cantilevered ⅓ the length of the joist). There is also a code limit to how long a joist any size can be cantilevered which must be taken into consideration as well.

If you are cantilevering for a deck, you will need to notch the cantilevered section 2 inches to allow the deck boards to sit below the level of the inside floor. This further reduces the strength of the joist.

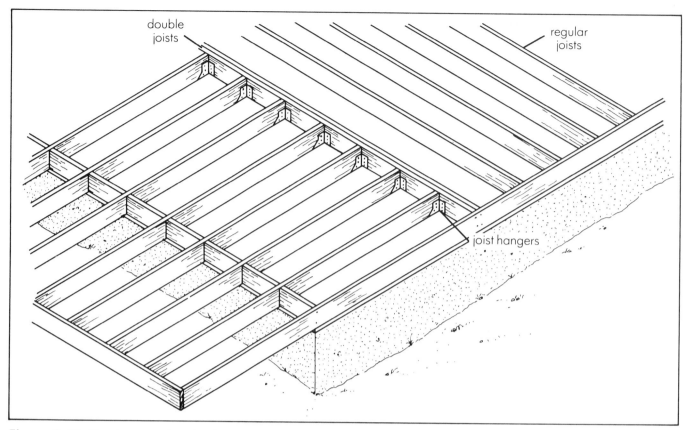

Figure 9-41. Cantilevering when the deck runs perpendicular to the regular joist.

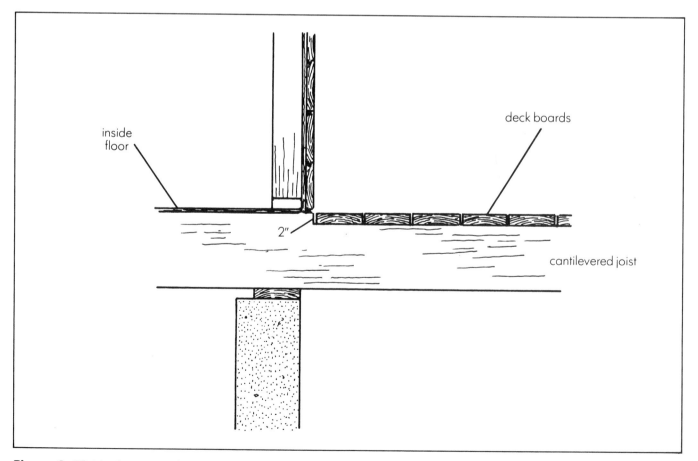

Figure 9-42. Notching a cantilevered joist to keep the deck slightly lower than inside floor level.

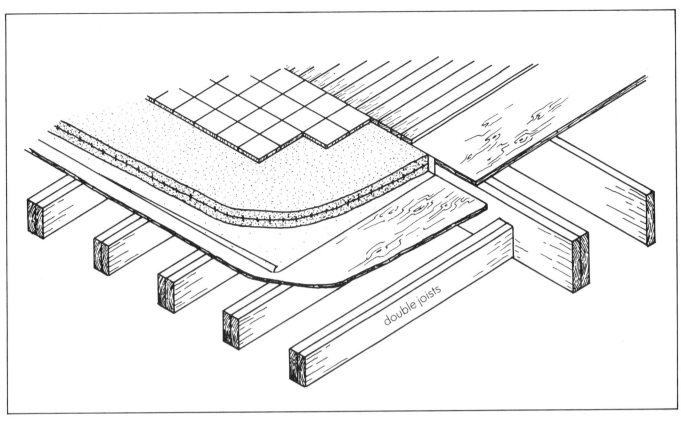

Figure 9-43. Changing the height of the joists to accommodate tiles being set on top of reinforced concrete.

Photo 9-17. A girder system using 4-inch-thick girders on 4-foot centers. A thicker subflooring, either 1⅛-inch plywood or 2-inch tongue and groove boards, is used. Note that underfloor utilities are already installed (optional in most areas).

Photo 9-18. Floor trusses being installed. They are lightweight, but can span greater distances than common floor joists.

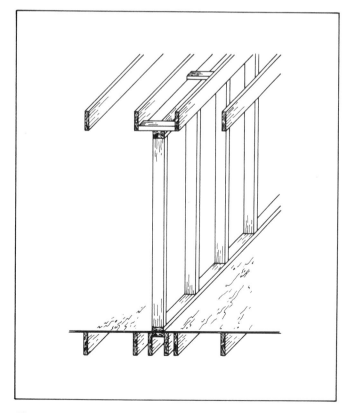

Figure 9-44. Manner in which the tops of walls are fastened to floor joists above and below.

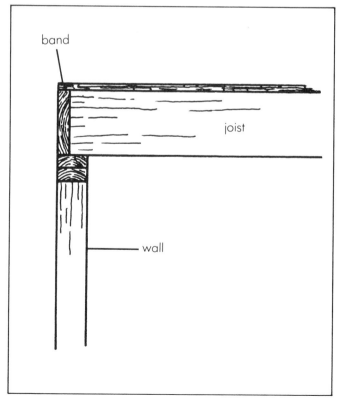

Figure 9-45. First-floor wall frame supporting second-floor joists.

Photo 9-19. Second-floor joists resting on the walls. Note that blocks are used at the end of the joists instead of a solid band.

Tile Floors

Diagram: Figure 9-43.

Sometimes in entrance halls and bathrooms, where tile on top of reinforced concrete is to be used as a flooring, you will need to lower the floor frame to accommodate the difference in height of this type of flooring. You will need to use either smaller joists or 4-inch stock or doubled joists to provide for this. If you are using standard thickness tile with mastic (adhesive) and no concrete base, usually no special adaptations are needed.

Girder System

Photograph: Photo 9-17.

Some builders choose to use a floor frame system made up entirely of 4-inch girders (usually 4×6s or 4×8s). This system requires girders every 4 feet supported by the standard posts; there are no joists. Because of the 4-foot span of the girders, you will need to use a thicker subfloor. 2-4-1 plywood (a 1⅛-inch-thick plywood) or 2-inch tongue and groove boards (2×6s or 2×8s) are used as subflooring. From what I know, this system is comparable in price and time to the standard floor frame system. You may want to ask a local builder who uses this system what they think. Some builders have switched to it.

Floor Trusses

Photograph: Photo 9-18.

As with roof framing, you can purchase pre-made trusses to be used as floor framing. These are made up of 2-inch material and plywood specially designed and built to be able to span long distances with lightweight materials. They seem to work pretty well. (See Photo 9-18.)

Second-Floor Joists

Diagrams: Figures 9-44, 9-45.

Photograph: Photo 9-19.

Floor framing for the second floor is almost identical to that of the first floor. Instead of bearing on foundation walls and girders, the joists bear on exterior and interior walls. The layout is done on the top of the wall's top plates and the band is nailed flush to the outside edge of the wall. Note the way parallel first-floor walls connect to second-floor joists.

Many designers or owner builders use large attractive timbers for second-floor joists and expose them to the downstairs. Usually 2-inch tongue and groove boards are used as a second-floor subfloor (and even as the finished floor). These make an attractive first-floor ceiling. Be aware when doing this that the noise transmission between floors is high.

Worksheet

ESTIMATED TIME

_____

_____

_____

_____

CREW

Skilled or Semi-Skilled

_____ Phone _____

_____ Phone _____

_____ Phone _____

Unskilled

_____ Phone _____

_____ Phone _____

_____ Phone _____

_____ Phone _____

_____ Phone _____

MATERIALS NEEDED

Joist stock, girder stock, posts and nails

ESTIMATING MATERIALS

_____ × _____ JOISTS _____ feet long: Quantity _____

Linear feet of HEADERS and BAND _____

Linear feet needed for BLOCKING _____

Quantity _____ of TAIL JOISTS at _____ feet long

_____ × _____ GIRDERS _____ feet long: Quantity ___

_____ × _____ POSTS _____ feet long: Quantity _____

_____ lbs. of 16d FRAMING NAILS (100 lbs. to frame a 1500 sq. ft. house.)

COST COMPARISON

_____ Store:

$ _____ per M = $ _____ total

_____ Store:

$ _____ per M = $ _____ total

_____ Store:

$ _____ per M = $ _____ total

_____ Store:

$ _____ per M = $ _____ total

PEOPLE TO CONTACT

If underfloor utilities need to be inspected before the subfloor goes on, you may want to contact any required subcontractors if you do not plan to do the work yourself.

Roofer _____

Phone _____ Work Date _____

Roofing materials supplier _____

Phone _____ Work Date _____

Sheet metal person _____

Phone _____ Work Date _____

Plumber _____

Phone _____ Work Date _____

Solar panel installer _____

Phone _____ Work Date _____

Daily Checklist

CREW:

_____Phone _____

_____Phone _____

_____Phone _____

_____Phone _____

_____Phone _____

_____Phone _____

TOOLS:

☐ framing hammers

☐ combination square

☐ hand saw

☐ power saw

☐ saw blades

☐ saw horses

☐ pencil

☐ steel tape

☐ nail belt

☐ sledge hammer

☐ safety goggles

☐ nail-pulling tools

☐ transit or hydro level

☐ chalkline

☐ 4-foot level

☐ extension cords

☐ junction box

☐ framing square

MATERIALS:

_____joists

_____joist stock for band, blocking, etc.

_____16d CC sinkers

_____metal fasteners

_____metal fastener nails

_____bridging material

_____bridging nails

DESIGN REVIEW: Just review to see if your planned joist installation system is the easiest and cheapest method. Would an added girder greatly increase the rigidity of the floor?

WEATHER FORECAST:

For week _____

For day _____

10 Subflooring

Photo 10-1. House with plywood subflooring being applied.

What You'll Be Doing, What to Expect

SUBFLOORING IS A SKIN or surface that covers the open floor joists. There are several different options in your choice of materials: plywood, 2-inch-thick (nominal) tongue and groove boards, particle board and 1-inch-thick (nominal) regular cut boards. This phase of construction is one of the simplest. It involves nailing and/or gluing the subflooring to the joists; an easy process with few possible mistakes. Though it can become monotonous, the work goes quickly. Since it is work that requires only one real skill, nailing, and little information, it is a good place to use your free or inexperienced help. Kids do well at this task. (See Photo 10-1.)

You will be working in a bending position for long periods of time so you may feel a strain. Finding the most comfortable positions to facilitate the flow of work can make a great difference in both the ease and speed of the work. As in other phases that require fastening large surfaces, like wall and roof sheathing, I would highly recommend using a pneumatic nailer (Photo 10-2). This tool, run by an air-compressor, can cut your application time by as much as 50 to 75% and is great when you are working with only 1 or 2 people. You can even rent a moving nailer mounted on wheels that you push while walking along, squeezing the trigger to drive the nails.

In this chapter we will introduce different types of subflooring and discuss plywood subflooring in detail.

Jargon

SUBFLOORING. Boards or panels that are installed over the floor joists and over which a finished floor will be applied.

SUBFLOORING NAILS. A deformed or annular shank nail (8d) that has a much greater holding power than a regular nail. These help prevent the floor from squeaking.

SUBFLOOR-UNDERLAYMENT. A plywood or particle board panel that is used both as the subfloor and underlayment.

TONGUE AND GROOVE (T & G). Boards that are cut in such a manner that their sides (and sometimes their ends) have a tongue or protruding strip on one side and a groove or continuous indentation on the other side.

2-4-1 PLYWOOD. A 1⅛-inch-thick plywood that can be used for subflooring when your supports are up to 48 inches on center.

UNDERFLOOR UTILITIES. All the different utilities that have wires, pipes, ducts, etc., that are installed below the subfloor in the crawlspace or basement. These include electrical and plumbing systems, heating and air conditioning, gas lines, solar pipes or ductwork.

UNDERLAYMENT. Sheets or particle board or plywood laid on top of the subfloor to provide a smoother surface for some finished flooring. This surface is applied after the house is built but before the finished floor is laid.

Purposes of the Subfloor

1) Provides a base for the application of the finished flooring
2) Adds rigidity to the floor frame
3) Provides a platform or surface on which the carpenters can lay out and construct the rest of the framing

Design Decisions

Here are the different types of subflooring materials available:

PLYWOOD (Figure 10-1). Plywood, either tongue and groove or butt edge, is the most common form of subflooring and perhaps the best. It goes on quickly and easily and adds a tremendous amount of rigidity to the floor frame. This is especially important in earthquake areas. Usually a ⅝-inch or ¾-inch CD exterior glue plywood is used. Exterior glue plywood is plywood that can get somewhat wet during construction, but cannot stay wet for long periods of time as can exterior plywood used for siding. Exterior glue plywood is also called exterior/interior plywood as it can be exposed during construction, but after construction it will be in the interior of the home.

Exterior glue plywood has knotholes on both surfaces. This is of little consequence since it will be covered by a finished floor. If, however, your finished flooring will be a thin linoleum that could press into these knotholes, a plug and

Photo 10-2. Using compressed-air nailing gun to apply subfloor.

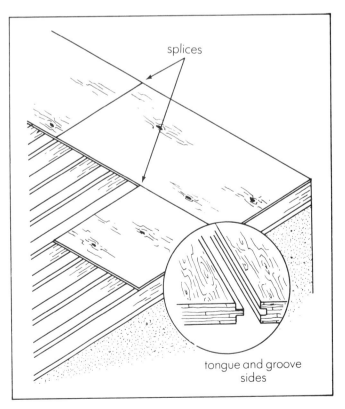

Figure 10-1. Plywood subflooring. Note that the splices are staggered.

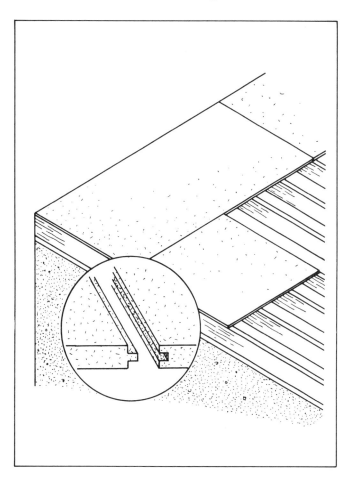

Figure 10-2. Particle board subflooring.

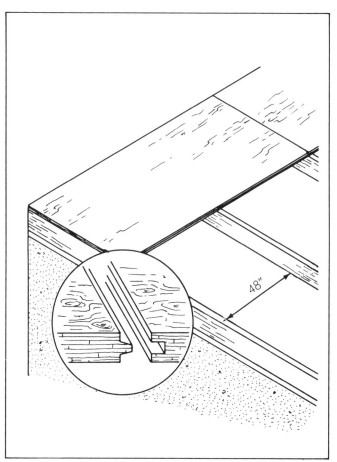

Figure 10-3. 2-4-1 plywood (1⅛ inch thick).

touch sanded (P & TS) grade is required, where the holes have been filled and sanded. (An underlayment grade can also be used.)

PARTICLE BOARD (Figure 10-2). This material is similar to plywood, but instead of being sheets laminated (glued) together, it is made of wood chips or particles and glue poured to form a sheet. This material is somewhat less expensive than plywood and is therefore becoming popular.

It does, however, have some drawbacks. The glue composition gives off a formaldehyde fume or gas, and if it gets wet, and stays wet (if a pipe were to burst while you were on vacation, for instance), it can totally dissolve. Some builders use it in bathrooms; if it only gets damp or a little wet it will hold up better than plywood, which can delaminate. It is also harder to nail through than plywood.

2-4-1 PLYWOOD (Figure 10-3). This is a 1⅛-inch-thick plywood that is used in the girder floor frame system (discussed in Chapter 9, *Floor Framing*, "Step 7") where the supports are 48 inches on center. This or a nominal 2-inch tongue and groove board is your only choice when framing members are on this great span.

NOMINAL 2-INCH TONGUE AND GROOVE BOARDS (Photo 10-3). Tongue and groove boards, either 2×6s or 2×8s, are sometimes used as subflooring. Like 2-4-1, they can span up

to 48 inches on center supports. Especially when used on framing members 16 or 24 inches on center, this material makes for a very stout floor.

When you are using large floor frame members for the second floor that will be exposed from the first floor below, and the subfloor will also be exposed from underneath, tongue and groove boards are often used. They look better from below than plywood, and sometimes you can even sand, stain and varnish the upper surface to make it into a finished floor as well. Be aware, however, that this type of system affords little sound control between floors.

Tongue and groove boards take a good bit more time to apply than does a plywood subfloor. The difference is applying many individual boards compared to large panels. You will need to check locally for the difference in the cost of the two systems for material and labor (if you are paying for labor). If you plan to expose one or both of the surfaces, buy "dry" lumber to avoid cracks caused by shrinkage.

NOMINAL 1-INCH COMMON BOARDS (Figure 10-4). Before the advent of plywood, this was the most common subflooring material. It consists of common 1×6 or 1×8 boards with regular surfaces and no tongue and groove, applied diagonally across the floor frame at a 45° angle for greater strength. This type of subflooring is much slower to apply and not nearly as strong as plywood; it is not recommended.

Photo 10-3. Tongue and groove 2×6 subflooring being applied. Large second-floor floor joists and the subflooring will be exposed from below.

What You'll Need

TIME   24 to 40 person-hours per 1000 square feet.

PEOPLE   2 Minimum (1 skilled, 1 unskilled); 3 Optimum (1 skilled, 2 unskilled); 6 Maximum (2 teams of 3 each).

TOOLS   Framing hammer, chalkline, power saw, hand saw, pneumatic nailer (optional), pencil, nail belt, tape measure, framing square, caulking gun, sledge hammer or maul, nail-pulling tools, saw blades, extension cords, plug junction box.

RENTAL ITEMS: Air compressor, nailing gun.

MATERIALS   With plywood subfloors I would use a tongue and groove plywood, at least ⅝ inch or thicker. A ¾-inch board makes for a stouter floor and is recommended. Use a CDX exterior glue plywood. In areas where the finished flooring will be a thin vinyl or linoleum or where the plywood subfloor will be the finished floor (e.g., laundry-room, workshop, unfinished areas, etc.), use an underlayment grade, plug and touch sanded (P & TS). Be sure when you buy the panels that they are all the same size. Some

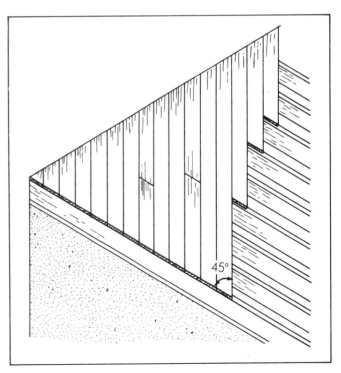

Figure 10-4. Nominal 1-inch (¾-inch) boards applied diagonally as subflooring.

manufacturers make them 48 inches wide, others 47½ inches wide. *You cannot mix the two; they all have to be the same size!*

Buy a high-quality subflooring adhesive and 8d subflooring nails. This should eliminate all squeaks.

ESTIMATING MATERIALS NEEDED

PLYWOOD: To estimate plywood divide the number of square feet of floor by 32 square feet (4-×-8-foot plywood sheet) and add 15% for waste.

ADHESIVE: Refer to the manufacturer's suggested use.

NAILS: Two pounds per sheet (CC sinkers 8d or threaded subflooring nails).

INFORMATION This chapter will deal in detail with applying plywood subfloors. For applying 1-inch common boards, consult some of the older housebuilding books written when this type of subflooring was prevalent. For more information on applying T & G boards, see section on 2-×-6-foot T & G in Chapter 14, *Roof Sheathing*, p. 327.

Be sure you meet all the specifications of your local codes for subflooring, e.g., plywood for subflooring must be graded, it must be the required thickness, etc.

PEOPLE TO CONTACT If in your area underfloor utilities need to be inspected, you may need to contact the plumber, electrician, solar installer and someone to install the sheet metal for the heating and air-conditioning.

INSPECTIONS NEEDED In many areas, especially those covered by the Uniform Building Code, you may need to get an underfloor utilities inspection before the subfloor is applied but after the floor framing is built. This means the following must be installed below floor level: water supply lines, both hot and cold water pipes, air conditioning and heating ductwork, waste-water pipes, electrical wires and gas lines. The inspector will want to inspect these before the subflooring is applied. If, however, you have a high crawl-space, the inspector may waive the inspection until after the frame is built. Getting a waiver, if you can, is usually an easier way to go.

Reading the Plans

Most plans will clearly state the thickness and type of sub-flooring to be used. Often they will not tell you whether it is tongue and groove or regular plywood. Tongue and groove is always recommended. You may decide to increase the thickness of the plywood called for in the plans in order to provide a stouter floor. Note that the plywood always is applied with its long (8-foot) dimension running perpendicular to the floor joists.

Your plans should show you any areas on the second or third floor where an open-beam floor system will be used, exposing the floor joists to the floor below. In these areas you may want to use 2-inch tongue and groove subflooring. Be sure there are no plumbing fixtures in these areas that need their drain or supply pipes protruding through the floor or they will be exposed from below.

Safety and Awareness

SAFETY: Be very careful with the pneumatic nailer.

Be careful not to fall between the floor joists, especially by stepping on an unnailed sheet of plywood that is hanging over a joist. This is a common accident.

Beware of back strain from bending over all day.

AWARENESS: Plywood always runs with its long dimension perpendicular to the joists. End edges must meet on the center of a joist. Stagger the splices in the panels. Best looking side faces up. Use glue and threaded nails.

Overview of the Procedure

Step 1: Check floor frame
Step 2: Spread the glue
Step 3: Lay the first course
Step 4: Lay the remaining sheets

Step 1
Check the Floor Frame

Margin of Error: Floor joist must be installed so that ends of plywood subfloor panels are at least within ¼ inch of center of joist.

Most Common Mistake: Joist not properly aligned.

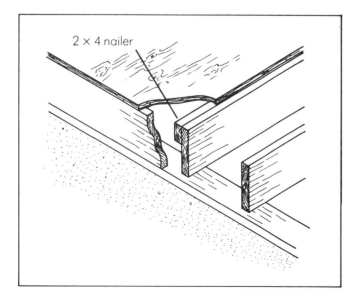

Figure 10-5. 2-×-4 nailer used to provide nailing surface where joists are off line.

Photo 10-4. Using a caulking gun to spread on subfloor adhesive.

Diagram: Figure 10-5.

Before starting on the subfloor, look over your floor joists. Do this both visually and with a tape measure. Walk along the system and sight down the joists. You should be able to tell if they are all running parallel. If you have accidentally placed one on the wrong side of the layout mark, you should be able to spot it with this check.

Starting at the end where your joist layout began (you will begin laying the plywood panels from this end), take a steel tape and measure from the outside edge of the first floor joist (where the first panel will start) to see if there is a center of a joist at every 4-foot interval. You have about a ¼-inch margin of error here. If there is not a center of a joist, determine why, and correct the situation. If only one or two joists are off, it is best to move them into place, or you can add a nailing surface by solidly nailing a 2×4 to the joist to catch the plywood edge.

Once you are sure there will be a center of a joist to catch the ends of the panels, you can begin to install the subfloor. If you are fairly certain that it will not rain before the house is weather tight, you can even install a floor insulation at this time. If there is a good chance of rain, wait until the roofing is on to protect the insulation.

Step 2
Spread the Glue

Margin of Error: Spread it wide, but don't be sloppy.

Most Common Mistake: Using wrong glue.

Photograph: Photo 10-4.

Many builders do not use subflooring glue; they just nail down the subfloor. Using glue, however, not only eliminates squeaks but also adds a tremendous amount of strength to the floor system as it ties it all together into one continuous piece. If you are planning to use wet lumber, instead of kiln-dried, this is even more important. As the lumber dries and shrinks, the nails can loosen up to half their holding power, causing the floor to squeak and lose strength. The use of glue is inexpensive and easy and I highly recommend it.

Be sure you are using subflooring adhesive. A novice builder I know used panel adhesive, which does not have as much bonding effect. The subfloor "pops" as it is walked on due to the sticking and then releasing of this less effective glue.

Spread approximately a ¼-inch-wide bead of glue over all the joists and headers that this first panel will cover. Use a

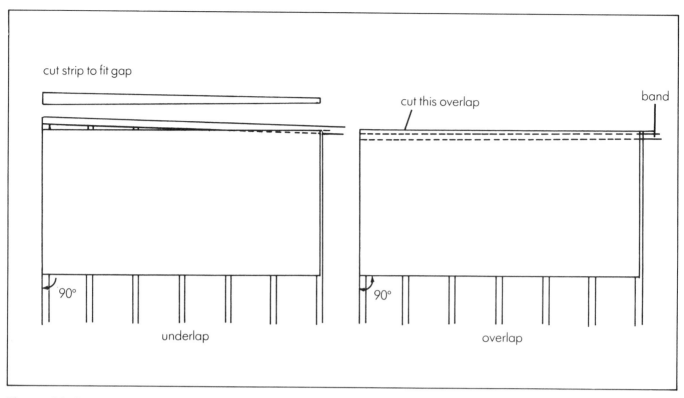

cut strip to fit gap

cut this overlap

band

90°

90°

underlap

overlap

Figure 10-6. Correcting for an off-square floor frame.

nice even bead. Just spread the glue over enough area to apply 1 panel at a time, until you feel you have the subfloor installation system down fast enough to spread the glue for 2 panels at once. Remember to disengage the caulking gun and stick a 16d nail in the open end each time you finish using it, to stop the cartridge from leaking.

Step 3
Lay the First Course

Margin of Error: Must have at least a ½-inch bearing on joist at end of panel.

Most Common Mistakes: Pointing wrong side up; not applying exactly perpendicular to the joists; not facing the tongue towards the outside.

Diagrams: Figures 10-6, 10-7, 10-8.

After the glue is spread, you are ready to apply the first course. As you apply the first row, it is more important that it be applied exactly perpendicular to the floor frame than that it be aligned exactly along the edge of the band. If your floor frame was accurately installed, both these conditions, good alignment and panels running exactly perpendicular, should exist. If it is off, align the panel perpendicular with the joists and, if need be, cut a small piece of subflooring to fill in the gap left by improper alignment. You may want to pop some chalklines that run perpendicular to the joists to assist you in laying the panels if the floor frame is not running true.

Be sure that you are pointing the good (smooth) side of the plywood towards the sky and the bad side towards the ground. The tongue edge should face the outside of the house as you apply, not the inside. You will need to pound on the edges of the plywood in order to get the tongue and grooves to interlock and it is always best to be pounding on the groove edge.

After the first panel is laid into place, use your chalkline and pop lines across the panels directly above the center of the joists. Use threaded 8d nails, nailing every 12 inches in the field and every 6 inches along the edge. Nail while the glue is still wet. Do not nail a few nails to hold the sheet in place and then return later to complete the nailing process. Nail it well at this time so that the panel presses against the joists and spreads the glue.

Some joists, you'll find, are simple field joists; no edge of any plywood sheet will break on them. Other joists will be used to support the ends of the panels. Since the joints are staggered in adjacent courses, one course's field joists may be the adjacent course's end joists. It is best to leave all end joists unnailed until the adjacent course has been installed, in case you need to push an end joist forward or backward a little to align its center with the end of the plywood panel.

Now continue to lay the rest of the course you have been working on. It is usually best to lay one entire course at a time, though this is not mandatory. As you lay the remaining sheets in each course, be sure their inside edges are in a straight line (a string stretched taut helps here) and be sure the ends of the panels fit tightly.

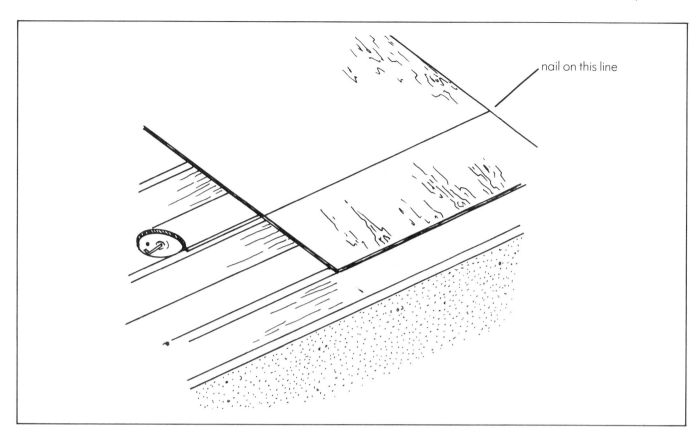

Figure 10-7. Popping a chalkline to mark the center of a joist on a plywood subfloor panel.

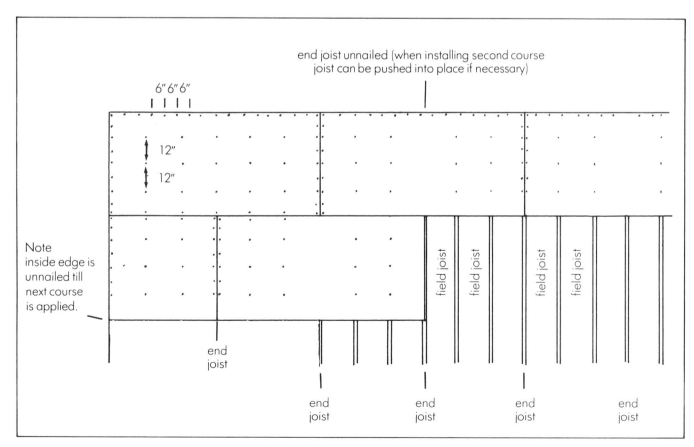

Figure 10-8. Subflooring application showing the nailing pattern. Note that the staggered splices and end joists are left unnailed until the adjacent course is laid.

Photo 10-5. Laying remaining sheets. Note the plywood has been cut for underfloor utilities. (Not required in many areas.)

Photo 10-6. Subfloor further along.

Photo 10-7. Using sledge and block of wood to tightly join T&G plywood.

Step 4
Lay the Remaining Sheets

Margin of Error: Snug fits between panels at all edges (except where spacing is recommended for possible swelling); edges of sheets aligned in a straight line — ¹⁄₁₆ inch.

Most Common Mistakes: Sheets not fitting together well or properly aligned; not staggering the joints during the application.

Photographs: Photos 10-5, 10-6, 10-7, 10-8, 10-9.

You are now ready to lay your next course. The ends and the edges of the panels should ordinarily fit tightly, but there is an exception to this. In many areas of the country it is recommended that you leave a ¹⁄₁₆-inch gap between the ends of the panels and a ⅛-inch gap at their edges. This is to ensure that should the panels get soaked during construction and swell, they will not buckle. If you will be able to close up the house (put roof and walls on it) before the subfloor gets soaked, such as in areas of the West where there are long, dry building seasons, or if you are willing to cover the subfloor with plastic in case of rain, the gap can be eliminated.

In order to get the tongue and grooves to interlock (assuming you are using T & G, not butt-end plywood), you will often have to force the tongue into the groove. Use a sledge hammer and a block of wood to prevent dinging the edge of the plywood. Sometimes this will take a little effort, but the sheets must fit snugly. The sheets are often manufactured to have a small surface gap even when the tongue has bottomed out in the groove. Don't bust a gut trying to close that gap; when you can feel the tongue hit bottom, stop hammering.

Photo 10-8. Marking "wild ends" of subfloor before cutting in place.

As you get to the end of the floor frame, the last sheet can either extend out to be cut after it is nailed or be cut before it is installed. Don't forget to apply the glue before nailing and be sure the good side is pointing up. As you start the second course, begin with a half (4-foot-long) sheet. If you begin every other course with a half sheet, this will automatically stagger your joints throughout the system. Continue applying the rest of the sheets in the manner described. If your last course will not take a full sheet, you will have to rip sheets to fit.

Photo 10-9. Finished subfloor.

Photo 10-10. Second-floor subfloor applied with area cut for stair entrance.

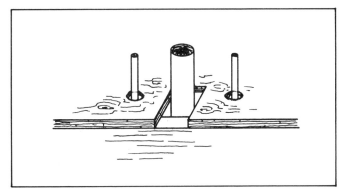

Figure 10-9. Notching subfloor to fit around underfloor utilities.

Special Situations

Subfloor Application and Underfloor Utilities

Diagram: Figure 10-9.

Photograph: Photo 10-10.

In some areas of the country, especially areas of the West, the building codes will require that you install all underfloor utilities *before* the subfloor is installed. This is to allow the inspectors to inspect the systems without having to crawl underneath the building. If this applies in your area but you have a high crawlspace that is easily accessible, often the inspector will waive this rule and allow you to erect the shell and later install the underfloor utilities. I think it is best to get the waiver if possible; installing the underfloor utilities before the subfloor has a few drawbacks:

1) You have to decide the exact location of your plumbing fixtures when only the joist are in place, and fix them in place;
2) Your subflooring is harder to apply when notches and holes will need to be cut to allow for the pipes to poke through;
3) The bottom plates of your walls will need to be notched to allow for these pipes;
4) You have to do your rough plumbing piecemeal; some at this stage, and the above floor part at a later stage.

Worksheet

ESTIMATED TIME

_____

_____

_____

_____

CREW

Skilled or Semi-Skilled

_____ Phone _____

_____ Phone _____

_____ Phone _____

Unskilled

_____ Phone _____

_____ Phone _____

_____ Phone _____

_____ Phone _____

MATERIALS NEEDED

Plywood, subflooring adhesive, nails.

ESTIMATING MATERIALS

PLYWOOD:

No. of sq. ft. of floor _____ divided by 32 sq. ft. (4-×-8 plywood sheet) + 15% waste

= _____ sheets

ADHESIVE:

One tube covers _____ sq. ft. of subflooring

Total no. of tubes needed _____

NAILS:

_____ pounds cover _____ sq. ft.

Total no. of lbs. needed _____

COST COMPARISON

_____ Store:

$ _____ per sheet = $ _____ total

_____ Store:

$ _____ per sheet = $ _____ total

_____ Store:

$ _____ per sheet = $ _____ total

_____ Store:

$ _____ per sheet = $ _____ total

PEOPLE TO CONTACT

If underfloor utilities need to be inspected, do any of the following need to be contacted about doing their work?

Electrician _____

Phone _____ Work Date _____

Plumber _____

Phone _____ Work Date _____

Sheet metal installer (Heating and Air Conditioning)

Phone _____ Work Date _____

Solar installer (for any pipes or panels or hybrid passive solar ductwork) _____

Phone _____ Work Date _____

Do any suppliers need to make deliveries before beginning?

☐ YES   ☐ NO

If YES, Who _____

Delivery date _____ Time _____

Inspector (if needed) _____ Phone _____

Daily Checklist

CREW:

_____ phone _____

_____ phone _____

_____ phone _____

_____ phone _____

_____ phone _____

TOOLS:

☐ framing hammer

☐ chalkline

☐ power saw

☐ hand saw

☐ pneumatic nailer (optional)

☐ pencil

☐ nail belt

☐ tape measure

☐ framing square

☐ caulking gun

☐ sledge hammer or maul

☐ nail-pulling tools

☐ saw blades

☐ extension cords

☐ plug junction box

RENTAL ITEMS:

☐ air compressor

☐ nailing gun

MATERIALS:

_____ sheets of PLYWOOD

_____ cartridges of SUBFLOORING ADHESIVE

_____ lbs. of SUBFLOORING NAILS

DESIGN REVIEW:

Type of subflooring to be used _____

_____

Where to use plug and touch sanded plywood _____

_____

Do builders in your area leave gaps between panels to avoid buckling?

☐ YES   ☐ NO

If YES, how much?

_____ at end _____ at edges

Thickness of plywood subflooring _____

Nailing pattern (refer to code book)

_____ " in the field _____ " at edges

Tongue and groove plywood or regular edge _____

INSPECTIONS NEEDED:

Underfloor utilities inspection needed?

☐ YES   ☐ NO

Waiver OK'd by _____

Inspector's Name _____

Phone _____

Time and Date of Inspection _____

WEATHER FORECAST:

For Week _____

For Day _____

If heavy rain is expected cover the subflooring with polyethylene, both the sheets that have been applied and those stacked and waiting to go on. This will prevent expansion and warping. On windy days be careful while lifting panels.

11 Wall Framing

Photo 11-1. House with first-floor walls in place.

What You'll Be Doing, What to Expect

WALL FRAMING (Photo 11-1) is a very exciting process in housebuilding. In a very short time, the form and skeleton of the house begin to take shape. Each day the house becomes more of a reality as its outline takes form against the sky and the trees. It is a process to be savored and enjoyed.

Though simpler than doing a complex roof, wall framing (Figure 11-1) is somewhat more difficult than floor framing, but floor framing has prepared you well for this experience. Since most of the framing pieces are 2×4s and 2×6s, there are few physical demands and little strain during this process. You will, however, need to be absolutely clear on the size and location of all your wall openings for doors and windows. This means that a lot of very crucial decisions are being finalized; decisions that will make a great impact on the way it feels to live in the house. Study well the information in "Special Situations" at the end of this chapter to acquaint yourself with all the variables in wall framing.

Jargon

BOTTOM OR SOLE PLATE. A horizontal base member to which the bottoms of the studs are nailed.

CAP PLATE (DOUBLE TOP PLATE OR DOUBLER). The second and uppermost horizontal member that is nailed to the top plate.

CRIPPLE STUDS (CRIPPLES). The shorter studs that are nailed below the window sills or sometimes above the headers.

HEADERS. Horizontal wood framing members that support and carry the load over a window or door opening in the wall frame. Usually 4×4s to 4×12s are used.

JAMBS. Vertical posts or pieces of a door or window frame.

PARTITION CHANNEL. A pattern of studs built into a wall to receive an intersecting interior wall at a 90° angle.

SILL (ROUGH SILL). The lowest horizontal member of a window opening in the wall, usually a 2×4 or 2×6 lying flat.

STUD. One of a series of wooden wall framing members, either 2×4s or 2×6s.

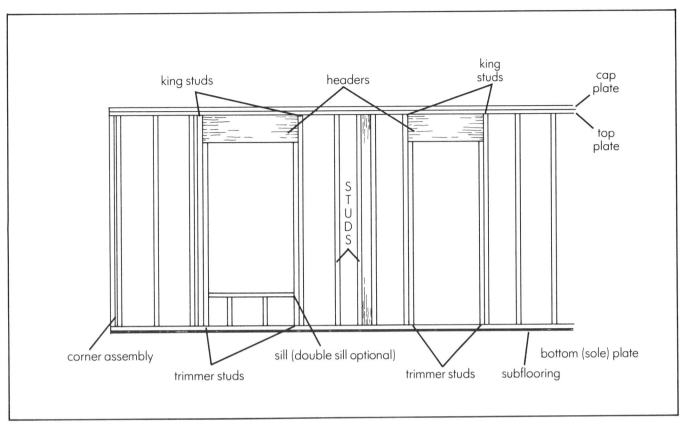

Figure 11-1. The components of a wall frame.

TOP PLATE.   A horizontal member to which the tops of the studs are nailed.

TRIMMER STUD (SHOULDER STUD OR JACK STUD).   The vertical wall framing member that supports the header and adds strength to the side of the opening.

Purposes of Wall Framing

1) Supports the load of the roof and carries this load to the floor frame, the foundation, and ultimately, the ground
2) Encloses the house and provides a framework on which to hang interior and exterior skins, doors and windows
3) Creates hollow cavities in which to place plumbing and an electrical system and insulation

Design Decisions

In the design classes at the Center, I have often said that a house that turns out looking exactly like its blueprints is a house whose design stopped growing during its most fertile growth period: the construction period. Many things become clear during this stage. The platform is complete and the actual location and sizes of the windows, doors and interior partition walls can now be evaluated.

Most builders build true to the blueprints, seldom making changes during this incredibly creative period. They are being paid to build it exactly as the plans read. But now is the best time for you to consider making changes. Will the view be better if a window is enlarged or moved? Will the room feel more intimate if a planned window is deleted; will the room feel safer if a door is moved? Locate the interior partitions on the floor and get a sense of the size of the room, now that you can visualize it. Stand where the wall is to be built and ask: Is this where I want this wall to be? Visit houses with rooms that have different height walls and ceilings. A wall 9 to 11 feet high gives the room a greater sense of space without much energy loss. The house will never be more pliable than at this stage; give it a chance to grow and expand on its original design.

If you are designing and building a passive solar house (in all but the most moderate climate I would advise building no other type of house) you may want to make a final check on your south-facing solar gain windows. This can be done with a "Solar Site Selector." This instrument (Photo 11-2) can give you a reading on how much sunlight will be entering the solar windows at different times of the day, at different times of the year. You may find that by moving an opening slightly to the left or right or up and down you can increase the amount of sunlight entering the house. If there are windows that are expected to admit sunlight during certain parts of the day for lighting purposes, e.g., sunlight in the morning in the kitchen, sunlight in the afternoon in the workshop, etc., you can observe the path of the sun to be sure it will indeed come through the openings. Remember to account for changes in the sun's height in the sky at different times of the year.

Photo 11-2. Using a site selector to check solar gain through windows on the southern wall.

Even after a wall is up, it is not too late to change it if it doesn't feel right. Granted, this will be harder to do; you will never want to re-do *any* of your work, especially when you are on a tight schedule. There were many times when I sensed that the decisions I made on window, door or interior wall placement did not work after I had erected them, but I did not want to go back and change them. Later, after the house was completed, I wished I had taken the 4 extra hours or even a full day to get it like I really wanted it. What are a few extra days of work, when you may have to live with the decisions for many years?

You may want to allot yourself a certain amount of time at the outset, say 20 working days, for re-doing any areas of the house you do not feel comfortable with after they are done. The house may take 20 days longer to finish (a small percentage of the overall time), but it will be what you really want, and isn't that what building your own home is all about?

I did a lot of design changing on the last house we built. The carpenters often threw their hands, and sometimes their hammers, up in exasperation at the idea of having to re-do their work. In the end I was glad I fought for the changes. The house pleases me and there are few compromises.

If you plan to change the location of an interior load-bearing wall, be sure any needed underfloor supports (double joists, girders, piers, etc.) are provided.

2x4s or 2x6s?

If you are planning to build a standard stud house, as opposed to a post and beam house, a dome, etc., one major decision you will have to make is whether the exterior walls should be constructed of 2×4s or 2×6s. (Sometimes the code will require the use of 2×6s in the first floor of a three-story structure, for instance.) More and more houses are

being built with 2-×-6 walls because they have the ability to hold more insulation (6 inches instead of 4 inches), reducing heat loss and, consequently, energy bills. Some builders are using 2-×-4 walls with an added layer of styrofoam insulation on the outside surface of the studs to increase the energy efficiency of the home (see Chapter 12, *Wall Sheathing*, p. 265).

The 2-×-6 wall increases your energy efficiency and gives the house a sense of greater strength since the walls are thicker. Also, their wider cavities make plumbing and electrical systems easier to install. Except for moderate climates where 2×4s are sufficient, we are encouraging our students to use 2×6 walls; we expect energy rates to continue to rise. Most window manufacturers now make windows to fit 2×6 walls, though the windows will cost more due to their wider jambs. Ask the energy-minded builders in your area what they recommend. Situations in wall framing having to do with 2-×-6 walls will appear at the end of the chapter under "Special Situations."

After you have made your design decisions, check and then *recheck* the sizes of all the openings (rough openings). Often builders build these openings to the wrong specs and it is not till much later, during the window and door installation stage, that they discover the error. Then it is much harder to repair. As you lay out the wall, go slowly and double and triple check your work; it will save many hassles later. Unless you are building very complex walls with many angles or slopes, wall framing is challenging but not too difficult once you get the hang of it.

What You'll Need

TIME 8 to 16 person-hours per 30 linear feet of wall 8 feet tall

PEOPLE 2 Minimum (1 skilled, 1 unskilled); 3 Optimum (2 skilled, 1 semi-skilled); 6 Maximum (2 teams of 3 each).

TOOLS Framing hammer, tape, sledge hammer, string, chalkline, power saw, hand saw, pencil, nail belt, combination square, framing square, plumb bob, 4- or 8-foot level, nail-pulling tool, com-a-long (hand winch), architect's rule, extension cords, plug junction boxes, saw blades.

RENTAL ITEMS: Ladders or scaffolding (if you are working on high walls); com-a-long (hand winch) to straighten walls after they are up. (They cost $30 to $40.)

MATERIALS You will need to choose whether to use standard "wet" lumber or kiln-dried lumber. Kiln-dried lumber (Photo 11-3) is somewhat more expensive and a little harder to nail. It does, however, have several advantages that I believe make it worth the added money and energy. This has all been discussed in Chapter 9, *Floor Framing*, under "Materials."

You will, of course, need standard studs. For conventional building, with interior skins made of 8-foot panels of drywall or paneling, the stud length is commonly 92¼ inches, not 96.

The reason for this, as shown in Figure 11-2, is that to determine the length of wall studs in a typical 8-foot wall, you must allow 1½ inches for the bottom plate, 1½ inches for the top plate and 1½ inches for the cap plate. Since the ceiling board rests on the wall boards, ¾ inch, including a margin of error of ¼ inch, is added to the wall height to accommodate an 8-foot board. If from the top of the cap plate to the lower edge of the bottom plate it's 96¾ inches and you subtract the 4½-inch allowance for plates, the stud becomes 92¼ inches.

If you are not planning to use standard 8-foot panels, you can order different length studs in increments of 2 feet beginning at 6 feet. If your wall is 12 feet high or over, consider building 2 walls. For instance, with a 12-foot wall, you can build a lower 8-foot wall supporting a higher 4-foot wall. (See Photo 11-4. Note that the cap plate of the lower wall is the bottom plate of the upper wall.) It is difficult to locate true, straight studs (Photo 11-5) over 12 feet long; the 2-wall process eliminates the need for this.

You will also need plate stock. These are long 2×4s (or 2×6s, if you are building a 2-×-6 wall) that will be used for bottom plates, top plates, and cap plates. These need to be true, straight boards as long as possible (12-foot boards will do, 14- or 16-foot are better, if you can find them). They must be straight, or the walls will be crooked.

You will need header stock which can be built up from 2-inch material or in many areas can be ordered in one solid piece, 3½ inches wide.

You will need nails, 8d, 10d and 16d CC sinkers (cement-coated sinkers). Steer clear of box nails or vinyl-coated nails. Box nails have thinner shafts than common nails, and vinyl-coated nails do not have the holding power of CC sinkers. In earthquake, tornado or hurricane areas, metal fasteners may be required or advised. If you are using wet lumber you may want to consider framing with threaded nails for added holding power.

ESTIMATING MATERIALS NEEDED

STUDS: Figure the number of studs needed per linear feet of wall using 12-, 16- or 24-inch centers. Add studs for the corners, king studs, trimmers, partition channel assembly, sills, cripples and blocking. Add 15% for waste.

PLATES: Use 10-foot, 12-foot or 14-foot lengths as needed to cover linear feet of wall. Add 15% for waste.

HEADERS: Exact linear feet needed. Remember, headers are 3 inches longer than rough opening widths.

NAILS: 1 pound of 16d sinkers per linear foot of wall, and ¼ pound of 8d sinkers per linear foot of wall.

INFORMATION This chapter deals with standard stud construction for both 2-×-4 and 2-×-6 walls. In most houses that are stud constructed, this will be all that is needed. If you have some tricky or very complex wall-framing situations, you may want to consult with a local builder before beginning. Be sure you check all the door and window rough-opening sizes with your manufacturer's catalogue.

Photo 11-3. Kiln-dried studs.

Photo 11-4. A 12-foot-high wall erected with a standard 8-foot-high frame plus a 4-foot-high frame.

PEOPLE TO CONTACT If you have made any *major* changes from your original set of plans, contact the inspection office to notify the governing bodies and get their OK. Minor changes such as changing the size or location of doors or windows often do not need to be approved.

INSPECTIONS NEEDED Usually no inspections are needed before or during this stage of construction, unless you have made major changes.

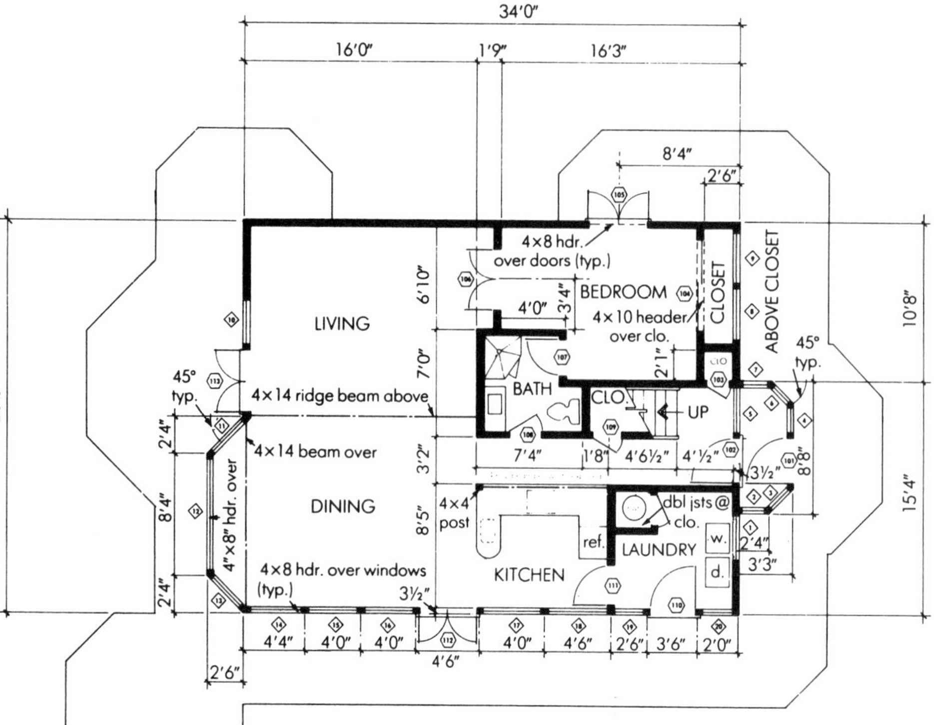
34'0"
16'0"
1'9"
16'3"
8'4"
2'6"
4×8 hdr. over doors (typ.)
BEDROOM
4'0"
3'4"
6'10"
4×10 header over clo.
CLOSET
ABOVE CLOSET
LIVING
7'0"
2'1"
45° typ.
45° typ.
4×14 ridge beam above
BATH
CLO.
UP
4×14 beam over
7'4"
1'8"
4'6½"
4'½"
3½"
3'2"
DINING
4"×8" hdr. over
4×4 post
8'5"
dbl jsts @ clo.
ref.
LAUNDRY
w.
d.
2'4"
3'3"
8'8"
KITCHEN
4×8 hdr. over windows (typ.)
3½"
26'0"
2'4"
8'4"
2'4"
10'8"
15'4"
4'4"
4'0"
4'0"
4'0"
4'6"
2'6"
3'6"
2'0"
4'6"
2'6"

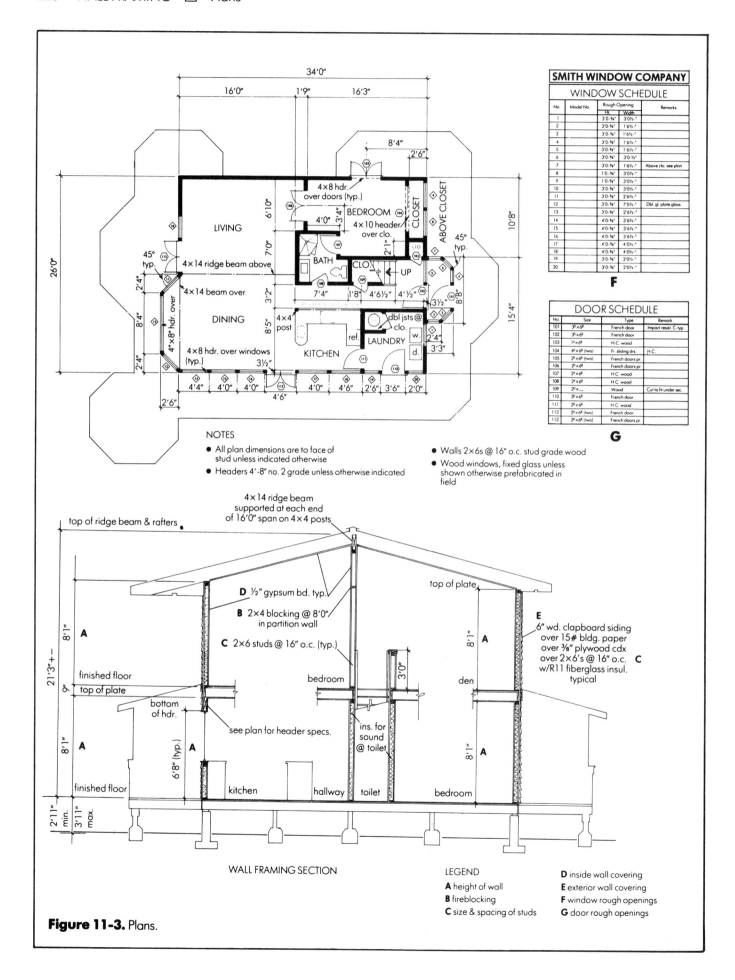

SMITH WINDOW COMPANY

WINDOW SCHEDULE

No.	Model No.	Rough Opening		Remarks
		Ht.	Width	
1		3'0-⅜"	3'0½-"	
2		3'0-⅜"	1'6½-"	
3		3'0-⅜"	1'6½-"	
4		3'0-⅜"	1'6½-"	
5		3'0-⅜"	1'6½-"	
6		3'0-⅜"	3'0-½"	
7		3'0-⅜"	1'6½-"	Above clo. see plan
8		1'0-⅜"	3'0½-"	
9		1'0-⅜"	3'0½-"	
10		3'0-⅜"	3'0½-"	
11		3'0-⅜"	2'6½-"	
12		3'0-⅜"	7'0½-"	Dbl. gl. plate glass
13		3'0-⅜"	2'6½-"	
14		4'0-⅜"	3'6½-"	
15		4'0-⅜"	3'6½-"	
16		4'0-⅜"	3'6½-"	
17		4'0-⅜"	4'0½-"	
18		4'0-⅜"	4'0½-"	
19		3'0-⅜"	2'0½-"	
20		3'0-⅜"	2'0½-"	

F

DOOR SCHEDULE

No.	Size	Type	Remark
101	$3^0 \times 6^8$	French door	Impact resist. C-typ.
102	$3^0 \times 6^8$	French door	
103	$1^6 \times 6^8$	H.C. wood	
104	$4^0 \times 6^8$ (two)	Fr. sliding drs.	H.C.
105	$2^0 \times 6^8$ (two)	French doors pr.	
106	$2^0 \times 6^8$	French doors pr.	
107	$2^4 \times 6^8$	H.C. wood	
108	$2^4 \times 6^8$	H.C. wood	
109	$2^0 \times$ __	Wood	Cut to fit under sec.
110	$3^0 \times 6^8$	French door	
111	$2^0 \times 6^8$	H.C. wood	
112	$2^0 \times 6^8$ (two)	French door	
113	$2^0 \times 6^8$ (two)	French doors pr.	

G

NOTES

- All plan dimensions are to face of stud unless indicated otherwise
- Headers 4'-8" no. 2 grade unless otherwise indicated
- Walls 2×6s @ 16" o.c. stud grade wood
- Wood windows, fixed glass unless shown otherwise prefabricated in field

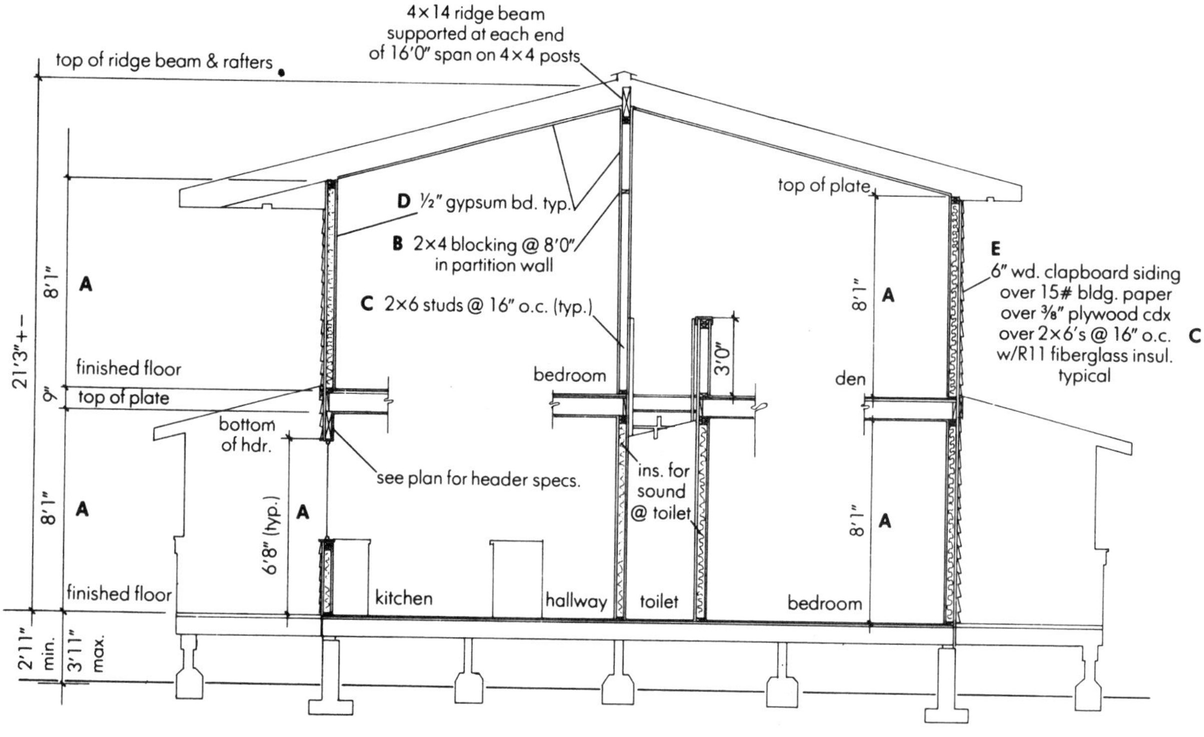
4×14 ridge beam supported at each end of 16'0" span on 4×4 posts
top of ridge beam & rafters
D ½" gypsum bd. typ.
B 2×4 blocking @ 8'0" in partition wall
C 2×6 studs @ 16" o.c. (typ.)
top of plate
E
6" wd. clapboard siding over 15# bldg. paper over ⅜" plywood cdx over 2×6's @ 16" o.c. w/R11 fiberglass insul. typical
C
8'1"
A
21'3"+–
finished floor
9"
top of plate
bottom of hdr.
bedroom
3'0"
den
8'1"
A
see plan for header specs.
ins. for sound @ toilet
8'1"
A
6'8" (typ.)
A
8'1"
A
finished floor
kitchen
hallway
toilet
bedroom
2'11" min.
3'11" max.

WALL FRAMING SECTION

LEGEND

A height of wall
B fireblocking
C size & spacing of studs
D inside wall covering
E exterior wall covering
F window rough openings
G door rough openings

Figure 11-3. Plans.

Reading the Plans

Your plans carry a tremendous amount of information about wall framing. (Figure 11-3.)They must be accurate and complete in regard to this information, unless you plan to be making many of the decisions at the job site. In this case many details can remain vague until you are ready to finalize them. The following is a review of details that will be listed in your plans.

2×4 OR 2×6: Your plans should call out which walls are 2×4s and which walls are 2×6s. The exterior walls may be 2×6s to provide added depth for insulation. The first floor of a three-story building may be 2×6s for greater structural strength. Some interior walls may be 2×6s to allow for plumbing pipes of larger diameter.

HEIGHT OF WALLS: Usually the elevations will note the heights of your walls.

LOCATION OF INTERIOR PARTITION WALLS: Be careful in this area; this is where many plan drawers make mistakes. The dimensions are often wrong, and the total of the distances called out does not equal the total in reality. Many times this mistake doesn't come to light until after you have erected the walls; the 3-foot-wide closet can only be 1 foot wide. Check all room/closet dimensions and total them up now. Take into account the width of each wall, i.e., the width of the studs (3½ inches or 5½ inches) and the wall coverings (½ to 1 inch).

This part of the plans will also tell you where to locate, in any wall, the partition channels for tying intersecting walls together at a 90° angle. Be aware of where the dimension given on the plans refers to, the middle of the partition wall or to one of its edges; a lot of mistakes are created in this area.

LOCATION OF DOORS AND WINDOWS: Your plans will show the placement of all the doors and windows. Often the exact measurements are not given; many times the location of a window is indicated by a measurement to the center line of the window. You may have to determine the exact measurements by measuring on the plan with an architect's rule and converting these measurements into dimensions on the proper scale. Remember, you can reconsider these locations at this time.

SIZES OF DOORS AND WINDOWS AND ROUGH OPENINGS: Your plans usually contain a separate door and window schedule with measurements for each door and window. See "Ordering Windows," in Chapter 17, *Windows*, p. 421. The schedule matches each number with a corresponding door or window size, or a manufacturer's model number. The rough opening sizes for the manufacturer's model numbers are listed in the window or door catalogue. The width is always given first, the height of the rough opening second. For windows and sliding doors you must know the manufacturer, the exact model and the corresponding rough openings.

For most common doors, both interior and exterior, the

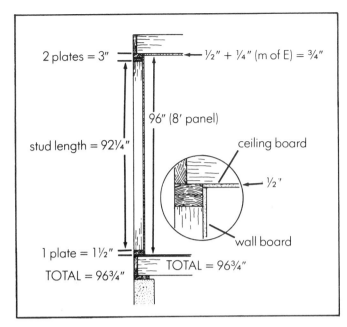

Figure 11-2. Determining the length of the common stud.

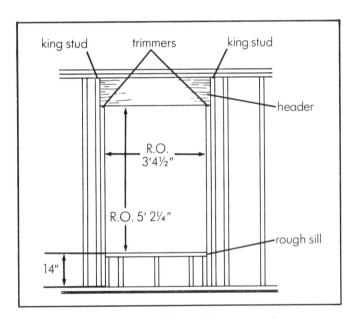

Figure 11-4. Example of from where a rough opening is measured.

plans will give only the width of the door; the height is often not mentioned because the height of common doors is 6 feet 8 inches. The width can range from 12 to 48 inches. These measurements are for the doors themselves, their rough openings are larger.

The width of a door's rough opening is the width of the actual door plus 2 inches. These 2 inches account for a ¾-inch finished jamb board on each side plus ½-inch adjustment room.

The height of the rough opening is the height of the door (6 feet 8 inches or 80 inches) plus a ¾-inch top jamb.

Door measurements are often written on the door schedule in 4 digits, e.g., 2068. This means the door itself is 2 feet, no inches, by 6 feet 8 inches, *not* 20 inches by 68 inches.

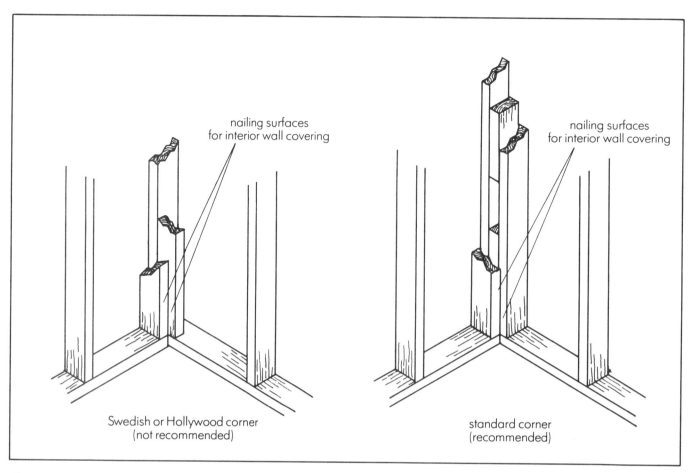

Figure 11-5. Two types of framing corner units.

Example of a typical window rough opening situation (Figure 11-4). Your manufacturer's catalogue calls for a window rough opening that is 3 feet 4½ inches × 5 feet 2¼ inches. Remember that the width comes first and the height next. The width measurement is between the inside edges of the trimmers. The height measurement is from the top of the rough sill to the bottom of the header. In this particular example, the bottom of the window is called out in the plans to occur 14 inches above the subfloor.

SIZE OF FIXED WINDOWS: Fixed windows (ones that don't open) are considered straightforward and often are not keyed to a catalogue or schedule for size. The dimensions of the actual glass of fixed windows are often called out on the plans. Your rough openings are usually these measurements plus 2 inches added to the height and width to allow for the window jambs on all 4 sides (¾ inch thick) and a margin of error (½ inch).

SPECIAL SITUATIONS: Your plans should also indicate, and often detail, any special situations in regard to wall framing. Some of these possible situations may include: load-bearing posts or beams, specially sized headers, fireblocking or other blocking needs, special support for wall-hung toilets or sinks, let-in bracing, chimney and hearth framing, arches, framing for cabinets or ductwork, etc.

Photo 11-5. High wall using long studs. (The horizontal blocks are therefore fireblocking.)

Parts of a Wall

Before beginning the actual construction of a typical wall, I thought it might be best to review in detail the parts of a wall and how they are built. Not only will it assist you in learning the jargon better but it should de-mystify many things about wall framing itself.

CORNERS (Figure 11-5): Outside corners of the exterior wall frame are usually done in one of two methods. As shown in the figure the section is either made of 3 or 4 studs (or 3 studs with some blocks). This is not so much for strength as to provide a nailing surface for the interior wall covering. Blocks are used for center spacers rather than a full stud because the blocks help prevent decay should any moisture get into this area. Usually the entire corner section is assembled, and installed into the wall as it is being built on the floor.

PARTITION CHANNELS (Photo 11-6, Figures 11-6, 11-7): Where interior partitions intersect outside walls, good anchorage is required. Nailing surfaces must also be provided for the interior wallboard on both sides of the partition, as well as on the exterior wall. The figures show 2 typical partition channels. The channel shown in Figure 11-6 is the most common and recommended, even though it will require a few more studs. These channels are preassembled and placed in the wall as the wall is nailed together. The same channels are used at the intersections of interior walls.

DOORS AND WINDOWS (Figure 11-8, Photo 11-6): Probably the most complex part of a wall frame is framing the door and window openings. Actually this is not very difficult at all. You are simply making a hole in the rough frame to hold the finished window or door frame. As you can see, the typical rough opening is made of the following pieces: king studs, trimmers, headers, sills, and cripples.

TRIMMERS: The trimmers in Figure 11-8 are the most typical, though some builders choose to cut through their trimmers and insert the rough sill as shown in Photo 11-6. This tends to hold the sill tightly in place, but it can weaken the wall when the trimmer shrinks or if the cuts are not tight.

To determine the height of a trimmer, determine the distance from the bottom of the header to the top edge of the bottom plate. Trimmers for standard doors and windows in standard 8-foot walls are 81 inches high. This allows the proper height for the opening when using a 4-×-12 header (11¼ inches deep) with a standard 92¼-inch stud (92¼ − 11¼ = 81 inches). Note: on occasion, when dealing with very wide openings such as garage doors, trimmers are doubled or even tripled.

SILLS: Some builders use double rough sills because they will tend to stay straighter as the wood shrinks; the single sills may twist and pull the interior trim off alignment. In some areas if the rough opening is over a certain width, usually 4 to 6 feet, a double sill may be required by code. The

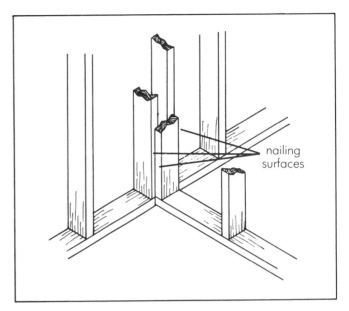

Figure 11-6. A partition channel.

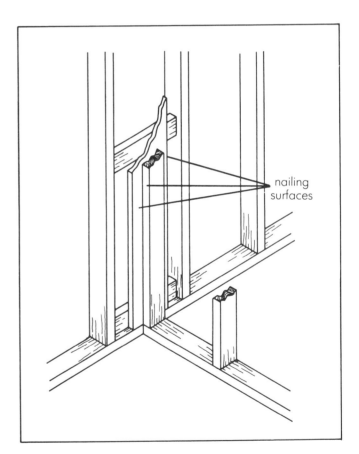

Figure 11-7. A partition channel.

length of the sill usually equals the width of the rough opening.

HEADERS: Headers can either be a solid piece of header stock (3½ inches thick, usually used in the West Coast areas), or a built-up header of 2 pieces of nominal 2-inch materials, or a trussed header.

Photo 11-6. The trimmers are cut and the rough sill is inserted.

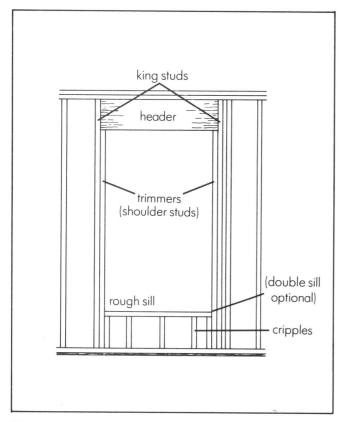

king studs

header

trimmers
(shoulder studs)

rough sill

(double sill
optional)

cripples

Figure 11-8. A typical window rough opening.

Types. Solid headers are most often seen on the West Coast (due to the availability of larger trees and therefore larger lumber stock). Header stock comes in heights from nominal lengths of 4 to 14 inches and is 3½ inches wide to match the width of a stud wall. They save the time needed to build headers out of pieces of lumber. They do warp and twist as they dry, however, more so than built-up headers.

The built-up header is another type commonly used. It consists of two pieces of nominal stock, sandwiched by pieces of ½-inch plywood (sometimes ⅜-inch plywood is used) to provide a header that is 3½ inches thick, the same as the stud wall itself. These pieces are nailed with 16d nails every 16 inches O.C. along each edge from both sides of the header.

Trussed headers (manufactured headers) are used where there is an unusually heavy load or where the span is unusually wide. These headers need to be designed by an architect or engineer and detailed in the plans.

Sizes. The length of the header is always the width of the rough opening plus 3 inches, assuming you are using single trimmers. The size of a header is determined by the width of the opening (this will affect the amount of load carried by the header) and whether the header will be supporting a floor above it or just a ceiling and roof. All exterior walls require full-size headers even if the wall is not a load-bearing wall (e.g., a gable wall). Interior walls will be discussed below. Local building codes usually have requirements for the sizes

HEADER SPANS

size of header	supporting one floor, ceiling, roof	supporting only ceiling and roof
2 × 4	3 feet, 0 inches	3 feet, 6 inches
2 × 6	5 feet, 0 inches	6 feet, 0 inches
2 × 8	7 feet, 0 inches	8 feet, 0 inches
2 × 10	8 feet, 0 inches	10 feet, 0 inches
2 × 12	9 feet, 0 inches	12 feet, 0 inches

the headers need to be. Above is a table that shows normal requirements.

Often, however, nominal 12-inch (11¼-inch) headers are used throughout for all window and door openings no matter what their width and loading. This is done for several reasons. First, in using nominal 12-inch-wide headers you can be sure they'll be sufficient to span almost any size opening in a typical single-family dwelling (except perhaps for the garage opening). Second, by using the nominal 12-inch header tight up against the bottom of the top plate, there is no need for bothersome cripples to be installed above the header. Thirdly, the industry has worked it out so that using a nominal 12-inch-wide header leaves just enough room to install a pre-hung standard size door. Finally, using nominal 12-inch-wide headers for all openings automatically sets the tops of all doors and windows at the same level, giving the house a more symmetrical look. There is one drawback however, 12-inch solid headers tend to crack and warp more than smaller headers, throwing your trim off. Filling the entire space with wood, where there could have been insulation, also increases your energy use.

CRIPPLES: Cripples are provided above the headers (unless nominal 12-inch-wide headers are used) and below the rough sills. These cripples need to continue the 12-, 16- or 24-inch layout pattern of the standard studs in order to support the top plate and/or rough sill, and to provide a nailing surface should 2 pieces of plywood or wallboard join in the area of an opening. Cripples are cut and measured to fill the space from the bottom of the rough sill to the top of the bottom plate and from the top of the header to the bottom of the top plate. There needs to be a cripple at either end of the rough sill to support the sill at the ends. You can draw a diagram and work out these measurements once you know your rough opening sizes and their locations in the wall.

Safety and Awareness

SAFETY: Standard precautions should be taken with your power saw. Watch that the wall does not fall off the platform when you lift it into place. Be careful on ladders while nailing on the cap plate (second top plate).

AWARENESS: Check, check and triple check your rough opening sizes. Be sure stud faces are flush with plate surfaces and headers are flush with stud surfaces. Use good plate stock and your straightest studs around doors and windows.

Overview of the Procedure

Step 1: Check the plans carefully

Step 2: Chalk walls on the subfloor

Step 3: Cut the bottom and top plates

Step 4: Lay out the wall

Step 5: Cut or build studs, headers, trimmer-king stud combinations, rough sills, corners, partition channels and cripples

Step 6: Nail the wall together

Step 7: Temporarily square and brace the wall

Step 8: Lift the wall into place and nail it to the floor

Step 9: Brace and plumb the wall

Step 10: Nail on the cap plates

Step 1
Check the Plans Carefully

Most Common Mistake: Misreading rough opening dimensions.

Photograph: Photo 11-7.

Before beginning construction of each wall, check the plans for the following:

1) Check the R.O. sizes to make sure they're right. Is the window and door schedule accurate? (See "Reading the Plans.")

2) Check opening locations, especially for solar gain, view, etc. Note how location is called out (e.g., is measurement to center or edge of opening?).

3) Check location of 6-inch-diameter plumbing pipes to see if any 2-×-6 interior walls are needed. (See "Reading the Plans.")

4) Check to see if room dimensions on plans will work out in reality.

5) Are wall heights properly called out?

Once all your decisions are finalized, the rough opening measurements thoroughly checked and the solar gain verified, you are ready to chalk the location of the walls on the subfloor. During framing, keep the plans on a table, near to where you are working, and check them at each step.

Photo 11-7. Using an architect's rule to check plans. Often plan dimensions are incorrect; check before building.

Step 2
Chalk Walls on the Subfloor

Margin of Error: ¼ inch off proper location.

Most Common Mistakes: Locating the walls in inaccurate locations; not checking all dimensions after all walls are chalked.

Diagram: Figure 11-9.

Photograph: Photo 11-8.

It is recommended that you locate and mark all the walls, both interior and exterior, on the subfloor with chalklines before you begin the actual wall construction. Many professional builders do this, and I highly recommend it for the novice builder. It gives you a chance to really see where the walls will be erected. All the measurements can be checked at this time to be sure they match the plan dimensions. You will get a better feel of the size of the rooms you have designed and can decide if they still feel right to you.

Another important reason for chalking the exterior walls is to make sure they are erected straight. If you use the edge of the floor platform as a guide, and align the outside edge of the bottom plate with the edge of the platform, the wall can come out crooked if the platform is crooked or bowed. By using a chalkline, you can be sure the wall is straight even if the platform is off.

To set your chalklines for your exterior walls (do these first), measure in at each end of the wall 3½ inches (5½ inches if you are using 2-×-6 walls), make a mark, and pop the chalkline between these 2 marks. If the walls are over 20 feet long, I recommend that you make a mark midway as well. Have a person at each end of the wall and a third at the center, pressing the line to the floor. The middle person then pops the chalkline by picking it up and releasing it first on one side of the center and then on the other.

Once all the exterior walls are chalked, do the interior partition walls. To locate the partitions, measure in from the outside edge of the platform the distance indicated on the plans. Be clear as to which point on the walls the dimensions are measured for, the inside edge, the outside edge or the middle of the wall. There is usually a small arrowhead at the end of the line on the plan, which calls out the dimensions by pointing to an edge or the middle of the wall. Measure in from the platform edge to each end of the wall, make marks and pop a line between these marks. Then locate the other edge of the wall 3½ inches from this line (5½ inches if using 2-×-6 walls), and pop another line. These 2 lines will mark where the 2 edges of the bottom plate will rest. Check the chalklines by measuring from another point of reference, i.e., the other end of the house, to be sure everything is working out. If you used, for instance, the north edge of the platform to locate a wall running east to west, measure again from the south edge of the platform to ensure the proper dimensions.

Once you have chalked all the walls, check and recheck their locations; make sure all the dimensions are working out according to plan. Re-evaluate your design decisions if necessary. *Now* you can start constructing the walls.

The choice of which wall to build first is not a random one. The main consideration in choosing the order in which the walls are built is the availability of open platform space; you need to leave space to build the last walls on the platform, after the first walls are raised. The exterior walls are usually built before the interior partitions. Some builders do not even build the non-load-bearing interior partitions until the house is dried in. If you have a large fiberglass tub/shower enclosure or any other large object that can not pass through door openings, you need to set that in its proper room before building walls around it. I once forgot about this in a house I was building and had to tear out part of a wall to install the tub enclosure.

In any case, consider the sequence of building the walls and remember to leave room on the platform to build all the walls. Usually the longest exterior walls are built and set into place; then the shorter exterior walls are completed in descending order, the shortest being built last; and finally the interior walls are finished, again in descending order.

Some builders erect 2 parallel walls 2 inches farther apart than called for. (The walls hang over the edge of the platform slightly.) These walls are then temporarily nailed and braced. This allows more working room to place the walls that fit in between. Once the in-between walls are built and permanently set, the 2 outer, parallel walls are moved into their proper places and nailed to the subfloor.

Step 3
Cut the Bottom and Top Plates

Margin of Error: ¼ inch off proper measurement.

Most Common Mistakes: Not using very straight stock; not cutting top plate so it breaks at the center of a stud.

Diagrams: Figures 11-10, 11-11.

Photographs: Photos 11-9, 11-10.

Having chosen the order or sequence of building your walls, you are now ready to build your first wall. All walls are built horizontally on the platform and then lifted into place. The first step is to cut the bottom plate and top plate (but not your cap plate, the uppermost of the 2 top plates). After these are cut you will lay out the wall on these plates by spreading them the appropriate distance apart and placing the adjoining pieces between them. Everything is then nailed together and the wall is lifted into place. All of this will be discussed in the remainder of this chapter.

In cutting your plates, you need to use very straight stock. If your plates aren't straight, your wall won't be either. Order good straight plate stock 12 to 14 feet long. Usually if it is longer than this it will have some warp or bow. Ask your lumber yard; they might have some straight 16- or 18-footers. Check each piece to be *sure* it is relatively straight.

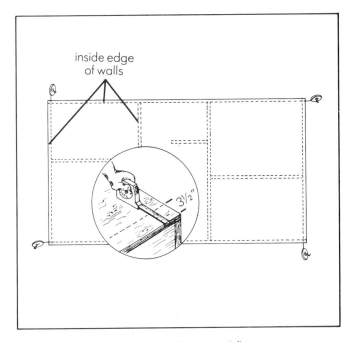

Figure 11-9. Walls being chalked on subfloors.

Photo 11-8. Snapping a line on the subfloor to mark the inside edge of a wall's bottom plate.

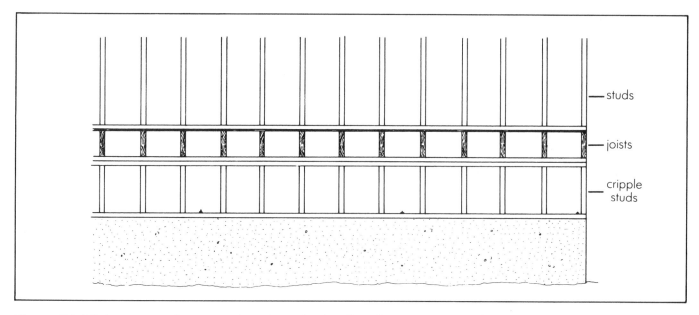

Figure 11-10. Lining up studs, joists and cripple studs. This allows for easy passage of the plumbing, electrical, and heating systems.

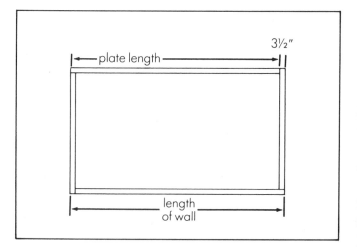

Figure 11-11. The bottom and top plate lengths are usually equal to the length of the wall, less the width of a bottom plate (3½ inches).

Photo 11-9. Top plates meeting over a stud. (Cap plates can meet anywhere, but must be at least 4 feet from any splice in the top plate.)

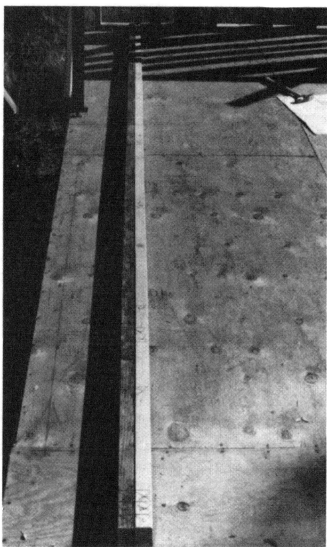

Photo 11-10. Top and bottom plates being marked along edges.

This can be done by using a taut piece of string on the plate to spot any curve along its length, or by setting the plate on its edge on the subflooring and looking for any gaps created by a bow. A plate that is bowed out of true by ½ inch or less is acceptable since no board is totally true.

Since the walls are generally longer than 14 feet, you will need to use several pieces of plate stock to make up the entire plate. There are a few rules when doing this.

1) Avoid very short pieces, under 6 feet.

2) Two separate plate pieces can join anywhere in the bottom plate.

3) The splices in the bottom plate stock, however, must allow at least 4 feet between any splice in the top plate.

4) Any splice in the top plate stock must break in the center of a stud, not in between studs, to ensure the strength of the plate.

5) Splices in the cap plate can occur between studs but must be at least 4 feet from any splice in the top plate.

You can begin piecing together the entire bottom plate, but the top plate is added as you go along. In this way you will know where your studs are located before cutting the top plate stock and can plan it so it breaks at the center of a stud. Generally, the total length of each plate equals the total length of the wall less 3½ inches (5½ inches for 2-×-6 walls), the allowance for overlapping 2 walls where they meet. When the end of a wall goes to the end of the platform on one side, its other end stops short of the platform edge on the other side and is overlapped by the intersecting wall, which extends to the edge of the platform.

Cut and piece together the entire bottom plate, cutting and placing the top plate for the first section only. Don't worry about any door openings in the bottom plate at this time. As you progress in the layout, the top plate will be added as you reach the end of each section. Cut each section of top plate so that it breaks in the center of a stud. (Be careful to be at least 4 feet from any splice in the bottom plate.) Repeat this process until you reach the end.

Another option is to figure out where the studs will be located (the center of a stud will be at every 4- and 8-foot interval to join plywood and drywall boards) and cut all the top plates at once so the splices always fall on the center of a stud. Lay the top and bottom plates together, joining them with a few double-headed nails or clamps, and do the layout on both plates at the same time. Now the top and bottom plates are totally ready to go. I prefer this method because it allows for a continuous flow of work from the beginning of the layout to the end.

Step 4
Lay Out the Wall

Margin of Error: Exact.

Most Common Mistakes: Not dropping ¾ inch for first layout mark; inaccurate measurements; not using straight plates; not breaking top plate at center of stud; not following a continuous unbroken 16-inch (or 12-inch or 24-inch) layout; using framing square for layout.

Diagrams: Figures 11-12, 11-13, 11-14.

Photograph: Photo 11-11.

As with the floor frame layout, the wall frame layout is a process by which you mark the exact locations of the studs, cripples, king studs, trimmers, corners and partition channels on both the bottom plates and the top plates. It is the most crucial part of wall construction, and the one where most of the mistakes are made—so *go slowly* and do good, accurate work.

Lay the plates out on the floor on edge (the 1½-inch side); you will be marking this side. The ends must be exactly flush. Then, join the top and bottom plates together with double-headed nails or clamps so they do not move apart during the layout. You can nail them to the floor temporarily to stabilize them. You will be marking both plates at the same time and the plates need to be in exact position, right next to each other, so the marks will correspond precisely once they are separated.

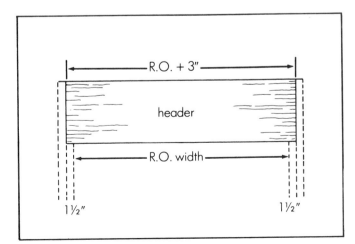

Figure 11-12. The length of a header is equal to the width of the R.O. + 3 inches.

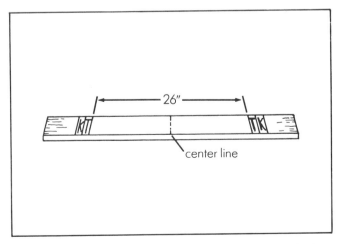

Figure 11-13. A layout pattern stick for a 24-inch door (26-inch-wide rough opening).

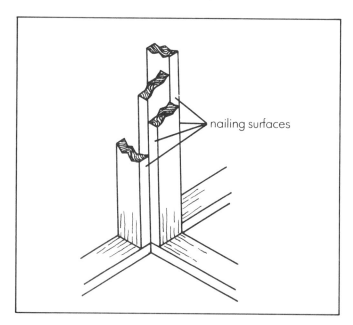

Figure 11-14. Partition channel.

Photo 11-11. Laying out a wall on plates. (It's best to use only one tape to make measurements accurate.)

Here are the symbols you will be using for the different pieces.

> X = Standard stud
> T = Trimmer
> C = Cripples above doors or windows and below windows
> K = King stud

There are several different sequences for doing the layout. I have chosen what I think is the most effective one to teach here. First, decide on which end of the wall to begin the layout. This may or may not be optional. If you are using a knee or cripple wall, you need to begin the wall layout from the same end of the building where the knee wall layout began. In this way, the studs in the knee wall will line up exactly with the studs in the wall above it. This is important when you apply the siding and sheathing.

If there are no knee walls, and the floor joists run perpendicular to the wall you're making, you may want to line the studs up directly over the floor joists. (This assumes that the on-center layout for the joists is the same as that for the studs.) This is of value if you plan to run any ductwork for your heating or cooling systems from under the floor up into the walls. It leaves a continuous unobstructed channel in which to run the ductwork. It also makes running any plumbing pipes or electrical wires easier.

After you have determined from which end you plan to begin, you are ready to start the layout. Let's assume you are using a 16-inch on center stud placement. Measure in from the end of the plates, assuming they are flush with the edge of the floor, 15¼ inches. (The reason for this is explained in Chapter 9, *Floor Framing,* "Step 5".) Using a sharp carpenter's pencil, make a clear mark and drive a nail at this point. Hook the end of steel tape over it (a 16- to 25-foot tape is recommended) and run the tape along the plates. Make a *clear* mark at every 16-inch interval, until you have marked the entire length of the plates. You may need to stop to cut and add new top plate sections if you have not already done so.

You now have marks at every 16-inch interval. Not all of these will be full studs, however. In the door and window openings they will be cripples (except for doors with 12-inch-wide headers, where the cripples are eliminated altogether). After marking one plate, draw a line across both the plates at each mark, but do not mark the line with an "X" yet (indicating stud); you can't be sure which marks will indicate full studs and which cripples yet. (See Figure 11-1.) At the end of the wall where you began the layout, mark the 3 full studs that will make up the corner section, where an end extends to the edge of the platform. The opposite end, which is set in from the edge of the platform, will have only 1 full stud.

Now you are ready to locate the door and window openings and the partition channels. In our example (Figure 11-3), there is a window with a rough opening of 3 feet 4½ inches wide × 5 feet 2¼ inches high. The center of the window is 64 inches from the end of the wall, it has a full 12-inch header above it, and the cripples below are whatever length needed to fill the space from the bottom of the sill to the top of the bottom plate. There is also a door opening for a 3-foot door, so the door itself is 3068 or 3 feet wide by 6 feet 8 inches high. The door's R.O. will therefore be 3 feet 2 inches wide × 6 feet 10 inches high. The center of the door opening is 12 feet from the end of the wall.

To locate the trimmers and king studs for the window, hook your tape at the end of the plates and put a mark at 64 inches, the center of the window. This represents the center of a 3 foot 4½ inch opening, so measure on each side of this mark ½ this distance or 20¼ inches. These 2 marks represent the inside edge of the trimmers. There will be 2 studs, the trimmer and the king studs, *to the outside* of each of these marks. Mark a "T" for trimmer and a "K" for the adjoining king stud. Now mark all the studs between the 2 trimmers with "C" for cripples. The door marking process is just the same, but without the cripples below the sill, though there may be cripples above the headers. Repeat the process for any doors or windows. Check your measurement once or twice, or better yet, have someone else check your work and tell them to look for any mistakes.

Many professional builders use jigs and patterns to make their work go easier. You can make patterns like these for your wall construction and layout, and it will make your work easier, quicker and more accurate. Make a pattern stick for your door layout as shown here. Use a good straight 1×4 for your pattern stick. You can make a pattern for all the common-width doors that you will be using. The pattern shows the centerline, the location of the trimmers and king studs for each width door. Then you can just mark the centerline on your plate and match this with the centerline of your pattern for the door width called for and draw on your plates the locations of the trimmers and king studs. Similar patterns can also be made for the locations of the kings and trimmers for common windows.

At this time check your openings to make sure they all have nailing surfaces for the trim. (See "Special Situations—Providing Nailing Surfaces for Trim.") Mark for any interior partitions. Make sure to place the center of the partition channel where the center of the wall will be. You can ascertain this location by measuring the chalklines on the floor or referring to your plans.

Photo 11-12. Cutting many studs to length in one cut. Standard-length studs in 8-foot walls need no cutting.

Photo 11-13. Eyeing for straightness of studs. The straightest studs are used for headers and trimmers.

Photo 11-14. Cutting a one-piece header, a 4×6 (3½ inches × 5½ inches), is shown. Often 4×12s (3½ inches × 11¼ inches) are used for all headers.

Photo 11-15. Nailing the corner assembly together.

Step 5
Cut or Build Studs, Headers, Trimmer-King Stud Combinations, Rough Sills, Corners, Partition Channels and Cripples

Margin of Error: ⅛ inch for measurement; ⅛ inch joined flush.

Most Common Mistakes: Boards cut to wrong size; using warped or bowed pieces around openings.

Photographs: Photos 11-12, 11-13, 11-14, 11-15, 11-16.

You are now ready to cut all the different components that will make up the wall. Collect all the full studs you will need. In a typical non-custom home you will be using standard-size studs that are 92¼ inches long (see "Materials" for explanation). If you are using another stud length or if you did not order the standard-length stud, you have to cut the studs you will need for the wall. You can cut several studs at a time for greater speed and accuracy. Use the straightest ones around your openings, for trimmers and king studs. This is very important; it makes nailing the final trim to these studs much easier.

HEADERS

If you are using solid 3½-inch-thick headers, you will simply need to cut them to the proper size, the width of the rough opening plus 3 inches. If you are using a typical hand power saw, you will need to make 2 cuts, one from each side, as the blade will not cut 3½ inches deep. Be sure not to leave much of a burr or it will throw your measurement off. If you are building headers, be sure they are nailed with all edges flush.

TRIMMER-KING STUD COMBINATIONS

Use the straightest studs you can find when building trimmer-king stud combinations. Be more concerned about a bow than a warp, because a warp can be nailed out. The king is a full-size stud. The trimmer is cut to whatever length is needed to support the header. In our example, we have a standard wall height, 96¾ inches from the subfloor to top of the cap plate, and a 6 foot 10 inch rough opening height (81 inches) for the door and for the window. The trimmer length, therefore, is 81 inches. This is a standard trimmer height in stud construction if your headers are 11¼ inches wide. Nail the trimmer to the king studs using 16d CC sinkers driven at a slight angle every 16 inches, alternating edges of the studs. Be sure that the king and trimmer are flush all along their edges and at each end. (See "Special Situations: Providing Nailing Surfaces for Trim" at the end of this chapter.)

Photo 11-16. Driving 16d nails at an angle.

Photo 11-17. Separating the plates and placing wall pieces between plates. Note the underfloor utilities in the foreground.

ROUGH SILLS

Rough sills are cut the width of the rough opening unless you plan to break the trimmers for the sills (see "Parts of a Wall: Trimmers"). In this case the sills are cut the width of the rough opening plus 3 inches (1½ inches for each trimmer). If you are using a double sill, cut 2 pieces the same size.

CORNERS

Prepare the corners for the wall by nailing full-length studs together (unless you are using blocking as a filler). Nail these together with 16d CC sinkers every 16 inches (some codes allow every 30 inches). They must be flush on all surfaces. (See "Parts of a Wall: Corners," p. 229.)

PARTITION CHANNELS

Partition channels are made from full-length studs nailed together. All surfaces must be flush. (See "Parts of a Wall: Partition Channels," p. 229.)

CRIPPLES

Where there are cripples above and below openings, they will need to be calculated and cut to length. Be sure there is one for each 16-inch layout mark. There needs to be one on each end of the sill also, to support it (unless the sill penetrates the trimmers). In our example (Figure 11-3), the cripples below the window sills are 28½ inches long. This is a good place to save lumber by using your short scrap pieces of wood. Bracing may be required; see "Special Situations: Bracing."

Step 6
Nail the Wall Together

Margin of Error: Exactly flush where pieces come together.

Most Common Mistakes: Insufficient nailing; studs not nailed flush with plates; headers not nailed flush with king studs; putting the partition channel in backwards.

Photographs: Photos 11-17, 11-18, 11-19, 11-20.

It is relatively simple to nail the wall together *if* your layout marks are accurate and your pieces are the proper size. Plan to build the wall as close to its final location as possible to avoid having to transport it too far. Start by separating the 2 plates a little more than a stud's length apart and placing the different pieces—the partition channels, full studs, corners, etc. —in position. Sometimes you need to nail a plate to the floor or brace it against something to prevent the wall from moving while you are nailing it together. Two people can nail, one at the top plate and one at the bottom plate, in unison, to prevent the wall from jumping around. This can be tricky, though.

Drive 2 nails (16d CC sinkers) through each plate at the ends into every stud. This applies to *each* stud in the corner assemblies, partition channels and trimmer-king stud combinations. Drive 2 nails into each cripple through the sill and plates. Nail the headers in place. Be sure, as you nail, all adjoining pieces have their edges flush with each other; no stud should protrude from the plates nor a header from the king studs. This can cause problems further down the line. Remember too that there is only *one* way the partition channel can be placed: with its center stud receiving the intersecting wall.

Photo 11-18. Nailing the wall together.

Photo 11-19. Nailing in cripples at the end of a rough sill.

Photo 11-20. A stud not nailed flush with the plate.

Step 7
Temporarily Square and Brace the Wall

Margin of Error: Square within ½-inch measurement of diagonals.

Photographs: Photos 11-21, 11-22, 11-23.

Once the wall is nailed together, it is advised that you temporarily square the wall and brace it. This makes lifting easier and provides the wall with some rigidity during moving. To square the wall, stretch a tape measure across diagonally, adjusting the wall until they are equal. Then nail a 1×4 or 1×6 diagonally across the outside surface of the wall to hold it square. The brace should extend from plate to plate. Check to see if each opening is the proper dimensions and if it is square at this time; they are a lot harder to fix when the wall is up.

Photo 11-21. Squaring a wall by checking for equal diagonals.

Photo 11-22. Let-in bracing notched at the top and cap plates. Note the cripples above the header at the left.

Photo 11-23. Let-in bracing installed before the wall is in place.

Photo 11-24. Blocks installed on the outside to keep the wall from slipping off the platform.

Step 8
Lift the Wall into Place and Nail It to the Floor

Margin of Error: Exact alignment with chalkline.

Most Common Mistakes: Wall falling off platform; plate not nailed flush to chalkline; not enough people to lift the wall; nailing the bottom plate in doorways.

Photographs: Photos 11-24, 11-25, 11-26.

The wall can now be raised and set into its proper place. One person for every 10 to 12 feet of wall is necessary to stabilize the wall and make lifting and carrying easier. When the wall is in position, start at one end and drive 1 16d CC sinker through the bottom plate into the subfloor about every 16 inches. If possible, drive the nails into the band joist or any joists below the subfloor. Be sure the inside edge of the bottom plate is aligned exactly with the chalkline. Do not nail the bottom plate at any of the door openings; the plate in these areas will be cut out later. You may want to nail some blocks to the outside edge into the rim joist to stop the wall from slipping off the platform. I once dropped a 22-foot wall off a second story. Fortunately no one was down below. Often a good sledge hammer (the "gentle persuader") comes in handy here to tap the wall into alignment with the chalk-

Photo 11-25. Lifting the wall into place.

Photo 11-26. Nailing the bottom plate to the subflooring and joists.

Photo 11-27. Plumbing a wall with a brace nailed to a cleat on the floor. Note the level being used at the wall.

Photo 11-28. Using a block and string to test the straightness of the cap plate.

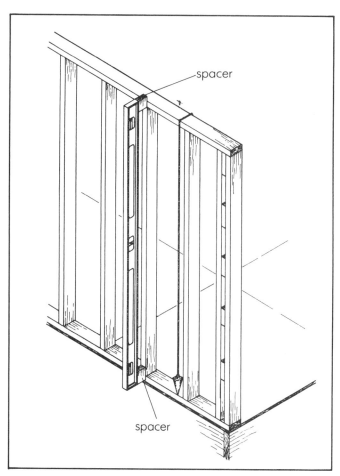

Figure 11-15. Using an 8-foot level with spacers or a plumb bob to plumb the wall.

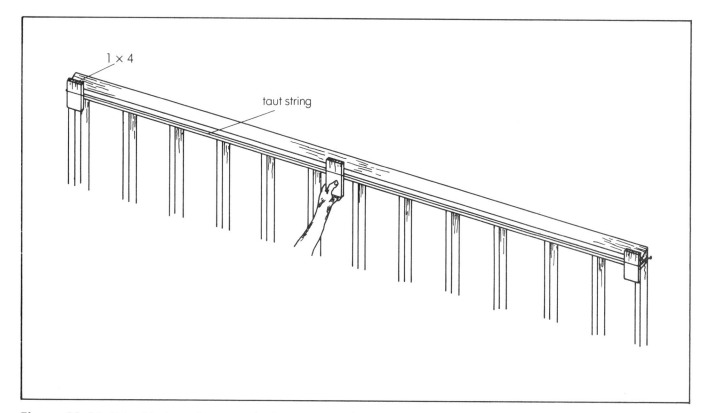

Figure 11-16. Using blocks and string to check straightness of the top plate.

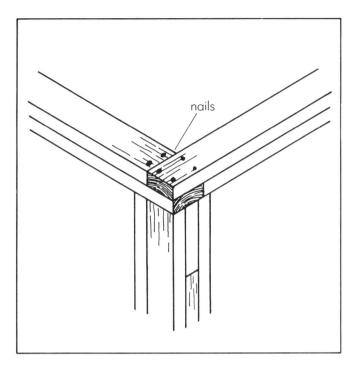

Figure 11-17. How cap plates interlock at corners.

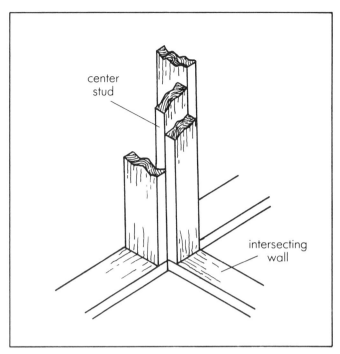

Figure 11-18. Note the location of the center stud.

lines. Always start nailing at one end and work your way to the other. If you have 2 people, 1 at each end, nailing towards the middle, you may trap a bow in the middle that you will never get out.

Step 9
Brace and Plumb the Wall

Margin of Error: Exactly plumb.

Most Common Mistakes: Using an inaccurate level; using a 4-foot level on an 8-foot-high wall; not nailing braces securely.

Diagrams: Figures 11-15, 11-16, 11-17, 11-18, 11-19.

Photographs: Photos 11-27, 11-28.

The wall, now nailed to the floor, will need to be braced and plumbed in all four directions. This is a rather simple task but a very important one. Do not rush through this stage. A few extra minutes will save you hours down the road.

The wall is braced by running diagonal 2-×-4 braces from the top of the wall to the floor where they are nailed to a temporary cleat. Erect a brace every 10 to 12 feet and secure it well. Drive double-headed nails through the brace into the cleats and the tops of the studs. The 2 ends of the wall can be braced with braces running to the outside edge of the building.

It is best to use an 8-foot level with spacers at each end to ride over the irregularities in the stud. A 4-foot level cannot do this and is, therefore, not advised. If you do not have a long level, a plumb bob can be used with equal accuracy.

Work your way down the wall, placing a brace every 10 to

12 feet. To allow for more building space on the platform, you can brace to the outside. Once you have finished, check the top plate for straightness before nailing on the cap plate. You can do this by running a string along the top plate over two 1-inch spacer blocks at either end of the top plate. By measuring the distance from the string to the top plate all along the wall, you can determine how straight it is. Where it is bowed in or out, use your braces and push or pull against the wall to try to straighten it. When it is as close to straight as you can get it and still plumb, nail a brace at the point where the bow was to hold it straight. If you have used good straight plate stock, this should not be very hard to do. If you haven't, you will have to do the best you can; it will probably never be completely straight.

After the wall is plumbed in this direction, check the plumb in the other direction by placing the level on the flat (3½-inch) side of each stud. Rather than adjusting for each stud, try to define a pattern in the wall as a whole, and ascertain where the top of the wall needs to be pushed to your right or your left. In these areas you will have to push against the top or hit the end of the wall with a sledge.

A trick that I have found to work very well is to use a com-a-long or hand winch to slowly crank the wall back into position. This ingenious tool is very useful where a slow, steady, yet powerful force needs to be applied. Attach the com-a-long to the bottom of the wall and the cable end of the winch to the top of the wall at the opposite end. Then slowly draw in the cable until the wall is plumb in this direction. Nail some diagonal braces temporarily across the inside surface of the wall and release the com-a-long. The braces will hold the wall in place until the siding and sheathing are on. Keep checking the wall until you are sure it is plumb in all four directions and the top plate is straight.

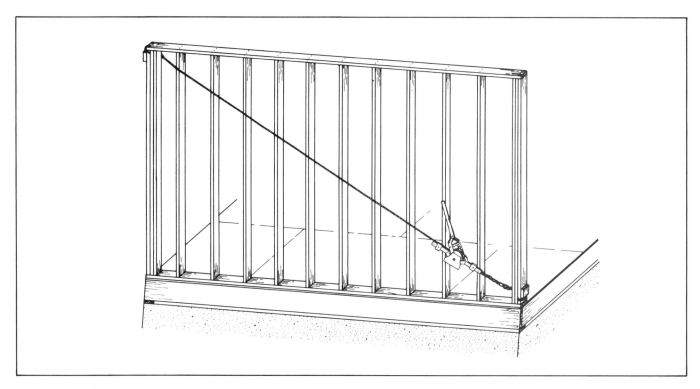

Figure 11-19. Using com-a-long to plumb a wall.

Photo 11-29. Nailing on the cap plate after the top plate is straightened.

Step 10
Nail on the Cap Plates

Margin of Error: Exactly flush with top plate.

Most Common Mistakes: Bowed stock; nails not over studs; falling off the wall; splinters in your rear end.

Photographs: Photos 11-29, 11-30, 11-31.

Steps 1 to 9 are repeated for each wall. When all the walls are up and well braced, you are ready to nail on the cap plates. This is done in a way that it interlocks, and thus strengthens, the intersecting walls. The cap plates lap each other at the corners, and intersect where interior walls meet exterior walls.

Use good, straight stock for these plates. Secure the 2 plates together with two 16d CC sinkers over each stud. By placing the nails over the studs, they will never be in the way of drill bits when you have to drill holes for plumbing and electricity later. Be sure that the edges are flush with the edges of the top plate, and that the cap plates fit tightly to make a strong interlocking joint.

Photo 11-30. A partition wall cap plate intersecting an exterior wall cap plate.

Photo 11-31. Aligning the cap plate nails over the studs to keep them out of the way when drilling holes for the plumbing and electricity.

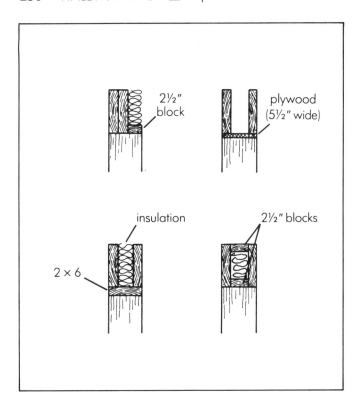

Figure 11-20. Headers for 2-×-6 walls.

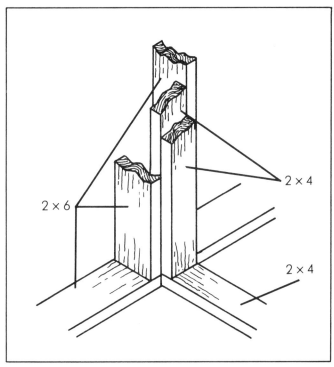

Figure 11-21. A partition channel for a 2-×-4 wall intersecting a 2-×-6 wall.

Special Situations

Unlike floor framing, which is usually fairly straightforward, wall framing and roof framing can have many variables. In this section, I will discuss some of the most common ones, but even this will not be a complete study of the subject. If there are any other situations that you are confronted with, refer to another resource—a builder, book or class that may help you.

2-x-6 Framing
(Figures 11-20, 11-21)

As mentioned earlier in the chapter, there are reasons that make using a 2-×-6 exterior wall frame desirable. In addition to energy benefits, the thicker wall panel provides for a quieter house and gives the house a greater sense of strength. In many areas of the country, a state energy-efficiency tax credit (as well as the existing federal credit) may allow you to deduct from your taxes the difference in cost between building a 2-×-4 wall and a 2-×-6 wall. There will be some added expense for using larger wood studs and switching to 6 inches of insulation (but this may still be cost effective—see *Wall Sheathing,* Chapter 12, "Design Decisions"), and door and window frames will also have to be extended for the thicker wall cavity (most companies do this for a small price increase). There can be some savings, however, if you can switch from a 16-inch center required for 2-×-4 framing to a 24-inch center for 2-×-6 framing. Be careful though, an increased span can require you to go from an interior drywall covering of ½ inch to ⅝ inch. This will definitely cost more.

Should you decide to go with a 2-×-6 wall frame (or if you happen to have an interior 2-×-6 wall to accommodate large plumbing pipes), there are a few things you will need to know. For the most part, 2-×-6 framing is the same as framing a 2-×-4 wall with slight adaptations for partition channels, corner assemblies and headers.

Partition channels: Partition channels for a 2-×-4 wall are built with the 2 side pieces from 2-×-6 stock and the center piece is a 2×4. If the intersecting wall is a 2×6, the center piece is also a 2×6.

Corner assemblies: Remember that the corner assemblies of a wall need to provide both interior and exterior nailing surfaces for the interior and exterior wall coverings. If we only used three 2×6s nailed together, as we do in 2-×-4 framing, the corner assembly's total thickness would only be 4½ inches. This is not enough for the 5½-inch intersecting wall used in 2-×-6 framing. To provide the needed nailing surfaces, 2-×-6 corners are built with two 2-×-6 side pieces sandwiching a 2×4. This gives the corner a total width of 6½ inches (1½ inches for each side, 3½ inches for the center 2×4), and allows for a 1-inch nailing surface on the inside.

Headers: Headers must also be adjusted for the thicker wall panels. They will still need to be the code required depth, but their construction will change for the thicker wall.

Post and Beam Construction
(Photos 11-32, 11-33, Figures 11-22, 11-23)

Earlier in the book, I discussed post and beam construction (see Chapter 2, *Types of Construction*). Even if you are not planning to build a post and beam home, you may want to use some large exposed timbers in a few places for aesthetic

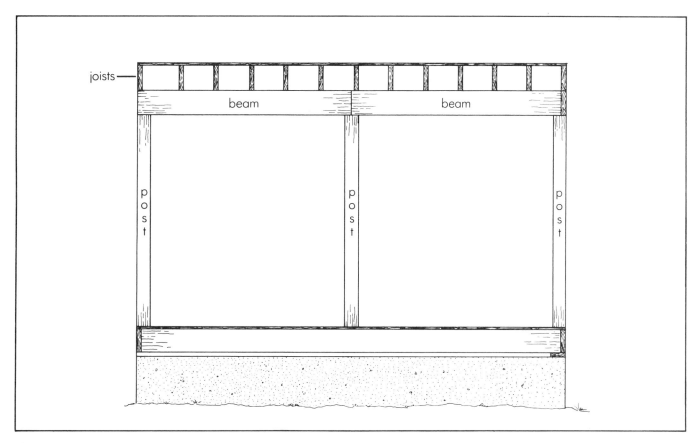

Figure 11-22. Posts and beams being used to support the second-floor frame.

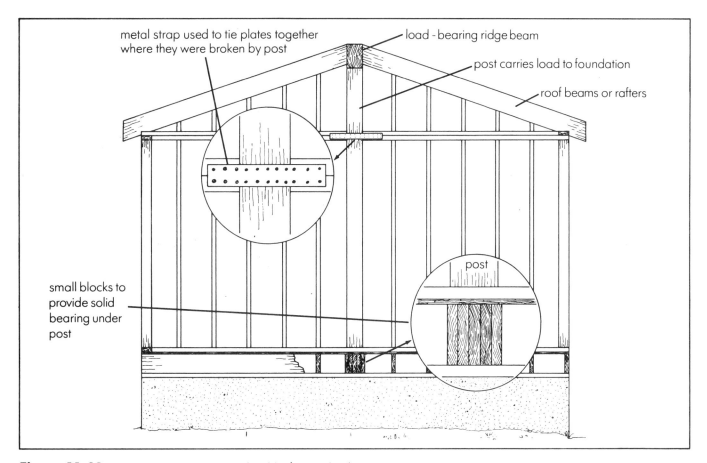

Figure 11-23. Using a post to support a load-bearing ridge beam.

Photo 11-32. Post and beam construction. The load is carried by vertical posts and horizontal beams.

Photo 11-33. A post installed in stud wall to support a roof ridge beam. Note the metal strap used to join plates where they are broken by posts.

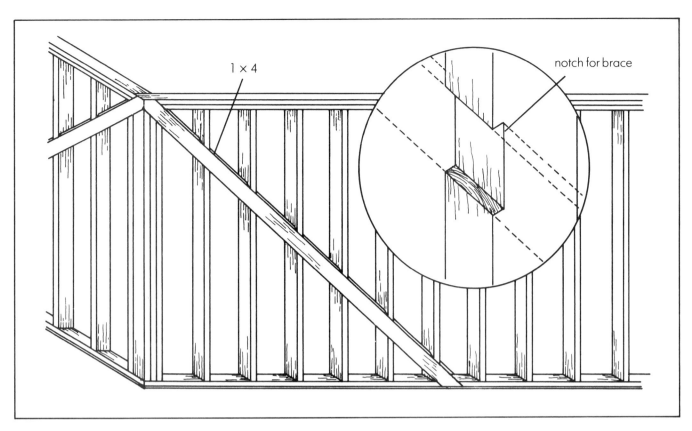

Figure 11-24. Let-in bracing.

purposes and to give the house a greater "sense of strength." Some special structural situations, such as a low sloping roof or large open areas, may also call for some post and beam framing to support the loads.

Post and beam construction is very common in a house that calls for an "open plan" design with large rooms and few interior partitions. The large post can be partially buried in the wall, leaving it partly projecting into the interior. The horizontal large beams can be partially or totally exposed. If your design is like this for the first floor, the load of the second floor frame can be carried by a post and beam situation as shown in the diagram. Another common situation where post and beam construction can be used to carry the load is in a low sloping roof panel where the ridge board has become a load-bearing beam.

In both the above situations, special foundation piers may be needed directly below the posts to help carry the load to the ground. If a vertical post must penetrate the top and cap plates, a flat metal strap is often required to join the two sections of plates together.

Walls over Standard Height

In almost all building there will be walls that are taller than the standard stud length. These can be dealt with in one of two ways. Either long studs can be purchased so that the taller walls can be built with single-length studs, or the walls can be built in 2 sections.

If walls are 10 to 12 feet high and you have access to good straight studs of that length, it is usually easier to build the wall from single-length studs. Often the code may require some horizontal fireblocks for safety and strength. If these are angled walls running from a high point to a low point, the upper end of the stud will have to be cut at an angle. The ends of the top and cap plates of the sloping walls will also be cut at an angle and will not fit as neatly with intersecting walls.

If walls are over 12 feet high, straight studs are hard to find and you may want to consider using 2 walls, a standard height wall and a smaller wall on top of this one. Build the lower wall first, but leave off the cap plate as this will be the upper wall's bottom plate. Brace and plumb the lower wall. Construct the upper wall with its bottom plate, a top plate and a cap plate. Lift it on top of the lower wall and nail through its bottom plate into the lower wall's top plate. Then brace and plumb this wall.

Bracing
(Figure 11-24, Photos 11-34, 11-35)

For rigidity and strength (so the wall frame does not collapse like an accordion) some type of bracing is needed for the walls. You won't need bracing if you are using plywood sheathing, plywood siding of intermediate density, fiberboard, or individual board sheathing applied at a diagonal. (See *Wall Sheathing,* Chapter 12.) If you are not using one of these types of sheathing, you probably will need wall bracing. There are two common types of wall bracing, metal straps and let-in bracing.

Metal bracing is relatively new on the market and may not pass all codes for bracing. It is quick and easy to apply and it

Photo 11-34. Metal diagonal bracing.

Photo 11-35. Metal bracing notched into studs.

only requires that you cut a small notch in each stud and plate. I am not sure of its strength as compared to let-in bracing. You may want to check the codes and ask builders about its strength before using it.

Let-in bracing is a process of cutting notches in the studs and plates and inserting a 1-×-4 brace diagonally. You nail it to each stud with two 8d nails. Each brace should start at a top corner and head down at a 45° angle. If a door or window opening is in the way, it is better to increase the angle rather than break the brace into pieces.

To install, temporarily tack the brace across the studs and plates at the best possible angle and draw lines for notches across the studs and plates on both sides of the brace. Remove the brace and set your saw cut for a depth of ¾ inch (the thickness of the nominal 1-inch brace). Cut on both lines and make an extra cut in the middle. Do this at each stud and plate. Using a chisel remove the wood to a depth of ¾ inch between the cuts. Make good cuts so the brace will be snug. After all notches are complete, cut the brace the proper length and with the proper angles at each end, and insert it into the notches. Nail at each stud and plate with two 8d nails. You can use glue for added strength.

Nailing Surfaces for Ceiling Boards
(Photo 11-36, Figure 11-25)

After you have finished your wall framing for each floor, walk from room to room and think about ceiling boards. Imagine how the ceiling boards will run and be sure there is a nailing surface at each edge. Along the edges where ceiling

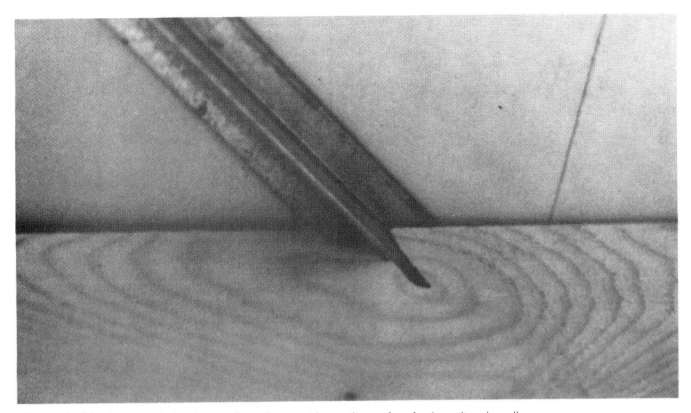

Photo 11-36. A board nailed to the top of a wall to provide a nailing surface for the ceiling drywall.

boards will run parallel to the ceiling joists or the second-floor framing, special blocking will be needed. Refer to your plans to pre-visualize what floor and ceiling framing will go in on the second floor.

Wall Running Parallel to Floor Frame or Floor Above
(Figure 11-26)

If a first-floor interior partition wall is running perpendicular to the floor joists of the second floor, you can toenail the cap plate into the floor joist where the wall crosses the joists. This will help stabilize the wall. If, however, the first-floor wall is running parallel to the second-floor joists, you will either have to arrange special joists on either side of the wall or block every 4 feet between 2 standard joists to provide a nailing surface to stabilize the wall.

Fireblocking
(Photo 11-37)

In the past, fireblocking (blocks placed horizontally between the studs) was used to prevent fires from spreading inside a wall cavity and creating a chimney effect in the space between the studs. Now we fill this cavity with insulation that is fireproof, and use fire resistant drywall gypsum board as an interior wall covering. Top and cap plates also prevent fires from spreading from floor to floor. Because of this, fireblocking is often not used. Consult your local codes to see if it is still required. Some areas require it if the wall is over 10 feet high for strength as well as fire protection. If it is not required, omit it.

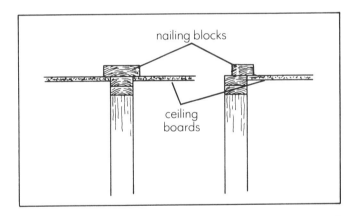

Figure 11-25. Using nailing blocks to nail the edges of ceiling boards.

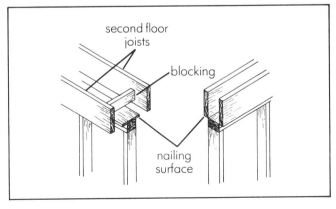

Figure 11-26. Two framing methods for second-floor joists to stabilize a first-floor wall that is running parallel to the joists.

Photo 11-37. Installing fireblocking on a 2-×-6 wall.

Photo 11-38. Sheathing a wall before the erection (courtesy of Heartwood Owner Builder School).

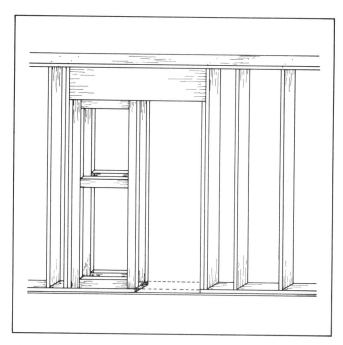

Figure 11-27. Framing for pocket door. Note the total opening must be twice the size of the actual door opening.

Sheathing or Siding the Walls Before Erection
(Photo 11-38)

Some builders sheath or side their walls while they are still on the platform and then lift them into place. I would not recommend this for the novice builder; it allows for less adjustment for mistakes or inaccuracies. It also takes more people to lift and move the wall. If you plan to go this route, check with a local builder as to things you should be aware of before you try it, e.g., that the floor is level, without irregularities; that studs are plumb after wall is lifted; that sheathing protrudes below the sill plate, etc.

Pocket Doors
(Figure 11-27)

Pocket doors are doors that slide and disappear into a wall. They create a pleasing effect and save space since no floor space is required for their swing. If you plan to use a pocket door, you can buy pocket door frames in different sizes from your lumber company. They are common and easy to install. Remember that the rough opening for a pocket door will need to be at least twice the size of the door itself, one space for the door passage and another for an empty pocket for the door to slide into. Always use good quality rollers with pocket doors so they will slide easily.

Wall Sinks, Tubs, Toilets, Toilet Paper Holders
(Figure 11-28)

Sometimes special pieces are inserted between studs for the following fixtures: wall sinks, built-in tubs, wall mounted toilets (details for these mounts are described in the manufac-

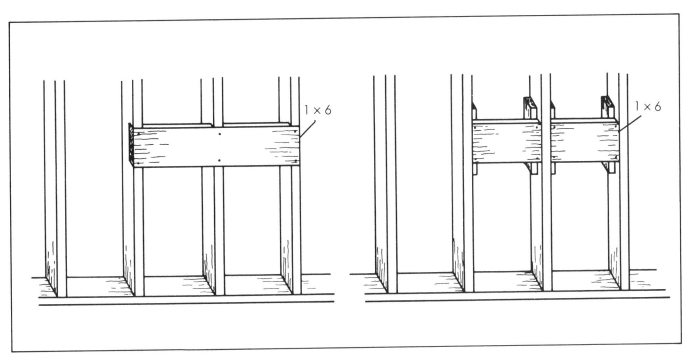

Figure 11-28. The backing for wall mounted fixtures or toilet paper holders.

turer's spec book), toilet paper holders, toothbrush holders, paper towel holders, etc. If you think back on all the times you searched a sheetrock wall for a solid framing member to hang something on, you may want to walk the house and anticipate this by adding an occasional interior horizontal framing mount. Be sure to mark where you have placed all backings so you can locate them after the sheetrock is up.

Building Walls Around Rough Plumbing Stub Outs
(Figure 11-29)

As mentioned in Chapter 10, *Subflooring,* "Special Situations," some codes require that the underfloor utilities be placed in before the subflooring is installed and the walls built. If this is the case in your area, and you can't get the variance, you will need to drill or notch all your bottom plates to allow for the plumbing pipes. If I need to drill or notch the plates, I do it after the wall is built, but before it is erected.

Nailing Surfaces for Vertical Siding
(Figure 11-30)

If you are using individual board vertical siding (such as cedar T & G or board and batten) and especially if there is no plywood sheathing, you will need to provide some horizontal nailing pieces between the studs so the boards can be nailed along their lengths. A piece every 30 inches is recommended. These pieces can be 2×4s installed flat, rather than cross-wise, like fireblocking. In this way you can run batt (rolled fiberglass) insulation directly over them rather than having to cut the insulation at each block. These nailing pieces also make installing your plumbing and electricity easier.

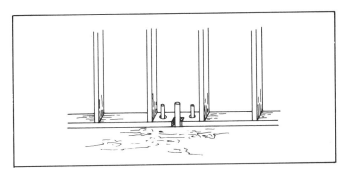

Figure 11-29. Notching or drilling for previously set underfloor utilities.

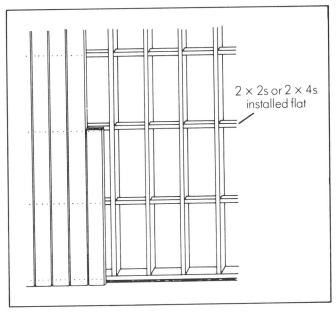

Figure 11-30. Nailing blocks for vertical siding.

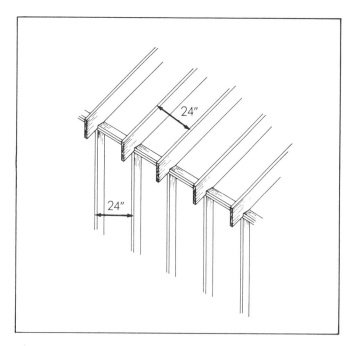

Figure 11-31. Rafters directly over studs on 24-inch centers.

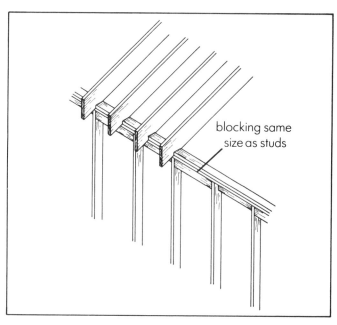

Figure 11-32. Alternate method of blocking when rafters fall between 24-inch o.c. studs.

Photo 11-39. The window and door are built so close together here that trimming will be difficult. No channels are provided to run the wires through so the studs are notched.

An Alternative to Placing Rafters Directly over Studs on 24-Inch Centers
(Figures 11-31, 11-32)

If you are using 24-inch centers for your wall studs, most codes require that the rafters be placed on 24-inch centers and fall directly over the studs. If they occurred between studs, the top (and cap) plates would be too weak to support the weight. If you don't want to match all the rafters and studs, you can place blocks at the very top of the studs, touching the bottom of the top plate. These blocks are installed on edge and alternate flush with the inside and exterior edges of the studs. The blocks are made from the same stock as the wall studs, 2×4s for a 2-×-4 wall, 2×6s for a 2-×-6 wall.

Providing Nailing Surfaces for Trim
(Photo 11-39)

As you build the walls, be sure that there will always be enough space and solid nailing for any window or door trim. This is especially important in walls that have a door right near one end of the wall. If it is built too close to the intersecting wall there will be no room for trim unless you are using a very thin trim. Adding an extra stud to the outside of the king stud can prevent problems. Trim can also be a problem when windows are built right next to each other or by a door. In this case they can share a common king stud if your trim will be working out OK. Be thinking "trim" as you take a look at all the openings.

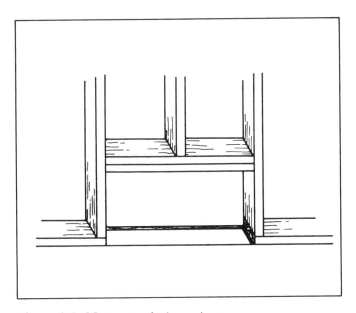

Figure 11-33. Framing for large ducts.

Ductwork
(Figure 11-33)

Most ducts are made to fit between 16-inch center studs. If one of your ducts will not fit, use the method shown in the diagram to accommodate the duct.

Metal Fasteners

In earthquake, hurricane or tornado areas, you may want (or you may be required) to use metal fasteners. It adds strength for holding the house together should any unusual forces be applied. In a catastrophe it could make a difference, and they are easy and inexpensive to install. Special fastener nails are provided with the fasteners.

Soundwall

Noise has a greater impact on our living space than most of us realize. My whole body relaxes when our washing machine finally cuts off. Fortunately our 11 year old took up the clarinet, but it could have been the drums. We lose a lot of peace of mind and sense of privacy due to unwanted noise. Walls (as well as floors and ceilings) can be constructed to minimize this. You can use a sound deadening board, which is easy and inexpensive to install, underneath the wall covering. Also you can use 2-×-6 plates with offset 2-×-4 studs on 12-inch centers to help reduce the passage of the sound waves from room to room (remember to use ⅝-inch drywalls, not ½-inch drywalls, on 24-inch centers). Willis Wagner's *Modern Carpentry* (Goodheart and Wilcox, publisher, 1973) has an excellent chapter on sound control.

Closet and alcove framing

You may want to consider using 2 × 2s, 2 × 3s or 2 × 4s flat to frame closets and alcoves. This will allow more interior space and less space required for the wall panel. For plates you can rip 2 × 4s into 2 × 2 plates. Check with your codes first.

Continuous headers

If your code will allow it, you may want to consider using a continuous header in the wall. It makes wall construction quicker and easier than the usual individual headers IF the wall has a lot of openings. It also allows you to easily change your mind about your door and window decisions. On one house I changed my mind so many times, I though about leaving a map of all the covered up windows and door openings for the buyers in case they ever wanted to add them back in.

Worksheet

ESTIMATED TIME

_____

_____

_____

CREW

Skilled or Semi-Skilled

_____ Phone _____

_____ Phone _____

_____ Phone _____

Unskilled

_____ Phone _____

_____ Phone _____

_____ Phone _____

_____ Phone _____

_____ Phone _____

MATERIALS NEEDED

Lumber, nails, metal fasteners (optional).

ESTIMATING MATERIALS

Number of STUDS needed _____ × _____ ft. long

= _____ total linear feet

PLATE STOCK:

Linear feet of total walls (interior and exterior) _____

Double this (top and cap plates) + 15% waste _____

HEADER STOCK:

Exact linear feet needed. Remember, headers are 3 inches longer than RO widths.

NAILS:

_____ lbs. of 16d per 1000 sq. ft. = _____ lbs.

_____ lbs. 8d per 1000 sq. ft. = _____ lbs.

(Usually 100-150 lbs. of 16d per 2000-sq.-ft. home)

METAL FASTENERS: No. _____

COST COMPARISON

_____ Store:

$ _____ per M board feet 2×4 (or 2×6) = $ _____

_____ Store:

$ _____ per M board feet 2×4 (or 2×6) = $ _____

_____ Store:

$ _____ per M board feet 2×4 (or 2×6) = $ _____

PEOPLE TO CONTACT

If you have made any major changes, contact inspection office.

Name _____ Phone _____

Changes approved _____ Date _____

You may want to contact the following people to let them know their work may soon be coming up, if you are planning to contract it out:

Plumber _____

Phone _____ Work Date _____

Electrician _____

Phone _____ Work Date _____

Solar Panel Installer _____

Phone _____ Work Date _____

Roofer _____

Phone _____ Work Date _____

Insulation Contractor _____

Phone _____ Work Date _____

Window Supplier _____

Phone _____ Work Date _____

Door Supplier _____

Phone _____ Work Date _____

Roofing Supplier _____

Phone _____ Work Date _____

Siding Supplier _____

Phone _____ Work Date _____

Skylight Supplier _____

Phone _____ Work Date _____

Daily Checklist

CREW:

_____Phone _____

_____Phone _____

_____Phone _____

_____Phone _____

_____Phone _____

TOOLS:

☐ framing hammer

☐ tape

☐ sledge hammer

☐ string

☐ chalkline

☐ power saw

☐ hand saw

☐ pencil

☐ nail belt

☐ combination square

☐ framing square

☐ plumb bob

☐ 4- or 8-foot level

☐ nail-pulling tool

☐ com-a-long (hand winch)

☐ architect's rule

☐ extension cords

☐ plug junction boxes

☐ saw blades

RENTAL ITEMS:

☐ Ladders or scaffolding may be needed if you are working on high walls.

☐ A com-a-long or hand winch may be used to straighten the walls after they are up. (These can be purchased for $30-40.)

MATERIALS:

_____ studs

_____ ft. header stock

_____ lbs. 16d CC sinkers

_____ lbs. 8d CC sinkers

_____ metal fasteners (optional)

DESIGN REVIEW:

Be sure that all your window and door location and sizes have been reconsidered before beginning this stage of construction.

1. Be sure each choice is a definite "Yes, that is exactly how I want it to be."

2. Take out your solar site selector (if you have access to one) and be sure any windows or glass doors that are supposed to allow sunlight in for solar or design reasons will function properly.

3. Decide on a window manufacturer and obtain their rough opening sizes before beginning. Any effort you put in improving your design at this stage will pay off a hundredfold later.

INSPECTIONS NEEDED:

Usually no inspections will be needed before or during this stage of construction.

WEATHER FORECAST:

For Week _____

For Day _____

12 Wall Sheathing

Photo 12-1. Wall sheathing applied to a building. Usually the sheathing is applied after the first and second floors are framed. This sheathing was applied early to accommodate the needs of a summer program class.

What You'll Be Doing, What to Expect

WALL SHEATHING consists of boards or panels that cover the wall frame like a skin. They not only provide rigidity and strength, but may also provide a nailing surface for the exterior siding. Sheathing also minimizes air and moisture leakage through the wall panels and adds an element of stoutness to the wall.

Most codes do not require that the walls be sheathed. Generally wall bracing (see Chapter 11, *Wall Framing*, "Special Situations: Bracing," p. 253) is allowed. Even in many earthquake areas it is not required, though I would highly recommend using plywood sheathing in such areas; it is your house's main protection in the event of a major earthquake. Though not required, many builders use sheathing for the advantages it provides.

The different materials used in sheathing include plywood sheathing, plywood siding, three types of fiberboard, rigid insulation and individual nominal 1-inch boards. There is a

complete discussion comparing these different types of sheathing materials later in the chapter.

Wall sheathing, like subflooring and roof sheathing, is a simple, but very monotonous, procedure in housebuilding. Since it will later be covered by the finished exterior siding (unless you are using plywood siding for sheathing), looks, accuracy and the quality of your cuts and fits are not important. You can pay minimal attention and still do a good job. There are few mistakes to make and it goes quickly. Physically, it is a little strenuous since you will be lifting and holding large plywood panels and doing a lot of nailing. It is somewhat easier if you are using fiberboard panels, individual board sheathing, or rigid insulation. If you have a high wall, it is even more physically demanding and you may want to consider using scaffolding.

Keep in mind that this is a rather formative stage that the house is going through. It is similar to entering puberty; kind of awkward and lanky. You are not really sure what the final product will look like. All the rough openings are gaping holes and the material itself is rather unattractive. You may think: "I've created a monster." Indeed, you may have, but this is

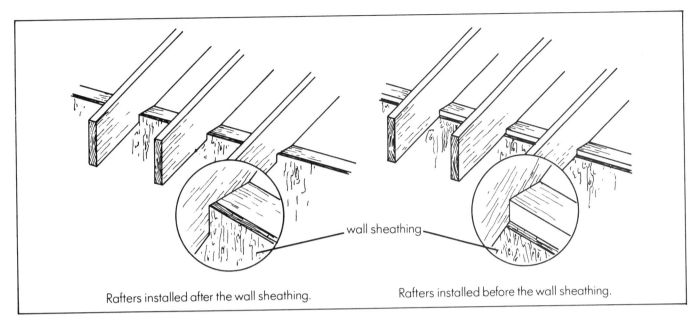

Rafters installed after the wall sheathing.   Rafters installed before the wall sheathing.

Figure 12-1.

really not the stage to judge. It is the most unappealing stage the house goes through.

Wall sheathing can be applied before or after the roof framing, or before or after the roofing is applied. If the siding and sheathing must go between the rafters, the rafters must be put on before the sheathing (Figure 12-1). If there is a need to protect the building from the elements, the roofing can be applied before the sheathing. If you are building for an extended period, you should finish the sheathing and roofing as soon as possible. (Photo 12-1.)

Jargon

ANNULAR-GROOVED NAILS (THREADED NAILS). Nails with grooves on their shanks to provide greater holding power.
END-MATCHED BOARDS. Boards with a tongue or groove at the ends as well as along their sides or edges.
EXTERIOR GLUE PLYWOOD. Plywood that can get wet for a period during construction and not delaminate.
FIBERBOARD. A sheet or panel made of wood fiber impregnated with a waterproofing chemical.
NAIL BASE. A board or panel that can be used as a nailing surface for attaching exterior siding.
SHEATHING. A covering, structural or non-structural, consisting of panels or boards and attached to the exterior of the studs or rafters.
SHIPLAPPED. Panels of boards that have been rabbeted to form a lap joint between adjacent pieces.
TEXTURE 1-11 (T-ONE ELEVEN). A plywood exterior siding that serves as the finished exterior siding and sheathing at the same time.
TONGUE AND GROOVE (T&G). Boards or panels with a tongue protrusion along one edge and a groove indentation along the other.

Purposes of Wall Sheathing

1) To provide rigidity and strength to the wall frame
2) To provide a nailing base for the exterior siding
3) To reduce air and moisture leakage

Design Decisions

A decision as to which material you will use for sheathing needs to be made at this point. Here is a review of the different types of sheathing available.

CD EXTERIOR GLUE PLYWOOD: Since its introduction into the construction industry, plywood has become the most common type of sheathing. It is also by far the strongest material. Usually a CD (refers to quality of ply) exterior glue plywood is used. It can get wet during construction and still not become delaminated, and since it will be covered by the exterior siding, a CD quality surface is all that is needed. Though it is a little costlier than some of the other materials, it takes little time or effort to install. Code requirements in most areas are as follows:

Thickness of Plywood	Inches on center for studs	Direction to studs
5/16 inch	16 inch	perpendicular or parallel
3/8 inch	16 inch	parallel only
3/8 inch	24 inch	perpendicular only
1/2 inch	24 inch	perpendicular or parallel

Most builders use ⅜-inch or ½-inch panels. I recommend the ⅜-inch-thick panels, as the added thickness of the ½-inch does not really add greater strength. The strength is more a function of the nailing and fastening process. As the chart shows, it is not as strong if it is applied horizontally because a

Photo 12-2. Black fiberboard is often used for wall sheathing. It has few structural qualities but reduces air and water infiltration.

great part of its 8-foot edge is left unnailed. This can be somewhat compensated for by placing blocking between the studs in order to provide a nailing surface for this edge. In areas where hurricanes or earthquakes are a problem, the sheathing is glued as well as nailed.

SHIPLAPPED EXTERIOR PLYWOOD SIDING: This is an exterior grade plywood that can withstand constant exposure to the weather. The final exterior ply of the plywood is of a higher grade fir or redwood that is meant to be exposed and serve as the finished exterior skin. No siding will be placed on top of it. It is applied similar to CD plywood sheathing with a few differences (see Chapter 16, *Siding,* "Plywood," p. 393). It is shiplapped to provide a better seal against water that will be running directly on the panels.

Most codes require that it be at least ⅜ inch thick, as there will be no sheathing underneath it. No additional wall bracing is required. Both ½-inch and ⅝-inch thicknesses are common. Most codes require that one ⅜ inch thick has at least 3 plys and can span 16 inches on center studs. Those ½ inch thick must have at least 4 plys and span 24 inches on center.

Shiplapped siding is becoming one of the most common types of siding used in the country. The advantages are that it is quick and easy to install. It does not, however, provide for a very stout wall, and the final finished plys have been known to bubble and delaminate. In material cost it tends to be rather expensive (though its labor application cost is rather low). If you purchase the lower-priced panel, it can have as many as 12 to 24 plugs in the finished surface, and unless you plan to paint over them it looks terrible. The better quality

panels without plugs are very expensive. Some panels require an additional sealant every few years, further increasing the cost.

FIBERBOARD (Photo 12-2): Fiberboard is a wood fiber impregnated with a waterproofing chemical. One side usually has a black tar-like surface and the other is brown. It has little structural strength and can easily be broken over your knee. Fiberboard sheathing is also called insulating board, but this is very deceiving as it has little insulating value. It is often used in homes that do not need sheathing for rigidity and structural purposes but need an air and moisture preventative and in some cases a nailing surface.

Fiberboard comes in three types: regular density, intermediate density and nailing-base density. Regular density is manufactured in both ½-inch and ²⁵/₃₂-inch thicknesses and comes in footage of 2×8, 4×8 and 4×9. Corner bracing is usually required on all fiberboard sheathing applied horizontally as well as on the ½-inch board regular density applied vertically. Corner bracing is described in Chapter 11, *Wall Framing,* "Special Situations: Bracing" (p. 253). Also ½-inch plywood sheets can be applied at all corners of the house. The space between these sheets is then filled in with fiberboard. Nailing of regular density is usually every 3 inches along the edges and every 6 inches in the intermediate framing members. A 1½-inch galvanized roofing nail is used for ½-inch thickness and a 1¾-inch for ²⁵/₃₂-inch thickness. Most manufacturers recommend leaving a ⅛-inch spacing between sheets.

Intermediate-density and nail-base sheathing are denser

Photo 12-3. Diagonally applied ¾-inch boards are seldom used since the arrival of plywood.

than the regular-density material and have more structural strength. The nail-base sheathing can be used when directly applying shingles to the wall using annular grooved nails. They are manufactured only in a ½-inch thickness and in panel footage of 4×8 and 4×9. These are usually installed vertically and require no corner bracing. The same nailing pattern and size apply for ½-inch-thick panels.

This type of sheathing will work OK in areas where no great shearing strength (ability to resist racking) is needed. As mentioned earlier, plywood sheathing or siding is recommended in earthquake, tornado, hurricane or high wind areas. Plywood sheathing, at a small increase in labor and cost, will generally make for a stronger house and stouter wall panel than fiberboard sheathing. It is worth considering no matter what your situation is.

INDIVIDUAL BOARD SHEATHING (Photo 12-3):   Before the advent of plywood, individual boards, nominally 1 inch thick and from 6 to 12 inches wide, were used for sheathing. They were butt end boards (with no tongue and groove or shiplap that ended over a stud), and were generally applied at a 45° angle for greater strength. Occasionally they were applied horizontally, in which case corner bracing was needed (diagonally, no additional corner bracing was needed). Boards with a 15% moisture content or less, usually a lower quality Number 2 or Number 3 common grade softwood, were used to reduce shrinkage and gaps between the boards. Three 8d nails were used at each stud for fasteners. Unless for some reason you have a cheap or free supply of 1-inch boards, I do not recommend using this type of sheathing. It does not provide the shearing strength of plywood and is labor and

cost intensive. It can, however, be used as a nail base for siding.

RIGID INSULATION (Photos 12-4, 12-5):   In the last few years, with the increased awareness of energy consumption in homes, builders have begun applying rigid insulation to the exterior edge of the studs and then applying siding over this. If a house needs greater strength, and rigid insulation affords almost none, plywood can be applied under the insulation. Both ½-inch and ¾-inch-thick panels of differing sizes are available with tongue and groove edges. The application, with flat-headed fasteners, is rather quick. It is good in colder climates as well as in very hot areas. Before making a decision, calculate the cost of adding this insulation and compare it to the cost of increasing your wall panels from 2×4s to 2×6s. If there's not much difference, I would recommend using the 2×6s. (You would seldom need a combination of 2×6s *and* rigid insulation sheathing.) The 2×6 provides a quieter, stouter wall panel while saving energy as well. When calculating the cost, add in the projected savings in energy to see what your pay-back period would be. Remember, your state and federal energy tax credit may apply.

GYPSUM BOARD SHEATHING: In some areas, where there is a threat of brush fires or other exterior fires, or where buildings are built very close to each other, the local codes may require that gypsum board (drywall) be applied on the exterior of the studs under the siding. This is to insure that, should the siding catch fire, the fire will not travel into the structural frame of the house. Check your local codes if you think this might apply.

Photo 12-4. Styrofoam wall sheathing comes in tongue and groove sheets.

Photo 12-5. Thin foil-faced boards are often placed behind brick. The boards offer little in the way of insulation but do help keep some heat out in summer.

What You'll Need

TIME   The time needed will depend on the complexity and height of the building as well as the number of window and door openings in the wall. All the nailing takes most of the time, and the times listed can be reduced if you are using a pneumatic nailer. 5 to 10 person-hours per 320 square feet (ten 4-×-8 panels).

PEOPLE   2 Minimum (both unskilled); 3 Optimum (2 nailers, 1 cutter, all unskilled); 6 Maximum (2 teams of 3, all unskilled).

TOOLS   Framing hammer, nail belt, chalkline, level, plumb bob, caulk gun (optional), power saw, pneumatic nailer (optional) pencil, tape measure, saw horses, framing square, nail-pulling tool, hammer tacker or staple gun, staples.

RENTAL ITEMS   Air compressor, scaffolding, nailing gun and ladders.

MATERIALS   You will be using ⅜- to ½-inch-thick plywood panels (⁵⁄₁₆ inch is often allowed but seldom actually used) in 4-×-8 or 4-×-9 sheets. CDX exterior glue plywood is used. Calculate how many sheets you will need to cover the house with a 10 to 20% waste factor. Do not use tongue and groove plywood; it is much harder to apply. Use 8d nails (CC sinkers) for fasteners and subflooring glue in heavy earthquake or hurricane areas. You will also need sheathing or builder's paper. This is either 15-pound felt or brown builder's paper.

ESTIMATING MATERIALS NEEDED

PLYWOOD: Determine the linear square feet of wall to be covered, divide by 32 square feet and add 10% for waste.

NAILS: 2 pounds of 8d CC sinkers per sheet of plywood.

SUBFLOORING GLUE: According to manufacturer's instructions.

BUILDER'S PAPER or 15-POUND FELT: According to number of square feet needed.

INFORMATION   This chapter will deal in detail with applying plywood sheathing. If you are using a plywood siding for sheathing as well, read this chapter and the section on plywood siding in Chapter 16, *Siding* (p.393). If you are using fiberboard sheathing or individual boards, refer to the section in this chapter comparing materials. If you are using rigid insulation, follow the manufacturer's instructions.

PEOPLE TO CONTACT   Unless you plan to do these things yourself, you will soon be needing the roofer, plumber and electrician. Contact them now. Suppliers will also be needed soon. Contact suppliers for siding, roofing, exterior doors and windows.

INSPECTIONS NEEDED   Most areas require an inspection of your sheathing-nailing pattern before it is covered by the sheathing paper and siding. Call the inspectors before the paper is applied and after the sheathing is up.

Reading the Plans

Usually your plans will clearly call for the needed sheathing. Generally you can find it on the elevations or cross sections. Sometimes no sheathing is called for and only diagonal bracing is indicated. You may want to consider adding plywood sheathing anyway. Some designers call out the minimal sheathing required (⁵⁄₁₆ inch); most builders upgrade this to ⅜ inch or ½ inch. If plywood siding is to be used without sheathing, its thickness (½ inch or ⅝ inch), the type of wood used (redwood or fir), its quality and its style will be called out in the elevations.

Safety and Awareness

SAFETY: Be very careful handling sheets on windy days, they can act as a sail. Be careful on ladders and scaffolding when handling the sheets, they are heavy and awkward. So that the power saw does not bind when cutting the panels, support the plywood properly.

AWARENESS: Be sure the sheets are plumb as they are applied and that they meet on a stud. Check that rough openings are the correct dimensions and square before beginning. In tall homes, stagger the end splices of the sheets. Be sure to overlap the crack between the top of the foundation wall and the bottom edge of the sill plate by at least 1 inch.

Overview of the Procedure

Step 1: Recheck the rough frame openings
Step 2: Apply the first panel
Step 3: Apply the remaining panels, cutting out doors and windows
Step 4: Apply the sheathing paper

Step 1
Recheck the Rough Frame Openings

Margin of Error: ⅛ inch off dimension; ⅛ inch off square in measurement of diagonal.

Most Common Mistake: Reading the window or door sizes incorrectly.

Before beginning to sheath any wall, check each rough opening size to be sure it is correct. Recheck to see if each opening is square or rectangular, either by using your framing

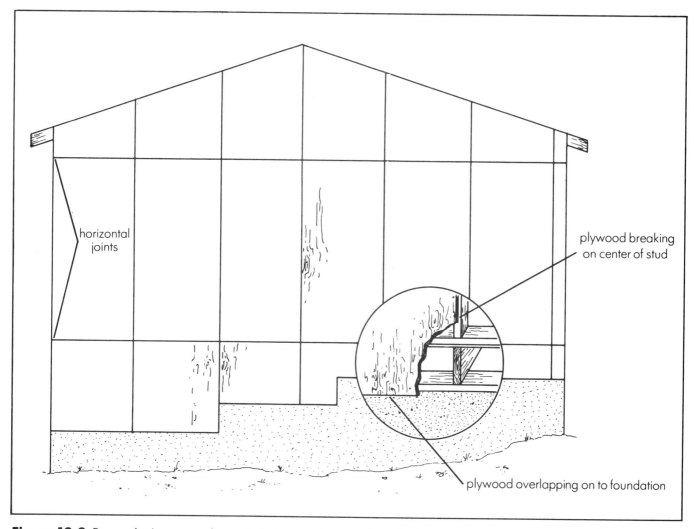

horizontal
joints

plywood breaking
on center of stud

plywood overlapping on to foundation

Figure 12-2. Pieces of siding are cut for use in the knee wall areas to keep the horizontal joints straight.

square or by measuring the diagonals of the openings (if it is a rectangle or square the diagonals will be equal). If you want to move or change an opening, it will never be easier than now.

Step 2
Apply the First Panel

Margin of Error: Exactly plumb.

Most Common Mistake: Not plumb.

Diagram: Figure 12-2.

Photographs: Photos 12-6, 12-7, 12-8.

You are now ready to apply your first sheet. You can begin on whatever wall of the house you want. Begin at the end of the wall, where you began your layout.

If you are using finished plywood siding (Texture 1-11) other factors must be considered. The sheathing will also be your finished siding, so you must accommodate for appearance. If the wall length is not exactly in 4-foot intervals, a

piece of siding will need to be ripped to finish out the wall. Decide on which end of the wall it will be least noticed and apply the ripped end there. The wall may be too high for one 8- or 9-foot-long panel to span. Decide where you would like the splice to be, low or high on the wall.

If there is a stepped wood knee wall in the foundation, and you are using plywood siding, begin at the lowest part of the wall and cut the lower panels. This will allow all the splices to line up across the wall.

After determining where to begin, check to see if there is a center of a stud at every 4-foot interval and that your corner stud assembly is plumb. (The first sheet should be applied plumb even if the corner stud is not exactly plumb.) Apply the first piece. It is important that this piece be very plumb (vertical); if it is out of plumb, all the others that follow will be out of plumb.

Usually the sheets stretch from 1 inch below the top of the foundation wall to the top of the cap plate. If the rafters have already been applied, the sheathing can come to the bottom of the rafters. Most often the sheets are applied vertically, though they can be applied horizontally with no wall bracing needed. Vertical application, however, is stronger.

Photo 12-6. Check to be sure walls are plumb in all directions before applying the wall sheathing or siding.

Photo 12-7. Be sure that the first course of sheathing you apply is level and plumb.

As you start to nail, start at one side of the sheet and work towards the other. If 2 people start at each side and work towards the center, it may create a bubble in the board. Use 8dCCsinkers or 8d annular groove nails every 6 inches around the edges and every 12 inches on the intermediate studs. It is easiest to pop a chalkline on the plywood to locate the center of each stud to make nailing easier. Apply a ¼-inch-wide bead of subflooring glue at this time if you are planning to use any.

Step 3
Apply the Remaining Panels, Cutting Out Doors and Windows

Margin of Error: Exactly plumb; within ½ inch to edge of opening, but do not extend sheathing into opening.

Most Common Mistakes: Extending sheathing into opening; inadequate nailing.

Diagram: Figure 12-3.

Photographs: Photos 12-9, 12-10.

After the first sheet is plumbed and nailed into place, continue with the rest of the sheets. Some builders leave a ¹/₁₆-inch gap at the ends of the panels and a ⅛-inch gap at the edges to avoid bubbles in the panels caused by expansion due to moisture rain. Unless you are planning to leave the sheathing uncovered for a long period of time (several months), I do not think it is really needed. Water runs more quickly off the walls than the subflooring or roof sheathing and is not as great a

Photo 12-8. Using a chalkline to locate the studs for nailing.

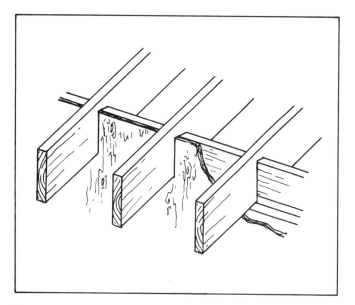

Figure 12-3. Plywood siding is cut between the rafters where siding will not be concealed by the soffit.

problem. Making sure the sheets are still plumb, nail flat against the studs. Try to use sheets as large as possible.

Openings for doors and windows can be handled in one of two ways. Some builders nail the sheathing on, covering over all the rough openings. Then from the inside, holes are drilled at the corners of each opening, and the sheathing covering the opening is cut out from the inside with a reciprocating saw. Chalklines representing the opening can also be popped on the outside of the sheathing and the sheathing cut out with a power saw. As in cutting any wood with a power saw, set the depth of your cut to penetrate about ¼ inch deeper than the wood.

I think an easier way is to temporarily tack the sheathing in place with 2 or 3 nails, press it tightly up against the studs and have someone mark the outline of the opening on the sheet from the inside. Then take the sheathing down, cut out the excess for the opening and permanently nail the piece into place. No matter what technique you use, be sure the sheathing does not overlap into the rough opening; this will reduce the size of the rough opening and the window or door will not fit unless it is later cut away.

Photo 12-9. Usually the wall sheathing is held or tacked temporarily in place and the openings are marked on the plywood from the interior. The sheet is then removed and the opening cut out.

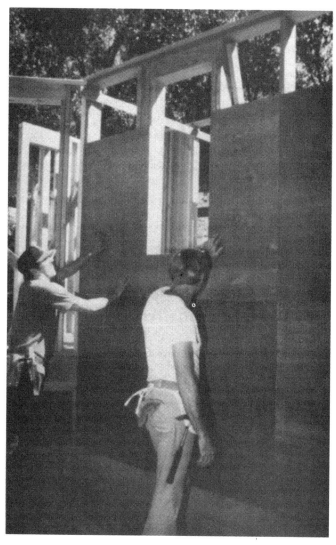

Photo 12-10. Nailing sheathing on the wall after the window opening has been cut out. Wall sheathing can be applied either vertically or horizontally.

Photo 12-11. Sheathing or plywood siding can be attached to the wall framing before it is lifted into place. This is not recommended for the first-time builder, since making adjustments for mistakes is easier when the sheathing is applied after the wall is up and plumb.

Continue with the rest of the sheets until the wall is entirely covered and then go on to another wall. If there is not a stud or cripple at the edge of each sheet, add an extra one. All edges of the plywood must fall on the center of a full stud or cripple.

Remember, as you apply the sheets, that the bottom plates in the door openings will be cut out (you can cut them now) and you should not cover this section of bottom plate with sheathing.

If the siding will run between the rafters, thereby exposing the overhangs of the rafters, the sheathing must also run between the rafters. Blocks will be needed between the rafters and flush with the outside edge of the cap plate to act as a nailing surface for the sheathing and siding. For more about these blocks see Chapter 13, *Roof Framing*, "Step 8."

Step 4
Apply the Sheathing Paper

Margin of Error: Just keep it flat.

Most Common Mistake: Not overlapping top piece over bottom piece.

Photograph: Photo 12-11.

Usually a sheathing paper is applied between the sheathing and the finished siding. If you are using Texture 1-11 plywood siding, this paper is applied directly over the studs. The purpose of the paper is to help prevent moisture and air leakage into the wall cavity and is essential in most good construction. The paper must be able to breathe to allow moisture to escape from the wall in the form of vapor, not water. The paper must have a perm rating of at least 6 (6 perms of water vapor can pass through it).

Generally a builder's paper or 15-pound felt is used. Builder's paper is 2 pieces of brown paper sandwiching a moisture-proof (but breathing) layer. Begin the paper at the bottom of the wall, letting each sheet overlap the one below it for a shingle effect. Tack it flat with a nailing gun or better yet a hammer tacker. The tacker is much easier to use and may be worth renting if one is not otherwise available to you. Overlap the paper 4 inches along the edges and 6 inches at the ends. Overlap the corners 6 inches and into the wall openings 2 to 3 inches.

Worksheet

ESTIMATED TIME

_____

_____

_____

CREW

Skilled

_____ Phone _____

_____ Phone _____

Unskilled

_____ Phone _____

_____ Phone _____

_____ Phone _____

_____ Phone _____

MATERIALS NEEDED

Plywood, nails, glue, sheathing paper.

ESTIMATING MATERIALS

PLYWOOD SHEETS:

Number of square feet of wall area _____

divided by 32 (4-×-8 plywood sheet, use 36 if

using 4-×-9 sheets) = _____

+ 10-20% waste _____ = _____ sheets

NAILS:

_____ pounds cover _____ sq. ft.

Total number of lbs. needed _____

ADHESIVE (optional):

One cartridge covers _____ sq. ft.

Total number of cartridges needed _____

SHEATHING PAPER or 15-LB. FELT:

One roll covers _____ sq. ft.

Total rolls needed _____

COST COMPARISON

_____ Store:

$ _____ per sheet = $ _____ total

_____ Store:

$ _____ per sheet = $ _____ total

_____ Store:

$ _____ per sheet = $ _____ total

_____ Store:

$ _____ per sheet = $ _____ total

PEOPLE TO CONTACT

You will soon be needing the roofer, plumbers and electricians, unless you are doing this yourself:

Electrician _____

Phone _____ Work Date _____

Plumber _____

Phone _____ Work Date _____

Roofer _____

SUPPLIERS:

Siding _____

Phone _____ Work Date _____

Roofing _____

Phone _____ Work Date _____

Doors (exterior) _____

Phone _____ Work Date _____

Windows _____

Phone _____ Work Date _____

Do any suppliers need to be called about daily deliveries?

☐ YES   ☐ NO

If yes, who _____

Daily Checklist

CREW:

_____ Phone _____

_____ Phone _____

_____ Phone _____

_____ Phone _____

_____ Phone _____

TOOLS:

☐ framing hammer

☐ nail belt

☐ chalkline

☐ level

☐ plumb bob

☐ caulk gun (optional)

☐ power saw

☐ hand saw

☐ pencil

☐ tape measure

☐ saw horses

☐ framing square

☐ nail-pulling tool

☐ hammer, tacker or staple gun

☐ staples

RENTAL ITEMS:

☐ air compressor

☐ nailing gun

☐ scaffolding

☐ ladders

MATERIALS:

_____ sheets of plywood

_____ lbs. of nails

_____ cartridges of subflooring glue (optional)

_____ rolls of builder's paper or 15-lb. felt

DESIGN REVIEW:

Check all rough opening measurements and re-evaluate all your design decisions regarding their size and location. After the sheathing is applied, they are much harder to change.

INSPECTIONS NEEDED

Most areas require an inspection of your sheathing-nailing pattern before it is covered by the sheathing paper and siding. Call the inspector before the paper is applied and after the sheathing is up.

Inspection called for _____

Time of inspection _____

Person contacted _____

WEATHER FORECAST:

For Week _____

For Day _____

13 *Roof Framing*

Photo 13-1. A house with all the rafters in place. The lower story rafters are cut at the wall line to accommodate a veranda roof that will be added later. Note that the lookouts off the front of the building are ready to receive the rake rafter.

What You'll Be Doing, What to Expect

THIS CHAPTER WILL deal with four different aspects of residential roof construction. The first section of the chapter will discuss various roof styles: flat, shed, gable, gambrel, and mansard. The second part will discuss the three main approaches to roof framing: site-built with standard materials, trusses, and large beam construction. The third part will deal with the actual step-by-step construction of a standard roof frame (simple gable roof). The last part of the chapter will deal with special situations in roof framing.

Roof framing can be anywhere from the simplest to one of the most complex types of framing; it depends on the complexity of the roof design. Complex roofs, with their many valleys, hips, ridges and intersections, can be a challenge even to a very experienced carpenter. Whereas simpler roofs, such as flat, gable or shed, can be easily mastered by the novice. My advice to you is, if you are building a complex roof frame (Photo 13-3), hire an experienced framer to serve

as your consultant or perhaps even work with you at the job site. It is one of those areas of building your own home in which paid professional help may be a good investment. You can accept the challenge of framing a complex roof for yourself, but be sure you understand the scope of the challenge.

Roof framing (Photos 13-1, 13-2) involves installing the roof framing members, ceiling joists, dormers and openings. It is mentally demanding; it takes a certain amount of comprehension of simple mathematical relationships as well as the ability to visualize the framing components. Physically, it is also a very demanding process. The roof ridges and rafters, usually the largest structural members used in the home, must be raised to the highest, and therefore the most dangerous, position. If ever your bad back will go out (if it hasn't already during the concrete pour), now is the time.

Safety should always be uppermost in your mind. Even a small fall can be disastrous. I was recently contacted by a lawyer from San Francisco who was representing a carpenter who had fallen only 9 feet and was paralyzed for life. *Be careful on the roof!*

Photo 13-2. A house with the large ridge beam sitting on top of 2 support posts. The first 2 rafters are in place.

Photo 13-3. Complex roof framing like this may require professional assistance.

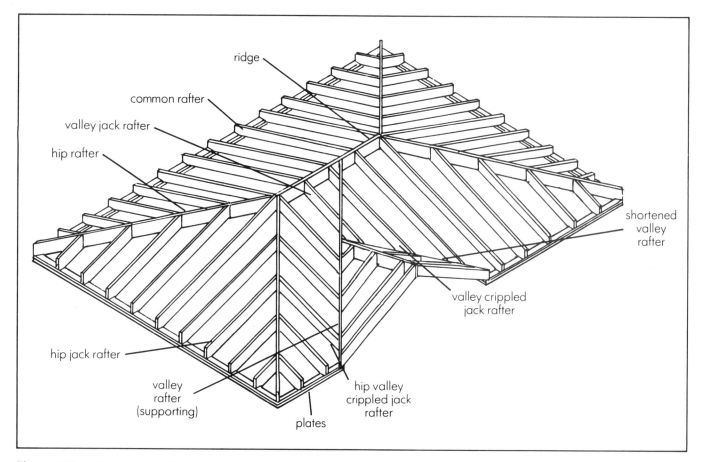

Figure 13-1. Names of roof framing components.

Jargon

Roof Framing
(Figure 13-1)

COMMON RAFTER.   An individual sloping member of the roof frame. Usually it runs at a right angle from the exterior walls and rests on the ridge.

CRIPPLE JACK RAFTER OR CRIPPLE RAFTER.   This rafter does not intersect the wall plate or the ridge. It is terminated at each end either by the hip or valley rafters. Both ends will be mitered where they intersect the valley or hip.

HIP JACK RAFTER.   A rafter running from the wall plate to a hip, where it is cut at an angle.

HIP RAFTER.   A rafter that runs diagonally (at a 45° angle) from the corner of 2 intersecting exterior walls to the ridge, to form the intersection of adjacent slopes of a hip roof.

HIP VALLEY CRIPPLE JACK RAFTER.   A cripple jack rafter that runs from a hip to a valley.

RAKE RAFTER.   The outermost rafter that runs parallel to the roof slope and forms the overhang between the roof and the wall at the gable end.

RIDGE.   The highest horizontal running member of the roof frame; it aligns the rafters and supports them at their upper end.

SHORTENED VALLEY RAFTER.   This rafter runs from the wall plate to the supporting valley rafter, where 2 sloping roof sections intersect and their ridges are at different elevations.

SUPPORTING VALLEY RAFTER.   The longer rafter that runs from the wall plate to the ridge, in a "valley" or hollow of 2 intersecting, sloping roof surfaces, where the 2 ridges are at different elevations.

VALLEY CRIPPLE JACK RAFTER.   A cripple jack rafter that runs from a valley to a valley.

VALLEY JACK RAFTER.   A rafter that runs from the ridge to the valley rafter.

VALLEY RAFTER.   A rafter running from the ridge to the wall plate in a "valley" or hollow of 2 intersecting, sloping roof surfaces, whose ridges intersect at the same elevation.

Rafters
(Figure 13-2)

BIRDSMOUTH.   This is the cut near the lower edge of the rafter where the rafter sits on top of the wall plate. It is made up of 2 cuts: the vertical plumb cut and the horizontal seat cut.

PLUMB CUT OR RIDGE CUT.   The cut of the rafter where it meets the ridge.

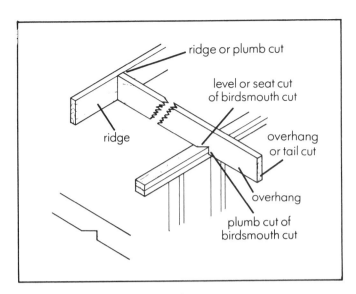

Figure 13-2. The parts of a rafter.

TAIL CUT OR OVERHANG CUT.   The cut at the lowest edge of the rafter.

TAIL OR OVERHANG.   The part of the rafter that extends past the exterior wall line.

Layout
(Figure 13-3)

In almost all layout principles, we are relating the cuts and dimensions of the rafters to 3 sides of a right triangle. Some common terms are as follows.

LINE LENGTH.   The hypotenuse of the triangle. The sloping distance measured along the top edge of the rafter, along the slope, from parallel lines drawn perpendicular from the rafter, at the centerline of the ridge, to the outside edge of the plate.

PITCH.   This term describes the incline of the roof as a ratio of the vertical rise to the total run of the entire roof. It is expressed as a fraction, e.g., a roof that rises 6 feet over a total wall span of 30 feet has a pitch of $^6/_{30}$ or $^1/_5$.

RISE.   The vertical side of the triangle. The altitude or total vertical distance the rafter rises as measured from the top of the wall plate to the top edge of the rafter.

RUN.   The horizontal side of the triangle. The total distance of the rafter as measured horizontally from the outside of the wall to the centerline of the ridge.

SLOPE OR THE CUT OF THE ROOF.   The incline of the roof as expressed in a ratio of its vertical rise to its horizontal run. The horizontal run is always expressed as 12 (for 12 inches) and the vertical rise varies. For example a roof that rises 3 inches for each horizontal 12 inches has a 3 in 12 slope.

SPAN.   The distance between the outside edges of 2 opposite walls.

Unit Measurements
(Figure 13-4)

UNIT LENGTH.   The measurement of the hypotenuse formed by a right triangle where one side is the unit rise and the other is the unit run.

UNIT RISE.   The vertical rise or distance the rafter extends in a vertical direction for every one foot it runs horizontally.

UNIT RUN.   Unit run is always one foot. This is the distance measured along the horizontal (level) plane.

Purposes of the Roof Frame

1) Protect the building from the weather
2) Provide the base onto which the roofing materials are applied
3) Contribute to the design of the home
4) Parking space for Santa's sleigh

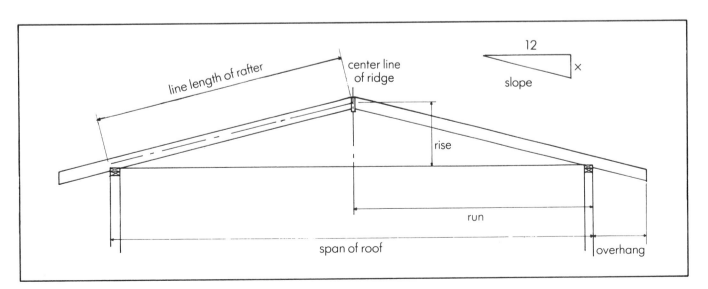

Figure 13-3. Terms used for a rafter layout.

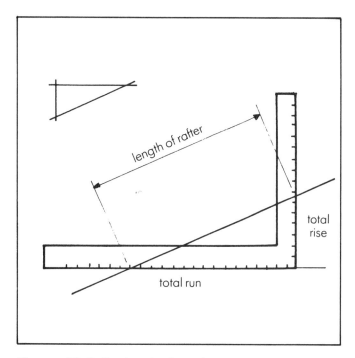

Figure 13-4. The length of a rafter as measured along its slope.

Design Decisions

In roof framing, perhaps more so than in any other part of housebuilding, there are many design decisions to make. These include two major decisions: the style of the roof and the type of members to be used; as well as minor decisions, including the type of fasteners used, the way in which the rafters intersect the ridge, etc.

Before you begin to order your materials, you will have to make several design decisions. Most of these may have been made during your initial design process and may be incorporated into your plans. Others may not have occurred to you until now. You may also want to reconsider any previously made design decisions to see if they are still in your best interest. The following sections will deal with three main areas: first, the style of the roof; second, the type of materials to be used; and third, we will consider some smaller design decisions that need to be clear at this stage. The list here is rather extensive, but it is important that you understand all the possible choices for each situation before you begin to build the roof frame. Otherwise your choices will be made for you, and they may or may not be the best ones.

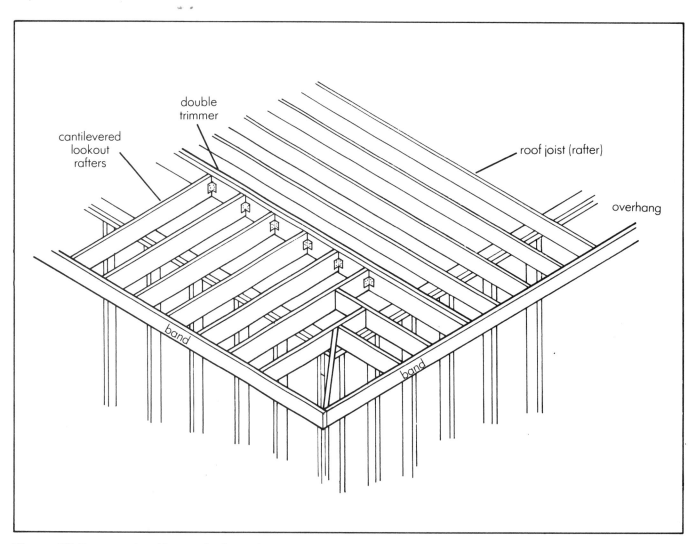

Figure 13-5. A typical roof framing for flat roofs.

Photo 13-4. Shed roofs running in different directions.

Style of the Roof

There are many different styles or designs of roofs. Each one has advantages and disadvantages. Your choice here is very critical for two main reasons: design and cost.

Probably no other design decision that you make will have more impact on your total exterior design as the style roof you choose. It is the most visual exterior element of the house, and thereby the most important, as far as the design is concerned. The same exact house, fitted with different style roofs, will appear very different. As you drive around your neighborhood, visualize the houses with different roofs or just different size overhangs, and you will get a sense of the impact a roof has. Be sure you have accurately visualized your dream house with its roof in place. Do some drawings and then try on different "hats" to see how much each changes the character of the house. Put a hat on your own head to see how different it makes you look.

You will want to consider your roof as an umbrella for the home. It should not only protect the interior but the exterior walls as well. If the overhangs prevent rainwater from running down the exterior walls, your windows and doors will never leak, and your painting and siding will be easier to maintain. Aesthetically, broader overhangs often look better also.

The design of the roof will affect the cost of the roof in money, time and energy. The design of the roof will affect the amount of roofing materials needed, as will the slope of the

roof, and may also affect the size and dimensions of the roof frame that is needed. It's true that simple roofs go up more quickly than complex roofs; but before you throw out your much-loved complex roof for the less-loved simple roof, determine how much more the desired roof will cost. It may be only several hundred dollars and may be well worth the money. It may cost several thousand dollars of an already stretched budget and cause you to consider some redesigning. Just be sure you are not making decisions, especially major ones, with incomplete data.

FLAT ROOF (Figure 13-5). A flat roof is perhaps the easiest, cheapest and quickest to build, but the hardest to seal. The roof rafters serve also as the ceiling joists and all needed insulation must be placed inside the rafter cavity, since there is no attic. As with all roof framing members, be sure flat roofs meet all local codes (e.g., snow can gather easily on these types of roofs).

Actually many flat roofs are not truly flat; they incorporate slight slopes to drain water. The flat roof will have to be sealed with a continuous membrane roofing material. Hot tar is the most common, and this usually must be professionally applied.

Overhangs are created by simply extending the roof rafters past the edge of the building, where the rafters are running perpendicular to these walls. When the rafters are running parallel to the walls, overhangs are created as shown in the diagram.

Photo 13-5. A simple off-center gable roof.

Photo 13-6. An off-center gable roof frame.

Photo 13-7. Two gable roofs on a multi-level house.

SHED ROOF (Photo 13-4). A shed roof is also very simple and quick to build. It spans 2 walls, one higher than the other. This type of roof has become very popular in modern architecture. Solar designers seem to have forgotten that other styles even exist. Unless you plan to drop a ceiling to form an attic, the insulation will go between the rafters (or on top of the exterior sheathing) and the rafters can serve as the ceiling joists as well. Since a shed roof is, in essence, half of a gable roof, the information in this chapter will enable you to build a shed roof.

GABLE ROOF (Photos 13-5, 13-6, 13-7). The gable roof is probably the most common roof style in the United States. It is formed by 2 roof planes meeting at a ridge. This ridge can either be centered over the house or offset to one side. Like the shed and flat roof, it is quick and simple to build. It can either be combined with an attic or the ceiling boards can be attached to the sloping rafters to create a cathedral ceiling. Usually ceiling joists or collar beams are used to take the stress off the rafters (the "A" splitting factor from the downward thrust of the roof), so the deletion of these must be considered from a structural standpoint. (See pp. 289–290, "Special Considerations: Ceiling Joists" and "Size, Spacing and Span" in this chapter.)

INTERSECTING GABLE ROOF (Photo 13-8). Though this roof can become complex, it is also rather attractive. It consists of several gables, each similar to the standard gable, intersecting and running in different directions.

THE HIP ROOF (Photo 13-9). In this type of roof, the ridge does not run the full length of the house; it stops short, and hip rafters extend diagonally from the end of the ridge to the corners of each wall. Though this roof can be difficult to build, a simple hip roof is well within the reach of a novice owner builder.

GAMBREL ROOF (Photo 13-10). The gambrel roof is very popular in many areas of the country. It consists of a top ridge with a slope being broken on each side by a second ridge, usually half way down the slope. The advantage of having this type roof is that it allows more head room upstairs and you can use your attic as a second floor.

MANSARD ROOF. This is a French style roof that is somewhat popular in this country as well. All 4 sides have a double slope; the lower slope is nearly vertical, while the upper slope is almost flat.

Types of Roof Construction

Most houses built today incorporate one of three different types of roof construction: standard materials, trusses or large beams. Each of these systems has its advantages and disadvantages, and your decision should be well considered. The decision between using either standard framing materials or trusses is mostly based on time, labor and money factors. Once installed, these two systems look the same. The decision to use large beams (e.g., exposing large beams to the interior) is usually a design decision made for aesthetic reasons.

Photo 13-8. An intersecting gable roof.

Photo 13-9. A hip roof. Note how the exterior wall facing us protrudes past the gabled-roof section.

Photo 13-10. Gambrel roofs, though more difficult to construct, offer beauty as well as greater head room on the top floor.

STANDARD MATERIALS. For many houses, the rafters, ridges, valley rafters, etc. are all constructed at the job site with standard framing lumber. This lumber can be as small as a 2×4 or as large as a 2×14. Until the introduction of trusses, all roofs were built this way, and many builders still prefer this way to using trusses.

Whereas trusses need to be ordered a few weeks ahead of time, site-constructed framing systems are built the day they are installed, allowing for last minute changes and avoiding mistakes made by the truss companies. Another advantage is that, if you are not paying for labor, site-built systems may be somewhat cheaper than trusses. You can also use larger materials in site-built framing than are used in trusses. Plus the quality of construction is controlled by the builder not the factory. This chapter will explain in detail how to build a roof frame using this method.

TRUSSES (Figure 13-6, Photos 13-11–13-14). Roof trusses are units which incorporate a series of structural members that are used to carry the load of both the roof and the ceiling to the outside walls. The structural members are usually 2×4s or 2×6s. Most trusses consist of a bottom chord, which also serves as the ceiling joist; a sloping member or top chord, which serves as the rafter; and diagonal connecting pieces, called webbing, that transfer the weight from one member to the next. These pieces are connected by metal truss connectors or plywood gusset plates, which are nailed

and glued. Often the bottom chord will have a slight camber or crown (about ½ inch over 24 feet) to compensate for the stress applied over the span. Most trusses are built in truss factories from engineering calculations. You may be able to build your own at the site, but check with the building inspector first.

Trusses are almost always used in tract development. Though the advantages are not as great in a single home project, trusses are becoming popular in this area as well. You order trusses from the factory to your individual specifications, and the factory then builds them for you and delivers them to the site. Usually the delivery is made with a special truck that can place the trusses on top of the walls. Trusses can be built for almost any roof, except for very complex designs. You must be sure to provide the truss factory with the proper information to build your trusses. I worked on a job where $1800 worth of trusses were built 1 inch too short. The moment of realization was not a pleasant one. The factory will want to know the span, the slope, the overhang, soffit details, etc. Just be sure you are having the correct trusses built.

There are several advantages to using trusses. Obviously there is a savings in time and labor, since much of the labor of roof framing is done by the factory. Once delivered, the roof framing erection process is much quicker. Trusses can often span greater distances than standard rafters, making load-bearing interior walls unnecessary. This, in turn, means you

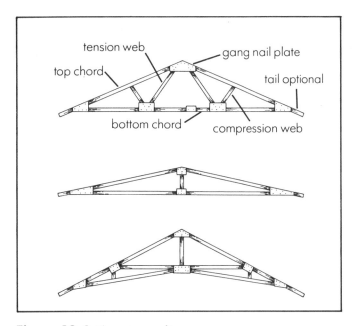

Figure 13-6. Three types of trusses.

can get the roof on faster, protecting the house from the weather, and then finish the interior walls. Trusses allow a greater freedom in design, since load-bearing walls are no longer necessary in certain locations. Be sure to inspect the trusses you have received from the factory to see that they have kept their standards of quality high. I have seen trusses with split, and even broken, members or with the fasteners hanging loose.

Trusses are rather easy to erect. They are usually delivered to the site and placed, point down, on the walls in the order they are to be installed. Usually 3 people are needed to handle the installation, and in the case of larger trusses 4 or more people may be needed. After the layout marks are drawn on the cap plates of each wall, the trusses are rolled into place. One person works at each wall while a person down below, using a long 2×4 or guide, holds the truss upright until it is secured. After it is nailed to the plate on both walls, a board is nailed across the sloping top chord to temporarily hold it plumb and in place. These temporary supports are removed when the sheathing is applied.

Photo 13-11. Trusses ready to be erected.

Photo 13-12. Trusses can be made for many different style roofs. Trusses for a mansard roof are shown here.

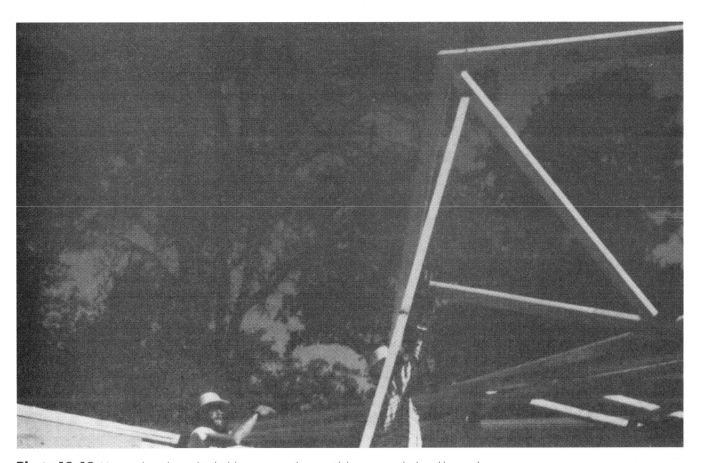

Photo 13-13. Using a long board to hold trusses in place until they are nailed and braced.

Photo 13-14. Trusses being delivered to the top plate of a garage. Most lumber companies use a truss truck which carries the trusses on a high rack for easy unloading. Trusses are then rolled off into place.

Photo 13-15. Large timber roof construction often requires several people or a crane. Be careful at this stage—accidents are not uncommon.

LARGE TIMBER ROOFS (Photos 13-15–13-18). Many houses use large beams for the roof frame. The beams are usually taken from 4-inch or 6-inch stock, 6 to 20 inches wide. Glulams (several boards laminated into a large beam) are often used. Sometimes these larger timbers are used because the loads applied to the roof (usually snow loads) are so great that the smaller, common framing lumber would not carry them. Other times large timbers are used just for the ridge beam because it has become a load-bearing ridge (see next section, "Ridge Beam or Board").

More commonly, however, these larger, more expensive beams are used in place of rafters because you are planning to expose them to the interior of the house and appearances are important. Visually, 4×8s on 4-foot centers are much more attractive than 2×10s on 16-inch centers. Usually 2-inch-thick decking is applied as the roof sheathing and this also is exposed to the interior. The insulation, usually rigid, is applied over the sheathing. If you are planning to use this type of roof construction, be sure you have compared the cost of it to that of the standard roof system with drywall ceilings. Often, with the more expensive large timbers, 2-inch decking, rigid insulation, etc., the roof can be quite costly. (A compromise would be to use a large ridge beam that is partially exposed to the interior; and standard rafters, with drywall attached directly to the interior of sloping rafters, creating a cathedral ceiling.)

For the most part, the principles used in building a roof frame with 2-inch materials are the same as in using 4-inch or 6-inch materials. All the cuts and layouts are the same, only

Photo 13-16. Large ridge beams and beams are being used instead of 2-inch rafters to create an exposed beam ceiling.

Photo 13-17. A crane is being used here to place large roof timbers. This might well be worth the $100 to $200 per day fee. It is safer, easier and quicker.

Photo 13-18. The ends of ridge beams and boards can be specially cut if they are going to be visible from the exterior.

the spacing is different. This chapter should give you all the needed information to build a large timber construction roof.

Aside from the aesthetic value of exposing the large timbers, the advantages are more subtle. They have to do with a sense of strength or the character of the home. Though it is subtle, its effect is profound. Modern houses are becoming lighter and lighter—small wall framing members, lightweight trusses in the roof, thin wallboards, hollow core doors, etc. In fact, stud construction was called "balloon construction" when it was first introduced. The old-time carpenters, who were used to post and beam construction, laughed and said the house would be so light that it would blow away like a balloon. The use of large framing timbers, especially when exposed in the walls and roof, lends the character and integrity that is missing in many modern houses. Solid doors, plaster walls, hardwood floors, tile, and good windows can also add to this. Visit, if you can, similar houses, one with large exposed framing members and the other without. See if for you the houses seem any different.

Aside from the expense of the large timber roof frame, the other great disadvantage is the availability of materials. On the West Coast, where there are still large trees (though they too are quickly diminishing), most lumber yards carry a large stock of timbers, of different lengths and dimensions. On the East Coast the selection is not so great. If they have to be specially ordered, there can be a time delay and greater expense, and the material shipped is often very wet, twisted and unattractive. As it starts to dry it twists and warps to a point that it is highly noticeable once the building is completed.

I have had more fights with my lumber suppliers over this than any other issue. I inform them that the timbers will be exposed to the interior and that I can use only the highest quality material. When it is delivered, before I allow it to be unloaded or pay the bill, I check it carefully for knots, straightness, delivery damage, checks, splits, twists, bad areas, etc., and, if I do not like it, I send it back (or use it at the gable where it will be covered). Be very clear with your supplier, especially if it is a special order, that it must look good enough to expose. If you are going to expose the overhang area to the outside, this must also look good. If possible, buy "dry" stock; it will not shrink and twist after it is installed.

Special Considerations

RIDGE BEAM OR BOARD (Photo 13-19). In most homes the ridges are not load bearing. They simply act to align and stabilize the top part of all the rafters. Though they are usually made of 2-inch-thick stock, the code will often allow 1-inch stock to be used, since they bear no load. In some cases, however, this ridge bears the load of the roof. It then becomes a ridge beam that must be calculated according to its span, loading, type of wood, etc. Usually the non-load-bearing ridge board becomes a load-bearing ridge beam when the slope is low (3 inches in 12 inches or less, for example), the span of the rafters is broad, or the loading, due to roofing materials, snow, etc., is great. Sometimes it may be a com-

Photo 13-19. Roof rafters hung from a load-bearing ridge beam.

bination of these factors. (Often the hip and valleys will need to be sized accordingly as well.)

If you are using a large ridge beam with standard rafters, you may want to expose the ridge to the interior for aesthetic purposes. If you are using large timbers as rafters as well, you will definitely want to expose the entire roof frame. But if only the large ridge is being used, the standard rafters can either be set on top of the ridge or hung from the ridge.

OVERHANGS, SOFFITS AND FASCIAS (Photo 13-20, Figure 13-7). Before laying out and cutting your rafters, you need to determine the type of fascia and soffit you plan to use, as well as the size of the overhang. The soffit is the underside part of the overhang of the rafters. The fascia is the board(s) nailed to the outer edge of the rafters. As you can see in the diagrams and photo, the manner in which the soffit is to be built and the way in which the fascia is to be attached will have a bearing on the way in which the end of the rafter is cut. Also the method of installing the "rake" rafters must be determined before beginning (see "Step 6" of this chapter).

CEILING JOISTS. The size and location of the ceiling rafters, as well as the manner and frequency in which they are attached to the rafters, may affect your roof framing. You may, for instance, be able to avoid collar beams (see "Span" in this next section) if you are attaching ceiling joists to the rafters. The spacing of the ceiling joists also needs to be coordinated with the spacing of the rafters to allow some of the joists to be nailed to the rafters. For more on ceiling joists and their installation see "Special Situations: Ceiling Joists" in the back of this chapter.

Photo 13-20. When a wide gutter is attached to the band rafter, no fascia board is needed.

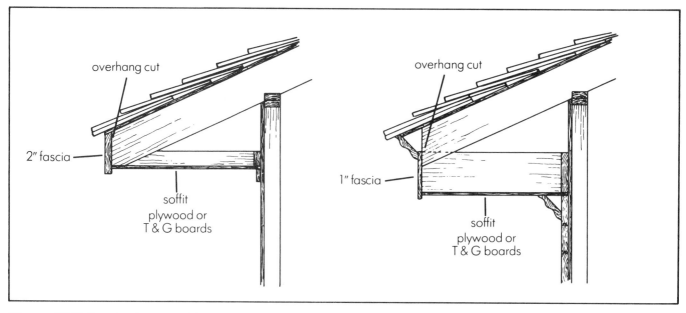

Figure 13-7. Details of two possible types of fascia and soffit.

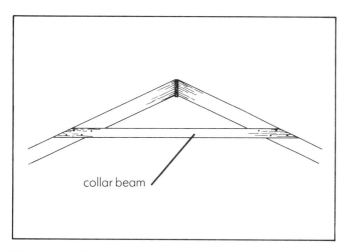

Figure 13-8. A collar beam or tie is used to hold rafters together and reduce the span of the rafters.

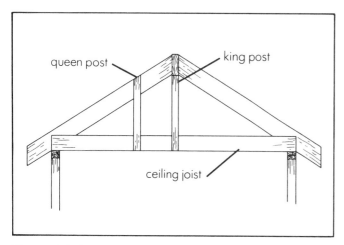

Figure 13-9. Using king and queen posts to reduce the span of the rafters.

SIZE, SPACING AND SPAN. The size, spacing and span of the roof rafters are all related to each other. And these are all related to the roof load (snow, wind, roofing materials, etc.). All these factors are coordinated so that certain size rafters, spanning certain distances, according to certain spacing will effectively carry a certain load. Usually there are a number of different combinations of these factors that will work for any one roof. Let's discuss each one of these factors now.

Size: Rafter sizes vary from 2×4s (rare) to 2×14s, with 2×10s and 2×12s being most common. If insulation is to be placed between the rafters, the rafters must be wide enough to provide for the thickness of the insulation. Other than this consideration, the size you use depends on what' you will need to carry the weight over the span at that spacing. The larger sizes are a little more expensive, but tend to be better quality boards.

Spacing: As with all stud construction, rafters are usually spaced on 12-, 16- or 24-inch intervals, with 12 and 16 inches being most common. As with sizes, whatever will work is fine. You will want to be sure the spacing is coordinated with the ceiling joists, so that every second or third ceiling joist can be nailed to a rafter to help reduce the outward thrust of the rafters on the wall.

Span: (Photos 13-21, 13-22, Figures 13-8, 13-9.) Rafter span, the distance between supported points of the rafter, affects both size and spacing. There is an upper limit that standard materials can span. In order to reduce the total span of the rafters, different techniques are used: a modified post and beam situation is placed either in the attic area or in the interior; or a load-bearing wall is used to break the span.

Collar beams can also be used to reduce the span of the rafters. They are attached across opposing rafters to reduce the outward thrust of the roof load, which wants to split the "A" formed by the gable rafters. They also serve to reduce the distance between support points on the rafters.

King or queen posts, usually made up of 2×4s running

Photo 13-21. Post and beam construction like this can be used in the attic area to break the span of a rafter. Two pieces are used to create the rafter, which is less expensive than one large, long rafter.

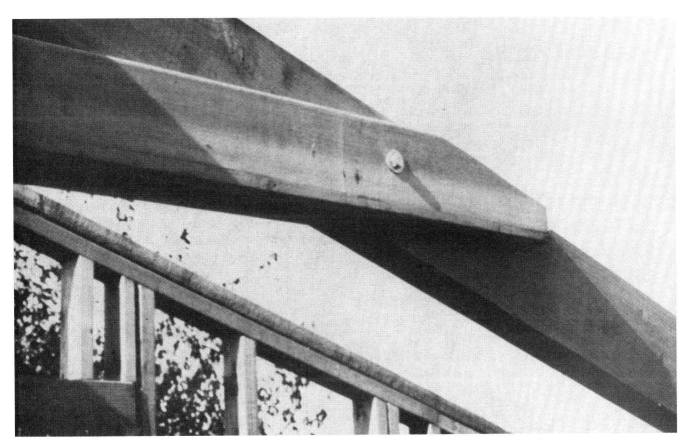

Photo 13-22. Using bolts to attach a double collar beam where they will be exposed to the inside.

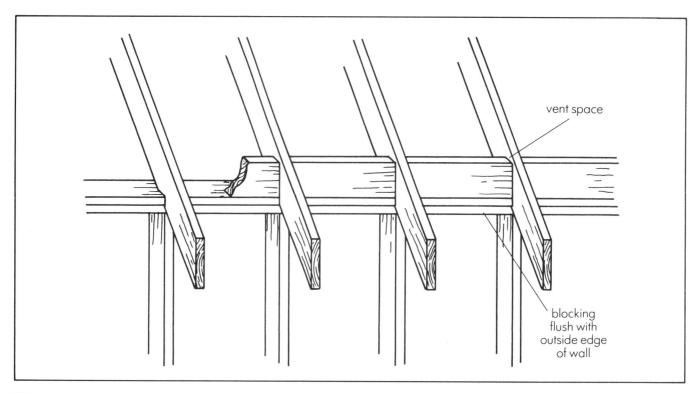

Figure 13-10. Vent space is shown above the blocking between the rafters at the wall.

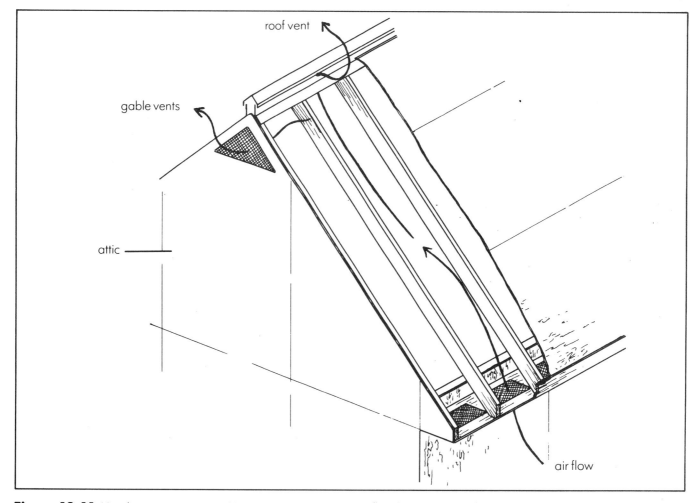

Figure 13-11. Ventilation in attic areas. Vents at eaves create an air flow between the rafters into the insulated area.

from the rafters to the ceiling joists, can also be used to reduce the span and transfer some of the load of the rafters to the ceiling joists and then over to the wall.

VENTILATION (Figures 13-10, 13-11). If you plan to incorporate an attic into your house, it must be ventilated. This not only allows hot air to be ventilated out in the summer but also avoids dampening of the attic insulation in the winter. The dampening is caused by condensation in the air in the attic. At a certain temperature, called the dew point, all the water molecules in the air will liquefy into water. This temperature often occurs in the attic or wall section of the house. Ventilating the attic allows this moisture to constantly be dried. Vents are usually placed in the gable walls of the attic area and are often built into the roof itself. It may also be wise to allow air to the floor from the overhang by placing vents between the rafters at the wall plates, as shown. Check with builders in your area to see if these are needed.

If you are planning to place the insulation between the rafters and then install the ceiling directly to the bottom edge of the rafter to create a sloping ceiling, you may want to ventilate between the rafters as well. This will allow air into the insulated area to dry any moisture build-up. It lowers the insulating properties of the roof section by only a negligible amount. Allow air to flow from the eaves to the ridge between each set of rafters as shown.

Be sure that the blocking between the rafters does not block the air flow, either at the wall, the ridge, or along the span. If you plan on including vents, make sure that the roof insulation stops 1½ inch short of the ridge.

Building a Roof Frame With Standard Materials

What You'll Need

TIME   The time you will need to do a roof frame is hard to generalize; it really depends on the size, complexity and height of the roof. Also how much thinking time you will need is hard to judge. A simple gable roof for a 2000-square-foot house could be completed in 3 or 4 days by two novice builders. A complex roof frame, with intersecting gables, valleys, hips, etc., could take 5 to 10 days.

PEOPLE   2 Minimum (1 skilled, 1 unskilled); 3 to 4 Optimum (2 skilled, 1 to 2 unskilled); 3 to 4 Maximum (all skilled).

TOOLS   Framing hammer, level, tape, pencil, saw horses, combination square, framing square, stair gauges, safety goggles, nail-pulling tools, nail belt, extension cords, power saw, saw blades, hand saw.

RENTAL ITEMS: Scaffolding or ladders, safety harness.

MATERIALS   As mentioned early in the chapter, your roof frame can be trusses, large timbers or standard rafters. The rest of the chapter will deal with roof construction with standard materials. To do the roof frame you will be using framing grade lumber, usually 2×10s or 2×12s. Aside from the rafter stock, you will need framing materials for the ridges, hips, and valleys (often one dimension wider than the rafters); blocking; bracing (temporary); band rafters and ceiling joists. Also you will need nails (8d and 16d) and any fasteners you plan to use.

ESTIMATING MATERIALS NEEDED

Since the rafter stock is usually rather expensive, you will want to order as close to what you really need as possible. Count all the rafters needed (don't forget the rake rafters and gable rafters). Order a few extra for blocking (use the worst pieces for this). Calculate the lengths of the hips, valleys and ridges, including the overhang lengths, and order these. Try to stay as close to the standardized 2-foot intervals in length as you can. It's a shame to have to order an 18-foot length for a 16-foot-2-inch rafter, but the scrap can be used for blocking. Order any ceiling joists at this time, too.

INFORMATION   This chapter should provide you with all the information you need to build a gable or shed roof. To build a complex roof may require added information. There are a few books available just on roof framing which may be available at a technical book store. I would advise you to consult with a good framer for complex roofs and even consider paying them to work with you.

If you are using large timber rafters instead of common framing materials, almost everything is the same as shown in this chapter, except that the stock is thicker and the spacing greater. You may want to buy a rafter table book for layout of the rafters. These books, which are readily available, supply layout information on common rafters, valleys, hips, etc.

PEOPLE TO CONTACT   If you do not intend to do the roof yourself or if you decide you need help, a roofer must be contacted at this time. Aside from the crew you will use, the roofing materials supplier and a plumber should be contacted. (A plumber is needed to install the vent pipes, that protrude through the roof, before the roof framing begins.)

INSPECTIONS NEEDED   The roof framing is inspected at a later date, during the general framing inspection, which occurs after the sheathing is installed. You may, however, want the inspector to inspect the frame as soon as you are finished, before you put on the roof sheathing. If there are any mistakes in the framing, it will be easier to repair before the sheathing is in place. (I once had a house totally enclosed, roofing—everything, when the inspector told me I needed a double ridge because of a loading factor.)

Reading the Plans

Most plans will give you the basic information about building the roof framing but often not many of the details. These are left for the builder to work out at the job site. So don't be surprised if you are standing around scratching your head for long periods of time. Many of the roof framing situations need to be worked out at the site, where it is easier to visualize

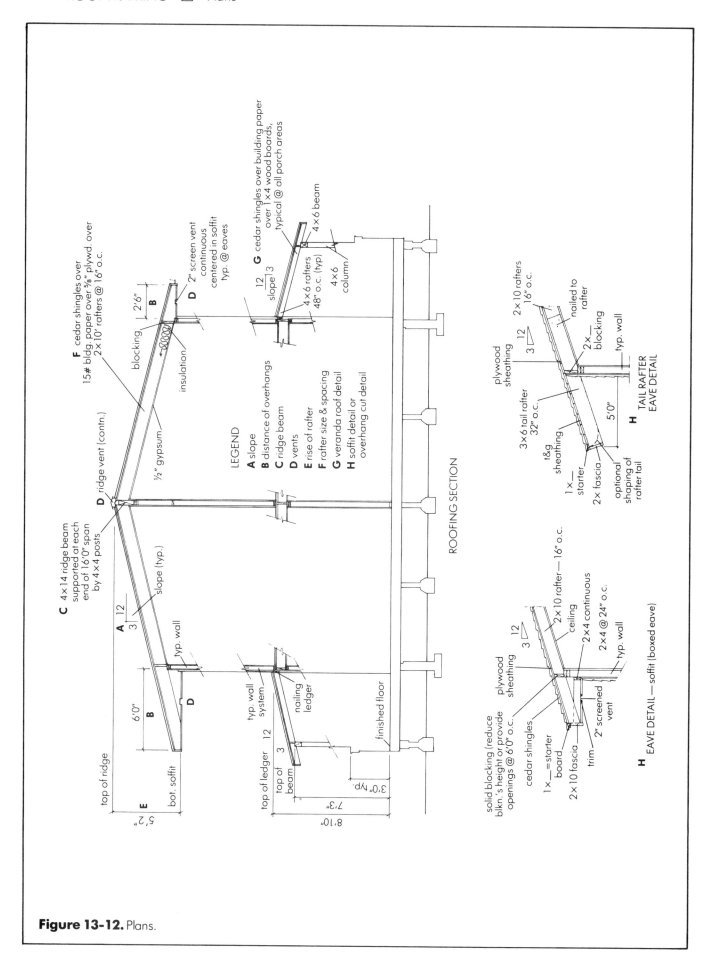

Figure 13-12. Plans.

Photo 13-23. The use of scaffolding during roof framing makes the work easier, quicker and safer.

and take measurements. The plans (Figure 13-12) should however include the following:

A) Slope
B) Size of rafters
C) Spacing of rafters
D) Length of overhang
E) Soffit detail for end cut of rafter
F) Location and size of ceiling joists
G) Size of ridge and any bearing beams
H) Location of any collar beam
I) Loading weights, including live and dead loads

Safety and Awareness

SAFETY: Roof framing is the most dangerous of all the building processes (except for estimating). There is little stable support while working in a high location. Also you are working with long, heavy framing members and the chances for a fall or dropping something on somebody is high. Move at a slow, safe speed and always be safety minded. Use scaffolding (Photo 13-23); it is your best insurance. Be careful where you are stepping and always be sure no one is standing below the work unless absolutely necessary.

Even with all this, there may be situations that just call for bravery (though not stupidity). I remember having to hold an 18-foot 4×8 in the air while sitting on top of a band rafter, looking 32 feet down, as we tried to nail it into the band.

Scaffolding was impossible, so we used a safety harness, which, fortunately, wasn't needed.

AWARENESS: Like all framing, roof framing is not highly exacting (¼ inch margin of error). The main thing to be aware of is that all the pieces fit into each other in the proper way. This can be demanding in complex roofs. Be sure to place the rafters on the correct layout marks, and that the end cuts at the overhangs are in a straight line.

Overview of the Procedure

Step 1: Select a pattern piece
Step 2: Lay out the pattern rafter
Step 2A: Do layout for the ridge cut
Step 2B: Do layout for the odd unit
Step 2C: Do layout for the full units
Step 2D: Do layout for the birdsmouth cut
Step 2E: Do layout for the overhang
Step 2F: Adjust for the ridge
Step 2G: Cut the pattern rafter
Step 3: Check the pattern rafter
Step 4: Lay out the top plates and ridge
Step 5: Install the first sets of rafters and the ridge
Step 6: Install rake or fly rafter
Step 7: Install the band rafter
Step 8: Install blocking (if needed)

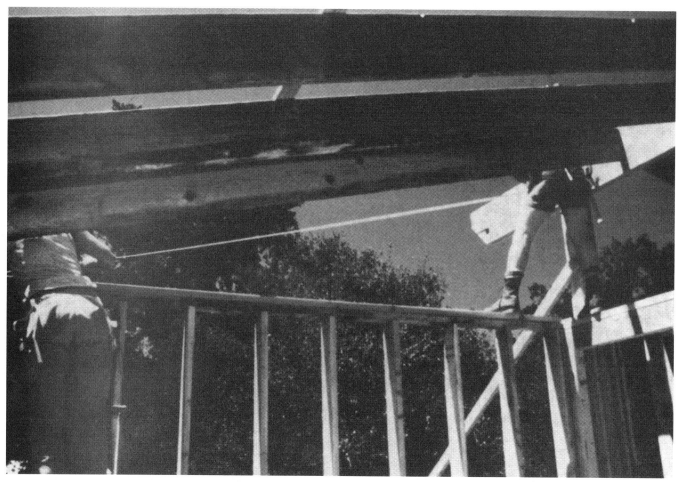

Photo 13-24. If the ridge or the walls are not straight or parallel, you will need to measure for each rafter instead of using a pattern rafter.

Step 1
Select a Pattern Piece

Margin of Error: ¼ inch off true over a 12-foot length.

Most Common Mistakes: Not using your best piece for the pattern; not allowing for deviations in the wall of ridge beam.

Photograph: Photo 13-24.

After digesting all the information in the beginning of this chapter (and I assume you have), and ordering all the proper materials, you are now ready for that glorious moment when you begin the roof framing—the most difficult of all framing. If you think you're not up to the task, step back and look at what you've already done. Then roll up your sleeves and go to work.

Most houses require a series of rafters that need to be cut equal lengths. In a gable (or shed) style roof, these rafters will be running from the ridge to the wall plates. When cutting a series of identical rafters, it is best to cut a pattern, test it and then use this pattern to mark the others. That is what we will be doing.

Before you begin (and this is crucial), you need to check the

walls to be sure that, indeed, all the rafters resting on them will need to be identical. If the tops of the walls where the rafters will rest are not parallel and straight, the distance from the ridge to the wall will vary along its length. If the walls cannot be straightened, and the variance is over ¼ inch, you may have to adjust your cuts as you go along to be sure that the rafters rest flat on the wall plates and tight against the ridge. Often walls can be forced out or pulled in to correct any problems.

If you have already installed a large load-bearing beam to serve as the ridge (see "Ridge Beam or Board" in the beginning of this chapter), you need to check this as well. Sometimes large timbers are warped, and the warps cannot be pulled out. This results in a ridge that does not run in a straight line, and, consequently, the distance from the ridge to the wall is not constant.

If at all possible, correct the warp in the beam. (You may have to winch it in place.) But if it cannot be corrected or if the walls are very crooked, you will need to make adjustments as you cut the rafters. This could involve measuring for each rafter, as shown in the photo. This is when you start realizing that there are no shortcuts: What you do at one stage of construction, affects what you do at later stages.

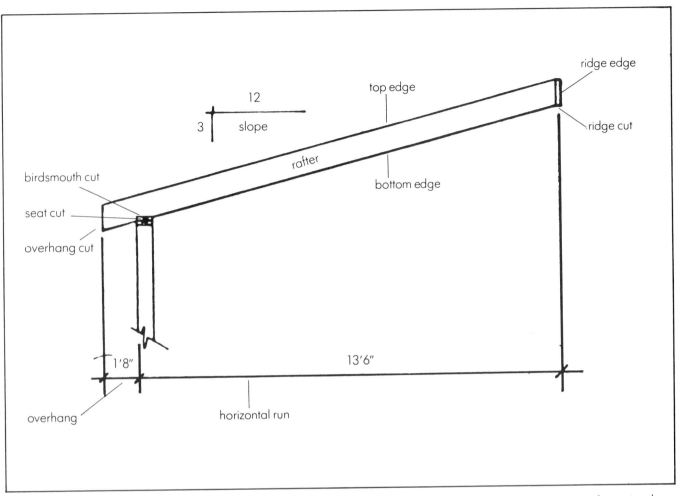

Figure 13-13. Example rafter. The run, the overhang and the slope are all necessary to step off a pattern piece to determine the ridge, birdsmouth and overhang cuts and their angles.

Step 2
Lay Out the Pattern Rafter

Margin of Error: Exact.

Most Common Mistake: Not laying out an accurate pattern.

Diagram: Figure 13-13.

Because of the length of this step, I have broken it into several sub-steps. In our example, which we will use in all the steps, the rafter will have a 13-foot-6-inch run (the horizontal distance from the outside edge of the wall to the centerline of the ridge), a 3 in 12 slope, and a 1-foot-8-inch overhang (the horizontal distance from the outside edge of the wall to the edge of the overhang cut). This is all you need to know to lay out the pattern, and your plans should contain all of this information.

Remember that the layout you are doing (the stepping off process), on the pattern piece is for the purpose of locating 3 cuts: the ridge cut, the birdsmouth cut and the overhang cut; and to determine their angles. This pattern will, in turn, be used to mark the cuts and angles on the rest of the rafters.

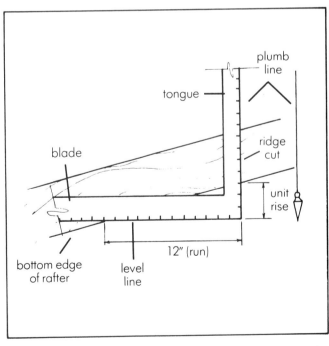

Figure 13-14. Note that the ridge cut will be plumb when the rafter is installed at its proper slope.

Photo 13-25. The square in place to mark off a rafter for a 3 in 12 slope roof. Note the unit rise, 3 inches, is set on the tongue (shorter arm) at the bottom edge of the board; and the unit run, 12 inches, is set on the blade (longer arm), also at the bottom edge of the board. Be sure you are using the numbers on the same side of the blade for both.

Step 2A
Do Layout for the Ridge Cut

Diagram: Figure 13-14.

Photograph: Photo 13-25.

After you have chosen your pattern piece, lay it flat between 2 saw horses. Even though you will be laying out the pattern in this position, you will still need to be able to visualize the rafter in place. Sometimes, if I get confused, I hold up the ridge end during the layout to visualize it.

Place the framing square on the pattern piece, as shown in the photo, with the shorter arm (the tongue) within an inch or two of the ridge end (the end that will eventually be against the ridge) of the rafter. Align the framing square so that the unit rise (3 in our example) is aligned with the bottom edge of the rafter on the tongue, and the unit run (always 12 inches) is aligned with the bottom edge of the rafter at the blade (the longer arm of the square).

Now draw a line along the outside edge of the tongue, close to the ridge end of the board. This will be the ridge or plumb cut. (Actually, we will be adjusting this cut later on to allow for half the thickness of the ridge board. See "Step 2F.")

Some builders do their alignment on the top edge of the rafter instead of the bottom; both will work. Be sure you are very accurate; use a sharp pencil, so you have thin precise marks. Use the numbers imprinted on the outside edge of the square, not those stamped on the inside. Patented clips, called stair gauges, are available that will maintain the square in position as you lay out the pattern.

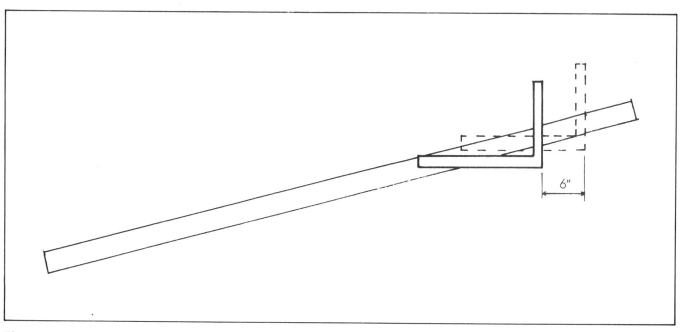

6"

Figure 13-15. Stepping off the odd unit.

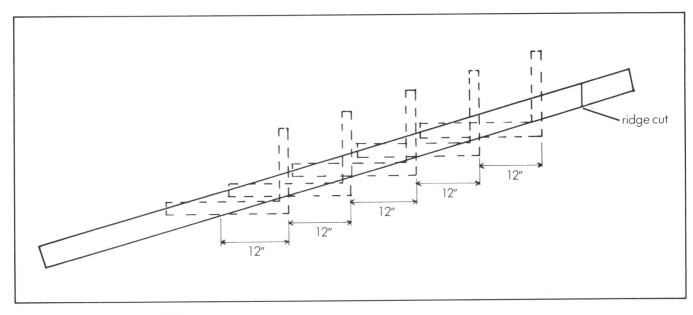

Figure 13-16. Stepping off full units.

Step 2B
Do Layout for the Odd Unit

Diagram: Figure 13-15.

After you have drawn the ridge cut, you are ready to step off the horizontal run, with a 3-inch rise for every 12 inches, in order to ascertain the next cut, the birdsmouth cut. First, you must measure the odd unit. This is a unit of measure on the horizontal run that is less than one foot along the horizontal. Since in our example, there are 13 feet 6 inches in the horizontal run, there is one odd unit measurement of 6 inches. Measure this odd unit (6 inches) along the blade (longer edge) of the square, which should still be in its original position on the stock. Mark the bottom edge of the stock at 6 inches. (Use a straight edge to transfer this point from the outside edge of the square to the board.) As always in the rafter layout, be precise.

You have now accounted for 6 inches of the 13 feet 6 inches that you must measure between the centerline of the ridge and the plumb cut of the birdsmouth (the horizontal run).

Step 2C
Do Layout for the Full Units

Diagram: Figure 13-16.

Photographs: Photos 13-26, 13-27.

Now begin to step off the full units (12-inch units). Slide the square along the stock until the 3-inch mark on the tongue of the square is on this 6-inch mark you have just made on the stock. Make another mark on the board where the 12-inch mark of the blade is crossing the bottom of the stock. Move

Photo 13-26. After the odd unit is marked off, move the square down the rafter to the odd unit mark and mark the rafter at the 12-inch mark on the blade.

Photo 13-27. Stepping off the full units. Continue down the board, placing a mark along the blade at the 12-inch mark, and then transferring the unit rise mark on the tongue (the 3-inch mark) to this point.

your square down again, until the 3-inch mark on the tongue is aligned with this 12-inch mark. Make another mark here and you have just stepped off another 12 inches along the horizontal run, for a total of 2 feet 6 inches. Transfer the tongue to this mark and continue to do this for the rest of the full units in our example 13. The last 12-inch mark that you make will correspond with the building line (the exterior edge of the wall), since it is exactly 13 feet 6 inches along the horizontal run from the beginning point at the ridge. This point also corresponds with the plumb cut of the birdsmouth.

Step 2D
Do Layout for the Birdsmouth Cut

Photograph: Photo 13-28.

After laying out all the full and odd units, you are ready to do the layout for the birdsmouth cut — where the rafter will rest on the wall. The birdsmouth is actually made up of 2 cuts: the seat or level cut, which is the horizontal (level) cut where the

rafter rests on top of the wall providing a good nailing surface; and the plumb cut, where the rafter "hooks" the exterior edge of the wall, preventing the walls from pushing out.

The seat is usually 3½ inches (5½ inches, if it is resting on top of a 2-×-6 wall), and the plumb cut is whatever is created by the seat cut being 3½ inches. To form this cut, turn your framing square over and align one arm along the last full unit mark. (The unit mark should extend, as shown in the photo, across the entire width of the rafter.) With the 3½-inch mark on the other arm of the square crossing the bottom edge of the rafter, draw a line running perpendicular from the last full unit line, toward the ridge end, to the 3½-inch mark. This line represents the seat cut of the birdsmouth.

Step 2E
Do Layout for the Overhang

Photograph: Photo 13-29.

After the birdsmouth cut layout is completed, you are ready to do the layout for the overhang and the overhang cut. You will have to ascertain what cut you plan to use at the end (see p. 289, "Overhangs, Soffits and Fascias" at the beginning of this chapter). In our example, there is a 1-foot-8-inch overhang (measured along the horizontal run) and a plumb cut for the overhang cut.

To lay this out, place the square on the top edge of the rafter, instead of the bottom edge. Once again, adjust the square so that the tongue (which is towards the overhang) crosses the top of the rafter at 3 inches, and the blade (which is towards the ridge) crosses the top edge at 12 inches. Be exact. (Remember to use the numbers on the outer edge of the square.) As shown in the photo, starting at the plumb cut of the birdsmouth (or building line, or last full unit mark, they are all the same), measure the odd unit, in our example 8 inches, and mark at the tongue. Then, move the 12-inch mark on the blade to this mark and draw a line along the outer edge of the tongue. You have now moved 1 foot 8 inches along the horizontal to the overhang plumb cut. In just a few pages I will show you how to check your layout using a rafter table book or the tables imprinted on the square.

Step 2F
Adjust for the Ridge

Diagram: Figure 13-17.

Photographs: Photos 13-30, 13-31.

Before we start to cut our pattern rafter, we need to make an adjustment at the ridge. If you remember, all our measurements have originated at a point that corresponds with the centerline of the ridge board. Obviously, the rafter does not extend this far. The rafter begins short of the centerline by

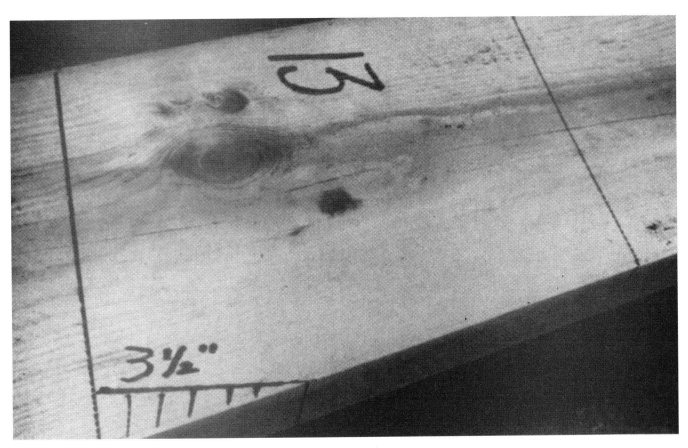

Photo 13-28. Rafter marked for the birdsmouth cut.

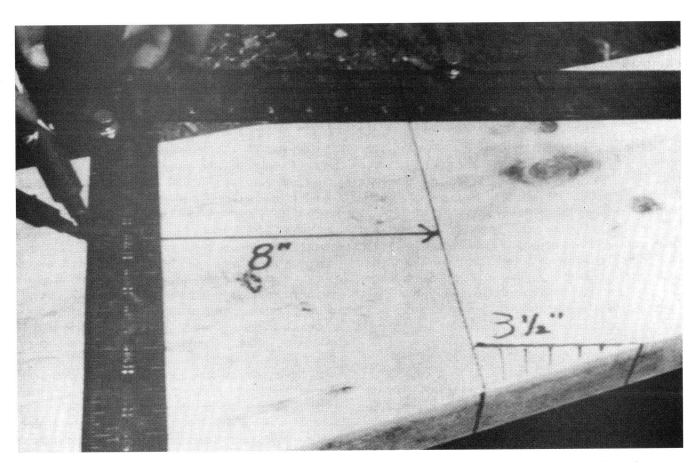

Photo 13-29. Marking the odd unit for the overhang cut (8 inches). Note that the square is now on the top edge of the rafter.

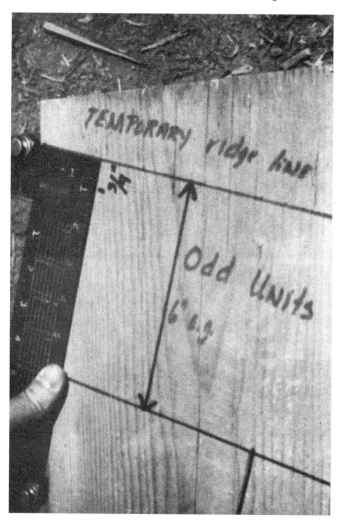

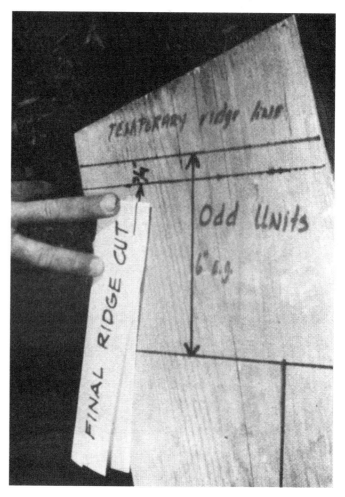

Photo 13-30. Marking the ¾-inch adjustment for the final ridge line.

Photo 13-31. Adjusting for the ridge cut involves laying out a new ridge cut line that is ½ the thickness of the ridge (along a horizontal plane), towards the overhang. In this example the ridge is 1½ inches, so the cut is ¾ inch towards the overhang.

half the thickness of the ridge board, which is short of our original mark. (In our example, we are using a 1½-inch-wide ridge board, so half the distance is ¾ inch). We therefore have to bring our first ridge mark ¾ inch (measured along the horizontal run) back towards the overhang.

To do this, simply place the square against the original ridge mark as shown in the photo, and measure along the blade ¾ inch and make a new mark. Then draw a line parallel to the original ridge mark at this point. This is your final ridge cut layout line. Pat yourself on the back, check your layout (using the tables) and you are ready to cut.

Using the Rafter Tables

Your framing square is a rather amazing tool. Though it seems simple, it holds a wealth of knowledge; somewhat like the great pyramids of Egypt, where, behind a mass of stones, many answers to the universe lie. The framing square can lay out rafters or stairs; measure jack rafters or valley rafters; calculate board feet of lumber; shine your shoes and take out

the garbage. There are books written just on the use of the square, but alas, this is not one of them.

For our purposes, I will show you how to lay out the common rafters using the square. You can then use either method, the square or rafter tables, or the step-off method just shown. Most builders step off the rafter with the framing square, and then check it with the tables.

The square can be used because it will show you the "length of the common rafter, per foot of run." What this means is that for every 12 inches of horizontal run, there is a certain length measured along the slope, for a given roof slope. This is the hypotenuse formed by a triangle using the unit run (12 inches, in our example) as one side and the unit rise (3 inches, in our example) as the other side. If you know what this hypotenuse (the unit length) is for a given slope, you can multiply that number times the total run, using whole numbers and decimals (in our example, 13 feet 6 inches of the total run would be 13.5; if it were 8 feet 8 inches it would be 8.666, etc.). Use a pocket calculator for the arithmetic involved in this process.

The unit length measurement can be obtained from a rafter

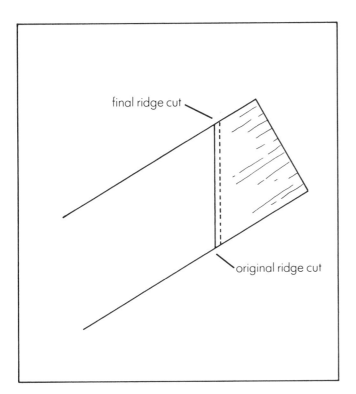

Figure 13-17. To adjust for ½ the thickness of the ridge board, the ridge cut is moved back from the original layout line, towards the overhang.

table book. In these books there are tables that give the unit lengths for all slopes and runs. These tables are also stamped on the blade of the square. Under the full inch numbers stamped on the blade are other series of numbers. The top series is unit length or, as written on the square, "length common rafter per foot run." One of these numbers times the total run would give you the total length along the slope from the building line (the outside edge of the wall) to the centerline of the rafter, as measured along its slope. Looking at the square under 3 inches (since we are using a 3 in 12 slope in our example), we find the number 12.37. Multiply this number times the total run, 13.5 (13 feet 6 inches), and get $13.5 \times 12.37 = 166.995$ inches (166.995/12), which in turn equals 13.91 feet or 13 feet 11 inches. This is the total length along the slope.

You can use this measurement to check your layout of the pattern. The distance on the pattern rafter from the overhang cut to the centerline of the ridge (the original, not final, ridge cut) should be the same as this measurement. If you are off more than ⅛ inch, check to see if you have measured wrong or stepped the rafter off incorrectly.

Photo 13-32. The birdsmouth after cutting. For a 2-×-6 wall, the cut would be 5½ inches.

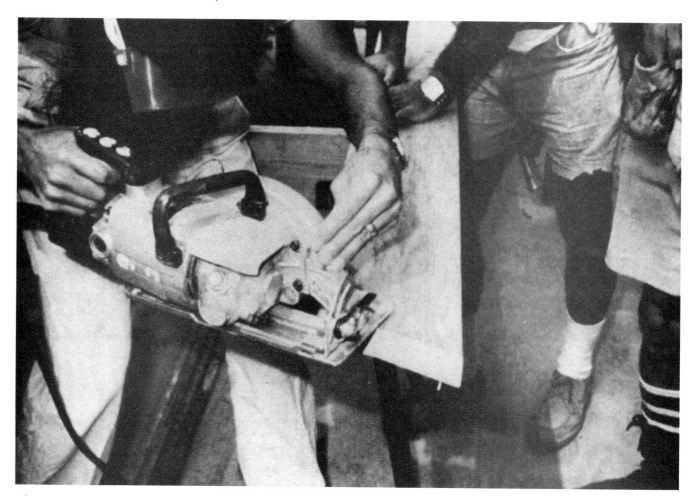

Photo 13-33. Cutting the ridge cut.

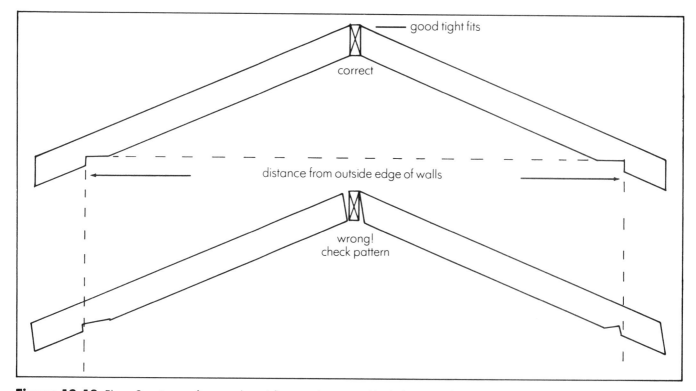

Figure 13-18. Place 2 pattern rafters on the subfloor, with a scrap block the same thickness as the ridge between them, to see how they fit.

Step 2G
Cut the Pattern Rafter

Photographs: Photos 13-32, 13-33.

If the calculated measurements are the same as those used for the cutting lines you made on the pattern, you can cut your pattern rafter. Do this very exactly because the other rafters will be cut from this pattern. Make your cuts at the ridge and overhang with a power saw, and at the birdsmouth, use a hand saw and power saw. You may want to let the overhang end of the rafter run "wild" and cut all the overhang cuts once the rafters are in place (see "Step 7").

Step 3
Check the Pattern Rafter

Diagram: Figure 13-18.

Aside from checking the pattern rafter with the tables, there is one last easy way of checking the rafter before you go and cut all that expensive stock. From the pattern rafter, cut one more rafter. Place these 2 rafters on a flat surface (preferably, on the subfloor) in opposing places, as if they were forming the gable roof. Place a scrap block the same thickness as the ridge between them and spread them out so that the distance between the plumb cuts of the birdsmouths equals the distance between the outside edges of the 2 walls (the roof span). When the rafters are spread like this, they should meet squarely and tightly at the mock ridge board, with no gaps between the rafters and the ridge pieces. If there are gaps, check to see where you are off before cutting more rafters.

Photo 13-34. Lay out the rafter locations on both the ridge and the wall plates. Be careful of splinters and falling.

Step 4
Lay Out the Top Plates and Ridge

Margin of Error: ¼ inch off proper layout location.

Most Common Mistakes: Not allowing for the rake (fly) rafter; placing the "X" on the wrong side of the layout mark; not allowing for any openings; falling off the wall.

Photograph: Photo 13-34.

Once you have ascertained that you have cut an accurate pattern and that the walls (or ridge beams) are not out of alignment to the extent that special measurements are needed, you can begin to cut all the remaining identical rafters. Always use the original pattern piece as the pattern, or you will be compounding errors by switching patterns. You are now ready to place the layout marks on the walls and ridge board in the same way a layout was done for the walls and floors.

When you begin your layout be sure you have accounted for any rake (fly) rafters. Actually your layout should begin from the outer edge of this rafter. Since these rafters are not yet in place you will have to calculate their eventual location, and lay out the other rafters from there.

Usually builders straddle the wall and lay out the top plates from this position. The greatest drawback to this method, needless to say, is getting splinters in the worst places. You may want to use a ladder. Mark your layout on the cap plates on a 12-, 16- or 24-inch center (wider if you are using large timbers for rafters).

Now lay out the ridge board on saw horses. Any joints in the ridge (where 2 pieces of ridge stock come together) must occur at the centerline of a rafter. To do this, you will have to cut your ridge pieces as you lay them out to make them work.

You will also be laying out ceiling joists on the walls at this time, if you are using them. Install them before the rafters; they can be used as a platform in erecting the roof framing. If you are not using any ceiling joists, seriously consider using scaffolding. I would advise it.

Photo 13-35. The ridge board is in place and permanently attached at the gable wall.

Photo 13-36. After the rafters are cut from the pattern, lean them against the wall near where they will be installed.

Photo 13-37. Installing the first set of rafters.

Step 5
Install the First Sets of Rafters and the Ridge

Margin of Error: ¼ inch.

Most Common Mistakes: Splice in ridge not occurring in the centerline of a rafter; placing the rafter on the wrong side of the layout mark; falling off the roof.

Diagram: Figure 13-19.

Photographs: Photos 13-35, 13-36, 13-37, 13-38, 13-39, 13-40, 13-41.

You are now ready to start erecting the rafters and ridge. In some buildings, the ridge will already be set permanently in place if the gable wall has been built (see photo). This makes things somewhat easier; the ridge is in place and stable. Often, however, gable walls (walls formed in the triangle created by the gables of the roof rafters) are not built until the roof rafters are in place. In this case, you will need to temporarily hold the ridge in place as you nail in the first few sets of rafters. If there are enough people around to help, they can just hold the ridge in its proper location. Or, you can build temporary supports for the ridge, as shown in the diagram, to hold it in place until the rafters can secure it.

After the ridge is secured, lean the rafters against the walls, each one near its layout mark. Starting at one gable end (use straight rafters at the gable), position one person at each wall and one person at the ridge. The workers at the wall should slide their rafters over the wall towards the ridge, and the worker at the ridge can hold these rafters against the ridge until the people at the wall toenail the rafters (use 10d) into the cap plates of the wall.

After the first 2 rafters are toenailed at the walls, the person at the ridge can nail them to the ridge board; one rafter can be face nailed (use 16d), but the other rafter will need to be toenailed (use 10d). As you install the rafters, alternate face nailing and toenailing the rafters, e.g., toenail the first one on the left hand side of the ridge, then the next one on the right hand side). Nail in every fourth set of rafters to stabilize the ridge before going back to fill in the other sets. Be sure, as with floor framing, that you install the rafters with the crown up so it will eventually settle down. Plumb and brace the assembly as you go; the ridge must remain straight.

If any metal fasteners are needed for hurricane, earthquake or other structural reasons, install these. In most cases they will be used where the rafters meet the wall and/or where the rafters meet the ridge, as shown in the photos. Check with your inspector to see if any metal fasteners are needed. In earthquake or hurricane areas, I would recommend them; they are inexpensive and quick to apply, but still offer protection in the event of a hurricane or earthquake.

If you are using a ridge beam instead of a ridge board, it will be permanently in place before you begin to install the rafters. This makes roof framing somewhat simpler. The beam rests on posts, usually built into the walls. These posts often require foundation piers directly below because they carry the weight of the roof. The rafters are either set on top of the ridge beam or hung from it.

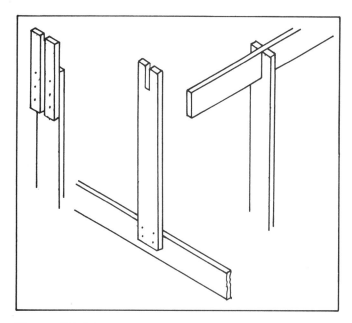

Figure 13-19. Temporary braces can be built to support the ridge rafters.

Photo 13-38. Rafter-to-wall metal fasteners are not common, but they are a good idea in earthquake or hurricane areas.

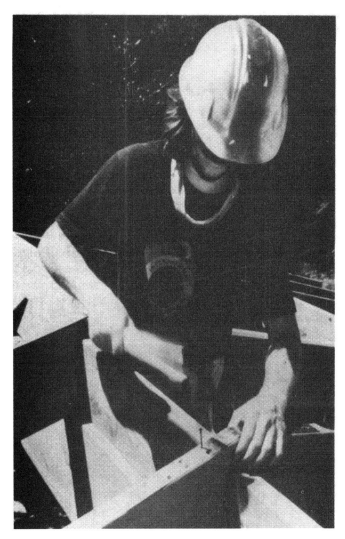

Photo 13-39. Installing metal plates over the ridge to connect corresponding rafters in a set is not always required, but it is a good idea in earthquake country.

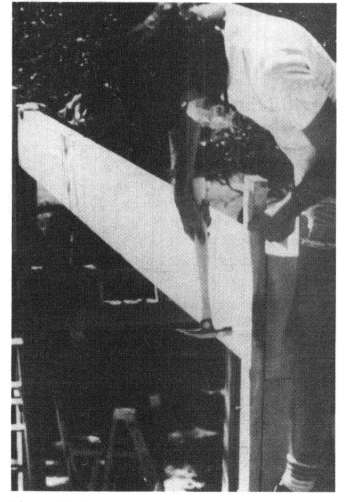

Photo 13-40. Nailing a load-bearing ridge beam to support posts. The 2×4s being nailed on either side of the beam are temporary supports that keep the beam from falling off the posts during construction. Ridge boards are much smaller in dimension than this ridge beam.

Photo 13-41. Installing rafters on top of a load-bearing beam.

Step 6
Install Rake or Fly Rafter

Margin of Error: ¼ inch.

Most Common Mistakes: Not providing adequate support; not installing rafters straight and aligned; falling off the roof.

Diagram: Figure 13-20.

Photographs: Photos 13-42, 13-43, 13-44, 13-45.

After all the common rafters are in place, you are now ready to install the rake or fly rafters. These rafters run parallel to the slope of the roof and create the overhang between the gable wall and the exterior edge of the roof (see diagram). There are several ways these cantilevered rafters can be supported, and you need to determine which method you are going to use before starting the roof framing.

One method for installing the rake rafters is as shown in the diagram and photo. This involves "lookouts" (usually the same dimension lumber as the rafters) being installed between the rake rafter and the last common rafter. As you can see there is no gable rafter and the gable wall serves as a fulcrum to support the weight of the rake rafters.

Another method involves installing the gable rafters, notching them, and then installing 2×4s flat running from the rake rafters to the rafter immediately inside the gable rafter.

The final method does not involve lookouts at all. In this case, the rake rafters are attached only at the ridge beam and the band rafter (see photo). You will need to be sure that the

Photo 13-42. Lookouts are being used to support a suspended rake rafter. Note that there is no gable rafter used here (rafter at the gable wall).

Photo 13-43. These "lookouts" are supporting the rake rafter. Note how the gable rafter is notched and the "lookouts" are inserted.

Photo 13-44. Attaching the rake rafter can be very dangerous. Do not take any chances.

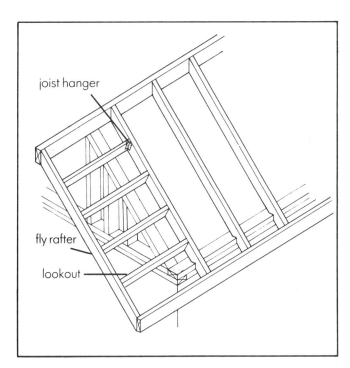

Figure 13-20. Using lookouts to support the rake rafter. Note there is no gable rafter.

Photo 13-45. In large timber roof framing, often the rake rafter is one piece, attached at the large ridge beam and band rafter.

rake rafter is properly sized; there are no supports to break the span. This is the easiest process if you can build the rake with one piece of stock. You will also need to install the band rafter before the rake in this case, since the rake is attached to the band at one end.

No matter what process you use, choose good stock for the rake rafters. Be careful installing these rafters, for nowhere is roof framing more dangerous than when you are leaning over the side of the building to install the rake rafters. Be sure they are installed in a straight line parallel with the gable wall. Check for straightness with a string to be sure it is true. With the sky as a backdrop, any discrepancy or wave in the board will be easily noticed from below.

Step 7
Install the Band Rafter

Margin of Error: ¼ inch.

Most Common Mistakes: Band rafter not installed straight; falling off the roof.

Photographs: Photos 13-46, 13-47.

After all the common rafters are installed, you can apply the band rafter or fascia. You can do this before or after the rake rafters (see "Step 6") are installed. The band rafter or fascia is the board that is nailed to the end of the common rafters at the overhang cut. The size and details of installing this board will depend on your soffit design (see p. 289, "Special Considerations: Overhangs, Soffits and Fascias" at the beginning of this chapter). Depending on the design, you will either be

using a 2-inch board (band rafter) or a 1-inch trim board (fascia) or both. Usually if you are installing the 2-inch-thick band rafter, this will later be covered with a more attractive 1-inch fascia as trim. Often a high-quality 2-inch band rafter is installed, painted or stained and left as the finished fascia.

In our example, we will be installing a 2-inch band rafter. The overhang cut at the end of the rafter is a plumb cut and the band rafter will be installed in a plumb position. The most important thing to remember when installing this rafter (as with the rake rafter) is that it must be installed straight. Any wave in the rafter should be easily noticeable against the sky backdrop, when viewed from the ground. Getting the band rafter straight can be difficult. Often all the overhang cuts are not aligned (in fact I have seldom known them to be). This may be due to a sloppy cut, or a wall or ridge that's not straight.

I highly recommend allowing the overhang ends of the rafters to run wild and cutting the overhang cuts after all the rafters are in place. Whether you are going to cut them in place or check them after they are cut and installed, run a string along the upper edge of the common rafters to see if the ends are all on line (mark them if they are running wild). They need to be straight within ⅛ inch maximum; no waver should be detected. Be very exacting with your cuts here. One of the first things I always do, to check a builder's attention to detail, is to check the straightness of the band rafters.

Once the string is in place, make marks on the rafters where you plan to make the overhang cuts. It there is one rafter shorter than all the rest, all the rafters will need to be cut back to this rafter (you can shim this rafter if it will not be exposed from below). With the square, mark the new overhang cuts. Using the power saw, as shown, make accurate cuts along these marks.

Photo 13-46. The band rafter is applied only after you have checked to be sure that the overhang cuts of the rafters are all in a straight line. Be sure this board is put on straight and level.

Photo 13-47. Cutting the overhang cuts after the rafters are installed is often suggested or required, if they are not all aligned. Wear safety glasses to protect your eyes, unlike the photo.

This is one of the most difficult cuts to make in framing. You are holding the saw in an awkward position and there is sawdust flying in your eyes. If possible, try to climb on top of the rafters and cut in a downward direction. In any case, be sure you are in a secure position and make an *accurate* cut. If you are cutting from below, wear safety goggles and a hooded sweatshirt and gloves. There is nothing more uncomfortable than spending the rest of the day working with sawdust down your back from cuts you made early in the morning. The gloves and hooded sweatshirt will also come in handy when you start to install insulation; another enjoyable task.

Step 8
Install Blocking (If Needed)

Margin of Error: ¼ inch.

Most Common Mistakes: Omitting blocking where needed; totally blocking rafter space where air passages are needed for ventilation; cutting blocking too small, too long or crooked; falling off the roof.

Photographs: Photos 13-48, 13-49, 13-50, 13-51.

As in floor framing, blocking may be required in the roof framing. Blocking, which would occur at the wall line and every 8 feet along the length of the rafters, adds rigidity to the roof frame and keeps the rafters plumb. Check with your plans or local inspector to see if blocking is required.

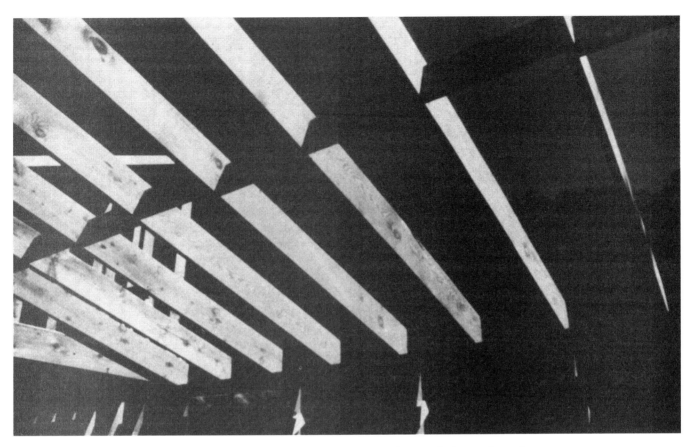

Photo 13-48. If you need to ventilate an attic area, screen vents can be installed every few rafters.

Photo 13-49. Blocking is often installed every 8 feet. Note that the blocks are smaller than the rafters to allow for ventilation.

Photo 13-50. Installing blocking at the wall line is done either after all the rafters are in place or as they are installed. Note that the blocks are aligned with the outside edge of the wall and do not reach the top of the rafter, to allow for ventilation.

Photo 13-51. If rafters are resting on top of the ridge, instead of hung from it, you will need to install blocking at the ridge.

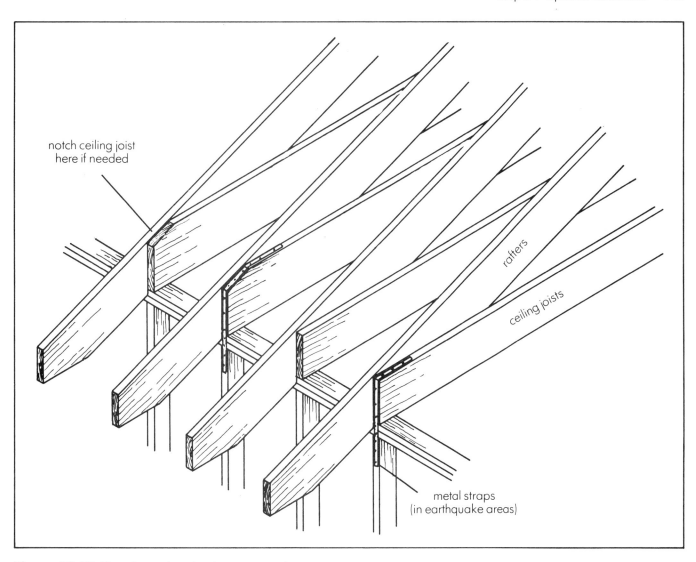

notch ceiling joist
here if needed

rafters

ceiling joists

metal straps
(in earthquake areas)

Figure 13-21. The relationship of ceiling joists to rafters.

Blocking is often installed at the building line (the outside edge of the wall) between the rafters. If a ventilation area is needed between the rafters (see, p. 293, "Special Considerations: Ventilation" at the beginning of this chapter), be sure that the blocks do not totally fill the rafter cavity; this will block ventilation. A 1½-inch gap at the top should be left for air flow. This gap will later be covered by a screen.

Install the blocks *flush with the outside edge of the wall.* If the interior wallboards are going to run up between the rafters on the inside (as in an exposed beam ceiling), you will need to provide a second nailing surface for the interior wall flush with the inside edge of the wall as well.

Blocks may be required along the length of the rafters as shown in the photo. Usually these are placed every 8 feet and, again, do not totally block the passage if air is to ventilate between the rafters. Stagger the blocks so they can always be face nailed, and be sure they are cut the proper size and straight so that they do not push the rafters out of alignment. Also check to see if the rafter is cupped or warped or twisted before installing the blocks. If any are, adjust the blocks to be sure the rafters stay on line.

Special Situations

Ceiling Joists
Diagrams: Figures 13-21, 13-22, 13-23, 13-24.

Many homes will require ceiling joists on which to hang the ceiling materials. If you are using a truss roof frame, the bottom chord of the truss will serve as the ceiling joists (see p. 283, "Design Decisions: Trusses" at the beginning of this chapter). If you are building the roof frame at the job site, you will need to cut and install your own ceiling joists. These ceiling joists are similar to the floor frame, but are lighter and have no band around the outside. They are usually 12 inches and 16 inches on center and are sized to carry a 20-pound dead or live load. Their size varies according to span, load and type of wood, but if you ever plan to use the attic as a living space or to store heavy objects, be sure to size the joists to handle these loads.

Ceiling joists can be supported in a number of ways. If the span is not too great or the loads are not too heavy, you can simply span the joists from wall to wall. If the span needs to

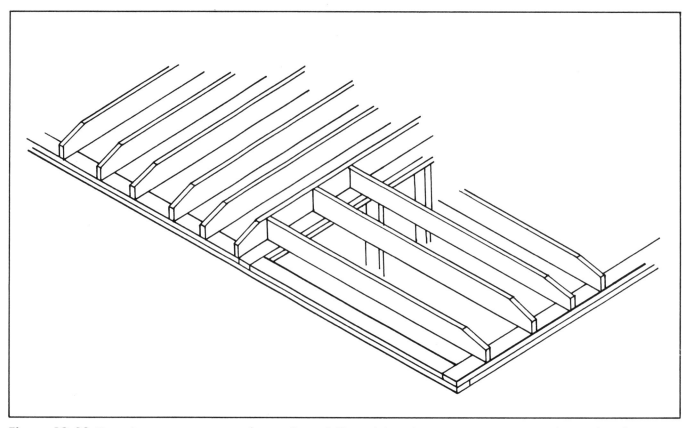

Figure 13-22. The ceiling joists are spanning from wall to wall. The ends have been cut so as not to protrude past the rafters.

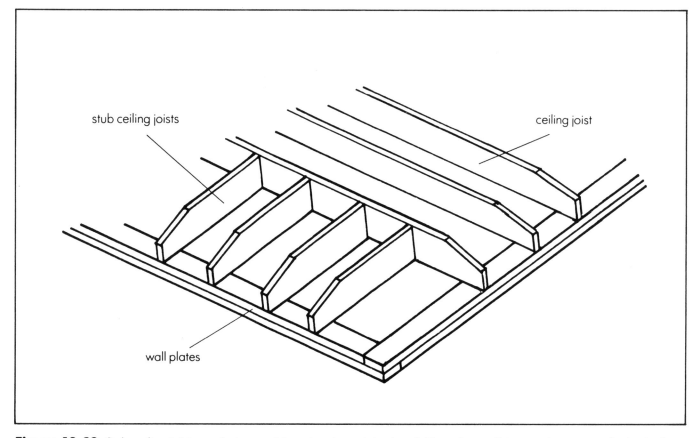

Figure 13-23. Stub ceiling joists are being used here in a low-pitched roof. The rafters will be running perpendicular to the regular ceiling joists.

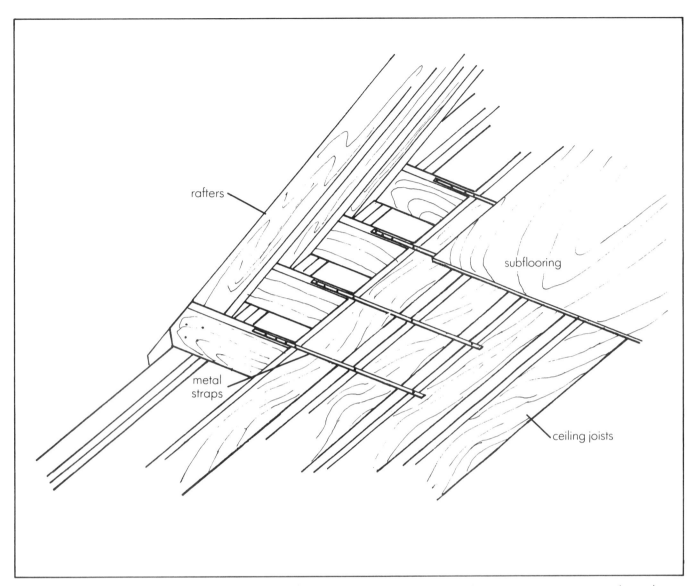

Figure 13-24. Metal straps are being used to tie the rafters to the subflooring when ceiling joists are running at a right angle to the rafters.

be broken by a support, you can build a post and beam into the building on which the joists can rest. Also you can hide this post and beam in the wall and ceiling, if this better suits your design. You can also use a "stongback," which is a beam in the attic fastened to the ceiling joists from above. This is often "retro-fitted" into existing ceilings that later have to be strengthened to carry a greater load.

Ceiling joists are usually cut to the slope of the roof where they intersect the rafters. These can often be rough cuts done with a hatchet, since they will be hidden in the attic. If you are using a very low sloped roof, you may need to use "stub ceiling joists" to replace the ceiling joists that are parallel and adjacent to the wall on which the rafters rest (see diagram). Otherwise the ceiling joist closest to the wall will be in the way of the rafters resting on the wall.

If the rafters run at right angles to the ceiling joists, stub joists and straps (metal) can be used to tie the rafters (and walls) to the attic floor diaphragm. These help to reduce the pressure created by the rafters to split the walls apart.

If any walls on the story below the ceiling are running parallel to the ceiling joists, the tops of these walls can be stabilized by placing blocks between the ceiling joists flush with the bottom edge of the joists. The cap plate of the wall is then toenailed into these blocks. Some ceilings may also need to be blocked (or bridged) as in floor framing. Again this adds rigidity to the ceiling and keeps the joists straight. Ask the inspector if it's needed, or if you ever plan to carry much of a load on the joists, install these now.

When there are openings in the ceilings (fireplace, chimneys, stairs, skylight shafts, etc.), they are framed identical to floor frame openings. You may not be required to use double trimmer joists or headers, however, if the ceiling does not carry a large load. Remember that most codes require an access to the attic (often built into the closet ceiling). This access is at least 22 inches × 30 inches with a clear height of at least 30 inches in the attic.

Photo 13-52. To frame for a skylight is similar to framing an opening in a floor.

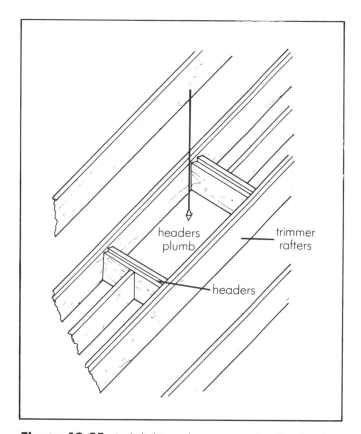

Figure 13-25. A skylight or chimney opening. The headers and trimmers are single or double, depending on the opening sizes and the roof loading factors.

Openings in the Roof Frame
Diagram: Figure 13-25.
Photograph: Photo 13-52.

Openings in the roof framing, created to allow for skylights or chimneys, are constructed in a similar way to those in the floor frame. If the opening can be formed within the normal span of 2 rafters, thereby eliminating the need to cut any rafters, only headers need to be added at the proper locations. Indeed, many skylight manufacturers do build their skylights to fit between the rafters. If a rafter needs to be cut, however, you may need to double the trimmer rafters and the headers, depending on the size of the opening and the load of the roof. The headers can be installed either plumb or square with the rafter, whichever way the design or manufacturer's requirements dictate.

As you create the opening, be sure that it is the proper rough opening as called for by the manufacturer's dimensions for that skylight. If you are building your own skylights, be sure that they match the rough opening you have provided in the roof framing. Check also the locations of these openings with respect to the fireplace chimney or the place you need light, to be sure they are properly located. You would not want to build a rock chimney and then find the opening was not aligned.

Never attach a masonry chimney to the frame. The chimney, due to its weight, will settle faster than the building, and if you attach the frame to it, it will pull the frame down. I saw this in an older house; the floor actually dipped at the hearth

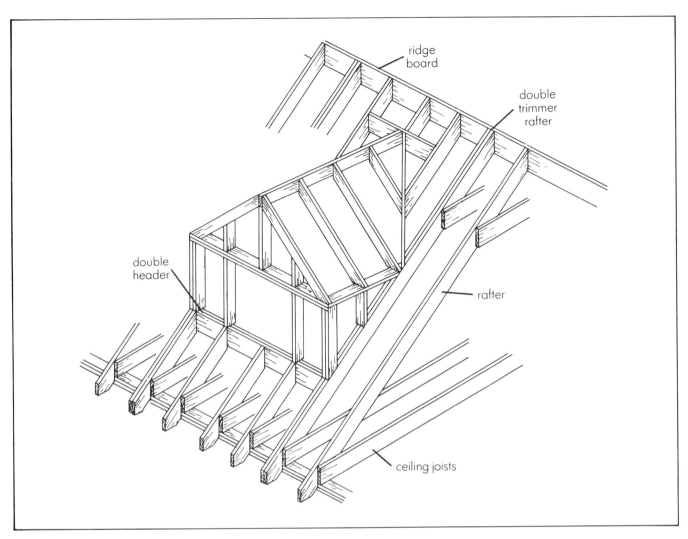

Figure 13-26. The framing of a roof dormer.

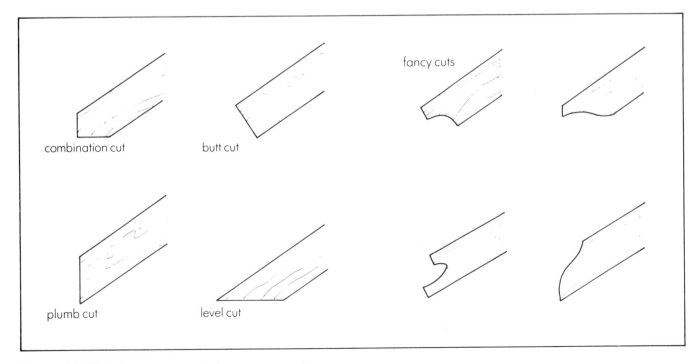

Figure 13-27. Different ways in which overhangs can be cut.

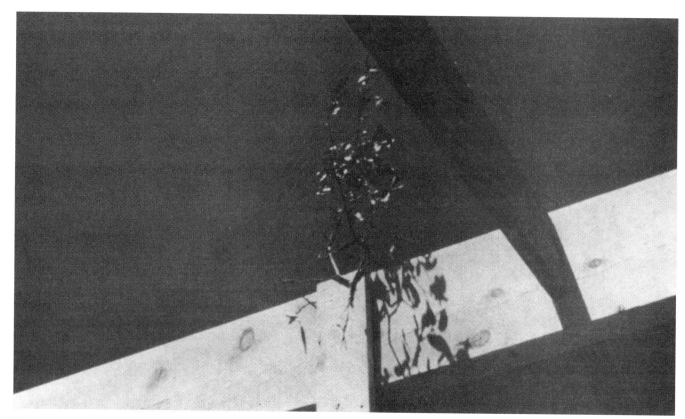

Photo 13-53. A Japanese tradition is to attach an evergreen twig to the top of the highest framing member as soon as it is erected and to celebrate. Not a bad idea to regain your perspective on how far you've come at this stage.

because they had attached the floor frame to the fireplace foundation. Leave at least a 1½-inch airspace between the frame and the chimney for fire safety (more, if required locally).

Dormers
Diagram: Figure 13-26.

A dormer is a framed part of the roof projecting from a sloping area of the roof, that contains a vertical window unit. They are often built into attic areas or upstairs rooms with sloping ceilings to allow for ventilation, light and more head room. They are also considered appealing from both the interior and exterior. An opening for a dormer is created in the same way as those for skylights or chimneys with the dormer framing built on top of the frame, as shown in the diagram. They can be built on either shed or gable roofs.

Fancy Cuts
Diagram: Figure 13-27.

Often fancy cuts are made on a ridge or at the end of the rafters when they are to be exposed to the exterior of the home. Your imagination is the limit. You would be surprised at what a difference these cuts make aesthetically, and they only take a few extra minutes per rafter. A little detailing like this really gives the house a lot of character. Be sure your design does not catch and hold water. If the rafters are exposed to the weather, use a weather-proof sealant on them.

Twig at the Top
Photograph: Photo 13-53.

The Japanese have a very nice custom. When the last roof framing member is in place, they attach the twig of an evergreen tree to the highest member and celebrate the completion of the frame of the house. I suggest you also build in ceremonial activities to your house construction. They allow you time to step back and get in touch with the beauty and joy of building your own house. It is also a chance to be thankful for the abundance that allows all this to happen.

Worksheet

ESTIMATED TIME

_____

_____

_____

_____

CREW

Skilled

_____ Phone _____

_____ Phone _____

_____ Phone _____

Unskilled

_____ Phone _____

_____ Phone _____

_____ Phone _____

_____ Phone _____

_____ Phone _____

COST COMPARISON

_____ Store:

$ _____ per 1000 board ft. rafter stock, $ _____ total

_____ Store:

$ _____ per 1000 board ft. rafter stock, $ _____ total

_____ Store:

$ _____ per 1000 board ft. rafter stock, $ _____ total

_____ Store:

$ _____ per 1000 board ft. rafter stock, $ _____ total

PEOPLE TO CONTACT

Roofer _____

Phone _____ Work Date _____

Roofing materials supplier _____

Phone _____ Work Date _____

Plumber _____

Phone _____ Work Date _____

MATERIALS NEEDED   Rafter stock; framing materials for ridges, hips and valleys; blocking, bracing and band rafter material; ceiling joists and nails.

ESTIMATING MATERIALS

RAFTERS: Size: _____ Length: _____ Number: ____
(include blocking)

RIDGES: Size: _____ Length: _____ Number: _____

VALLEYS: Size: _____ Length: _____ Number: ____

HIPS: Size: _____ Length: _____ Number: _____

CEILING JOISTS: Size: _____ Length: _____

Number: _____

Daily Checklist

CREW:

_____ Phone _____

_____ Phone _____

_____ Phone _____

_____ Phone _____

TOOLS:

☐ framing hammer

☐ level

☐ tape

☐ pencil

☐ saw horses

☐ combination square

☐ framing square

☐ stair gauges

☐ safety goggles

☐ nail-pulling tools

☐ nail belt

☐ extension cords

☐ power saw

☐ saw blades

☐ hand saw

RENTAL ITEMS:

☐ ladders

☐ scaffolding

☐ safety harness

MATERIALS:

Rafter stock _____

Hips _____

Ridges _____

Valley _____

Blocking _____

Band rafters _____

Ceiling joists _____

Nails _____

Fasteners _____

DESIGN REVIEW:

Are you sure where you want that skylight?

INSPECTIONS NEEDED:

Inspector's name _____

Phone _____

Time and date of inspection _____

WEATHER FORECAST:

For week _____

For day _____

14 *Roof Sheathing*

Photo 14-1. Skip sheathing before the ends are cut.

What You'll Be Doing, What to Expect

ROOF SHEATHING is the skin or surface that covers the roof rafters. There are different choices in materials: plywood, 2-×-6 T&G boards, skip sheathing, 1-×-6 butt-end boards, etc. We will discuss each one of these and then, in detail, describe how to apply three of them: plywood sheets, skip sheathing and 2-×-6 T&G boards.

Like wall sheathing and subflooring, roof sheathing is rather straightforward and boring work to do. There are few opportunities for mistakes and since it will all be covered by the roofing material, appearance is not a concern. In some ways it is easier than wall sheathing, because you are nailing on a flat, near-horizontal surface, and the sheets do not have

to be supported while the first nails are being driven. There are also fewer openings to cut out. However, since you are working high on the roof, the materials have to be lifted up high and your pace is slower because of the increased danger.

Sheathing can become very tiring since there is a lot of repetitive nailing from a bending position. Stretch a few times each hour to relieve the pressure from your back. Knee pads are highly recommended. As with wall sheathing and sub-flooring, you may want to consider using a pneumatic nailer in these areas.

This stage goes quickly and easily. The view from your working position can often be restful. Though somewhat physically demanding, the mental demands are few and it may be a good time to use some of those inexperienced friends who promised to come help.

Jargon

EAVES. The part of the roof that extends beyond and over the walls of the building.

END MATCHED BOARDS. Boards with tongue and groove on their ends as well as their edges.

EXPOSED BEAM CEILING. A ceiling where the roof beams or rafters (usually 4 inches thick or more) are exposed to the living space below as a design effect.

ROOFING FELT. Asphalt impregnated paper used as a moisture barrier and added protection below the roofing.

ROOF LOAD. The weight or load on the roof from the combined weight of the sheathing, roofing materials, snow and wind.

ROOF SHEATHING. A structural covering consisting of boards or panels that are attached to the exterior surface of the rafters and on which the roofing is nailed.

SELECT DRY DECKING. Tongue and groove boards, 2×6s or 2×8s, used for roof sheathing, especially in areas where the sheathing will be exposed to the living surface below. It is select, meaning there are few knots; and dry, meaning it has a low moisture content, reducing shrinkage to a minimum.

SKIP SHEATHING OR SPACE SHEATHING. Butt-end boards, 1×4s or 1×3s, applied to the exterior surfaces of the rafters with a designated space between each board. Used for rigid roofing materials only (e.g., wood shakes, metal, tile, etc.).

SOFFIT. The underside area, where the roof projects to the outside of the walls of the building.

TONGUE AND GROOVE BOARDS. Boards that are cut in such a manner that they have a tongue or protruding section on one side and a groove or continuous indentation on the other side, and each panel locks into the one adjacent to it.

2-4-1 PLYWOOD. A 1½-inch-thick plywood that can be used for roof sheathing when the rafters are up to 48 inches on center.

UNDER LAYMENT. A layer of roofing felt below the roofing.

WEATHER EXPOSURE. The part or area of the shake or shingle that is exposed to the weather.

Purposes of the Roof Sheathing

1) Nailing surface for the roofing material
2) Adds strength and rigidity
3) Carries the roof load to the rafters

Design Decisions

Your choice of roof sheathing materials depends on a few different factors: roof loading, type of roofing, use of exposed ceilings, and spacing of the rafters. Read carefully the following section and narrow down your choices.

Factors to Consider in Selecting Sheathing Materials

ROOF LOADING: The roof load is the combined weight or load of the sheathing, roofing, snow and wind. The codes also factor in additional weight for people and materials on the roof during re-roofing. The heavier the roof load, the thicker or stronger the sheathing must be for any rafter span. In areas of high snow loads, both the thickness of the sheathing and the size of the rafters are increased. Check your local codes for the roof-load factor in your area to determine the sheathing thickness and the rafter size and spacing for your house.

TYPE OF ROOFING MATERIAL: (Photo 14-1.) Flexible or pliable roofing materials, such as asphalt shingles, rolled roofing, built-up hot tar roofing, etc., must have solid sheathing; they are too soft and pliable to span a gap. Rigid roofing materials, such as tile, wood shakes or shingles, metal, etc., can span a gap and skip sheathing (boards or slats with spacing between them) can be used. This type of sheathing is less expensive, and quicker and easier to install. In some areas, skip sheathing is required for wood shakes or shingles to provide adequate ventilation and reduce the risk of the shakes or shingles rotting.

EXPOSED BEAM CEILING: If you plan to expose your roof beams or roof framing structure to the interior of the room below, you will need a sheathing that will have at least one finished side, the side that will serve as the finished interior ceiling.

SPACING OF THE ROOF RAFTERS OR BEAMS: The greater the spacing between the roof supports, the stronger the sheathing will need to be in order to span the distances. Often, exposed beam ceilings are 4-inch-thick beams on 4- to 8-foot centers. Situations like this will demand the use of either 2-4-1 plywood or 2-inch T&G boards for roof sheathing.

In the following discussion on the different types of roof sheathing, decide which ones can be used in your specific situation, and which one of these is the best choice.

Types of Roof Sheathing

PLYWOOD SHEATHING: (Photo 14-2, Figure 14-1.) Unless wood shakes or shingles are used, plywood roof sheathing is most commonly used. It is relatively low in cost and it is quick and easy to apply. Usually an exterior glue plywood that is water resistant is used for protection during construction. Some builders even paint over exposed exterior edges, at the eaves, to prevent exposure to any wet weather. In homes where there are no exposed ceilings or rafter spacings over 48 inches, this is used.

The chart on page 327, from the 1976 Uniform Building Code, gives the regulated thickness of plywood, given the span, the use of blocking or not, and the load capacity ("total" is for both live and dead weight). Check with your local codes to see if these apply in your area. The Identification Index refers to the numbers stamped on each piece of plywood. The first number refers to the maximum rafter spacing; the second, the maximum joist spacing.

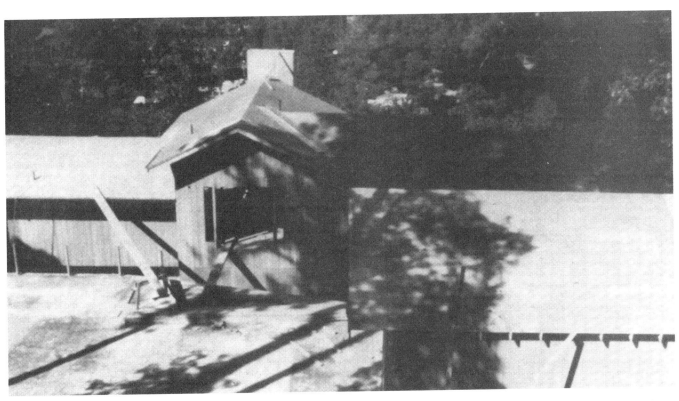

Photo 14-2. Plywood roof sheathing is most commonly used if you are not using a rigid roofing material. It is not only strong, but it is quick and easy to apply.

Photo 14-3. Individual 1-inch-thick boards are seldom used for sheathing since the introduction of the stronger plywood. (Photo courtesy of Heartwood Owner Builder School, Washington, Massachusetts.)

Photo 14-4. This is an insulated roof sheathing panel. It has a foam center with a particleboard exterior sheathing and a drywall interior siding. (Photo courtesy of Heartwood Owner Builder School, Washington, Massachusetts.)

Photo 14-5. A closeup of insulated roof sheathing.

Panel Identification Index	Plywood Thickness	Maximum Span (inches)		Load (lbs. per sq. ft.)	
		Edges Blocked	Unblocked	Live	Total
12/0	5/16	12	—	150	155
16/0	5/16, 3/8	16	—	75	95
20/0	5/16, 3/8	20	—	65	75
24/0	3/8, 1/2	24	16	50	65
30/12	5/8	30	26	50	70
32/16	1/2, 5/8	32	28	40	55
36/16	3/4	36	30	50	55
42/20	5/8, 3/4, 7/8	42	32	35	40
48/24	3/4, 7/8	48	36	35	40

Even in roofs with exposed beam ceilings, plywood can be used. 2-4-1 Plywood is a 1⅛-inch-thick plywood that can span roof beams with a 4- to 6-foot on center span (unless there are heavy snow loads or other loads). This material can be bought with one surface finished so that it looks OK from below. (I recently saw, in a student's home, where she had used a plywood siding, with a rough-sawn finished side, as roof sheathing exposed to the interior. I was pleasantly surprised by its attractive appearance, as I seldom like plywood ceilings.)

The minimum thickness plywood used for roof sheathing is ⁵⁄₁₆ inch. Usually, however, roof sheathing is between ⅜ inch and ¾ inch thick. The sheets are always applied perpendicular to the rafters and need to span at least 3 rafters for strength. This type of sheathing may be used with wooden shakes and shingles. In some areas the shakes or shingles are nailed directly to the plywood. In other areas, where greater ventilation is required, 1×2s are nailed to the sheathing with spaces between each slat, and the shakes or shingles are nailed to these.

2-INCH TONGUE AND GROOVE BOARDS: This type of siding is very popular in areas where there is an exposed beam ceiling. It is also used in flat or low-pitched roofs. It can span a 4- to 8-foot span, depending on the roof load, and has an appealing look when viewed from below. Selected for its few knots, it is called "Select Dry Pine Decking," as it is often of pine, and has a low moisture content to reduce shrinkage. Cedar, redwood and fir are also available. The most common size is 2×6 or 2×8, though 3×6 and 4×6 are also sometimes used. The 4×6 can span from 10 to 12 feet.

Usually this type of decking or sheathing is "V" grooved. This means that as the tongue and groove are fitted together, the edges on one surface meet at a V shape, rather than flush. This "V" groove helps hide the inevitable gaps that will form (even dry boards will eventually shrink). The groove is faced downward towards the living area, making a very attractive ceiling. Insulating is usually done by nailing rigid styrofoam sheets to the exterior side of the sheathing. Roofing is sometimes nailed through this insulation and into, but not totally penetrating, the decking. This is not always a satisfactory application, especially in high wind areas. In this case, another layer of plywood sheathing is nailed over the decking and the roofing is nailed to this. With wood shakes or shingles, slats are often nailed to the insulation to provide the proper ventilation and nailing strength.

This type of exposed beam ceiling can become expensive, for the following reasons: the high cost of the heavier roof timbers, the more expensive Select Dry Decking, the more expensive insulation (rigid insulation is more expensive than the same "R" factor in fiberglass insulation), the added labor to apply the individual boards and finally, the possible need for a second layer of sheathing. Before deciding whether it is worth the added expense or not, calculate the exact additional costs. Decisions made without complete data can be very dangerous in housebuilding.

SLATS OR SKIP SHEATHING: Skip sheathing (also called space sheathing) is used with various types of rigid roofing materials. These include wooden shakes and shingles, tile, metal roofs and slate. (Solid sheathing should be used in areas with blowing snow.) All of these materials have the rigidity to span the space left between the boards. Either 1×4s or 1×3s are nailed perpendicular to the rafters. They are spaced so that their "on center" distance equals the weather exposure of the roof shingles to be used. Common low-quality softwoods are usually used. The sheathing is quick and easy to apply. It cannot be used, however, if it is to be exposed to the view from below, in which case, you would see a not too appealing collage of slats, shingles and tar paper. In garages or utility or storage spaces you may want to consider exposing it to view.

SQUARE EDGE BOARDS: (Photo 14-3.) Before the advent of plywood, common square edge boards were used as sheathing. These are usually nominal 1×6s or 1×8s. Their end splices need to meet over a rafter and each board has to span at least 3 rafters. Boards that extend over the edges of the roof to the rake rafters have to span at least 4 rafters. End splices also need to be staggered by at least 2 rafters. Usually it is best to use stock that has a 12% moisture content or less so as to minimize gaps. These boards are slower to apply than plywood and not as strong, so I would suggest using plywood, unless you have a very inexpensive source of square edge boards.

NEW PRODUCTS: (Photos 14-4, 14-5.) There are several new products available in roof sheathing material. One type is a solid board bonded together by layers of Kraft paper. The boards come in sizes up to 2×16 feet. Another product is an insulate board that uses a composite material with some insulating properties. It usually comes in 2-×-8-foot panels, 2 inches thick, and has tongue and groove edges. Both of these products have various types of finished surfaces, and can be used as an exposed beam ceiling.

Another type of sheathing I have seen effectively used at the Heartwood Owner Builder School in Massachusetts is a thicker sheathing that has a layer of drywall for the interior facing, a pressboard for the exterior and a center core of 3 inches or so of foam insulation. The advantage of this type of sheathing is that it provides the sheathing, insulation and interior finish in one application. Check with your local suppliers for more info on these and other new types of sheathing.

Now, decide which sheathing is best for you. The rest of this chapter is in three parts: installing plywood sheathing, installing tongue and groove 2-inch decking, and installing skip or space sheathing. Proceed to the appropriate section.

Installing Plywood Sheathing

What You'll Need

TIME 4 to 8 person-hours per 320 square feet (ten 4-×-8 panels).

PEOPLE 2 Minimum (1 semi-skilled, 1 unskilled); 3 Optimum (1 semi-skilled, 2 unskilled); 6 Maximum (2 teams of 3).

TOOLS framing hammer, power saw, hand saw, metal tape, chalkline, framing square, pencil, nail belt, knee pads, ladders, combination square, safety goggles, nail-pulling tools, saw blades, extension cords, terminal plug boxes, staple gun or hammer tacker, staples.

RENTAL ITEMS: Pneumatic nailer, scaffolding, safety harness.

MATERIALS You will be using a ⅜-inch, ½-inch, ⅝-inch or ¾-inch-thick plywood. Usually an exterior glue plywood is used in cases where it may rain before it is covered. If a plywood is used that does not contain an exterior glue, be sure that you seal any edges that will be exposed so that it does not delaminate. Unless you plan to expose the interior surface, a CD grade will do. Usually a straight edge plywood is used with metal clips (discussed later), though some builders use T&G boards with no clips.

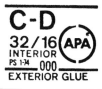

Figure 14-1. A typical stamp on a piece of plywood. "C-D" refers to the grade of veneer of the panel faces. "Interior" designates the type of plywood. "Exterior glue" refers to the type of glue used.

ESTIMATING MATERIALS NEEDED

PLYWOOD: Divide the number of square feet of roof surface by 32 square feet (4-×-8 panel), and add 15% for waste.

NAILS: Calculate the number of pounds per square feet needed and multiply times the number of square feet to be covered.

TAR PAPER: Enough to cover the surface of the roof in square feet.

INFORMATION For the most part, this section will give you all the information needed to apply typical plywood roof sheathing.

PEOPLE TO CONTACT Contact your roofer (if you are using one) and your roofing supplier to let them know you will be needing them soon. Contact a sheet metal person if you are using one to provide any flashing on the roof. Check with the plumber to be sure all the proper plumbing vents are through the roof before you begin the application of roofing materials (the sheathing can be placed before).

INSPECTIONS NEEDED Before the roof sheathing is covered with roofing paper, the inspector may want to inspect the sheathing for proper nailing. Check your local codes.

Reading the Plans

Usually the plans will indicate the type and thickness of the roof sheathing to be used. Plans should also show details of any special situations, such as slats nailed to plywood sheathing or 2-×-6 decking. You may want to reconsider any decisions made by the designer; be sure you agree with their choice of materials and thicknesses.

Safety and Awareness

SAFETY: Be very cautious when working on the roof; it is very dangerous. A fall from the roof, especially through the rafters, can cause severe injury. Workman's compensation insurance is almost 50% higher for roofers than for carpenters; this gives you a sense of the dangers involved in working up high. Do not lift panels to the wind—they will act as a sail.

AWARENESS: Keep the panels perpendicular to the rafters, stagger the joints, and be sure you do adequate nailing.

Overview of the Procedure

Step 1: Temporarily nail on the first (eave) course
Step 2: Check and permanently nail on the first course
Step 3: Nail on the remaining courses
Step 4: Do sheathing at ridges, hips and valleys
Step 5: Apply the tar paper

Step 1
Temporarily Nail on the First (Eave) Course

Margin of Error: Perpendicular to rafter — *exact*.

Most Common Mistakes: Falling off the roof; aligning course with edge of roof rather than perpendicular to rafters; plywood wrong side up; nailing intermediate rafter that becomes the next course's edge rafter.

Diagram: Figure 14-2.

Begin applying the sheets from the eaves (bottom) of the roof and work your way up to the peak. The last course at the peak may need to be ripped if the roof length is not in 4-foot intervals. If the overhangs are going to be exposed to view from below, with no soffit, you may want to consider using plywood with a finished surface and exterior glue where it will be exposed at the eaves and overhangs.

The main thing to watch for here is to see that the eave line is indeed exactly perpendicular to the rafters. Often it is best to only temporarily nail this course, until you have checked it out. If the eave line is not exactly perpendicular to the rafters, be sure to adjust the edges of the plywood until they are. It is more important that the edges be perpendicular than that they be flush with the edge of the roof at the eaves. A tapered piece can be cut to fill in if needed.

Check the rafter ends (or tails) also, to be sure they are all on a straight line. Often they are not. You may need to pop a straight line and recut them, as explained in Chapter 13, *Roof Framing,* "Step 7." If there will be no soffit to hide any discrepancy, it is important that these rafters be cut straight before the sheathing is applied.

Don't forget to leave some overhang at the eaves if a fascia board is to be added on later, unless your plans call to cover the edge of the plywood with the fascia. Face the better surface of the plywood towards the sky. In areas where rain is a threat during construction, you may want to leave expansion gaps of ⅛ inch at the edges, ¹⁄₁₆ inch at the ends. Some builders double these spaces in very wet or humid areas. You will occasionally have to cut a panel in order to make up for these gaps as you go along.

Step 2
Check and Permanently Nail on the First Course

Margin of Error: Same as for "Step 1."

Most Common Mistake: Same as for "Step 1."

After the first course is up and temporarily nailed, check it out to be sure that the sheets are running perpendicular to

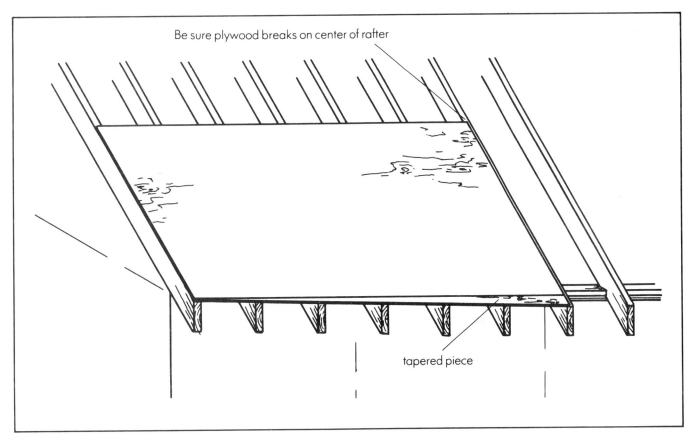

Be sure plywood breaks on center of rafter

tapered piece

Figure 14-2. Note that a tapered piece has been used at the eaves. This is done to keep the inside edge of the plywood on the center of the rafter.

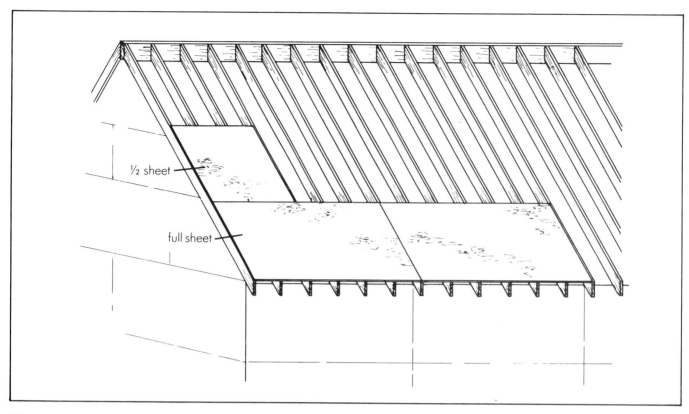

Figure 14-3. Use half sheets of plywood for alternating courses to be sure that joints are not aligned.

the rafters and their ends are falling on the centers. Make any needed adjustments. Once they are in line, nail them, using 8d CC Sinkers or annular groove nails, 6 inches along the edges and 12 inches in the intermediate rafters. As you nail, do not nail the field rafter that will become the next course's end rafter (see Chapter 10, *Subflooring*, "Step 3" for further explanation). This allows you to move the rafter around a little if you need to. As in subflooring and wall sheathing, pop chalklines across the sheets to locate the centers of the rafters as you nail.

Step 3
Nail on the Remaining Courses

Margin of Error: Nail end of panel within ¼ inch of the center of the rafter.

Most Common Mistakes: Falling off roof; no gaps.

Diagrams: Figures 14-3, 14-4.

After the first course is properly set, you can begin nailing the remaining courses. They are applied as the first one, with a few things to remember. First, stagger your joints. You can do this by cutting panels in half and using a half sheet to start every other course. Then, remember to leave the gaps as you go along, if they are needed. Some areas require the use of metal plywood clips to support the panels along their edges between the rafters. You can avoid this either by using a

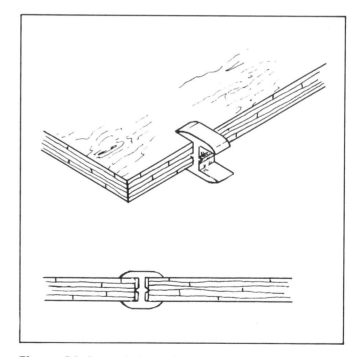

Figure 14-4. Metal plywood clips.

⅛-inch-thicker plywood than required, by using tongue and groove panels, or by blocking under the panel, i.e., nailing blocks between the rafters. Continue to work your way carefully up the roof, checking for alignment and end support as you go.

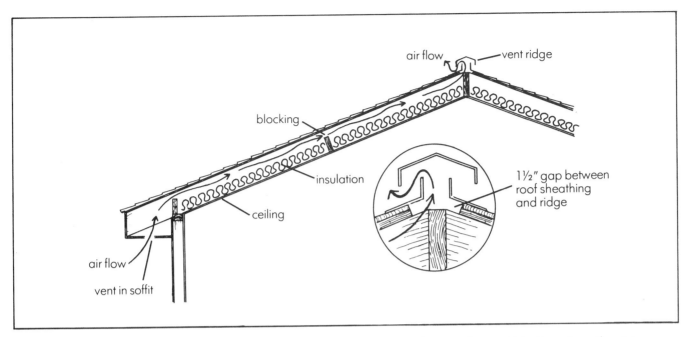

Figure 14-5. Example of how to allow ventilation through the roof panel. Note how smaller-sized blocking allows the air to pass from the soffit to the ridge.

Step 4
Do Sheathing at Ridges, Hips and Valleys

Margin of Error: ¼ inch.

Most Common Mistake: Falling off the roof.

Diagram: Figure 14-5.

Photograph: Photo 14-6.

Where the panels meet at the hips, ridges and valleys, no special beveled or angled cuts are needed, unless the sheathing will be exposed from underneath. Otherwise the sheathing needs to be nailed down flat and tight and the sheets should touch, but no special treatment is needed.

If you are using a ridge vent to ventilate the roof panel, leave a 1½-inch gap on both sides at the ridge to allow a continuous flow of air.

Step 5
Apply the Tar Paper

In order to make the house weather tight as well as protect the sheathing from the weather and possible expansion and buckling, it is advisable to apply the tar paper (roofing felt) as soon as possible after the sheathing is on. The roofing felt, also called underlayment, provides additional weather protection for the roof, and places a barrier between the shingles and the wood.

Usually the material used is a 15-pound (15 pounds per 100 square feet) roofing felt. This is a breathable material. You do not want to use a non-breathing material, which will act as a vapor barrier and trap moisture below—it can possibly rot the sheathing. Follow instructions for felting in accordance with the type of roofing to be used.

Many builders and roofers apply a metal drip edge if they are using asphalt shingles. These are applied both at the eaves and along the slope. Drip edges are usually ready-made and are of 26 gauge galvanized steel with a top flange of 3 to 4 inches. At the eaves, the paper is placed *over* the drip edge; at the edge of the slope, the paper is placed *under* the drip edge.

Photo 14-6. A plywood roof sheathing being applied. Note how all the joints are staggered.

Installing Tongue and Groove 2-Inch Decking

What You'll Need

TIME   15 to 25 person-hours per 500 square feet.

PEOPLE   2 Minimum (1 skilled, 1 unskilled); 3 Optimum (1 skilled, 2 unskilled); 6 Maximum (2 teams of 3 working on 2 sections of the roof).

TOOLS   Framing hammer, framing square, combination square, power saw, hand saw, metal tape, pencil, nail belt, knee pads, ladders, safety goggles, sledge, nail-pulling tools, saw blades, extension cords, terminal plug boxes, staple gun or hammer tacker, staples.

RENTAL ITEMS: Scaffolding, ladders, safety harness.

MATERIALS   Usually this sheathing is called select dry decking (Figure 14-6). It's done with fir or pine or the more expensive redwood or cedar. Use the type of wood that is "V" grooved on one surface, or all your gaps will be highlighted. Be sure that the material is dry, or it will shrink. Since in most applications, it will also serve as the interior ceiling, it is important that the material look good on the "V" groove side, showing no discoloration, yard marks or scratches. Usually 2×6s or 2×8s are used (occasionally thicker stock is used). I recommend the 2×8s; the boards are likely to be straighter and therefore easier to apply than

2×6s. Do not use anything wider than 8 inches; it will tend to cup.

When you order, take into account a waste factor of 10 to 15%. As you estimate the materials, remember the boards are 5½ inches, not 6 inches, wide. Order the longest stock possible. Check the stock for quality while it is in the yard or on the delivery truck, *before* they dump it on your site.

ESTIMATING MATERIALS NEEDED

DECKING: Each board foot will cover one square foot. Add 15% for waste.

NAILS: Estimate the number of pounds per square foot.

TAR PAPER: Estimate the number of feet that will cover the roof surface.

INFORMATION   This chapter should give you everything you need in the way of information to apply this type of sheathing.

PEOPLE TO CONTACT   Contact your roofer (if you are using one) and the roofing supplier to let them know you will be needing them soon. Contact a sheet metal worker if you are using one to provide any flashing for your roof.

You may also need to call your plumber back (unless you are doing your own plumbing). All plumbing vents, as well as electrical weatherheads, must be run through the roof before the roofing is applied.

INSPECTIONS NEEDED   Before you cover the roof decking with tar paper, the inspector may need to inspect the nailing pattern of your sheathing. Check your local codes.

Reading the Plans

The plans will simply state that you are using tongue and groove decking. Actually, there is little else you need to know before you install it. You may want to change the overhang areas, as mentioned in "Step 2," if you are going to enclose these with soffits.

Safety and Awareness

SAFETY: Be very cautious in doing any work on the roof; it is very dangerous. A fall from the roof, especially through the rafters, can cause severe injury. Workman's compensation insurance is almost 50% higher for roofers than it is for carpenters, which should give you a sense of the dangers involved in working up high.

AWARENESS: Because the decking will be your finished ceiling, inspect the boards well before you apply them. Be care-

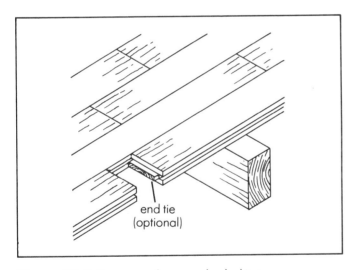

Figure 14-6. Tongue and groove dry decking.

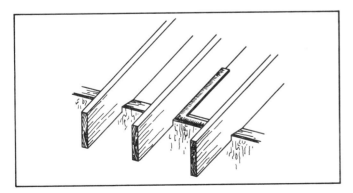

Figure 14-7. Checking rafters with a framing square to be sure they are perpendicular to the wall.

ful of dings and dirt. Do not step on them before they are applied; a footprint on the ceiling looks rather strange. Be sure the boards are running in a good straight line—parallel with the walls. Hopefully this will make them exactly perpendicular to the rafters as well.

Overview of the Procedure

Step 1: Check rafter alignment
Step 2: Apply the first course
Step 3: Apply the remaining courses
Step 4: Make cut for the ridge
Step 5: Apply the tar paper

Step 1
Check Rafter Alignment

Margin of Error: Rafters ¼ inch off proper alignment.

Most Common Mistake: Allowing too great a margin of error.

Diagram: Figure 14-7.

Before beginning your application, check the rafters. Walk around the building to see if they are all running parallel to each other. Very often one was installed on the wrong side of the layout mark. Check to see if the rafters are running at true right angles to the walls they are resting on. If they are not, either they were installed improperly or the walls are not running parallel to each other. Try to determine what has gone wrong. If it is not easy to correct or if the discrepancy is not too great, you can let it go and learn to live with it. Chances are you will never notice it unless you are lying on the living room carpet, staring up at the ceiling.

If the rafters are not running perpendicular to the walls they are resting on, and you install the decking boards perpendicular to the rafters, the decking will not be running parallel to the walls. I think it is more important that the decking runs true parallel to the walls than it is for the decking to run true perpendicular to the rafters. The mistake will be less noticeable.

Check to be sure the rafter ends at the overhang are all on a straight line. Often they are not. You may need to pop a straight line and recut them, as explained in Chapter 13, *Roof Framing,* "Step 7." If there will be no soffit to hide any discrepancy, it is important that these rafters be cut straight before the sheathing is applied.

If the rafter ends are running in a straight line and this line is parallel to the wall the rafters are resting on, you can apply the first course of deck boards, which will be aligned with the edge of the rafters. If this is not the case, you will have to pop a line on the top of the rafters to mark where the first course will go.

If you plan to enclose the overhang areas with a soffit, you do not need to use the expensive T&G in these areas of the roof. Since this area will not be seen from below, you can use the less expensive plywood sheathing. The top of the plywood will have to be shimmed to make it flush with the top of the thicker decking.

Photo 14-7. Do not nail the butt end of 2 boards down where they meet over a rafter until the course above it is installed.

Step 2
Apply the First Course

Margin of Error: ¼ inch off proper alignment.

Most Common Mistakes: Course not running parallel with wall; installing boards upside down with "V" groove facing the sky (see "Types of Roof Sheathing: 2-Inch Tongue and Groove Boards" in the beginning of this chapter); not pointing the tongue downroof.

Photograph: Photo 14-7.

You are now ready to apply the first course. Choose good straight pieces for this course because all the other courses will be laid against this one. The rafters should be marked where the course will be laid (unless you are using the ends of the rafters as marks). Place the boards on the rafters with their "V" grooves facing down, not up. See that the tongues are facing downhill of the slope, and the grooves are facing uphill of the slope; you will be banging against the uphill sides of the boards and it is better to be banging against the grooves, which are not so easily damaged, than the tongues.

Nail the first course on with 16d nails, 2 at every rafter. Apply pressure, where needed, to bring the board in a straight line. You may not want to nail in the uphill nail at each rafter until after the next course is installed. Nailing it here can bring the decking so tight against the rafter that it will be difficult to bring the tongue of the next course into the groove. You may not want to nail in any of the nails at the ends of the boards where the 2 pieces of stock come together. This allows you to move these ends, making sure their grooves line up, so the tongue of the next course can be easily inserted.

Usually the T&G is cut flush with the outside edge of the rake or overhang rafters. Then a trim or fascia board is applied to cover this raw, exposed edge. Before you apply your sheathing, be sure you understand how you plan to finish out the edges of the roof. If you do not know, leave some overhang that can be cut later, when you decide.

Step 3
Apply the Remaining Courses

Margin of Error: ¼ inch off alignment; ⅛ inch crack between boards.

Most Common Mistakes: Boards not fitted tightly enough; boards upside down; dirt or dings on underneath (exposed) side.

Photographs: Photos 14-8, 14-9.

Continue applying each course up to the top of the roof. As with the first course, avoid nailing the upper edges or ends until the adjacent course has been installed. Every now and then go below to see if you are looking good—the boards are still running parallel to the wall.

Photo 14-8. Be sure to use a "beater block" so as not to damage the tongue or groove of the decking during installation.

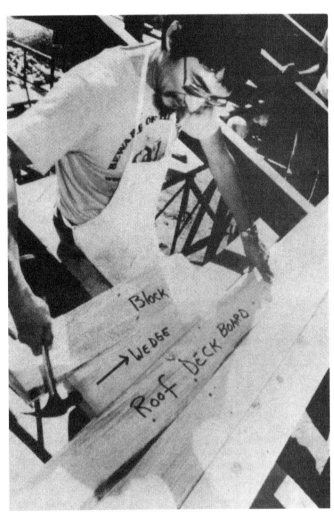

Photo 14-9. Using a temporarily nailed block and a wedge to force a bowed roof deck board tight against the lower one.

One of the hardest things about applying this type of sheathing is getting the pieces to fit tightly along the edges. Because almost all boards have a bow, if the pieces are not fit tightly together, you will see the cracks from below. Usually a sledge is used to pound these boards into place. A scrap piece of T&G is used so you will not be hitting against the groove of the piece you are installing. This works to some degree, but bowed boards still keep popping out after they are knocked home with the sledge.

One of our teachers taught me a great way to bring the boards together. Temporarily nail a block of wood to the rafter 2 to 3 inches above the unnailed board. Insert a precut wedge block between the T&G stock and the block you just nailed. Using your framing hammer (or sledge for difficult pieces), tap the wedge until it pushes the unnailed piece of decking tightly to the lower nailed piece. It works, even for badly bowed pieces.

In most cases, with an exposed beam ceiling, the rafter or beams will have no blocking between them. Because of this, any warps or twists will prevent them from running straight and parallel to each other. If you do not straighten them, if you nail the decking on and trap them crooked, it will be very noticeable from below. As you apply the decking, you may

want to straighten the rafters and nail a temporary brace to the rafters to hold them straight until the decking is applied.

Step 4
Make Cut for the Ridge

Margin of Error: ⅛-inch gap between ridge pieces.

Most Common Mistake: Too great a gap.

Photograph: Photo 14-10.

At the ridge, you may need to make a special cut on the decking so that the ridge pieces fit together well. If there is a large ridge beam and the interior decking will not be exposed from below at the ridge, this will not be needed. But if the decking at the ridge is visible, you will want it to fit tightly together. To do this, you will need to bevel cut the upper edge of the last course of stock on both sides. If there are metal ridge straps running from rafter to rafter on the top edge (as in earthquake areas), you will have to drill through these metal straps to nail on the last few courses of decking.

Photo 14-10. The final boards at the ridge will need to be beveled.

Step 5
Apply the Tar Paper

In order to make the house weather tight as well as protect the sheathing from the weather and possible expansion and buckling, it is advisable to apply the tar paper (roofing felt) as soon as possible after the sheathing is on. The roofing felt, also called underlayment, provides additional weather protection for the roof, and places a barrier between the shingles and the wood.

Usually the material used is a 15-pound (15 pounds per 100 square feet) roofing felt. This is a breathable material. You do not want to use a non-breathing material, which will act as a vapor barrier and trap moisture below — it can possibly rot the sheathing. Apply the tar paper according to the type of roofing you are using. Secure the paper well with staples to be sure that it does not blow off before the roofing is applied. Top sheets should overlap the sheets below by 4 to 6 inches to create a shingle effect.

Many builders and roofers apply a metal drip edge if they are using asphalt shingles. These are applied both at the eaves and along the slope. Drip edges are usually ready-made and are of 26 gauge galvanized steel with a top flange of 3 to 4 inches. At the eaves, the paper is placed *over* the drip edge; at the edge of the slope, the paper is placed *under* the drip edge.

Installing Skip (Slat) Sheathing

What You'll Need

TIME   16 to 25 person-hours per 1500 square feet of roof area.

PEOPLE   1 Minimum (skilled); 2 Optimum (1 skilled, 1 unskilled); 4 Maximum (2 teams of 2).

TOOLS   Framing hammer, combination square, power saw, saw blades, hand saw, metal tape, pencil, nail belt, ladders, safety goggles, nail-pulling tools, extension cords, terminal plug boxes.

RENTAL ITEMS: Scaffolding, ladders, safety harness.

MATERIALS   Usually the least expensive grade of 1×4 is used. If you have access to used 1-inch material, this may be a good place to use it, even if it is wider than 4 inches and has to be ripped down. Try to buy pieces that are as long as possible (14 feet, 16 feet, 18 feet) so less cutting is required.

ESTIMATING MATERIALS NEEDED

In order to estimate the materials needed, you will have to know the weather exposure of your roofing shingles (see "Step 3"). Knowing this, you can then figure your on center distance of the slats, and from that you can calculate the number of courses you will need and the amount of material you will need.

INFORMATION   You may want to check with your roofing supplier as to the on center distance between slats they recommend. Otherwise this chapter tells you all you need to know.

PEOPLE TO CONTACT   Contact your roofer, if you are using one, and the roofing supplier to let them know you will be needing them soon. Contact a sheet metal worker, if you are using one to apply the flashing on your roof and the solar panel installer, if you're using solar panels.

Unless you are doing your own plumbing, call the plumber. All plumbing vents, as well as electrical weatherheads, must run through the roof before the roofing is applied.

INSPECTIONS NEEDED   None.

Reading the Plans

Most plans should simply indicate the use of slat or skip sheathing. If your plans call for wood shakes or shingles, you will probably be using skip sheathing (in some cases solid sheathing can be used).

If skip sheathing is being used on top of plywood or 2-inch T&G sheathing, the plans may even show it in detail. Many builders add skip sheathing on top of solid sheathing when using wood roofs (shakes or shingles), to ensure greater ventilation and thereby reduce the possibility of rot.

If you are using other types of rigid roofing, e.g., tile, metal, etc., you will probably be using skip sheathing; it is cheaper and easier to apply.

Safety and Awareness

SAFETY: This is a very dangerous procedure. As with all roof work, the height is a dangerous element, but unlike the solid types of sheathing, with skip sheathing you can step between the slats or step on the slat between the rafters and break through. *Always step directly on top of the rafters, or very close to one. Never step on a slat between the rafters.* Be careful also while cutting on the roof. Be sure both you and the board you are cutting are stable before proceeding.

AWARENESS: Be sure the slats stay at the proper spacing, by checking with a tape every 8 to 10 courses. Other than that, there is little to watch for.

Overview of the Procedure

Step 1: Check rafter alignment
Step 2: Apply the first course
Step 3: Apply the remaining courses
Step 4: Apply the ridge course

Step 1
Check Rafter Alignment

Margin of Error: Rafter ¼ inch off proper alignment.

Most Common Mistake: Allowing too great a margin of error.

Before beginning your application, check the rafters. Walk around the building to see if they are all running parallel. Very often they are installed on the wrong side of the layout mark. Check to see if they are running perpendicular to the walls they are resting on.

Applying skip sheathing is not an exacting procedure. The sheathing will not be exposed to the inside (as in the case of T&G) and the slats can be cut to any length; therefore, the locations of the rafter centers are not as vital as with the plywood sheathing. There is one situation to watch for, though. If you are planning to apply sheetrock panels or other fixed-length ceiling panels directly to the bottom side of the rafters, it is important that there be a rafter at all the right places to receive the splice in the ceiling panels. If this is the case for you, check this out with your tape as you apply the slats, and move the rafters around as needed. At times this may involve knocking out blocking in the rafter system and replacing it with different size blocking.

Unless the above situation applies, fix any major discrepancies, but otherwise don't worry too much about rafter alignment.

Step 2
Apply the First Course

Margin of Error: ¼ to ½ inch off proper placement.

Most Common Mistake: Falling off the roof.

Diagram: Figure 14-8.

Photograph: Photo 14-11.

After you have checked the rafter placement, you can begin to apply the skip sheathing. If you do not plan to use a soffit, which would enclose the underside area of the rafters where they hang over beyond the walls, you will want to deal with the overhang areas in a way that you will not look up and see the slats. That can be done by using solid slats or an attractive ¾-inch plywood (the same thickness as the slats) such as plywood siding. If a soffit is planned, it will hide these areas from view, so just your common slats can be used.

Apply the first course of slats so that the outside edge is flush with the outside edge of the band rafter. At the end of the roof, apply the outside end of the slat flush with the outside edge of the rake rafter. If you are using no soffit and using solid sheathing at the eaves and overhangs, apply the solid sheathing in the same manner. Be sure that all splices between slats occur in the center of a rafter.

Use two 8d nails where the slats cross each rafter. Avoid placing any cracks or large knotholes between the rafter spans. This creates weak spots.

Usually I bring a power saw up on the roof with me and make the few cuts I need up there. You can allow the slats to hang over the edges of the roof and then cut them all at once when you are finished applying the slats.

Photo 14-11. Nail the first course of skip sheathing flush with the edge of the band rafter. Be sure it is applied in a straight line.

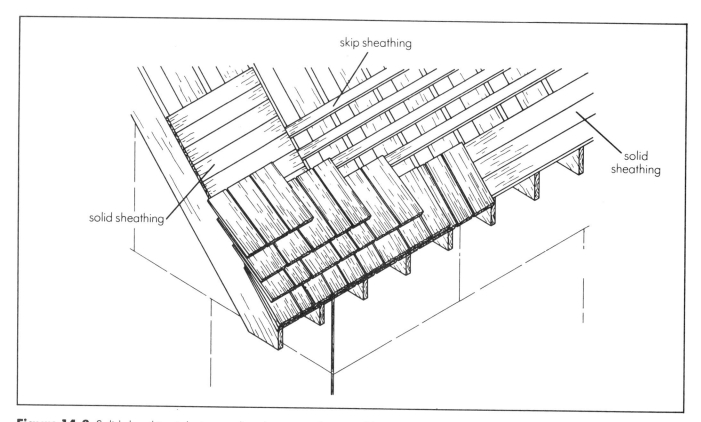

skip sheathing

solid sheathing

solid sheathing

Figure 14-8. Solid sheathing is being used on the eaves where it will be exposed from below.

Photo 14-12. After every few rows of skip sheathing, check your courses to be sure you are still on line.

Photo 14-13. Applying skip sheathing. The board in the middle is a spacer board and will be removed.

Photo 14-14. Skip sheathing as it is being applied to the roof. Note how the builder is stepping close to a rafter.

Step 3
Apply the Remaining Courses

Margin of Error: ¼ to ½ inch off alignment.

Most Common Mistakes: Getting off line; falling off the roof; sticking your foot through a weak spot.

Photographs: Photos 14-12, 14-13, 14-14.

Using the on center distances for the slats in relation to the weather exposure of the roofing, continue applying the remaining courses on the roof. If, for example, you are using a wood shake roof with a 10-inch weather exposure, the on center distance between slats would be 10 inches. The gap between two 3½-inch-wide boards would be 6½ inches. In some wood shingle applications, where there is a small weather exposure (3 to 5 inches), this formula does not apply. In this case the on center distance of the slats would make the slats butt together, thereby forming a solid sheathed roof with no ventilation for the shingles. Consult the suppliers in your area if you are using a wood shingle roof with a small weather exposure. If you are using a metal roof, consult the manufacturer's instructions on how to space.

Every 8 to 10 courses, take your tape out and measure from the bottom of the roofline to see if you are running true.

If not, make whatever adjustments are needed in the courses you have yet to apply. You do not need to tear up the courses you have nailed to correct discrepancies; exactness is just not that important.

Step 4
Apply the Ridge Course

Margin of Error: ¼ to ½ inch.

Most Common Mistake: None..

Photograph: Photo 14-15.

Be sure that there is a course at the top of the ridge. This course will be needed to provide a nailing surface for the last course of shakes or shingles that are to be applied to the roof. If you are using a ridge vent to ventilate the roof panel, leave a 1½-inch gap at the ridge, on both sides, to allow a continuous flow of air.

You may or may not have to apply an underlayment layer of tar paper. Wood shingles use no tar paper (roofing felt) at all. Wood shakes use shake paper applied in a very definite manner (see Chapter 15, *Roofing,* "Wood Shakes," Step 1, p. 355). Other rigid roofings may or may not use a tar paper underlayment; ask your supplier.

Photo 14-15. Skip sheathing completed on the roof.

Worksheet
Plywood Sheathing

ESTIMATED TIME

_____

_____

_____

_____

CREW

Skilled or Semi-Skilled

_____ Phone _____

_____ Phone _____

_____ Phone _____

Unskilled

_____ Phone _____

_____ Phone _____

_____ Phone _____

_____ Phone _____

_____ Phone _____

MATERIALS NEEDED

Plywood, nails, sealing glue, tar paper, staples, metal clips.

ESTIMATING MATERIALS

PLYWOOD

No. of sq. ft. of roof surface _____

divided by 32 sq. ft. (4-×-8 panel) + 15% waste =

_____ sheets

TAR PAPER

_____ sq. ft. roof surface = _____ rolls.

NAILS:

_____ lbs. cover _____ sq. ft.

Total no. of lbs. needed _____.

COST COMPARISON

_____ Store:

Cost per sheet _____ total $ _____

_____ Store:

Cost per sheet _____ total $ _____

_____ Store:

Cost per sheet _____ total $ _____

_____ Store:

Cost per sheet _____ total $ _____

PEOPLE TO CONTACT

Roofer _____

Phone _____ Work Date _____

Roofing materials supplier _____

Phone _____ Work Date _____

Sheet metal person _____

Phone _____ Work Date _____

Plumber _____

Phone _____ Work Date _____

Solar panel installer _____

Phone _____ Work Date _____

Daily Checklist

CREW:

_____ Phone _____

_____ Phone _____

_____ Phone _____

_____ Phone _____

_____ Phone _____

_____ Phone _____

TOOLS:

☐ framing hammer

☐ power saw

☐ hand saw

☐ metal tape

☐ chalkline

☐ framing square

☐ pencil

☐ nail belt

☐ knee pads

☐ ladders

☐ combination square

☐ safety goggles

☐ nail-pulling tools

☐ saw blades

☐ extension cords

☐ terminal plug boxes

☐ staple gun/hammer tacker

☐ staples

RENTAL ITEMS:

☐ pneumatic nailer

☐ ladders

☐ scaffolding

☐ safety harness

MATERIALS:

_____ sheets of PLYWOOD

_____ lbs. of NAILS

_____ rolls of TAR PAPER

_____ SEALING GLUE

_____ boxes of STAPLES

DESIGN REVIEW: Re-evaluate any skylight locations.

INSPECTIONS NEEDED:

☐ YES   ☐ NO

Inspector _____ Phone _

Time of inspection _____

WEATHER FORECAST:

For week _____

For day _____

If heavy rain is expected, try to cover the plywood sheathing after each day's work. Do not apply sheathing on windy days.

Worksheet
Tongue and Groove
2-inch Decking

ESTIMATED TIME

_____

_____

_____

_____

CREW

Skilled or Semi-Skilled

_____ Phone _____

_____ Phone _____

_____ Phone _____

Unskilled

_____ Phone _____

_____ Phone _____

_____ Phone _____

_____ Phone _____

MATERIALS NEEDED

T&G stock, tar paper, nails, staples.

ESTIMATING MATERIALS

DECKING:

_____ sq. ft. of roof surface + 15% waste = _____

NAILS:

16d CC Sinkers _____ lbs. per _____ sq. ft.

TAR PAPER:

_____ sq. ft. roof surface = _____ rolls.

COST COMPARISON

_____ Store:

Cost per bd. ft. _____ total $ _____

_____ Store:

Cost per bd. ft. _____ total $ _____

_____ Store:

Cost per bd. ft. _____ total $ _____

_____ Store:

Cost per bd. ft. _____ total $ _____

PEOPLE TO CONTACT

Roofer _____

Phone _____ Work Date _____

Roofing materials supplier

Phone _____ Work Date _____

Sheet metal person _____

Phone _____ Work Date _____

Plumber _____

Phone _____ Work Date _____

Solar panel installer _____

Phone _____ Work Date _____

Daily Checklist

CREW:

_____ Phone _____

_____ Phone _____

_____ Phone _____

_____ Phone _____

_____ Phone _____

_____ Phone _____

TOOLS:

☐ framing hammer

☐ framing square

☐ combination square

☐ power saw

☐ hand saw

☐ metal tape

☐ nail belt

☐ knee pads

☐ ladders

☐ safety goggles

☐ sledge

☐ nail-pulling tools

☐ saw blades

☐ extension cords

☐ terminal plug boxes

☐ staple gun/hammer tacker

☐ staples

RENTAL ITEMS:

☐ scaffolding

☐ ladders

☐ safety harness

MATERIALS:

_____ linear feet of T&G STOCK

_____ lbs. 16D CC SINKERS

_____ rolls of TAR PAPER

_____ boxes of STAPLES

DESIGN REVIEW: You may want to re-evaluate any skylight locations.

INSPECTIONS NEEDED:

☐ YES   ☐ NO

Inspector _____ Phone _____

Time of inspection _____

WEATHER FORECAST:

For week _____

For day _____

Beware of rain before decking is covered with tar paper. If the decking is applied and then gets wet, it will strain and buckle due to expansion. Do not apply on windy days.

Worksheet
Skip Slat Sheathing

ESTIMATED TIME

_____

_____

_____

_____

CREW

Skilled or Semi-Skilled

_____ Phone _____

_____ Phone _____

_____ Phone _____

Unskilled

_____ Phone _____

_____ Phone _____

_____ Phone _____

_____ Phone _____

_____ Phone _____

MATERIALS NEEDED

1×4s and nails.

ESTIMATING MATERIALS

SLATS:

_____ No. of courses _____ ft. long (for each course)

= _____ total linear feet + 20% waste.

NAILS:

_____ lbs. of 8d CC Sinkers.

COST COMPARISON

_____ Store:

Cost per bd. ft. _____ total $ _____

_____ Store:

Cost per bd. ft. _____ total $ _____

_____ Store:

Cost per bd. ft. _____ total $ _____

_____ Store:

Cost per bd. ft. _____ total $ _____

PEOPLE TO CONTACT

Roofer _____

Phone _____ Work Date _____

Roofing materials supplier _____

Phone _____ Work Date _____

Sheet metal person _____

Phone _____ Work Date _____

Plumber _____

Phone _____ Work Date _____

Solar panel installer _____

Phone _____ Work Date _____

Daily Checklist

CREW:

_____ Phone _____

_____ Phone _____

_____ Phone _____

_____ Phone _____

_____ Phone _____

_____ Phone _____

TOOLS:

☐ framing hammer

☐ combination square

☐ power saw

☐ saw blades

☐ hand saw

☐ metal tape

☐ pencil

☐ nail belt

☐ ladders

☐ safety goggles

☐ nail-pulling tools

☐ extension cords

☐ terminal plug boxes

RENTAL ITEMS:

☐ scaffolding

☐ ladders

☐ safety harness

MATERIALS:

_____ linear feet 1×4S

_____ lbs. 8D CC SINKERS

DESIGN REVIEW: You may want to re-evaluate any skylight locations.

WEATHER FORECAST:

For week _____

For day _____

15 Roofing

Photo 15-1. Roofing material has been delivered to the top of the roof and is stacked near the ridge.

What You'll Be Doing, What to Expect

APPLYING YOUR ROOFING (Photo 15-1) consists of installing a waterproof covering over your roof sheathing to protect the house from the weather. This covering is not only the roofing material itself but can also involve underlayment (roofing felt); flashing; ridge caps; vents; or flashing for plumbing vents, wood stoves and skylights.

There is a lot to consider before you choose any one type of roofing. Your choice will affect not only the appearance of your house, but your pocketbook and work schedule as well. In this chapter the different types of roofing and their advantages and disadvantages will be discussed in detail. We will also describe, in detail, how to apply two of the most common types of roofing: wood shakes and asphalt shingles. Each will be presented in separate sections.

For the most part, roofing is a rather simple procedure. There are a few key areas that are a little tricky; it is important that you understand how to roof these areas properly or you will have leaks. These areas include: valleys, hips, ridges, vent pipes, skylights and vertical walls intersecting sloping roofs. Other than these areas, roofing is straightforward and simple. Laying shakes or shingles on the broad uninterrupted areas of the roof is simple, and monotonous, and can cause back strain from all the bending. Being up off the ground with a good view, on the other hand, can often get you into a rhythmic, almost meditative, flow of installing the shakes or shingles; the work becomes enjoyable and almost effortless.

More than in almost any other area of housebuilding, you want to constantly strive for an efficient system in roofing that allows you to install the greatest number of roofing materials with the least amount of effort in the least amount of time. Go watch a few professionals at work. Note the way they sit, where the materials are in relation to where they are working, what their assistants do, what tools they have and what tricks they use. Often observing a professional for half an hour can save you many hours on your own roof. Speed and efficiency of energy use are essential here.

Be sure in whatever you do with the roofing: you do not want a leak! A leak can turn a cozy, rainy evening into a nightmare, and have you kicking your behind from here to Toledo for not doing it right the first time.

Jargon

BUNDLE. Term used for wood shakes or shingles, describing a stack of material banded together. Usually there are 4 or 5 bundles in a square.

BUTT END. The bottom end (lower end) of the shake or shingle.

EAVES. The lower edges of the roof.

FLASHING. A metal or composition material used on a roof around skylights, chimneys, vents, intersecting vertical walls, etc., or on a wall above doors and windows. The flashing is placed in such a way that it will shed rain water.

GABLE ROOF. A roof sloping on 2 sides, named for the triangular wall section, the gable, created by the slopes of each end wall.

HIP. The external roof angle formed by 2 inclined sides of a roof.

PLY. Layer.

PREFELT. To apply the roofing felt before the roofing material goes on.

RAKES. Sloping edges of roof that frame the gable.

RIDGE. The horizontal line at the top of the roof formed by the intersection of sloping roof surfaces.

RISE. The vertical distance covered by a rafter or roof.

RUN. The horizontal distance covered by a rafter or roof.

SLOPE. The angle of the roof.

SQUARE. Materials needed to cover 100 square feet of roofing.

STARTER COURSE. A layer of shingles that is installed directly below the first course on the roof, at the eaves.

VALLEY. The internal roof angle formed by 2 inclined sides of a roof.

WEATHER EXPOSURE. The section of any roofing material, shake or shingle, that is exposed to the weather (outside).

Purpose of the Roofing

To provide a weather tight covering over the house (as if you didn't know), protecting it from rain, wind, ice, snow, dust and sunlight.

Design Decisions

Selecting your roofing material is a rather important decision, involving a large amount of money, a lot of application time and a heavy impact on the looks of the home. You can better grasp the scope of the decision when you realize that during the life span of a normal roof, it will withstand 1,000,000 gallons of water weighing 5,000 tons; 50,000 hours of direct sunlight; 2 million miles of wind and ½ million degrees of temperature changes. Your choice affects other parts of the construction process also, e.g., a tile roof will need heavier rafters and sheathing material to hold the weight of the tiles. Read the following section and consider your choice well.

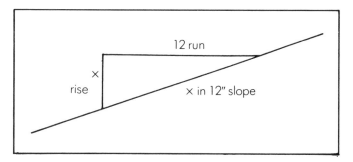

Figure 15-1. Roof slope represented in a triangle.

Factors to Consider in Selecting a Roofing Material

SLOPE: (Figure 15-1.) The slope of the roof refers to the angle or pitch of the roof. It is always written in relationship with the number of inches a roof rises over 12 inches (1 foot) of horizontal run. Therefore, a roof with a 4 in 12 slope rises 4 inches for each 12 inches of horizontal run. Your plans will indicate the roof slope by using a triangular symbol; 12 will be on the horizontal bottom side of the triangle with a variable number on the vertical side of the triangle representing the rise over 12 inches of run in number of inches.

The greater the slope, the steeper the roof. The steeper the roof, the easier it will be for water to run off the roof and, therefore, the less likely it will be to leak. Some roofing materials must have at least a minimum slope angle in order for them to work effectively. If the roof has too shallow a slope, rain can be blown under the roofing and it will leak. Other roofing materials (wood shakes and shingles, for instance) will vary in their weather exposure for different slopes, i.e., the amount of exposure used depends on the length of the shake or shingle and the angle of the slope. The following chart shows the minimum slope you can have for the various roofing materials.

ROOFING MATERIAL	MINIMUM ALLOWABLE SLOPE (in inches)
Hot tar built-up roofing	0 in 12 (flat)
Roll roofing (single coverage)	0 in 12 (3 in 12 maximum)
Roll roofing (double coverage)	1 in 12
Metal	2 in 12
Composite shingles (wind resistant)	2 or 3 in 12
Composite shingles (standard)	4 in 12
Wood shingles	3 in 12
Wood shakes	4 in 12 (7 in 12 in some rainy climates)
Slate	4 in 12
Tile	4 in 12

DURABILITY: Different roofing materials have different life spans. These life spans are not fixed; they can vary from area to area or even in the same areas, depending on whether they are in the shade or sun, etc. The greater the life span, the greater the cost. The cheapest roofing, single coverage roll roofing, has a life span of 5 to 8 years. The most expensive roofing, terne metal, has an indefinite life span. You pay for what you get. You will have to consider this within the framework of your overall budget. We would all like the best of everything, but can you afford it? If the roof with a longer life is $1500 more than what you planned, what are you giving up? Looking at it another way, if you use, say, a 20-year roof instead of the more expensive 30-year roof, and place the savings in a savings account, it may be that in 20 years you would have enough to buy 5 new roofs. If you can get 20 years out of your roof, that seems adequate.

COST: As mentioned above, cost is related to durability, for the most part. Galvanized metal roofs are an exception. They last a long time, but are relatively inexpensive. Costs for roofing can run into many thousands of dollars; shop well for the materials—this is an area where big discounts and savings can be found, especially if you are buying with cash or check.

Another consideration is the resale value of the house in relationship to the roof. Some very good roofs, e.g., galvanized metal or double coverage roll roofing, have a negative effect on the resale value. Others, terne metal and wood shakes or shingles, have a positive effect.

If you are paying for labor, you must consider labor cost in the choice of the material you plan to use. Metal roofs are very quick and easy to apply and therefore less costly, whereas wood shingles and shakes are more costly. Roll roofing is easy and composition shingles are somewhere in the middle.

Slope may also affect materials and application costs. If the roof has a steep pitch and is difficult to work on, the labor price may go up. On the other hand, shallower slopes may require greater quantities of materials, as in the case of cedar shingles. On a 4 in 12 slope, a 16-inch-long shingle can have 5 inches exposed to the weather; on a 3 in 12 slope, only 3¾ inches can be. You do not get as much coverage from the same shingle in this shallower slope, so you need more shingles (25% more). Because there are more materials to apply, the roofing workers will charge you more to do the shallower slope roof.

COMBUSTIBILITY: In making a choice of roofing material, combustibility of materials must be taken into consideration (e.g., wood shakes and shingles are highly combustible, while composite shingles, tiles, slate and metal are not). If you live in an area where there is a history of brush fires or you are building close to another home, this should definitely be a consideration. Your roof, and consequently your home, could ignite from an outside fire. Fire insurance premiums may be higher if you choose a material that's more combustible. In some areas, where brush fires have been a problem, wood roofs are not allowed. In some places it is required that roofs be treated with a liquid fire retardant. This is of limited value however, as it usually washes off in a few years.

NUMBER OF LAYERS: In applying most roofs, several layers of material, overlapping each other, are used in order to waterproof the roof. Water cannot flow up the back of a shingle or shake that overlaps the one beneath it. Lateral overlap, with staggered joints between the shingles or shakes, prevent sideway leakage. This overlapping process creates layers of roofing material at any one point on the roof. This is called coverage. Roofs can have double or triple coverage depending on the length of the shingle or shake exposed to the weather. When about ½ of the length of the shingle or shake is exposed, there is a double coverage (2-ply) roof; if about ⅓ of the length is exposed, there is a triple coverage (3-ply) roof.

APPLICATION: Some roofs are much easier to install than others. Some materials, such as terne metal or hot tar, need to be applied by professionals and can be quite costly. Consider the application differences when you choose your roof (this is discussed in more detail in the next section). Consider how the differences of application will affect your budget, time, pace, energy, etc.

APPEARANCE: Like siding, the roofing material you choose, in regards to both the type and the color, will have a major impact on the looks of your house. It can enhance it, have a neutral effect or detract from it. Look around your neighborhood at the different roofing materials and envision them on your house. Your choice in this matter can have almost as great an effect on the house as the design itself. After you have considered it well, your final decision should be a definite "yes," not a definite "maybe."

Types of Roofing

HOT TAR BUILT-UP ROOFING: (Photos 15-2, 15-3.) This type of roofing is most common on flat roofs and shallow slope roofs (1 or 2 in 12). It is used very often on stores and commercial buildings which use the flat roofs to support their mechanical equipment for heating, cooling, and sprinkler systems, etc. The roof consists of layers of roofing felt bonded together with hot tar or pitch. The first layer of 30-pound (30 pounds per 100 square feet) felt is applied directly to the solid sheathing and fastened with large-headed nails. This first layer, applied dry, keeps the melted tar from penetrating the joints in the sheathing. Between each layer of felt (3 to 6, depending on the quality of the roof) melted tar is mopped or sprayed on. On top of the last ply of felt, the tar is applied, and finally a layer of gravel, crushed stone or marble chips is placed. This allows you some choice in color and texture. At the edges of the roof, gravel stops are installed. Where a hot tar roof intersects a vertical wall, it is flashed.

Advantages: The greatest advantage of this type of roofing is its ability to provide a watertight membrane on a flat or low slope roof. Its life span is from 10 to 25 years depending on the site specifics, the local weather, the quality of the materials and the number of layers or plies used. It is moderately expensive. (Melted coal tar is better than asphalt. Find out which one your roofer plans to use.)

Disadvantages: The greatest disadvantage in hot tar roofing is that it is relatively expensive, since it usually needs to be applied by a professional. To do it yourself, you would need to use large equipment to melt the tar that would be

Photo 15-2. Hot tar is being pumped to the rooftop for application. The cord operates the on-off switch for the pump.

Photo 15-3. A hot tar roof after the final layer of gravel has been applied. Note how tar is being used to flash the vent pipes.

hard to locate at rental yards. Even if you could locate it, I would recommend against applying this type of roofing yourself. It is an incredibly messy job, and information on how to apply it is hard to find. It also has little fire resistance and leaks are hard to locate.

There are several alternative types of materials now available for flat or low sloped roofs. These include liquid rubbers you paint on, sealants you spray on and fiberglass roofs applied in the same way as hot tar roofs. There are even cold tar applications that require no melting of the tar. As with all products, their manufacturers tell you it is the best thing since sliced bread. Before you use any of them, talk to the people who apply them and the people who live in houses where they have been used for many years to see if there are problems.

ROLL ROOFING: (Photo 15-4.) Roll roofing is made from the same materials as asphalt shingles, only put together in a different size and form. It is made of long rolls of asphalt impregnated paper with an exterior coating of different colored mineral granules to help resist the elements. This material is also called mineral surfaced asphalt, SIS and salvage roofing. Because of its appearance, which is considered to detract from the structure, it is most often seen in rural areas or on out buildings. If this is not a problem for you, you may want to seriously consider it for your house because of its low cost and ease of application. It can be used for awhile and then covered over with a better material when your ravaged pocketbook has recovered.

This material comes in two different types: single coverage and double coverage. Both come in 36-foot rolls, 3 feet wide. The weight is from 55 to 90 pounds per square (100 square feet). In the single coverage type, the mineral granules cover all but the top 3 to 4 inches of the roll and it is applied with a 3- to 4-inch top overlap. This creates a single coverage which is good for 5 to 8 years. It weighs about 90 pounds per square. It is good on a 2 to 3 in 12 slope.

The double coverage roll roofing only has about ½ the width of the roll covered with mineral granules. This requires that you use a 19-inch overlap, creating a double coverage. This roof is good for 12 to 20 years.

Advantages: The advantages to this type of roofing are its low cost and ease and speed of application. It is also easy to maintain.

Disadvantages: It does not have the durability of some of the other roofings, which can affect the resale value of the home. It has poor fire retardant properties as well.

Warning: With either type of roll roofing, unroll the material a day before you plan to install it so it will expand and

Photo 15-4. Roll roofing being applied to the roof. Installation is quick and easy.

lie flat. Otherwise, it will buckle, even if you have installed it flat. Also do not use this material on a steep slope as its heavy weight can cause it to pull away from the nails.

METAL ROOFING— ALUMINUM AND GALVANIZED: (Photo 15-5, Figure 15-2.) Metal roofing comes in many forms and materials. In the old days you could buy small metal shingles and they were installed very much like wood shingles. Today, most metal roofs come in long sheets, 2 feet wide and up to 18 feet long. These long sheets make metal roofing one of the quickest and easiest types to install. Most panels have 2 crimps on either edge and a center crimp. The edge crimps are water guards, since water running down the roof will not work its way over both interlocking crimps. The center crimp, as well as the edge crimps, are for nailing. Metal roofing is one of the few types of roofing that has its nails exposed. Because of this, all nailing is done through the peak of the crimps, not on the flat areas where it would be more prone to leak. Because the metal expands and contracts,

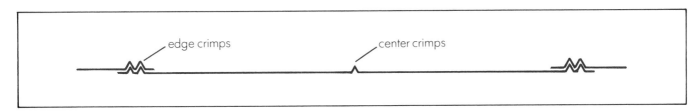

edge crimps   center crimps

Figure 15-2. The crimps in metal roofing interlock with adjacent pieces and make them weather tight.

Photo 15-5. Metal roofing on an owner built home.

Photo 15-6. A terne metal roof being installed. The cost of this roof makes it prohibitive for most residential construction.

Photo 15-7. A composite shingle roof with tile ridges and hips for accent.

threaded nails are used to prevent the nails from eventually working themselves loose. The nails have either a lead head that spreads out over the nail hole or a neoprene washer underneath the head that seals the hole.

The most common metal roofs are either galvanized or aluminum. Unless other factors are involved, I would go with the galvanized. It has been around longer (the old "tin roofs"). Certain tree saps falling on aluminum roofs can cause them to corrode. In the South where I was raised, tin roofs were very common. I have seen some 60 years old or older still in good shape.

A good quality galvanized roof will have at least 2 ounces of zinc per square foot. Be sure if you use a galvanized roof that you use galvanized nails and flashing. *Never allow galvanized metal to touch aluminum; they will cause each other to corrode.* Use at least one roofing nail per square foot. If more than one sheet is needed to cover the slope of the roof, overlap the sheets 8 inches if the slope is less than 3 in 12, and overlap 4 inches if it is more.

Advantages: The advantages of this type of roofing are its ease and speed of application. It goes on in a day or two. The materials cost approximately the same as composite shingle roofing. It also has a very long life span, 30 to 60 years. The sheets now come in various colors and patterns and with insulation in the roof panel; they are not noisy in a rainstorm. It is fire resistant.

Disadvantages: The disadvantages are that these roofs can affect resale value, since they are considered shoddy (mis-

takenly so), and sometimes they do not sit as flat nor trim out as well as you would like. Every few years you will need to crawl on the roof and drive the nails back in that have worked themselves out. It is also subject to damage by wind or falling objects.

METAL ROOFS—TERNE AND COPPER: (Photo 15-6.) These two types of metal roofs are much more expensive than their cousins, galvanized and aluminum. They are perhaps the most expensive of all roofs. (Terne costs six times as much as galvanized roofing or asphalt shingles, twice the cost of cedar shakes. Copper is $795 a square. Someone may steal your roof one day.) Terne metal is a copper-bearing steel. Both of these roofs will probably outlast the house. The terne metal roof on Thomas Jefferson's home in Virginia, the Monticello, is still the original one installed in the 1790s. Today only major commercial or institutional buildings use these types; only banks or governments have that kind of money to waste. Sometimes these roofs have to be put on with a special machine and professional labor since their application, using folded or soldered seams, is complex.

COMPOSITE (ASPHALT) SHINGLES: (Photos 15-7, 15-8.) This is the most common roofing used. Like roll roofing it is composed of asphalt impregnated paper and mineral granules, which line the exposed area. There are many different colors, qualities and styles on the market. They even make one that looks very much like wood shakes. The quality differs according to the weight per square (100 square feet).

Photo 15-8. A composite shingle roof on an owner built passive solar home.

Photo 15-9. A cedar shingle roof. Shingles lie flat, for a smooth, even finish.

Anything at 235 pounds or more is pretty good, and should last 15 to 25 years. Standard types are used with a slope of 4 in 12.

Wind resistant or wind sealed shingles have a dab of roofing cement under each shingle, along the bottom edge. When the sun hits these shingles after they are installed on the roof, the cement melts and sticks the shingle to the one below it. This type of shingle can be used on shallower slopes (3 in 12) since there is less likelihood that the wind can blow the water underneath the shingles. I would advise using them no matter what the slope; they cost only a little more.

The shingles are usually 3 feet long with 3 tabs per shingle. At the bottom of the shingles, there is either a butt cut or another decorative cut. There are many manufacturers to choose from. Stick with one of the larger, better known brands to avoid problems.

Advantages: This roofing is easy to install and comes in a large selection of styles and colors. It requires little maintenance and is easy to repair.

Disadvantages: It has little fire resistance.

WOOD SHINGLES: (Photo 15-9.) There is a difference between wood shingles and wood shakes. Wood shingles are machine sawn from cypress, redwood or, most commonly, cedar logs. They are uniform in length (16 inches, 18 inches, or 24 inches), tapered from ½ inch to about ¹/₁₆ inch and have random widths. They are very uniform and lie flat and smooth on the roof. Wood shakes, on the other hand, are rougher, thicker and more irregular. They are hand or machine "split" from logs, much like the shakes made by the pioneering immigrants of this country. They are from 24 to 36 inches long and taper from ¾ inch to about ¼ inch. The various kinds of shakes will be dealt with in the next section.

Wood shingles are most often made of cedar. Cedar has a low expansion rate, a high impermeability level and an even grain that allows the water to flow off easily. The rougher shakes require at least a 4 in 12 slope. Because of their uniformity, which allows them to lie flat, shingles can be used with a 3 in 12 slope. No tar paper or roofing felt is used in shingle application, because ventilating is more important.

Wood shingles come in different grades for roofs, walls, and starter courses of the roof. Many code areas require that shingles be approved shingles; they must have an approval label attached to each bundle. The weather exposure of shingles varies according to the length of the shingle and the slope of the roof. The table shows some common exposure lengths in relation to pitch and length, but check these with your local codes or suppliers before you use them.

ROOF PITCH	SHINGLE LENGTH (in inches)	EXPOSURE (in inches)
4 in 12 or Steeper	16	5
	18	5½
	24	7½
Less than 3 in 12 to 4 in 12	16	3¾
	18	4¼
	24	5¾

If you will note, almost always ⅓ or less of the shingle is exposed to the weather. This creates a triple coverage, which is what is recommended for this type of roofing (an exception to this is a 4 in 12 or more slope, using a 24-inch shingle).

Advantages: The value in the wood shingle is its beauty and uniformity as well as its prestige and resale value. It is also easy to install and has some insulating properties.

Disadvantages: This type of roofing is expensive, and for the money, does not really give you a long life span. It costs almost twice as much as composite shingles, with about the same life span (20 to 30 years), and almost as much as wood shakes, which have a longer life span (30 to 45 years). Installing this type of shingle is time-consuming and can be expensive if you are paying for labor. It is combustible and can be a fire hazard; in some areas it is prohibited for this reason.

WOOD SHAKES: (Photo 15-10.) As described in the section on wood shingles, wood shakes are split from cedar or redwood rounds. There are several different types of shakes. Taper split shakes are split at a taper and are irregular on both sides. Straight split shakes are straight with no taper. When these are sawn in two, they are called resawn shakes. Resawn shakes have one irregular side and one smooth side, allowing them to sit flatter on the roof than taper split or straight split shakes.

All of these shakes are split, not sawn, from logs. Splitting cleanly separates the wood fibers and leaves the cell walls

Photo 15-10. A shake roof is being applied. Shakes, being rougher than shingles, lend a handcrafted touch. Note the large flashing where the upstairs deck joists intersect the veranda roof.

Photo 15-11. A slate roof. Note the broken slate shingles—this is a major problem with these roofs.

Photo 15-12. A ceramic tile roof is being applied. The wave pattern in the tiles will allow them to interlock. Note that skip sheathing is used here.

intact. The wood consequently absorbs little of the runoff water. Most shakes are from heartwood, with little or no sapwood. The shakes come in random widths and the lengths are 24 or 36 inches. On the roof they have a hand-hewn irregular look which is desired by many people. It is often called the aristocrat of roofs because of its beauty and long life. This chapter includes a section on installing this type of roofing.

Advantages: The value in using wood shakes is their beauty and durability (30 to 45 years and sometimes longer). Shakes have a very high resale value and are considered prestigious. Because of the thickness of the wood, shakes do offer a small amount of insulating value. They are also easy to install.

Disadvantages: The disadvantages of shakes are their combustibility and relatively high cost. Cost-wise they are twice as much as composite roofing and, with installation, can run even more.

SLATE SHINGLES: (Photo 15-11.) Slate shingles have been used in many homes, especially in the New England areas where the quarries are located. Because of the cost, they are seldom used on new homes. If you can afford them, however, or know of a good supply of used slate, you may want to consider them. They are flat pieces of slate, usually black in color, that are uniform in length, width and thickness. They can weigh up to 850 pounds per square. There are 2 holes drilled in each tile and the shingles are attached with nails or wires. Usually the slate is ³/₁₆ inch thick and 18 or 24 inches long. They need at least a 4 in 12 slope and often require additional roof support to handle their added weight.

Advantages: It is a permanent roof that should outlast the house. Aesthetically it is appealing and has a great resale value.

Disadvantages: It is a very expensive roof, especially if you live very far away from the quarries. It is also a very time-consuming roof to install, because you have to be very careful. The slate shingles are prone to crack and break should a branch or something fall on them. Due to their dark color, they retain a lot of heat in the summer months.

TILE ROOFS: (Photos 15-12, 15-13, Figure 15-3.) There are now several types of tiles on the market. There is the old type of clay tile, the cement tile, the fiberglass tile and there are probably a few I have not heard of yet. Clay tile can weigh from 800 to 1600 pounds per square and costs more than concrete tile. Concrete tiles can weigh from 700 to 1000 pounds. The old type of clay tile is rather time-consuming to apply; it is brittle and you have to use a tile cement or mastic at the ridges and the eaves. The newer tiles have done away with that, but they still take more time to install than composite shingles or metal. For most tiles, some sort of interlocking or overlapping structure is used to prevent leaks.

Like slate, tiles may require additional roof support to carry their extra weight. Some of the newer cement-based tiles, however, require no additional support. This type of roofing comes in many different styles and colors and can be quite attractive and durable, lasting many years longer than composite shingles. The life spans differ according to the type you are using, but 30 to 60 years is the range.

Photo 15-13. A ceramic tile roof after installation. Tile is durable and the price is within reason.

Advantages: Tile is both beautiful and durable, and it adds to the resale value of the house. It comes in a wide choice of colors and styles and can be applied by the owner builder. Because of its fireproof quality, it is good in fire prone areas.

Disadvantages: Like any long-life roofing material, it is expensive. Prices vary according to what type you get, so check with your local suppliers. Tile is a labor-intensive product in regards to application. It is also brittle and can break should anything fall on it.

NEW MATERIALS: (Photos 15-14, 15-15, 15-16.) Aside from the common materials mentioned, there are many new materials on the market. Three are new types of composite shingles, new types of rigid shingles, new waterproof membranes for flat roofs, wood shakes joined together to make a panel, etc. All of these products have their advantages and disadvantages. Be careful, the sales people will only tell you their advantages and are often trained to give the proper response when asked about their disadvantages (it's called in sales language "overcoming objections"). Evaluate each product carefully and try to talk to a professional roofer, as well as someone who has been using the roof for several years, before deciding.

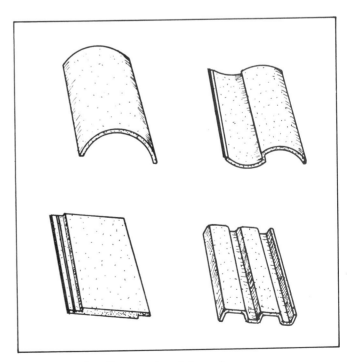

Figure 15-3. Different styles of tile roofing.

Photo 15-14. Sod roofs are becoming popular for earth-sheltered homes. You can grow your garden on your roof, which might be needed on small lots.

Photo 15-15. A thatched roof. Beautiful, but probably would not get by the local codes.

Photo 15-16. Your guess is as good as mine!

Applying Wood Shakes

What You'll Need

TIME The time varies according to the complexity of the roof. A roof with skylights, intersecting vertical walls, or a lot of hips or valleys will take longer than a simple straightforward roof with few special situations. It is also important to have your materials delivered to the rooftop. If they are set on the ground, more time will be required to carry the bundles up to the roof. Also your ability to get an even flow of work going is a big factor in how long it will take. Keep stepping back to see if there are ways to streamline your operations.

For a straightforward roof: 40 to 60 person-hours per 1000 square feet (10 squares).

PEOPLE 1 Minimum (semi-skilled); 3 Optimum (1 semi-skilled, 2 unskilled); 6 Maximum (2 teams, one on each side of the roof).

TOOLS Rubber soled shoes, knee pads, chalkline, power saw, tape, tin snips, ladders, extension cord, utility knife, staple gun or hammer tacker, staples.

Aside from these common tools, there are a few special tools that you will need to make the work go easier.

NAIL STRIPPER: (Photo 15-17.) A nail stripper is a small metal box filled with roofing nails that hangs from your neck

Photo 15-17. A nail stripper. These come in both right- and left-handed versions. They are good for applying siding as well.

Photo 15-18. The roofing hatchet comes with a wrist strap so it will not fall off the roof.

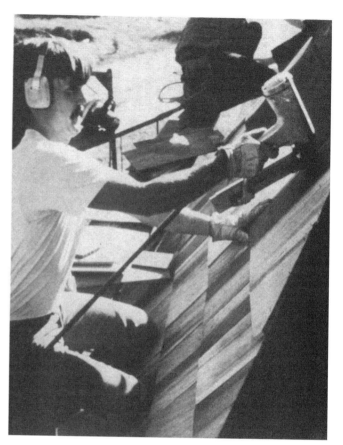

Photo 15-19. A compressed air gun makes roofing a lot quicker, if one is available to you. Note the use of ear phones.

Photo 15-20. A safety harness. A good idea for steep slopes or high roofs.

Photo 15-21. A roof jack to support the foot. Note how a support nail is driven under the butt line of a shingle.

Photo 15-22. A toe board. This is used like the roof jack, to provide support on a sloped roof.

with a strap around your back supporting it. The stripper causes the nails to all hang point down and in line, so you can reach across and pull several nails out at once, all with their heads up. Do not overload the stripper; put in a handful at a time and jiggle the stripper so the nails will settle. Strap the stripper tightly to your chest, so you can pull on it to get the nails. You can lubricate the nails or the discharge doors of the stripper with mineral oil for easier use. Wear a nail bag to hold extra nails to supply the stripper. Strippers come in both right-handed and left-handed models, for both composite shingle nail sizes and nails for wood roofs. Usually a roofing supplier or a good lumberyard will stock them. If not, they can be ordered from:

> Roofmaster Products Company
> P. O. Box 63167
> Los Angeles, CA 90063
> or
> South Coast Shingle Company
> 2220 East South Street
> Long Beach, CA 90805

ROOFING HATCHET: (Photo 15-18.) A roofing hatchet is used in applying wood roofs. There is a hatchet end that can be used to split the shakes or shingles to any needed width. It has a non-skid, waffle-surfaced head to grip the rough-surfaced galvanized nails. The roofing hatchet also comes with an adjustable pin, and notches on the handle, to easily measure the proper weather exposure for wood shingles or shakes during application. Keep the hatchet very sharp; dull cutting tools are very dangerous.

COMPRESSED AIR NAILER AND EAR MUFFS (Optional): (Photo 15-19.) Like sheathing and subflooring, roofing is an area where you may want to consider using a compressed air nailer. Many professionals do, and even if you have to rent one, it may be well worth your while. Get one with a rubber air hose for mobility. Be sure that the nails you use are right for your roofing and sheathing situation.

SAFETY HARNESS: (Photo 15-20.) On steep roofs, you may want to consider a safety harness for added protection. This can be fastened to something on the other side of the ridge from where you are working; should you start to fall, it will catch you. Attach it to a tree or stationary object. A friend of mine attached it to the bumper of his truck, and his wife came out and got into the truck to drive away. Fortunately she felt the tug before going very far, but can you imagine the way he felt when he heard the truck start!

ROOF JACK: (Photo 15-21.) An apparatus nailed to the roof to support your foot while working on the roof.

TOE-BOARD: (Photo 15-22.) A board attached to the roof to provide support to a person who is working on a steep slope.

LADDER JACK: These jacks allow you to support planks (2×10s, up to 9 feet long) from the rings of 2 ladders, up to 20 feet above the ground. The ladders rest against the house so that they are 2½ feet from the house, where you will be standing. Bracket arms holding the planks are levelled by adjustable braces. The planks should extend 1 foot past the arms. This can be used for working on the lower areas of the roof.

	SHAKE WEATHER EXPOSURE (in inches)					
Pitch	**No. 2 Blue Label** Length of shakes in inches			**No. 3 Red Label** Length of shakes in inches		
	16	18	24	16	18	24
3 in 12 to 4 in 12	3½	4	5½	3	3½	5
4 in 12 or steeper	4	4½	6½	3½	4	5½

ADJUSTABLE SCAFFOLDING: This is a type of portable scaffolding that has jacks to pump the platform up and down wooden uprights. Pumped up to 30 feet off the ground, they can be used to stand on as well as to transport materials up to the roof. Follow the manufacturer's directions for use.

LADDER HOOKS: These are used to hang a ladder from the ridge so that it can be used as a toe-hole while applying the roofing. They may be needed for steep roofs.

MATERIALS Shakes are graded according to their thickness. There are two basic grades: heavy and medium heavy (there are lesser grades also). The various types of cuts include taper split, straight split and resawn, as mentioned earlier under "Types of Roofing." Most builders are using a medium grade shake, which is thinner, of lower quality and has a shorter life span than the heavy shake that used to be the standard in the industry. In the past, 80% of the shakes produced were heavy shakes, and 20% were medium heavy. Today those numbers are reversed.

If you decide to order the heavy shakes to get a better roof, you may find that the heavy shakes delivered are also of low quality, come with many knots, are not very wide and have waves and irregularities. This is because most shake mills today are milling mostly medium-heavy shakes. Those pieces that are too irregular to cut into the thinner medium-quality shakes are thrown aside and used as heavy shakes, since they are thicker and may survive the milling process. If you are planning to use heavy shakes, which I advise because of their durability, try to locate a mill that is still milling a lot of heavy shakes, or at least look closely at the shakes before you buy them. Check to see if there are a fair number of wide shakes (12 to 18 inches), if there are too many knotholes, or if the shakes are straight and without a lot of splits.

Watch the newspaper (the classified ads) if you live in an area of the country where shakes are still milled (wherever cedar grows). Often there are small one- or two-person mills that sell shakes directly to the public at a greatly reduced cost. Some owner builders even buy the redwood or cedar rounds and cut their own shakes. A week's work doing this could save you several thousand dollars.

Shakes are usually graded according to their quality. Some areas require that only graded shakes be used. Number 1 Blue Label shakes are the best. Number 2 Red Label or number 3 Black Label shakes can sometimes be used with less weather exposure or on walls.

When you order your shakes, be sure the delivery is made to the rooftop, not to the ground. Find out if there is an extra cost per square for this service; argue against it, if there is. A supplier can deliver the materials to the roof via a scissors truck (Photo 15-23), conveyor belt or forklift. It saves you a lot of effort to have it placed on the roof for you. You can rent an electric ladder hoist (Photo 15-24) if your supplier doesn't have the equipment to deliver materials to the roof.

Aside from the shakes, you will need various types of flashing, nails and roofing felt. Flashing is used where vent pipes or chimneys penetrate the roof, around skylights, in the valleys and where vertical walls for dormers or other parts of the house intersect the sloping roof. These will be described in detail later in the chapter. The nails used are rust-resistant hot-dipped galvanized 6d or 7d nails. They are 13 gauge with $7/32$-inch-wide heads and are 2 inches long. Be sure you use galvanized nails to secure galvanized flashing, and aluminum nails to secure aluminum flashing. *Never mix the two; they will corrode.* The type of roofing felt you will need is 36-inch and 18-inch-wide, 30-pound roofing felt. You can buy shake felt already cut 18 inches wide or buy standardized 36-inch-wide rolls and cut them yourself.

ESTIMATING MATERIALS NEEDED

In estimating the materials needed to do your roofing job, you will need to estimate not only how much roofing material you will need, but also the roofing felt, metal flashing and nails, as well as any hip and ridge shakes. (See Photo 15-25.)

In order to estimate the number of squares (shakes that will cover 100 square feet of roof surface), you first must determine the square footage of the roof. Simple roofs can be easily estimated by multiplying their dimensions. More complex roofs are somewhat more difficult. A simple method is to determine the square footage of the house at ground level, then multiply this times a factor, the roofing coverage factor, related to the slope of the roof. If there are overhangs and eaves, you will have to account for these in your square foot measurements. The factors are as follows:

Slope	**Roofing Coverage Factor**
2 in 12	1.02
3 in 12	1.03
4 in 12	1.06
5 in 12	1.08
6 in 12	1.12
7 in 12	1.16
8 in 12	1.20

Add on 10% for wastage.

Once you know the number of square feet of the roof surface, you must relate this figure to the exposure of the

Photo 15-23. A scissors truck. Ask your supplier if one is available, and if the price includes rooftop delivery.

Photo 15-24. An electric ladder hoist. You may want to consider renting one if carrying roofing materials to the roof by hand is the only alternative.

Photo 15-25. Shakes and shake felt (18 inches wide) ready to be applied.

shake, in order to know what to order. (Note that for every 100 feet of valley, allow 2 extra squares of shakes.) A roof with a 10-inch exposure will need fewer shakes than a roof with a 7½-inch exposure. Your supplier will have conversion tables and can help you figure your needs once the roof surface in square feet is known.

In ordering shake felt, order enough to cover the square footage of the roof. You will need 1 or 2 rolls of 36-inch-wide felt for the first course, and the rest will be 18-inch-wide shake felt. You will need about 2 rolls of 18-inch-wide felt for each 120 square feet of roof surface.

To estimate the number of hip and ridge shingles you will need, measure the length of all the hips and ridges, convert that number to inches, and divide by the exposure of the shakes. One bundle will cover 16⅔ linear feet of ridge or hip. You can either build these yourself from common shakes or order special hip and ridge shingles.

Measure for all your flashing needs: center crimped valley flashing, flashing where a vertical wall intersects a sloping roof, step flashing for skylights, bottom and top flashing for skylights, flashing for the different sized vent pipes, chimney flashings and flashings for mounts for solar panels.

In nails, you will need about 2 pounds, per square of shakes, of hot-dipped galvanized or galvanized 6d or 7d nails of 13 inch gauge with 7/32-inch-wide heads, that are 2 inches long. Plus, you will need 2 to 4 pounds of 8d nails for starter courses, hips and ridges.

INFORMATION Shake application is rather routine and simple. The information given in this chapter should get you through the process with no trouble. If you have any unique situations that are not covered here or any other doubts, talk to your supplier before proceeding. Be clear in whatever you do; re-doing a roofing section is a hassle. For more information, write to:

> Red Cedar Shingle and Handsplit Shake Bureau
> 5510 White Building
> Seattle, Washington 98101

Be aware of the code requirements for roofing in your area. Most codes require that the starter course be doubled, that there be a 1½-inch overhang, that a ½-inch gap is left between the shakes, that joints be staggered (1½ inches between), that all protrusions through the roof, and all intersections, be properly flashed, and some areas require that you use inspected and marked shakes only.

PEOPLE TO CONTACT You may want to check with your plumbers and electricians (unless they are you), to be sure all plumbing vents and electrical weatherheads are in place. Check with the solar panel installer to be sure mounts are in place. Contact your insulation installer at this time to let them know that the house will soon be watertight and the insulation can be installed. Often it is cheaper to pay an insulation contractor to install the insulation than it is for you to just buy the materials. They buy the stuff by the train car loads. Check it out; it may save you from a lot of itching.

INSPECTIONS NEEDED Before the roofing is applied, usually an inspector will need to look at the way you nailed your roof sheathing, to be sure it meets code.

Reading the Plans

Most plans will simply indicate the type of roofing to be used; e.g., composite, wood shake, cement tile, etc. Seldom will the plan tell you the type or quality, unless the plans are well detailed. Some plans will say nothing about the type of roofing to be used but will simply leave this to the owner's discretion. Usually there is nothing indicated on the plans about the application procedure, unless there is a unique problem that the designer has detailed. Otherwise, it is assumed that the roofing installer has all the information needed to install the roof.

Safety and Awareness

SAFETY: As in roof sheathing or any work up high, applying the roof is very dangerous. Be extremely careful; accidents can happen before you even realize it. Do not step on the slats between the rafters, keep good balance and do not lean over the sides of the roof too far.

When using a ladder, lean it against the wall so that the bottom is placed at a distance from the wall equal to ¼ the height of the wall. Rest scaffolding and ladders on level, solid ground, or on a secure foot on one side. One person on a ladder at a time; someone should hold on to the ladder as you climb; and avoid overhead electrical wires. Stay off scaffolding or ladders in high winds or rain.

AWARENESS: Water, water, water! Everything you are doing is to stop water from leaking into the house. As a contractor, the worst call back I could ever have was for a leak (fortunately, there were few). Where was the leak coming from? How do I locate it? Do I have to strip off large portions of the roof to repair it?

Be sure you understand what you are doing before proceeding. It is all simple, but don't trust your common sense, and make no assumptions. You want to welcome that first rainstorm.

Also with long sections of repetitious work the system you use is very important in order to finish the job with the least amount of effort and time. Constantly be seeking quicker and easier ways to apply the shakes. One hour watching professionals and adopting some of their tricks may save you many hours on the roof.

Overview of the Procedure

Step 1: Prefelt the roof
Step 2: Apply the valley flashing
Step 3: Apply the starter course

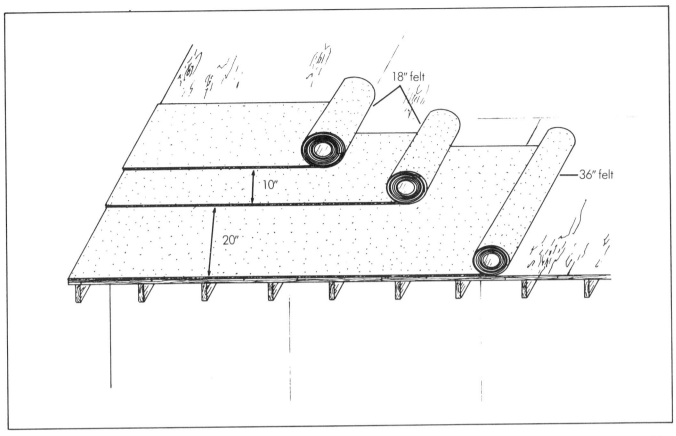

Figure 15-4. Prefelting using 36-inch felt for the first course and 18-inch rolls for all the successive courses. The first spacing is 20 inches and all the rest are 10 inches (assuming a 10-inch weather exposure).

Step 4: Apply shakes
Step 5: Flash for plumbing vents and chimneys
Step 6: Apply hips and ridges
Step 7: Flash for skylights
Step 8: Flash where vertical walls intersect the roof

Step 1
Prefelt the Roof

Margin of Error: Within 1 to 2 inches of prescribed distances and overlaps.

Most Common Mistakes: Not using 36-inch-wide starter piece; not spacing felt sheets at proper intervals.

Diagram: Figure 15-4.

Photographs: Photos 15-26, 15-27, 15-28.

Unlike wood shingles, which require no felt underlay because they lie flat and are therefore less prone to leak, the rougher shake roofs must have a felt underlay or they will leak. Usually a section of the roof is pre-felted before the roofing is applied and the shakes are inserted underneath the laid felt as you go along. Lay the felt in sections, as much as you can roof at a time, because the felt can be affected by the wind or sun if left uncovered for too long.

Use a 30-pound felt in both 36-inch-wide and 18-inch-wide rolls. The 36-inch-wide rolls are used only at the eaves and the 18-inch for the rest of the roof. Buy precut 18-inch-wide rolls of shake roofing felt. It is especially made for this application and is much easier than buying the 36-inch-wide rolls and cutting them yourself.

There is a certain system of prefelting, explained here, that must be followed to ensure that the roof will not leak. It allows for the upper 4 inches of each shake to be inserted under a piece of felt interlay. This will prevent water, should any run below the shakes, from penetrating the felt; the water will simply be carried back to the top of the shakes and then over the roof and into the gutters.

Begin at the lowest point of the roof (eaves) and apply a 36-inch-wide piece of felt so that it hangs over the roof ⅜ inch. This overhang will ensure that the water is delivered into the gutters. Apply this sheet smoothly, stapling it with a hammer tacker or a staple gun every 12 inches. Be careful not to rip the felt. Overlap all side laps 12 inches. At these overlaps, double back the ends of the felt 12 inches for added protection against water moving sideways under the shakes. Do this also around any openings, such as skylights, etc. Overlap in any valleys or over any hips by 6 inches.

After the first 36-inch piece is in place, you will be using 18-inch pieces of felt the rest of the way up the roof. The second piece of felt (the first piece of 18-inch-wide felt) begins 2 times the weather exposure above the bottom edge of the first course. If your weather exposure is 7½ inches,

Photo 15-26. Note how all the materials are stacked from the ridge down. This is because you will be starting from the eaves and working towards the ridge.

Photo 15-27. Prefelting the roof is the first step.

Photo 15-28. Roofing felt is overlapped at the hips, valleys and ridges.

then begin this second (18-inch-wide) course 15 inches above the bottom edge of the first course (36-inch-wide course). If the weather exposure is 10 inches, use 20 inches as the measurement.

Continue applying 18-inch-wide courses up the roof so that they begin the distance of the weather exposure above the bottom of the course below. For 7½-inch exposure, the courses are spaced 7½ inches; for a 10-inch exposure, the felt courses are spaced 10 inches. At the ridge, apply a sheet so that it overlaps both sides of the ridge.

Be careful working around the felt, especially if it is being installed over slats. Do not step on a slat between 2 rafters; step near the rafters. Cover about as much area as you plan to shake in a day. Measure up from the ridge every now and then to be sure you are going straight. If the roof intersects any vertical walls, run the felt 10 inches up the wall for added protection. Some builders use a drip edge at the sides of the roof to divert the water back to the center.

In snow-free areas, no felt interlay or underlay is required for straight split or taper split shakes if the weather exposure is less than ⅓ the total shake length (3-ply roof). In snow areas, you may want to use an extra 18-inch strip at the bottom so that it is about 8 inches up from the eave for 10-inch exposure, 6 inches up for 7½-inch exposure. This allows more of the first course of shakes to be inserted under the felt and allows you to cover the holes in the starter course.

Step 2
Apply the Valley Flashing

Margin of Error: 1 to 2 inches off proper location.

Most Common Mistakes: Using valley flashing with no center crimp; nailing aluminum flashing with galvanized nails or vice versa; placing nails in exposed section of flashing.

Diagrams: Figures 15-5, 15-6.

Photographs: Photos 15-29, 15-30, 15-31, 15-32.

If you have any valleys in your roof, you will need to apply the valley flashing before the shakes are applied, but after the underlay felt has been applied. Valley flashing is usually 20 inches wide, 28-gauge galvanized metal with a center crimp or ridge. This ridge prevents water running down one side of the roof, from running across the valley and under the shakes on the other side. The flashing usually comes in long sheets; you may need 2 or more pieces to run the entire length of the valley. Be sure the upper piece overlaps the lower piece (by about 8 to 12 inches), and not the other way around. I advise wearing gloves when handling this material, because it has very sharp edges.

Before applying the valley flashing, run a 36-inch-wide piece of roofing felt down the valley on top of the pre-felting, pressing it tight against the deck and stapling or nailing its edges. This is for added water protection. After this is done,

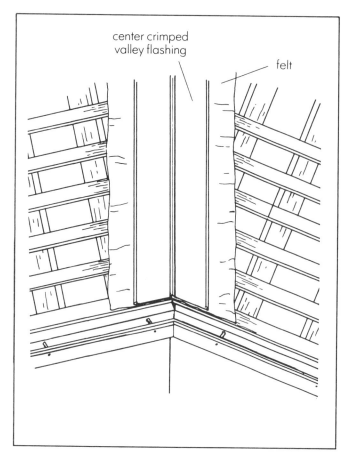

Figure 15-5. Valley flashing is applied on top of the felt layer.

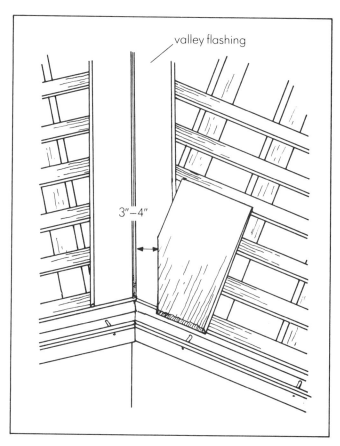

Figure 15-6. This shows how the shakes are cut at an angle to match the center crimp (the felt is omitted for clarity).

Photo 15-29. Valley flashing has been applied to this roof. Usually felt is applied first below the flashing. Note how the upper flashing overlaps the lower flashing.

apply the flashing, the lowest piece first. Nail with 1¼-inch galvanized nails (aluminum nails, if using aluminum flashing) about every 12 inches, an inch or so from the edges. Never nail in the exposed area of the flashing; this will cause a leak.

After the flashing is nailed in place, you need to pop a line with a chalkline where the shakes will fall on the flashing. The shakes taper in such a way that the exposed trough gets wider and wider as you go down the trough. At the peak, pop the line 3 inches from the center crimp on each side; and at the bottom, at a point so that the exposed flashing area expands at a rate of ⅛ inch per foot on each side.

Your shakes should be cut in such a way that their edges will line up along this line. Use wider shakes in the valleys. Lay them into the valley, scribe lines where to cut them and then cut and nail them in place. Always nail each shake 1 to 2 inches above the butt line of the shake to be applied on top of it. Be sure no nails are exposed. There should remain, at minimum, a 3-inch trough of exposed flashing on both sides of the center crimp.

If 2 valleys intersect at a ridge, fit the valley ends of the flashing snugly over the ridge and each other by making small cuts and bands along the edge. Cover with shakes.

You can paint the valley flashing, as well as any other flashing, to closer match it to the color of the shakes. First clean the metal with a solvent, and then apply a zinc-based primer coat. Spray on 2 coats (not one heavy coat) of rust-resistant metal paint.

Photo 15-30. Valley flashing is intersecting at the eaves on this roof.

Photo 15-31. Valley flashing with shakes laid 4 inches from the center crimp and cut at an angle, parallel to the crimp.

Photo 15-32. A finished valley flashing. The center crimp is to prevent water running down one slope from going under the shakes on the other slope.

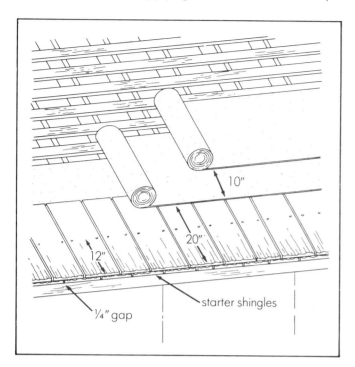

Figure 15-7. In applying shakes, note that the first 18-inch-wide felt begins 20 inches above the eave line. All other 18-inch-wide felt layers are 10 inches above the course below it.

Photo 15-33. Starter shingles are being applied with a 1½- to 2-inch (2 finger digits) overhang. Note how shingles are tucked under the shake felt.

Step 3
Apply the Starter Course

Margin of Error: ¼ to ½ inch off proper location.

Most Common Mistakes: Using rough shakes for starter course; not overhanging starter course 1½ inches; not leaving gap between shakes.

Diagram: Figure 15-7.

Photograph: Photo 15-33.

After you have prefelted a section of the roof and applied any needed valley flashing, you are ready to begin applying the shakes. The first course in shake application is different from the rest of the courses because it must be doubled to prevent water from running through the cracks of the first course (this is a code requirement). To accomplish this, the first course is a double layer with the joints staggered (at least 1½ inches in between), so that water will not penetrate the felt.

The lower course is made up of starter (thin) shingles, not shakes. These starter shingles are 15 inches long and lower in quality than most shingles. Place nails ¾ to 1 inch from the side edge and 1 to 2 inches above the butt line (bottom edge of the shake that will be on top). This ensures that no nails will be exposed, because they will be covered by the shake above. Be sure that you drive the nail just so it is flush with the shake; if you try to drive it deeper than this, you may break the shake. Use only 2 nails per shake. You will want to use the longer 8d nails for the second layer of the starter course, since they need to penetrate the lower layer. Leave a ½-inch gap between each shake or starter shingle to allow for expansion and remember to offset the joints in each succeeding course by at least 1½ inches (code requirements). Overhang the eaves 1 to 1½ inches, so that water drips into the gutters. Overhang the rakes about 1 inch.

The last 4 inches of the shakes (if you are using 24-inch-long shakes) will be tucked under the second course of roofing felt. Be sure in all your shake courses that the upper portion of the shake is tucked underneath the felt. This provides the necessary waterproofing. Work along the eaves and nail in both layers of the starter course. After this is done you are ready to start laying the remaining shakes.

Step 4
Apply Shakes

Margin of Error: Within ½ to 1 inch of proper location.

Most Common Mistakes: Using poor quality shakes; having improper weather exposure; not leaving a sufficient gap between shingles; not inserting top of the shake under the felt; not nailing above the butt line.

Diagram: Figure 15-8.

Photographs: Photos 15-34, 15-35.

After the starter course has been applied, you are ready to start laying the rest of the shakes up the roof. With standard shakes, the tapered end is up and the thicker end is down,

Photo 15-34. The second course is being applied. Note how the shakes overlap at the hip. These will be cut flush later.

Photo 15-35. The second course of shakes being applied. Note how the upper 4 inches of the shakes are tucked under the felt. The weather exposure here is 10 inches.

the rough side is always exposed and the smoother side is down. Work your way across the roof, applying shakes in an area that is comfortable for you without straining, and then move. The more efficient your system for applying shakes is, the quicker the job will go. Observe some professionals, before you begin, to see where they place the shakes, how they sit, how they hold the nails, how big an area they cover at once, etc. Having a system can make a great deal of difference in how much you get done, in how long a period of time, with how much effort.

As mentioned earlier, use two 6d nails for each shake, ¾ to 1 inch in from the side, 1 to 2 inches above the butt line (the bottom edge of the course above). Do not go higher or further in than this or the shakes might curl. Remember not to drive the nail too deeply into the wood, or the shake may crack. Use the head or handle of the shake hatchet as a guide for your weather exposure. An 18-inch-long shake usually uses a 7½-inch exposure; for one 24 inches long, use a 10-inch exposure. A superior 3-ply roof can be created by using a 7½-inch exposure with a 24-inch shake. There will be a slight increase in cost.

Be sure to leave approximately a ½-inch gap between shakes and offset the joints of succeeding courses by at least 1½ inches. Tuck the last 4 inches of the shake underneath the felt for waterproofing. Also save your wider shakes to be used at the edges and in the valleys. At the rakes, overhang the edge about 1 inch and nail as close to the edge as possible to prevent curling. Should a shake split while nailing, treat it

as 2 shakes, unless this disturbs the 1½-inch offset of joints, in which case, it will have to be removed.

Work your way up the roof. Should you encounter hips, ridges, valleys, skylights, plumbing vents, chimneys, etc., do these as you go, as explained in this chapter. You can measure the exact distance to the ridge and adjust the weather exposure to finish out evenly at the ridge. On steep roofs, you may want to build a toe-board out of a long 2×4 and shakes at 3-foot intervals. Be careful and work smoothly and without strain. Pace yourself well and enjoy the view.

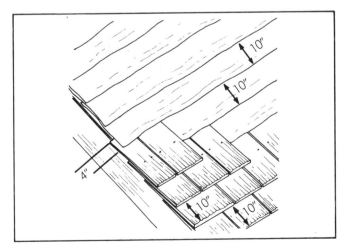

Figure 15-8. Note that 4 inches of each 24-inch shake are under the roofing paper.

Photo 15-36. The finished application of vent pipe flashing. Note how the flashing overlaps the lower shakes and is over-lapped by side and upper shakes.

Photo 15-37. Pre-made vent pipe flashings with rubber gaskets.

Photo 15-38. To roof around fireplace or wood stove chimneys, first apply insulated pipe assembly. Adjustable ''wings'' on the roof mount adjust for any slope.

Photo 15-39. After the roof mount is installed, continue to apply the felt. Install the chimney flashing, tucking the upper edge under the felt.

Photo 15-40. Continue to apply shakes, cutting the shakes that are directly below the chimney, as shown. Note that the flashing overlaps the lower shakes.

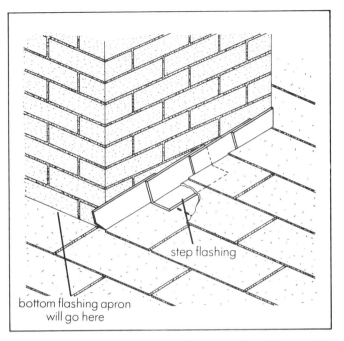

Figure 15-9. Note that the step flashing is placed against the vertical projection (chimney, skylight, etc.) and under the shake or shingle. It lies on top of the shake or shingle in the course below. (Counter flashing, embedded in the brick mortar, has been omitted for clarity.)

Step 5
Flash for Plumbing Vents and Chimneys

Margin of Error: ½ to 1 inch off proper location.

Most Common Mistakes: Placing upper shakes underneath the flashing; placing lower shakes on top of the flashing; not leaving a 1½-inch gap between shakes and the neck of the flashing; nailing in exposed parts of the flashing.

Diagram: Figure 15-9.

Photographs: Photos 15-36–15-42.

Most builders use pre-made plumbing vent flashing, though it can be made at the site. I highly recommend the pre-made units with neoprene gaskets, where the pipe pokes through the flashing. It ensures that no water will run between the vent pipe and the flashing into the house. They are made of galvanized metal and come in the common vent pipe sizes, 1½ inches, 2 inches, and 3 inches. Another model, made of lead, simply bends at the neck, into the opening of the pipe, to stop water from getting between the pipe and flashing.

There is a certain manner in which the flashing is installed that allows it to be interwoven with the roof in such a way that the roof will not leak. This procedure is the same for plumbing vents or chimneys. The sequence is as follows:

First, lay the shakes up against the protruding vent pipe. You may need to notch the shake to fit around the pipe. (Note that the pipe has been poked through the roofing felt.) Never apply the course of shakes above a pipe vent until after you have installed the flashing.

Photo 15-41. After the lower shakes are applied, install the shakes above and on the sides of the chimney. These shakes will overlap the flashing, as shown.

Photo 15-42. The finished application of the shakes around the chimney. Unlike the photo, it is best to leave at least 2 inches between the shakes and the edge of the flashing.

After the shakes are properly laid against the pipe, slip the flashing over the pipe so it overlaps the lower shakes. If possible, tuck the top part of the flashing under the building felt. If the bottom of the felt is too far above the top of the flashing for you to slip the flashing under it, cut a piece of felt 3 feet wide and slip one edge underneath the felt and the other over the flashing. Press the flashing against the deck and nail it to the deck along the sides and top where shakes will cover over the nails. Use flat-headed galvanized roofing nails (or aluminum if using aluminum flashing).

Now, continue laying shakes on both sides of the pipe. Leave a 1½-inch gap between the shake and the flashing neck to ensure drainage and to prevent debris from damming up.

Finally, lay the course of shakes immediately above the pipe. You will probably need to cut the shake right above the pipe so it fits. These upper shakes overlap the flashing. Leave a 1½-inch gap between the bottom of this shake and the top of the flashing neck.

You can now continue along the roof, applying shakes until you are finished or come to another flashing situation.

Wood stove chimney flashings

Wood stove chimneys protruding from the roof are flashed in a similar way, only on a larger scale. Special flashing and the insulated pipe that goes through the roof section are available for all different sized pipes. Often your stove manufacturer can provide these or will recommend a certain model. These are applied exactly like the plumbing vent pipes, unless the flashing manufacturer directs otherwise. Before making this final, be sure that your wood stove will be the required distance away from any interior wall. Don't forget to tuck the upper part of the flashing under the felt.

Step 6
Apply Hips and Ridges

Margin of Error: ½ to 1 inch off proper location.

Most Common Mistakes: Applying line of shakes crooked; not beginning with double starter course; using wrong weather exposure.

Photographs: Photos 15-43–15-52.

You can buy special hip and ridge shakes by the bundle. These are shakes approximately 5 inches wide that are

Photo 15-43. Pre-assembled ridge and hip shakes, as opposed to site-built ones, are advised.

Photo 15-44. Hip shakes being cut flush at the hip.

Photo 15-45. Shakes are cut flush at the hip and are ready for the hip/ridge shakes.

Photo 15-46. Applying hip/ridge shakes at a hip. Use the same weather exposure and a double layer for the first course. Pop a line 5 inches from the hip line as a guide to align the hip shakes.

Photo 15-47. A hip almost completed.

Photo 15-48. A hip completed. Note how straight the hip shakes have been applied.

Photo 15-49. Flashing where a hip and a corner of a second-floor section of the house intersect.

stapled together to form a peak or a ridge. If you don't want to buy these pre-made, you can make them yourself by cutting 5-inch-wide shakes and assembling them at the job site. For the added money, I recommend buying them pre-made, unless every dollar is important to you.

Hips

Before finishing off hips, remember to overlap the felt paper at the hips, about 6 inches. As the shakes are applied up both sides of the hip, they are scored and cut at the proper hip angle. Shakes that intersect the hip (as well as those at the overhangs and the valleys) should be wider shakes, since you will be cutting them. Score them with your hatchet and carry a power saw to the roof to cut them. After the shakes are cut back at the ridge, cut a 6-inch-wide piece of felt and run it the length of the hip for added protection. You are now ready to apply the hip shakes.

Applying hip shakes is the same as applying standard shakes. Use the same weather exposure and apply a double course at the eave. Alternate the joints where the 2 shakes meet in the assemblies, one on your left, the next on the right. It is best to pop a line 5 inches from the ridge on one

Photo 15-50. The shakes have been applied to the ridge. They will be cut flush before the ridge shakes are applied.

Photo 15-51. A ridge vent has been used, instead of ridge shakes, to ventilate this roof section.

Photo 15-52. Ridge shakes have been applied from both directions and a "saddle" has been placed where they intersect in the middle.

side, to use as a guideline, to be sure the line of hip shakes is running straight. Any waver or irregularities can be easily spotted. Some roofers cut back the lower shakes of the starter course so that they don't underlap the second row of shakes. This will allow a smoother hip. You will need to use your longer, 8d nails for hips and ridges since you need to penetrate more shakes. At the top of the hip they are cut to follow the roof line or a wall they may be intersecting.

Ridges

Ridges can be done in several ways, mostly depending on whether you plan to use a ridge vent or not. Many builders are now using a ridge vent (see Chapter 14, *Roof Sheathing*—Plywood Sheathing, "Step 4") in order to rid the roof panel of any possible moisture build-up. It is especially recommended in colder climates, though I would check with local, open-minded builders for their opinions. When in doubt, put one in (unless you have an exposed beam ceiling with no real roof panel); it couldn't hurt.

RIDGE VENTS: If you are planning a ridge vent, you will have to leave a 1½-inch vent space in the sheathing and the shakes to allow the air to flow. After you have done this, purchase a ridge vent in your desired color, either in plastic or metal, and nail it over the ridge. See if there are any special manufacturer's instructions as to its installation.

RIDGE SHAKES: If you are planning to use a common ridge with no ridge vent, apply your pre-made or job-site built ridge shake assemblies. It is best to begin the ridge shake application from the side opposite the prevailing winds, so that no water will be blown under the shakes. If the winds do not come from a consistent direction during the rainy season, you can start from both ends and meet in the center, covering the joint with a ridge shake to form a "saddle." As with hip shakes, begin your shakes with a double starter course. If the ridge line will be noticeable (and even if it won't), pop a guideline to keep it straight. Alternate the joints on the assemblies; one on your left, the next on your right. If you are using no vent space, cover the ridge with a 6-inch-wide piece of roofing felt, applied along the ridge and beneath the shakes (use 30-pound felt).

Step 7
Flash for Skylights

Margin of Error: ¼ to ½ inch off proper locations.

Most Common Mistakes: Not using step flashing; not overlapping step flashing onto bottom skylight flashing; not overlapping top skylight flashing onto step flashing; not using shop-made bottom and top flashing.

Photographs: Photos 15-53, 15-54, 15-55, 15-56.

Installing the skylights and their flashing properly is essential if they are not to leak. Most skylights are put on over a curb, or a box-like wooden protrusion coming from the roof, that is

somewhat smaller than the lip of the skylight frame. The frame is inserted over this curb to stop leaks. Most curbs are built to specification (usually ½ to ¾ inch less than the width and height of the skylight frame) at the job site from 2×6s. Some skylights are provided with their own curbs and flashing, but this is not common.

Most curbs for skylights are built directly over the rafters, which are on 16-inch, 24-inch or 48-inch centers. The roof will need to be framed in such a way that there is a rafter directly underneath each side of the curb in the area you want the skylight (see p. 318, "Skylights" in Chapter 13, *Roof Framing*). After the decking is applied to the rafters, build the curb and nail it to the decking directly above, and in line with, the rough opening you have left in the rafter framing. Check this opening, once again, to be sure it is the proper size. Then begin applying the felt and shingles. When the felt reaches the curb, run it up the sides of the curb, at the bottom and top and along the sides.

I advise that you get the bottom and top flashing made at a sheet metal shop. The flashing in these areas is very important, and soldering the corners is an added protection against leaks. It's wrapped around the curb a certain distance and comes up on the sides of the curb at both the top and the bottom. Many sheet metal shops have experience in making this type of flashing, and some places may even have it ready-made. If you cannot find it ready-made, give the metal shop the details on the slope of your roof, the type of roofing you plan to use and the width and height of the curb (outside edge to outside edge), and for a reasonable price ($10 to $25) they will make your flashing. You may want to order the step flashing at this time, too, rather than bending it at the job site.

Bring the shakes up to the bottom of the curb. You may need to cut some of the shakes directly below the curb for them to fit. After these shakes are applied, install the bottom flashing. This flashing must overlap the shakes below it in order to keep the water moving down the roof. Nail this curb with galvanized nails (aluminum nails, if you are using aluminum flashing). Place all nails into the curb or along the areas of the flashing that will be covered by shakes. Never nail in an exposed area.

You are now ready to begin laying shakes again. This is done as you have been doing the rest of the house, with one important difference: the use of step flashing. Step flashing is a separate piece of flashing, the same length as the shake or shingle, that is bent in an L-shape to extend up the sides of a protrusion on the roof, to lie flush with the bottom of the shake on the roof. It is used whenever a sloping roof with shakes or shingles intersects a vertical plane, such as a skylight curb, a brick chimney, a dormer, a vertical wall, etc.

Step flashing, cut the same length as the shakes, is inserted directly below each shake so that its bottom is flush with the bottom of the shake and its upper end is tucked under the roofing felt along with the shake. The flashing should extend 6 inches onto the roof, under the shake, and be within an inch or so of the top of the curb. It is nailed into the curb, using small rust-resistant nails, and nailed to the roof, with the nails that hold the shakes to the roof decking. Work your way up both sides of the roof, using the flashing interlaid

Photo 15-53. This curb is flashed and ready to receive the skylight. Note how the upper shakes overlap the top flashing and the top flashing overlaps the side step flashing.

Photo 15-54. Applying the lower section of the flashing for a skylight installation. This flashing was supplied by a manufacturer. Note the soldered corners.

Photo 15-55. A close-up of skylight flashing. Step flashing is being used. Note how the flashing always overlaps the piece below.

Photo 15-56. The finished installation of a skylight. Note the upper and lower flashing sections and the side step flashing.

with the shakes, until you reach the top of the curb. Always leave a 1½-inch gap between the curb and the shake for drainage.

At the top of the curb, insert the top flashing in such a way that the very top of the step flashing will be inserted *underneath the top flashing,* and the shakes above that will overlap onto the top flashing. Your concern is, once again, delivering the water from layer to lower layer until it is off the roof. Leave a 1½-inch drainage gap between the shakes at the top and the curb. After the shakes are applied at the top of the curb, begin your normal pattern of applying the shakes. You may have to cut some of the shakes so they will fit around the top of the curb.

Step 8
Flash Where Vertical Walls Intersect the Roof (optional)

Margin of Error: Flashing ¼ to ½ inch off proper location.

Most Common Mistake: Not flashing properly.

Photographs: Photos 15-57–15-63.

Flashing in this area is very similar to skylight flashing. I advise you to have the corner flashing made at your local sheet metal shop also. Step flashing is applied along the sides in the same manner as described in "Step 7." The siding and its under layer of tar paper are applied after the roofing,

overlapping the step flashing and the roofing felt that has been continued 6 inches up the side of the wall. As always, leave a 1½-inch gap between the shake and the wall and a 1½-inch gap between the bottom of the siding and the roof.

Photo 15-57. Pre-made soldered corner flashing can be bought at local sheet metal shops. These are recommended to ensure against leaks.

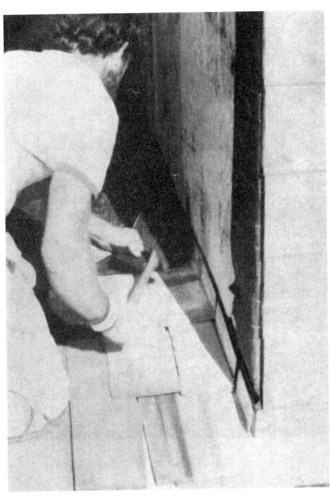

Photo 15-58. Step flashing always overlaps the flashing below it, to provide proper drainage.

Photo 15-59. Applying step flashing where a sloping roof intersects with a vertical wall. Note how the step flashing lies directly below a shake and is tucked under the felt.

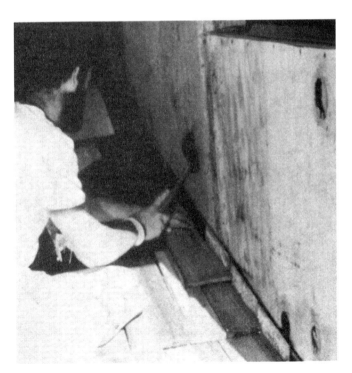

Photo 15-60. Shakes are applied directly on top of the step flashing and are tucked under the felt.

Photo 15-61. Pre-made corner flashing is in place here. Note how the flashing overlaps the shakes.

Photo 15-62. The roofing is almost completed.

Photo 15-63. Shakes as viewed after application.

Applying Asphalt Shingles
What You'll Need

TIME 25 to 40 person-hours per 1000 square feet of roof.

PEOPLE 1 Minimum (semi-skilled); 3 Optimum (1 semi-skilled, 2 unskilled); 6 Maximum (2 teams, 1 on each side of the roof).

TOOLS Rubber soled shoes, knee pads, chalkline, power saw, tape, tin snips, ladders, extension cord, utility knife, staple gun or hammer tacker, staples, framing hammer, nail stripper (Photo 15-17).

RENTAL ITEMS: Pneumatic nailer (Photo 15-19), safety harness (Photo 15-20), ladder jack, adjustable scaffolding and ladder hooks.

(For a description of the last 6 items, see "Tools" under "Applying Wood Shakes," pp. 359–362.)

MATERIALS Asphalt or composite shingles are made from wood or rag fibers, impregnated with asphalt and coated with gravel. There is also a lighter weight shingle with a fiberglass base, that is usually 3 feet long, 12 inches wide and sectioned into 3 tabs by two 5-inch-deep slots called cutouts. Most shingles have dabs of roofing adhesive on the surface, directly above the cutouts. This adhesive, heated by the sun, adheres to the shingle above it, and holds it in place in the event of high winds or curling due to heat.

Shingle quality is graded by the weight. A 232-pound shingle means that the shingles of this grade used to cover 1 square will weigh 232 pounds. The heavier the shingle, the higher the quality and the longer the life. The exception to this is fiberglass-based shingles, which are light in weight but still high in quality. A 232-pound-standard-shingle roof is considered a good roof.

In making a choice of the type of shingle best suited for your house, select carefully; it will make an immense difference to the appearance of the house. Shingles come in many different styles and colors to choose from. Buy all the shingles at the same time, and be sure they are from the same lot number; colors can vary slightly. Patterns can be formed on the roof through creative application of the shingles. You can, for instance, use 3 layers at every eighth course, with their bottoms flush, to create a ribbon course. This will give emphasis to the horizontal lines of the house. (You can test this effect by taking a picture of the house and then drawing the lines on the photograph). In hot, humid climates, you can buy shingles treated with an agent to help resist fungus and algae. Do not store the materials in the sun or allow them to get wet.

ESTIMATING MATERIALS NEEDED

You will need to estimate shingles, roofing felt, metal flashing and nails. To estimate the number of shingles you need, first determine the square footage of the roof surface and then the number of squares (the number of shingles that will cover 100 square feet of roof surface) needed. The roof surface can be calculated simply, by finding the square footage of the house at ground level and multiplying this times a factor, the roof coverage factor, related to the slope of the roof, given in the chart below. (Include all overhangs and eaves in the measurements.) Add on 10% for wastage.

Slope	Roofing Coverage Factor
2 in 12	1.02
3 in 12	1.03
4 in 12	1.06
5 in 12	1.08
6 in 12	1.12
7 in 12	1.16
8 in 12	1.20

Once you know the number of square feet in the roof surface, this must be used in relation to the shingle exposure you will use. Your supplier will have a conversion table to ascertain the quantity of shingles your roof will need.

Order enough shake felt to cover the square footage of your roof. You will need 36-inch-wide felt for the first courses on each side, and the rest will be 18-inch-wide felt.

You must measure for your flashing needs. Be sure to include all protrusions (vents, chimneys, etc.) through the roof, the center crimp valley flashing and flashing where vertical walls intersect sloping roofs.

In nails, you will need 2 pounds per square of shingles (12-gauge galvanized with ⅜-inch head, 1¼ inches long).

The number of hip and ridge shingles needed is determined by measuring the length of the hips and ridges, converting this measurement to inches and dividing by the exposure of the shingles.

INFORMATION Asphalt shingle application is one of the easiest procedures in housebuilding. There is little complexity or difficulty. The information given in this chapter should get you through with no trouble. If you have any unique situations that are not covered here or any doubts, talk to your supplier or a professional before proceeding. Be clear in whatever you do; you do not want leaks.

PEOPLE TO CONTACT People to contact at this time include the plumber and electrician, to be sure the plumbing vents and electrical weatherheads are in place; the solar panel installer, to be sure mounts are in place; and the insulation contractor, to let them know the house will soon be ready for insulation.

INSPECTIONS NEEDED Usually before the roofing and roofing felt are applied, the inspector will want to inspect your roof sheathing to be sure the nailing pattern meets code.

Reading the Plans

Most plans will simply indicate the type of roofing to be used; e.g., composite, wood shake, cement tile, etc. Seldom will the plan tell you the type or quality, unless the plans are well detailed. Some plans will say nothing about the type of roofing to be used but will simply leave this to the owner's discretion. Usually there is nothing indicated on the plans about the application procedure, unless there is a unique problem that the designer has detailed. Otherwise, it is assumed that the roofing installer has all the information needed to install the roof.

Safety and Awareness

SAFETY: As in roof sheathing or any work up high, applying the roof is very dangerous. Be extremely careful; accidents can happen before you even realize it. Keep good balance and do not lean over the sides of the roof too far.

When using a ladder, lean it against the wall so that the bottom is placed at a distance from the wall equal to ¼ the height of the wall. Rest scaffolding and ladders on level, solid ground, or on a secure foot on one side. One person on a ladder at a time; someone should hold on to the ladder as you climb; and avoid overhead electrical wires. Stay off scaffolding or ladders in high winds or rain.

AWARENESS: Water, water, water! Everything you are doing is to stop water from leaking into the house. As a contractor, the worst call back I could ever have was for a leak (fortunately, there were few). Where was the leak coming from? How do I locate it? Do I have to strip off large portions of the roof to repair it?

Be sure you understand what you are doing before proceeding. It is all simple, but don't trust your common sense, and make no assumptions. You want to welcome that first rainstorm.

Also with long sections of repetitious work the system you use is very important in order to finish the job with the least amount of effort and time. Constantly be seeking quicker and easier ways to apply the shakes. One hour watching professionals and adopting some of their tricks may save you hours on the roof.

Overview of the Procedure

Step 1: Prepare the roof deck: felt, flashing, and drip edge
Step 2: Apply the starter course
Step 3: Apply the first course
Step 4: Apply the remaining courses
Step 5: Apply hips and ridges

Step 1
Prepare the Roof Deck: Felt, Flashing and Drip Edge

Diagrams: Figures 15-10, 15-11, 15-12, 15-13.

Roofing Felt

Before applying the asphalt roofing, inspect the roof deck. Some roofers go so far as to fill any knotholes with filler, but most simply ascertain that the deck is clean and free of debris. Measure the deck itself to see if it is square and the proper dimensions. If it is not square, adjust the course of shingles or the lengths of shingles as they are applied to make any discrepancies less noticeable. Be sure that the deck is dry; if it has rained, allow one or two days' drying time before beginning. Once it is clean and dry, you can begin to apply the roofing felt.

Snap horizontal lines on the decking to align the upper edge of the felt. The first line is popped 35⅝ inches above the eave (the 36-inch-wide felt overlaps the eave ⅜ inch). All horizontal lines after that are popped at 34-inch intervals, since there is a 2-inch overlap between the courses. Begin by applying 15-pound felt at the eaves, overlapping the edge by ⅜ inch to ensure proper drainage into the gutters. The felt should be flush at the rake edges and have 4-inch side laps wherever 2 pieces of felt are joined. There should be 6-inch overlaps over all hips, valleys, and ridges as measured from their centerlines. Be sure the felt lies flat and smooth. Secure the felt with a hammer tacker, staple gun or pneumatic stapler, using enough staples to secure the felt against any possible wind damage.

Drip edges, metal flashing installed along the eaves and rakes, are often used. They extend in from the roof edge about 3 inches and keep draining water free from any cornice construction. In snow areas, eave flashing is also recommended. This is a cut strip of either smooth or mineral-faced roll roofing that is wide enough to extend from the edge of the roof to a point that is 12 inches inside the wall line. It is installed over the drip edge and the underlayment (felt). The lower edge of the strip is placed even with the drip edge.

Flashing

Roofing is basically a very easy operation in the housebuilding process. The only really complicated areas are where intersections occur with other roofs or walls, or where pipes or chimneys protrude through the roof. These areas must be properly dealt with or they will leak, and repair work is frustrating and hard to do. Be sure you clearly understand how to flash all these areas before you begin. Properly done, there will be no leaks; but one small mistake can cause a big hassle.

VALLEY FLASHING: A valley is where 2 sloping roofs meet. Heavy drainage from both roofs occurs at this point and proper flashing must be installed. Usually a heavy roofing felt (90-pound) is used as flashing and installed in a double layer: first an 18-inch-wide layer with the mineral surface down; and

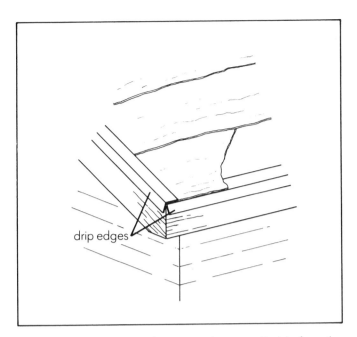

Figure 15-10. Drip edges are often installed before the roofing goes on. Note how the felt overlaps the bottom drip edge, but not the side edge.

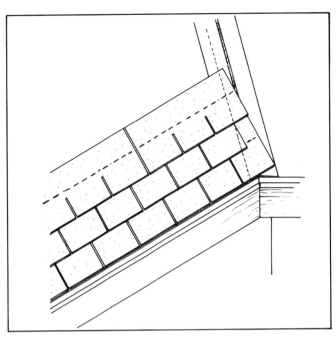

Figure 15-11. Note how shingles are cut at an angle to be parallel to the valley center crimp.

second, directly on top of this, a 36-inch-wide layer with the mineral surface pointing up. Overlap pieces about 12 inches and join them with roofing cement. Be sure a good, distinct valley is created.

Once both pieces have been installed, pop a chalkline in the center and 2 more on each side of this centerline. The exterior lines are the boundaries; the shingles will come only as far as these lines, thereby creating a waterway in the center. This waterway will be 6 inches at the ridge (highest point), expanding ⅛ inch per foot as it progresses down the roof. Your shingles will be cut to meet this line to create the valley.

Valleys are sometimes created using the shingles themselves as the valley flashing. These are called woven or closed-cut valleys and are commonly used in re-roofing work. I will not go into this type of valley flashing here, but you may want to ask your roofing supplier about its installation if you are interested.

VENT OR PIPE FLASHING: Where plumbing vent pipes protrude through the roof, the pipes must be properly flashed. Because this is done during the roofing process, all plumbing must be completed before the roofing is started. It is very difficult to install a vent pipe through roofing that has already been installed.

You can buy pre-fabricated vent pipe flanges, made of metal, for different sized pipes. First, lay the shingles up to the pipe itself, cutting the shingle to fit around the pipe. The flange is then installed over the lower shingles, and the remaining courses are applied around it, cutting the shingles slightly to fit near the pipe.

NOTE: For skylight flashing and flashing where the roof meets a wall, see "Step 7" and "Step 8" at the end of the wood shake section in this chapter. Be sure the upper shingles overlap the flashing.

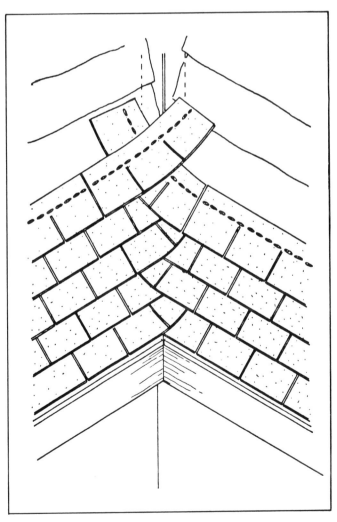

Figure 15-12. A woven valley created with interwoven composite shingles.

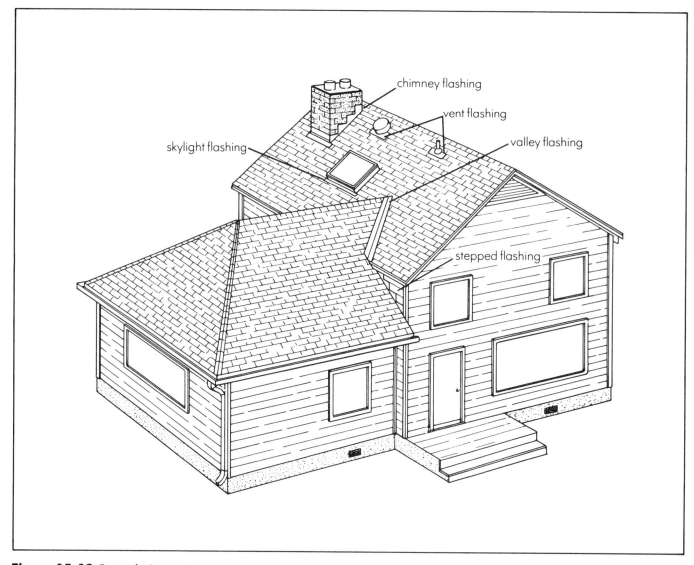

Figure 15-13. Parts of a house that require roof flashing.

Step 2
Apply the Starter Course

Margin of Error: ¼ inch off proper alignment with roof edge.

Most Common Mistakes: Leaving out starter strip; not overhanging roof with starter strip.

Diagram: Figure 15-14.

As with wood shakes, asphalt roofing makes use of a starter course of shingles at the eaves. This course fills in the spaces between the tabs of the first regular course, which is installed directly above it. It also adds extra protection to the roof.

Usually the starter course is a standard row of shingles installed in an inverted position, the tabs facing towards the ridge instead of the eaves. This course is 9 inches wide, and you will need to cut 3 inches off the tab end of the shingles (9-inch strips of roll roofing can also be used). Install these cut shingles (tin snips are used to cut shingles) with the wind seal down and the mineral surface pointing up. Cut 6 inches off the length of the first shingle to stagger the joints of the tabs of the shingles of the starter course with the first standard course of shingles. For proper drainage, overhang the eaves and rakes by ½ inch and leave a ¹/₁₆ inch space between the shingles. Use 4 nails 3 inches above the eaves, 1 inch and 12 inches in from each end. If you are using a drip edge, be sure the starter course is installed *on top* of the drip edge at the eaves and *underneath* the drip edge at the rakes.

Step 3
Apply the First Course

Margin of Error: ¼ inch off proper alignment and location.

Most Common Mistake: Not overhanging shingles at eaves and rakes.

Diagram: Figure 15-15.

The first course is installed directly on top of the starter course and is made up of full-sized shingles. Repeat the same

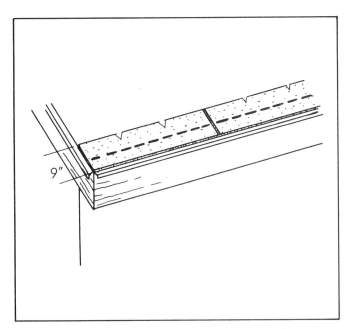

Figure 15-14. Shingles used for the starter course have their 3-inch tabs cut off.

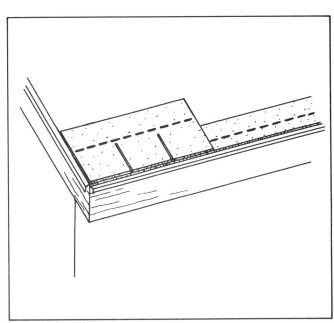

Figure 15-15. The first shingle of the first course being installed.

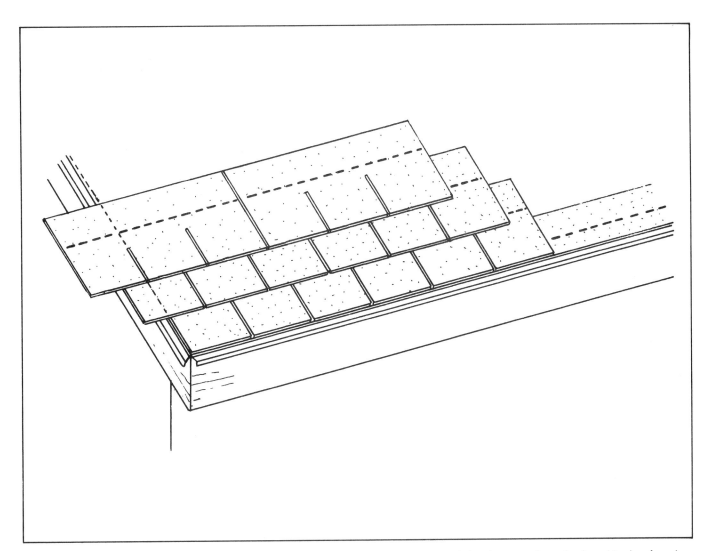

Figure 15-16. The first 3 courses have been applied. Note how an additional ½ tab has been cut from the first shingle of each progressive course in order to stagger the joints between the tabs.

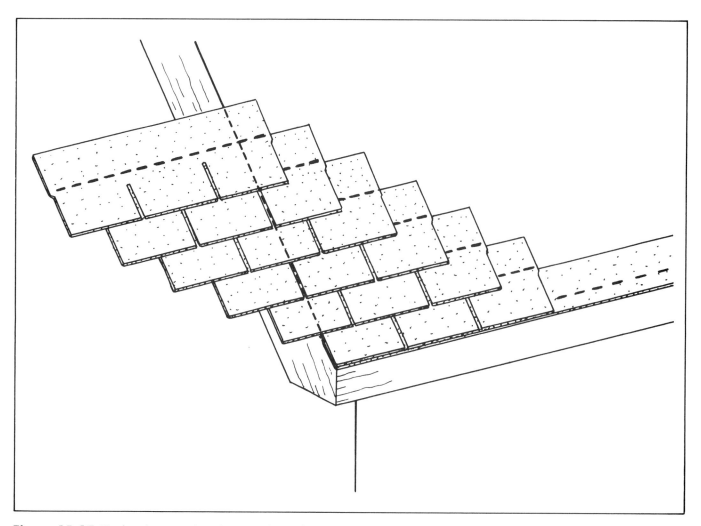

Figure 15-17. The first 6 courses have been applied. After this, begin the process again, using a whole shingle.

overhangs (½ inch) at the eaves and rakes, and space the shingles ¹/₁₆ inch apart. Use 4 nails per shingle, placed 5⅝ inches above the bottom line of the shingle (the butt line) or ⅝ inch above each cutout. Nail the shingles 1 inch and 12 inches from each end. Nails are usually 12-gauge galvanized roofing nails, 1¼ inches long (2 inches for some textured roofing). They usually have a ⅜-inch-diameter head and are nailed flush with the top edge of the shingle. To prevent buckling, start nailing each shingle from the end next to the shingle you just laid.

Step 4
Apply the Remaining Courses

Margin of Error: ¼ inch off proper alignment and location.

Most Common Mistakes: Off alignment; not staggering joints.

Diagrams: Figures 15-16, 15-17.

After the starter and first courses are applied, you are ready to install the rest of the roofing for the remaining courses.

This is monotonous, repetitious work, so always be asking yourself if there is an easier, quicker way to do it. Be careful not to strain your body or endanger yourself by a precarious perch. Relax and enjoy the view.

As in all parts of construction, the joints, in this case, the joints between the tabs, are staggered. You will want the joints (or cutouts, as they are called) to be centered over the tab of the course below. To do this, cut 6 inches off the first shingle of the second course, at the rake. The third course, cut 12 inches off (one full tab), the fourth course, 18 inches (a tab and a half), and continue to reduce the length by 6 inches for each subsequent course. After the sixth course, the sequence is repeated, with the seventh course starting with a full shingle. There are variations on this pattern if you want different visual effects, but the above is the most common.

Many roofers pop chalklines on the felt to make sure that the shingles are aligned, as they are installed, in straight vertical and horizontal lines. For the novice roofer, this is probably a wise thing to do; just to be sure. These lines are popped every 36 inches vertically down the roof and every 10 inches horizontally across the roof, thereby allowing you to align the top and side edges of the shingles along straight lines.

Usually roofers work a number of courses at a time. In the method we have been describing, usually 6 courses are worked at one time, in a fan-like pattern. Remember to allow $^{1}/_{16}$ inch between shingles, and to cut an additional 6 inches off the first shingle as each course progresses.

Step 5
Apply Hips and Ridges

Margin of Error: ¼ inch off alignment and location.

Most Common Mistakes: Ridge or hip shingles not in a straight line; installing ridges before hips.

Diagram: Figure 15-18.

After the roofing has been applied in the field areas and around the flashing for pipes and valleys, you are ready to complete the roofing process by applying the hip and ridge shingles. Always apply the hip shingles first, as these are overlapped by the ridge shingles, which are installed next. Hip and ridge shingles are simply one 12-inch tab of a standard shingle. They can also be purchased pre-fabricated, as hip and ridge shingles. If you are making your own, bend them in the middle to conform with the ridge or valley. In cold weather, you may need to warm the material up before bending it.

Now pop lines 6 inches on each side of the centerlines of the hips and ridges. These will be guidelines for the edges of your hip and ridge shingles. For the hips, start at the eaves with a double layer and work towards the ridge with the standard 5-inch exposure. Use 2 nails, 1 inch from each side and 5½ inches above the butt line, in each shingle. To finish hips where they intersect with the ridge, cut 4 inches up the center of a tab and nail this so that the unslit portion is nailed to the ridge and the 4-inch slit portion overlaps and is nailed to the top of the hip, covering the last hip shingles.

For the ridges, start from the ends and work towards the center, nailing a saddle where they meet. Overlap the tab portions where they come together in the middle. Finish off so that the last tab installed on the left leaves a 5-inch exposure of the last tab installed on the right. You will need to cut the left-hand tab to do this. Finally, install a tab, the saddle, over this joint with a nail in all 4 corners. Cover these 4 exposed nails with roofing cement as the final frosting on the cake. Pat yourself on the back, take a long look around (you might not be this way again for awhile), and your roof is completed.

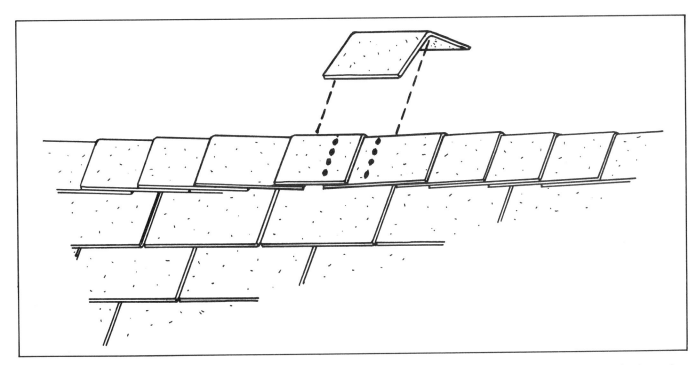

Figure 15-18. As the shingles from both directions meet at the center of the ridge, overlap their top portions. The last tab installed should be cut so that it leaves 5 inches exposed on the tab next to it. Over this place the saddle, the lower half of a cut shingle, and nail it at each corner. Finally, put roofing cement over each nail.

Worksheet
Applying Wood Shakes or Asphalt Shingles

ESTIMATED TIME

_____

_____

_____

CREW

Skilled or Semi-Skilled

_____ Phone _____

_____ Phone _____

Unskilled

_____ Phone _____

_____ Phone _____

_____ Phone _____

MATERIALS NEEDED

Shakes or shingles, felt, flashing, nails.

ESTIMATING MATERIALS

SHAKES OR SHINGLES:

_____ sq. ft. of house at ground level (including

overhangs and eaves) × _____ (the roofing

coverage factor related to the slope of your roof) =

_____ sq. ft. of roof surface. (Add 10% for wastage.)

_____ shake or shingle exposure

_____ no. of squares (as determined by your supplier's conversion table)

_____ + 2 squares for every 100 ft. of valley

_____ + squares for hip and ridge shakes or shingles

_____ = Total no. of squares.

FELT:

_____ sq. ft. of roof surface =

_____ rolls (36 inches wide)

_____ rolls (18 inches wide).

FLASHING:

_____ ft. of standard flashing

_____ no. of pieces of step flashing

_____ no. of pieces of vent pipe flashing, size _____

_____ no. of pieces of top flashing (skylights)

_____ no. of pieces of bottom flashing (skylights)

_____ no. of pieces of chimney flashing, size _____

_____ ft. of center crimped valley flashing

_____ no. of pieces of solar panel mount flashing,

size _____

NAILS:

_____ squares of shingles × 2 = _____ 12-gauge galvanized nails.

_____ squares of shakes × 2 = _____ lbs. 6d or 7d nails.

_____ lbs. 8d nails for starter course, hips and ridges.

COST COMPARISON

_____ Store:

Cost per square _____ total $ _____

_____ Store:

Cost per square _____ total $ _____

_____ Store:

Cost per square _____ total $ _____

Daily Checklist

CREW:

_____ Phone _____

_____ Phone_____

_____ Phone _____

_____ Phone_____

_____ Phone_____

_____ Phone _____

TOOLS:

☐ rubber soled shoes   ☐ nail stripper

☐ knee pads   ☐ roof jack

☐ chalkline   ☐ toe-board

☐ power saw

☐ tape   **RENTAL ITEMS:**

☐ tin snips   ☐ pneumatic nailer

☐ ladders   ☐ safety harness

☐ extension cord   ☐ ladder jack

☐ utility knife   ☐ adjustable (or
                             regular) scaffolding
☐ staple gun/hammer tacker
                           ☐ ladder hooks
☐ staples

☐ roofing hatchet (for shakes)

☐ framing hammer (for shingles)

MATERIALS:

No. of squares _____

No. of hip and ridge shakes or shingles _____

Pounds of nails _____

Rolls of 36-inch-wide roofing felt _____

Rolls of 18-inch-wide roofing felt _____

Linear feet of center crimped valley flashing _____

No. of pieces of step flashing _____

Top flashing for skylights:

No. _____ Size _____ ; No. _____ Size _____

Bottom flashing for skylights:

No. _____ Size _____ ; No. _____ Size _____

Vent pipe flashings:

No. _____ Size _____ ; No. _____ Size _____

Chimney flashing: Size _____

Solar panel mount flashing: No. _____ Size _____

DESIGN REVIEW:

For asphalt shingles, determine:

Color _____ Type _____

Lot no. _____

INSPECTIONS NEEDED:

☐ YES   ☐ NO

Inspector _____ Phone _____

Time and date of inspection _____

WEATHER FORECAST:

For week _____

For day _____

Avoid working on wet roofs; it is even more dangerous.

16 Siding

Photo 16-1. Siding, as it progresses up the wall. Note that the deck ledger board has been installed.

What You'll Be Doing, What to Expect

I HAVE ALWAYS loved putting up the siding. Finally, you are starting to look at a finished product. All the rough work you have been doing up till now, with the exception of the foundation, will be covered up. What you put up during the siding stage will be seen forever. Basically the process, applying the siding course by course from the bottom up, is a simple one. There are a few tricky areas, mostly at the beginning, around the doors and windows and at the corners, but they do not offer any real trouble — if you know what you are doing. The work goes easily, though not always quickly; there is little physical strain; and I have always found it to be one of the most satisfying of all the building processes. It is finish work, but not as demanding as trim or cabinets.

Because it is finish work that will be exposed, you need to

enter a new mind-set about the work you are doing. There is no great event to mark the transition from rough work to finish work (you may want to create one, however, a small party, a day off, etc.). Because of this, many people take the pace, attitudes and standards of quality into the siding that they used on the framing. Be careful not to do this. Each board must be separated and selected in order to use the most attractive boards in the most visible places (around entrance doors, at eye-level, on the street side, etc.). The less attractive or discolored boards can be used up high or down low, in areas of the house where there is little traffic, below decks, etc. Switch over to your lighter-weight finishing hammer, and be careful not to leave too many hammer marks (though many will disappear after a few rains swell them out, or a steam iron can be used to steam them out). Just keep in mind that what you do at this stage will be seen; it will be a part of your autograph — so keep your standards high and your pace efficient, but not rushed.

For high areas, I highly recommend using scaffolding rather than ladders. Many builders do not want to spend the time or money to rent and set up metal scaffolding. Ladders or benches will work well for one-story houses on relatively level lots, but if you have many high areas, I would recommend setting up scaffolding and doing all the high work at one time, e.g., fascias, siding, trim, paint, window installation, etc.

Keep in mind that the siding is a final protective skin of the house, as is the roof. You need to always be thinking about any possible water or air penetration. The weather tightness of the wall panels will depend on the quality of the siding application, trim and caulking. Flashing is also very important. It is often overlooked, done incorrectly or left out, or caulking is used instead. Apply a tight, unbroken flat under-skin of felt paper; flash and caulk properly; trim well; and your walls and house will be well protected.

Jargon

BEVEL CUT. To cut a board at an angle.

FINISH HAMMER. A lightweight (14- to 16-ounce smooth faced hammer used for finishing work.

FLASHING. A plastic or metal material used in wall and roof construction that prevents water from entering at the joints.

HOT DIPPED GALVANIZED NAILS (HDG). Rust-proof nails with a zinc coating.

SHIPLAPPED. Boards whose edges have been rabbeted to form a lap joint between adjacent pieces.

TONGUE AND GROOVE. Boards that are cut in such a manner that their sides (edges), and sometimes their ends, have a tongue or protruding strip on one side and a corresponding groove or continuous indentation on the other.

TRIM. The finish materials of a house, including the mouldings around openings (door and window trim), at the outside corners, around floors and ceilings, etc.

Purposes of the Siding

1) Cosmetic covering
2) Protection from weather
3) Adds shear and rigidity to the walls

Design Decisions

Your decision as to the type of siding (Photo 16-1) you plan to use is a crucial one. It will have a profound effect on the final exterior appearance of the home and can have even a more profound effect on your money and energy budgets. Many people make their decision without all the necessary data or thought. Decisions made in this way can often have devastating effects (as you may already know). Check your conclu-

sions carefully — be sure you feel good about your choice or don't act on it.

In making your decision, keep in mind the overall trend of housebuilding in this country. Most materials and techniques developed have been worked out for large volume builders, who build several thousand houses per year, or for small contractors, whose annual income often depends on the number of houses (not number of hours worked) they build per year. For the large builder, small changes or "improved" building techniques, that may save only a few dollars per home, can be used in several thousand houses and save many thousands of dollars. These changes often lower the quality of the home and are simply not worth the effort when building one house.

For the small contractor who is building several houses per year and needs to complete them all in order to get his 20% profit, time is of the essence. For these contractors, higher quality methods or materials may not be more expensive, but simply take more time and therefore slow up the building process. So they may opt for a quick method and lower quality. This may not be the wisest decision for someone building their own home.

When choosing your siding, you will find that some types of siding, such as plywood, aluminum, stucco, etc., can be installed very quickly. Other types, such as board and batten, vertical or horizontal wood, shakes or shingles, take more time to install, but add a certain quality or handcrafted appearance to the home (along with a week or two of extra building time). As you consider your decision about siding, evaluate each type and consider the following:

- ☐ Price
- ☐ Resale value
- ☐ Appearance
- ☐ Sound insulation
- ☐ Installation labor
- ☐ Repair
- ☐ Maintenance
- ☐ Combustibility
- ☐ Resistance to scratching and impact

The following is a discussion of various types of siding materials. The remaining part of the chapter will show in detail how to install horizontal board siding. I have chosen this type because it is one of the more difficult ones to apply. Actually stucco is perhaps the most difficult, but this is usually applied by professionals, and only used in areas where there is little or no freezing. Other sidings, such as plywood and aluminum, are very easy to install by simply reading the manufacturer's instructions.

Types of Siding

PLYWOOD SIDING: (Photos 16-2, 16-3; Figure 16-1.) I have begun with this type of siding, not because it is my favorite, but because it is becoming one of the most common. The reason for this is that it has one great asset: it goes up

Photo 16-2. A smooth-surfaced plywood siding.

Photo 16-3. Trimming out plywood siding is easy and can hide even large mistakes. The horizontal trim should be flashed so it won't trap water behind it.

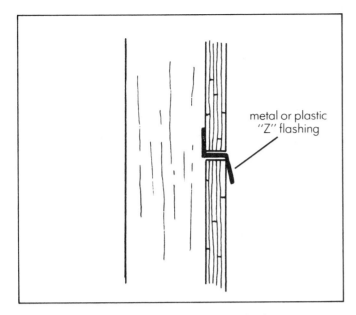

Figure 16-1. Flashing used with a plywood siding.

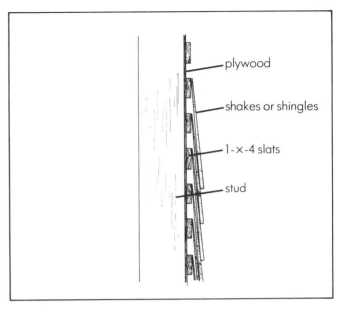

Figure 16-2. Here 1-×-4 slats are used over plywood sheathing. The shakes or shingles are then nailed to these.

quicker than any other type of siding, and as I mentioned in my introduction, this can be quite an inducement for many builders. It may be the best decision for you as well, especially if your time is very limited (though, in 20 years, you may wonder what difference a few weeks would have made).

Exterior plywood, often known as Texture 1-11 (Texture, one eleven) is an exterior grade panel made from thin sheets (plies) glued (laminated) together under pressure with waterproof glues. The interior plies are of mixed woods, but the final exposed exterior ply is Douglas fir, western red cedar, southern pine or redwood. The redwood and cedar plies are more expensive than the others and can run so high as to be the most expensive types of siding to use per square foot. The cheaper boards often have up to 12 plugs per panel (plugging for exposed knotholes), and can be used only if they are painted.

The panels come in a large range of textures and designs. They can come with grooves, to appear as different individual boards; rough or smooth; or straight, with no grooves. They are 4 feet wide and 8 to 10 feet long. Usually they are installed vertically. Horizontal application is sometimes used, but there is the danger of water getting between the plies into the joints. They also come as lap boards, 6 to 12 inches wide and up to 16 feet long. Thicknesses vary from 3/8 to 5/8 inch and are priced accordingly. They also come with various types of treatments: painted, primed, stained, untreated or treated with a water repellent.

The types of plywood siding also vary according to their use. Medium-density overlaid (MDO) is recommended for paint. The exterior surface is a resin-treated sheet, hot bonded to the panel to make a smooth surface for painting. 303 is for stain. It comes in many textures and patterns, often with tight knots or holes that must be filled with synthetic filler. It has a questionable look with the knotholes, but the clear stock is more expensive. Texture 1-11 has shiplapped edges and parallel grooves, and is a higher quality material.

All panels are applied with rust-resistant nails that penetrate at least 1 inch into the studs. Horizontal joints between sheets should be flashed with pre-fabricated "Z" flashing. Trim above doors and windows and other horizontal trim should also be flashed. Often for horizontal joints, a nailing surface between the studs needs to be provided.

Perhaps the greatest drawback of this product is that the boards can delaminate over time, especially if they are not properly maintained or sealed every 2 to 5 years. This siding can give a certain flatness or dullness to the appearance of the house, while the more attractive panels can be very expensive. It is not fire resistant.

Cost: Moderate to expensive.

Lifetime: 20 years to life of building.

Maintenance: Treated every 2 to 5 years.

Installation: Quick and easy; requires 2 people.

WOOD SHAKES AND WOOD SHINGLES: (Photos 16-4–16-8; Figure 16-2.) Wooden shakes and shingles are perhaps the most attractive and, correspondingly, the most time-consuming to install of all the sidings. Since each piece covers only a few square inches (compared to 32 square feet, with 4-×-8-foot plywood panels), application can take a *very* long time. One of our students, who covered his 3200-square-foot home with shakes, said towards the end that he bordered on insanity and/or enlightenment. Shakes or shingles can give a home anywhere from a classy look to a funky look. Be sure you have the patience, if you are working by yourself; or the money, if you are hiring the labor.

Most of the shakes or shingles are made from cedar or redwood, and occasionally from oak or sugar pine. Shingles are cut and shakes are split (see discussion in Chapter 15, *Roofing*, p. 355, "Types of Roofing"). Shingles come in various grades, Number 1 Blue Label, Number 2 Red Label, and other utility grades. The blue label shingles are considered the best. Red label shingles are often used on walls or as underlayment when double coursing in roofs. The weather

Photo 16-4. Shake siding is rougher than wood shingles. It is also more costly and time-consuming to apply, but lends a real handcrafted feeling.

Photo 16-5. Shingle siding is beautiful, but costly and time-consuming.

Photo 16-6. Cedar shingle siding can be bought in 8-foot strips for quicker application.

Photo 16-7. If you are willing to take the time, the siding can tremendously affect the looks of the house.

Photo 16-8. An adjustable jig for applying shakes or shingles. The horizontal bar is supported with 2 vertical pieces.

exposure for shingle siding varies, but it is greater than that of roofs, since weather protection is not as critical on a vertical surface. Shingles are usually 12 to 16 inches long with random widths, and are about ¼ inch thick at the thick end. They are sometimes cut with fancy, rounded or pointed or patterned ends for more decorative applications.

Shakes are thicker, rougher pieces split from the wood. They are usually 18 and 24 inches long and up to ½ to ¾ inches thick at the thick end. Sometimes shakes can be bought pre-bonded to 8-foot panels for quick application. Shakes can be installed over either solid or spaced sheathing.

Both shakes and shingles can be very expensive, especially if you are paying for labor. Be sure you compare prices of installing shakes or shingles with other types of siding before you decide. Wooden shakes and shingles are rather fire prone and are not suited to all areas. Termites can also be a problem. They are, however, beautiful, easy to work with, easy to repair and replace and can add to the resale value of your home.

Cost: Moderate to expensive.

Lifetime: 30 years to the life of the building.

Maintenance: In humid climates a fungicide is recommended; in other areas, oil every 5 years to preserve resiliency.

Installation: Easy, but very slow; can be done by one person.

Photo 16-9. Horizontal lap siding offers beauty and durability.

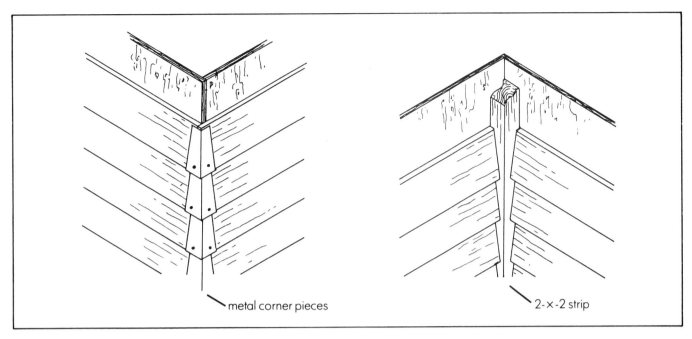

metal corner pieces

2-x-2 strip

Figure 16-3. Two different ways of trimming corners when applying horizontal siding.

WOOD BOARDS: (Photos 16-9, 16-10, 16-11; Figure 16-3.) Natural wood boards, milled from pine, fir, cypress, redwood, or cedar, are often used as siding. I personally favor this type of siding; it has the beauty and durability of natural wood, and a moderate impact on time and money budgets. Though its application is much slower than plywood, it goes on much quicker than shakes or shingles, and it can easily be applied by a novice. It adds beauty to the design and a high-quality look to the home.

The boards are usually ¾ inch thick, 4 to 12 inches wide and up to 20 feet long. They are applied horizontally, vertically or diagonally, depending on the type. They can be butt-cut, shiplapped, clapboard, beveled or tongue and groove. The price for this type of siding can vary from very inexpensive to very expensive. It all depends on 3 main factors: the type of wood, the quality of the wood and the number of milling processes required.

If redwood or cedar is used for the boards, the price will be high. Many people think these woods are used because they are resistant to rot. All woods will wear well if they can thoroughly dry after periodic wettings, and this is the case with siding. It is only when a wood gets wet and stays wet, as when it is near or in the ground, that a rot-resistant wood such as cedar or redwood is needed. Cedar and redwood are used for siding many times because they have fewer knot-holes and have straight, smooth grains. If good pine or fir is available, either one will be adequate for siding, provided it has a chance to dry after a rain. You may want to consider using one of these and save some money.

The quality of the wood you use will also affect the price. Clear lumber, boards with no knotholes, will be more expensive than lumber with knotholes. (Knotholes are often attractive on walls.) In some types of woods, the knots will fall out in time; in others they will remain in for the life of the house. Two kinds of redwoods are available: all-heart (all red), or

redwood containing some sapwood (white areas). The all-heart clear redwood siding is very expensive. You can also find dry or wet lumber for siding. Dry lumber is more expensive, but it will not shrink and leave unsightly cracks. Usually a relatively dry siding is recommended, unless you are using a clapboard or board and batten siding, where the boards overlap sufficiently to dry without showing cracks. Be on the lookout for good deals: lower quality boards that are still attractive.

The number of milling operations that a board must pass through before the final product will also affect the price. Trees in the forest are the cheapest lumber you can buy; for comparison sake, say $50 per 1000 board feet. Once it is cut into logs and trucked to a mill, it runs $80 per 1000 board feet. After it's rough cut in the mill, it's about $180. When it's finished and planed in the mill, it's $280; dried in the kiln, $360; picked for clear (no knots), $800; and finally, run through a special process to shiplap or tongue and groove the edges, it is $1500 per 1000 board feet. So if you are planning to use a high-grade redwood or cedar dry lumber siding, you will pay a lot for the materials. If you are paying for installation, it will cost a good bit more.

The way to use wood boards, if you are on a tight budget, is to use wet rough-cut lumber (in a board and batten or clapboard style), or the lower grades with knots, or some sapwood (redwood), and apply it yourself. Application is simple for a novice; it will be thoroughly explained later in this chapter.

The boards can be run horizontally, vertically or diagonally. The main consideration is the looks of the house. Running the boards vertically will give the house a greater sense of height; running them horizontally will tie the house more into the ground plane. Most styles of board siding can be used in any of the three positions, except the butt-cut boards, which must be used vertically in a board and batten style, and the

Photo 16-10. Board and batten siding is inexpensive and offers a handcrafted look.

Photo 16-11. Individual shiplapped boards have been applied diagonally. This application adds shear strength to the wall.

Photo 16-12. Aluminum siding.

Photo 16-13. Aluminum siding being applied.

beveled boards, which are most often used in a horizontal position. The shiplapped and tongue and groove boards can be used in any of the three directions (if using tongue and groove horizontally, be sure to point the tongue, *not the groove,* up or the groove will trap water).

Cost: Inexpensive to expensive.

Lifetime: 30 years to the life of the house.

Maintenance: Often no maintenance is needed; with redwood, seal it immediately after application to prevent darkening; other types of wood may require an occasional treatment to keep resiliency.

Installation: Moderately time-consuming, but easily mastered by the careful novice.

HARDBOARD SIDING: Hardboard is manufactured from wood chip pulp, which is compressed and bonded under heat and pressure. It is often made to appear like real wood boards or even stucco. It comes in many styles and textures and costs less than the expensive real wood sidings, but not particularly less than the inexpensive ones. It comes in panels like plywood or individual boards and is often primed or painted, and sometimes stained. It takes paint well and has no knotholes. It needs to be kept out of any direct contact with the ground because it rots easily. It is available in tongue and groove or shiplapped.

This siding does allow you the appearance of wood without the cost of the expensive real wood siding. If you can locate a moderately priced real wood siding, however, I would advise that over hardboard; it looks better, and I am more trustful of nature-made materials than manufactured materials for siding. Hardboard can delaminate and buckle over the years.

Cost: Moderate.

Lifetime: 20 years to lifetime of building, depending on the maintenance; finishes are guaranteed for 5 years.

Maintenance: Edges must be sealed upon installation and all siding painted shortly after; repaint every 3 to 6 years.

Installation: Rather quick and easy and well within the scope of the novice builder.

ALUMINUM, VINYL AND STEEL: (Photos 16-12, 16-13.) Metal and vinyl types of siding have become rather popular in many areas of the country. They are often used over an existing siding. Like many things nowadays, they seldom stand up to the manufacturer's claims (whereas with real *wood* siding, the Manufacturer never lets us down). These types of siding are usually molded or extruded panels that simulate lap siding (and sometimes shakes or shingles). Vertical panels are made 8 inches wide with battens, to simulate board and batten siding. The standard length is 12 feet 6 inches with other styles available.

Except for vinyl, these types of siding are fireproof; something to consider, if fires are a problem in your area. Usually

Photo 16-14. Stucco is often used in Florida, California and the Southwest, where freezing is not a problem.

Photo 16-15. Wire mesh is stretched across the house before the stucco is applied. In this case, the felt paper and wire mesh are fastened together at the factory.

Photo 16-16. Applying a brick veneer. Laying bricks is a challenge to the novice.

they won't rust, peel or blister or, needless to say, rot. The siding does dent and scratch easily though, and the aluminum can corrode near salt water. The biggest problem, however, is that the pre-finished surfaces can get chalky after a few years and start to buckle, and this is hard to repair or repaint.

Cost: Inexpensive to moderate.

Lifetime: 40 years to lifetime of home.

Maintenance: Vinyl — none; aluminum and steel — hose off every year and repaint according to manufacturer's instructions; paint all scratches on steel siding to prevent rust.

Installation: Easy and quick.

STUCCO: (Photos 16-14, 16-15.) Stucco is a cement or mortar-type material made from three parts fine sand, one part Portland cement and water. The stucco is prepared at the site and then applied wet over wire laths in 2 or 3 coats. It can be finished smooth or rough, or in various textures and patterns. Pigment can be added directly to the mixture, or the stucco can be painted after drying. This type of siding is found mostly in moderate or temperate climates (Florida and California), because in areas where there is freezing weather, any water in the wall will freeze and cause the stucco to crack.

Stucco can be an attractive, durable siding. Its main drawback for owner builders is that it is difficult to master its application the first time around. Because of this, it is most often done by professionals, which can be costly (though still within reason). You may want to try applying some on a mock-up or outbuilding first, to see if you can learn to do it yourself. It provides a seamless, even surface that is easy to repair. It is fireproof and durable, and surprisingly, has good

shear strength. It can easily be molded to rounded or curved surfaces. Its greatest drawbacks are the cracking, due to either freezing or the house settling, and the fact that it may require a professional to install.

Cost: Moderate.

Lifetime: Life of the building.

Maintenance: None, except for repainting every few years.

Installation: Difficult and time-consuming for the novice.

BRICK: (Photos 16-16–16-19.) Brick siding is still commonplace in many areas of the East. You seldom see it in the West, and almost never in the earthquake areas of California (for apparent reasons). Brick is a veneer that covers and is attached to either a wood frame or concrete block walls. It is expensive and difficult to apply, but it gives a permanent veneer that is both attractive and maintenance-free. If you have a knack for this type of work and the patience, you may want to try to do it yourself, otherwise it is best left to professionals.

Many people think the brick veneer is one of the structural components of the house. Actually, the bricks are not structural and bear no load; they simply are a veneer. In the past they have been considered a mark of quality, and in many cases this is still true. If you plan to use this type of siding, you must adapt the foundation so it can support not only the house but the brick veneer as well. This is often done by using a 12-inch-wide concrete block with 4 inches of it extending to the outside of the wall line to support the 4-inch-wide bricks. Consider the expense and labor before deciding on this high quality siding.

Photo 16-17. A brick veneer being applied.

Photo 16-18. Brick veneers are often used only on the side of the house facing the street.

Photo 16-19. This is not a real stone house, it's only a stone veneer over a frame house.

Cost: Expensive.
Lifetime: The life of the house.
Maintenance: None.
Installation: Difficult and time-consuming; usually done by professionals.

Horizontal Siding

What You'll Need

TIME   16 to 32 person-hours per 30-foot wall, 8 feet high.

PEOPLE   2 Minimum (1 skilled, 1 unskilled); 3 Optimum (2 skilled, 1 unskilled — 2 nailers, 1 cutter); 6 Maximum (2 teams of 3 each).

TOOLS   Finishing hammer, ladders, utility knife, staple gun or hammer tacker, staples, measuring tape, hand saw (fine toothed), power saw with finishing blades, extension cords, pencil, nail belt, level, caulk gun, combination square, chalkline, sabre saw (if needed), bevel square for odd angles, copycat (if needed), nail punch.

RENTAL ITEMS: Scaffolding (Photo 16-20), transit.

MATERIALS   In this chapter, we will be doing horizontal, shiplapped wood board siding. You will need enough 15-pound felt to cover the entire building, enough siding materials (include a 10 to 15% waste factor; more if using a poor-quality stock), flashing materials, trim materials, caulking and hot-dipped galvanized nails. (See the Design Decisions section of this chapter, "Wood Boards," for a discussion of the different types of lumber and their quality.)

Clear, shiplapped redwood or cedar siding can be very expensive ($1100 to $1500 per 1000 board feet). I have often located a lower grade cedar siding with tight knots, that seldom fall out, for about $600 per 1000 board feet. Shop well, because the siding for a home can run to several thousand dollars. You need to shop prices before deciding. Be sure that the supplier you buy from will continue milling your type of siding, in case you did not order enough. You don't want to be looking for a match for your siding when you are 90% finished.

Store and protect the siding boards very well; you don't want to lose any to the elements. Siding is much more expensive than framing lumber, and therefore should have few, if any, unusable pieces. Check the stock as it is unloaded and do not accept any pieces that have been damaged or marred, such as from strapping, marking crayons, greying or weathering. You are paying for boards that will be exposed — they should be in good shape. Unload the truck by hand if there is any danger that the boards could be damaged by dumping.

Trim boards should be 1×4s that are of dry, knot-free stock. Use cedar, if the trim will be left unpainted, and fir or pine if you are going to paint the trim.

Always buy hot-dipped galvanized siding nails. Siding nails

Photo 16-20. Though it takes a little time and money to set up, scaffolding makes all exterior work on second floors easier and quicker.

have smaller heads so they are not so visible. There are a few other types of non-rusting nails, but hot-dipped galvanized nails are recommended. Electro-plated galvanized nails often rust, and aluminum nails bend easily.

ESTIMATING MATERIALS NEEDED

SIDING BOARDS: To calculate the amount of siding boards you will need, you must first figure out how many square feet of walls need to be covered. Deduct for door and window openings and add 10 to 15% for waste. Order the longest boards you can get, or try to calculate the number of boards for each length you will need. If you have a 12-foot wall and a 16-foot wall, order 12- and 16-footers, etc. Take your time and work out a figure that comes close to your needs; the material is expensive.

TRIM: Trim boards are often dry, high-quality stock and therefore expensive. Try to estimate the exact amount of trim you will need and order the pieces in the proper lengths, to keep waste to a minimum. Include inside and outside corners, door and window trim, gable trim, etc. Also remember that many doors and windows come with their own trim, and you do not need to order trim for these.

NAILS: Figure 2 pounds of nails per sheet (4×8).

INFORMATION   This chapter should provide you with all the info you need to complete the siding. For unique trim situations, consult an expert or use your problem-solving abilities.

PEOPLE TO CONTACT   If you have opted to do the siding before the roofing, this is a good time to contact your roofer or roofing supplier. A plumber and electrician and sheet metal people, for ductwork, may need to be contacted also.

INSPECTIONS NEEDED   If you are installing a plywood sheathing below the actual board siding (this is recommended, especially in earthquake, hurricane or tornado areas), you must get the nailing pattern of the sheathing inspected before covering the sheathing with felt and siding boards.

Reading the Plans

The plans will usually indicate what type of siding is to be used. Also they may indicate the type of trim. If a certain window manufacturer's model numbers are listed, you can check to see if the windows are pre-trimmed or have been ordered that way. Complex trim situations are sometimes detailed on the plans also. The plans should indicate any sheathing and may indicate the felt underlayment. (See Chapter 17, *Windows*, p. 421, "Materials: Trim," for trim applied before siding.) Other than this, everything else about siding is worked out at the job site.

Safety and Awareness

SAFETY: Unless you are dealing with heights, there is little danger involved in handling this type of siding, other than the common hazards at any construction site.

AWARENESS: Appearance should be of utmost importance as you apply the siding. Remember, this is finishing work you are doing and will be seen in the final product. As you apply, be aware of the following:

  1. Horizontal lines must meet at the corners where 2 walls come together;
  2. The best looking boards should be placed in visible areas;
  3. The color matches in adjoining areas;
  4. Hot-dipped galvanized siding nails are used;
  5. Flashing is applied above all horizontal trim pieces where water will be running down the wall;
  6. There is a felt underlayment;
  7. Your siding is always level and aligned, especially on either side of an opening;
  8. No dents, scratches or marks have been inflicted on the wood.

Photo 16-21. Be sure to separate your siding boards according to length, quality and color.

Overview of the Procedure

Step 1: Inspect and select the stock
Step 2: Pop a straight line to align the first board
Step 3: Apply the felt paper and locate and mark studs
Step 4: Apply the first course
Step 5: Apply successive courses
Step 6: Side and flash around openings
Step 7: Apply the trim

Step 1
Inspect and Select the Stock

Most Common Mistakes: Mixing colors; using poor-quality stock in highly visible areas.

Photograph: Photo 16-21.

Once the siding stock has been delivered to the site, spend some time selecting the good pieces and if necessary, separating the boards into different color shades. I have seen in one shipment of cedar siding colors vary from red to brown to white, and we had to separate the colors for various walls (often the difference in color diminishes with weathering). Separate out badly bowed pieces or pieces with flaws that are too extreme. Often, in the process of placing the siding, you will nail up a piece before noticing the bad knothole or ding in

it. The unacceptable pieces can be cut up and used where shorter pieces are needed, to cut down on waste. Put these in a separate pile and stack the good full-length pieces according to color and quality.

Select some of the straightest pieces to use for the first course, which needs to be really straight. Be on the lookout for especially beautiful pieces, ones with unique grain patterns or coloring, and use these around the main entrance doors at eye-level.

Step 2
Pop a Straight Line to Align the First Board

Margin of Error: Exactly level and in the exact location.

Most Common Mistake: Sloppy work.

Diagram: Figure 16-4.

Photograph: Photo 16-22.

Before applying the tar paper, it is best to snap a chalkline on the foundation to use to align the bottom of the first board. This line is usually about one inch below the top of the foundation. Bringing the felt paper and the siding down to this mark ensures that no water can run behind the bottom of the siding onto the top of the foundation and wet the sill plate. If the top

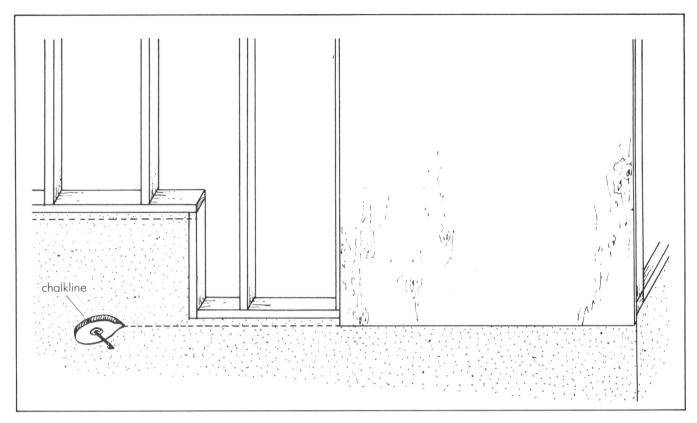

Figure 16-4. Pop a line for the first course about 1 inch below the top of the lowest level of the foundation, or ¼ inch below plywood sheathing at that level.

Photo 16-22. Scribing a level line on the foundation wall to serve as a guide for the first course.

Photo 16-23. Applying the tar paper before the siding. Always overlap the higher courses onto the lower ones.

of the foundation is exactly level (and it seldom is), you could simply measure down one inch at either end of the foundation and snap a line between these 2 points. Most likely the foundation will be somewhat off level, so I would recommend using a transit (as described in Chapter 4, *Layout,* p. 77) to locate exactly level points. Be *very* exact; off-level boards are highly visible on the side of a building.

If you have a single-level foundation, where the top of the entire foundation wall is the same level, you can simply snap level lines around the entire foundation, one inch below the top. If, however, you have a stepped foundation, where the top of the foundation is at several different levels, start at the lowest section of the foundation wall and pop a level line one inch below the top of this section. Then, as the boards are applied, you will have to cut the pieces that break at a step. Do the same at each level.

Step 3
Apply Felt Paper and Locate and Mark Studs

Margin of Error: Apply felt smooth and approximately level; mark studs within ¼ inch of its center.

Most Common Mistakes: Not providing adequate overlaps at side or edges of felt; eliminating felt entirely; not overlapping felt courses so that the top overlaps the bottom.

Photographs: Photos 16-23, 16-24.

You are now ready to apply the felt. Usually 15-pound asphalt or rosin paper is used. Sometimes a brown paper called building paper is used. Whatever is used, it must breathe, in order to allow any water vapor that has built up inside the wall cavity to escape to the outside. Therefore, this paper should be water resistant, but not waterproof. Do not use a non-breathing surface such as polyethylene sheets or other types of plastic.

Do a good job as you apply this moisture barrier. Avoid punching holes or tearing the paper. Apply it flat and smooth. Begin at the bottom with the bottom edge ¼ inch above the line you popped, so the line is still visible when you apply the first course of boards. Apply the first (lower) course of paper and then overlap subsequent courses so that the one above overlaps the ones below.

Remember to get the nailing pattern of the sheathing inspected before you cover it with paper. In some areas the paper may be omitted. Inquire about the local building practices. Use a staple gun to affix the paper to the studs, using enough staples to keep it smooth and prevent any possible wind damage. Overlap the edges about 4 inches and where the ends of 2 sheets meet, overlap them 6 inches. Wrap one of the pieces about 6 inches around the inside and outside corners. Work up as high as you can and, after the ladders or scaffolding is erected, you can continue up higher.

After the paper is up on a section that you plan to side, locate and mark the centers of the studs, so you will know where to nail. This is done by simply snapping a chalkline down the centerline of the stud on top of the felt paper.

Photo 16-24. Before installing the siding, locate the center-line of studs on the felt with a chalkline.

Photo 16-25. Be very careful, when applying the first course, to use the straightest boards and check closely for level.

Step 4
Apply the First Course

Margin of Error: Exactly level.

Most Common Mistakes: Using a bowed board; not starting the first board below the top of the foundation.

Diagram: Figure 16-5.

Photographs: Photos 16-25, 16-26, 16-27, 16-28.

With the building paper or asphalt paper in place, you are now ready to apply the first course. These boards should be of very straight stock because the next course will be nailed directly against them. If there is a bad bow in these first boards, it will cause the other boards to bow. This is really true of the whole siding application process, and you should not use badly bowed boards at all, though they need not be particularly attractive boards unless they will be easily seen.

Before applying the first board, there are a few things you need to know about installing any wood board siding.

1. All vertical joints, where 2 boards meet at their ends, must be beveled and the joint must occur over a stud for solidity. Butt cuts will not work—they show cracks very easily. These joints are crucial to the quality of the job, because if they are badly cut or fitted, there will be a large crack showing. Using a jig and setting the saw blade at a true 45° angle is the best way to ensure good cuts. Shown here is a jig you can build at the job site. Bevel cuts are hard to make, but with a good straight jig and an accurate blade angle of 45°, it can be done. If a radial arm saw or table saw is available, it can be used to get an even better cut.

2. The vertical joints should be staggered on the wall so that joints in successive courses are offset by 4 to 8 feet; joints in alternating courses will line up. Keep this in mind when you order your board lengths to eliminate waste.

3. Use 3 hot-dipped galvanized siding nails at each stud if the boards are 6 inches wide or more. You may want to place these 3 nails in a straight line for appearance sake.

With all this in mind, nail on your first course. Be sure the bottom edge is flush with the line popped on the foundation wall. Place a level on the board to be sure the board is level before you permanently nail it on.

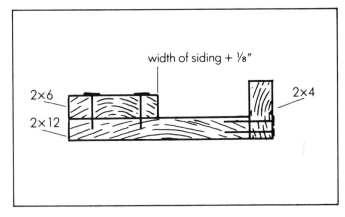

Figure 16-5. A jig used to cut siding.

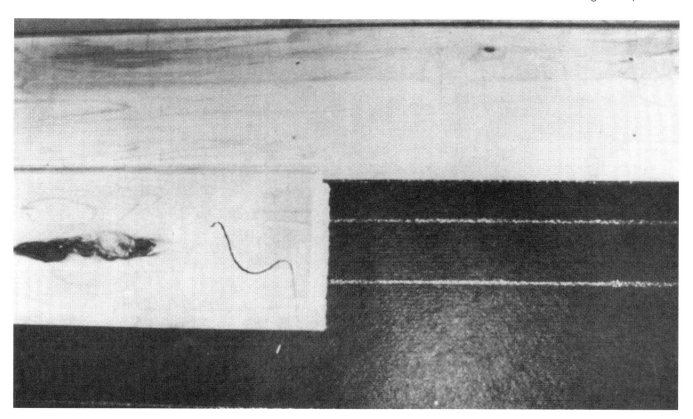

Photo 16-26. Bevel cut the ends of the boards and be sure they meet over a stud. Be careful to avoid irregularities, like those shown here, in highly visible areas.

Photo 16-27. Using a site-built jig and a power saw at a 45° angle to bevel-cut the ends of the siding boards.

Photo 16-28. To bevel cut siding boards without a jig takes a steady hand.

Photo 16-29. Siding being applied. Note how the boards, on both sides of the openings, are level with each other.

Step 5
Apply Successive Courses

Margin of Error: Within ⅛ inch of level.

Most Common Mistakes: Not pushing new top board all the way down on to the top of the lower board; pushing top board too far down, causing it to buckle outwards; nailing boards from both ends towards the middle, thereby trapping a bow; board off level.

Diagram: Figure 16-6.

Photograph: Photo 16-29.

You can now start to apply the successive courses. This is a rather simple process as long as you are careful not to make any of the common mistakes, and you follow the guidelines set forth in "Step 4." Just continue working your way up the wall. If you find you are starting to get off level, adjust your next few boards to return to level (the transit comes in handy here). Check your level every second or third course to see how you are doing. Remember to choose each board for its visual effect and work on getting "grade A" joints where 2 boards meet at their ends.

Some builders use a story pole to be sure the boards are staying aligned and level. This is a straight 1×4 or 1×6 marked very accurately at intervals equal to the width of the boards. This marked pole can then be held at each end of the wall with the bottom of the pole aligned with the level marks, and the top of each board is aligned with an interval mark. Where 2 walls meet at an inside or outside corner, be sure the horizontal joints of the boards on each wall line up, or it will look very strange.

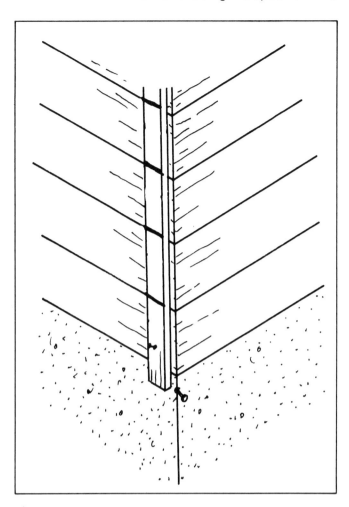

Figure 16-6. Using a story pole, with marked intervals, to keep individual boards level.

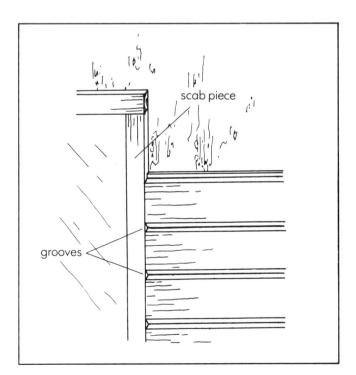

Figure 16-7. Using a scab piece to stop air infiltration through the openings left by the grooves.

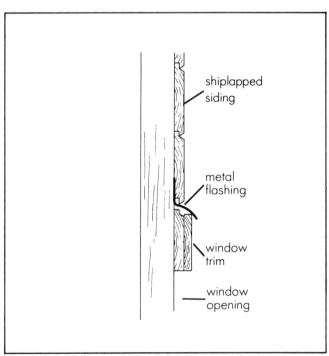

Figure 16-8. How flashing is applied to horizontal trim when using shiplapped siding.

Step 6
Side and Flash Around Openings

Margin of Error: ⅛ inch off alignment and level.

Most Common Mistakes: Omitted or improperly installed flashing; boards on either side of the opening getting off line.

Diagrams: Figures 16-7, 16-8.

Photographs: Photos 16-30, 16-31.

The only really challenging part of siding is around the door and window openings. These areas must be sided so that the boards in the same course on opposite sides of the opening are aligned with each other. If these boards get off line, the first full piece installed above the opening will not fit properly. There may be some fancy cutting needed wherever a full-size piece is notched to fit around the top or bottom of an opening. To be sure the pieces on opposite sides are level, simply make level marks on each side of the opening, using a level to represent the top of each board.

One-inch-wide strips, that run from the top to the bottom of the opening on the vertical sides only, are used as wind stops. These scab pieces prevent air from penetrating the small V-shaped grooves created where the shiplapped boards meet along their edges. They will be covered, in turn, by the wider exterior trim. Many builders eliminate this step and use pre-made plugs or caulking. I prefer the strips.

Flashing must also be installed above all door and window openings where water will be running down the wall. If the openings are well under the overhangs of the roof and no water will get above them, the flashing can be eliminated. This was the case with the opening in the photos.

Photo 16-30. Note how the horizontal siding below the window is notched to fit around the opening. The strips on the sides of the window are to stop air infiltration through the grooves in the siding.

Photo 16-31. Be sure to continually check the level on the boards applied to each side of the openings. They must stay aligned.

Photo 16-32. The corners can be left rough until the trim is applied. Be sure that the horizontal lines of the grooves are aligned where the 2 walls meet.

Where any roofs are joined to the wall, flashing is necessary. The flashing here will be exposed above the trim, but it is unavoidable if you want good weather protection. Usually an aluminum or galvanized flashing cut from rolls of standard 12-inch-wide flashing is used. This can be painted with a bituminous paint for longer life and a more attractive appearance. Terne metal, a copper-bearing steel, can also be used. This is not only attractive, but will last the lifetime of the house as well (it is expensive).

Step 7
Apply the Trim

Margin of Error: ⅛ inch.

Most Common Mistakes: Sloppy work; no flashing on exposed horizontal pieces.

Photographs: Photos 16-32, 16-33, 16-34.

The siding is now finished and you are ready to apply the trim at the corners, around the openings and at the gables. Check to see if your doors and windows are coming pre-trimmed (most windows are pre-trimmed). If they are, you might want to coordinate the rest of the exterior trim with the trim provided by the manufacturer.

Use high-quality dry stock for the trim and be sure all the trim pieces around any one opening are color-matched. Be sure the trim is applied level or plumb and has tight-fitting joints. These joints can be either mitred or butt-cut.

Photo 16-33. This corner has been finished with 1-×-4 trim.

Photo 16-34. The house after the siding has been applied and the deck added.

Worksheet

ESTIMATED TIME

_____

_____

_____

CREW

Skilled or Semi-Skilled

_____ Phone _____

_____ Phone _____

_____ Phone _____

Unskilled

_____ Phone _____

_____ Phone _____

_____ Phone _____

_____ Phone _____

MATERIALS NEEDED

Siding, felt, flashing, trim, caulking, nails.

ESTIMATING MATERIALS

SIDING BOARDS:

No. of sq. ft. of walls _____

+ 15% _____ = total _____ sq. ft.

No. _____ 8-foot boards (No. of sq. ft. _____)

No. _____ 10-foot boards (No. of sq. ft. _____)

No. _____ 12-foot boards (No. of sq. ft. _____)

No. _____ 14-foot boards (No. of sq. ft. _____)

No. _____ 16-foot boards (No. of sq. ft. _____)

No. _____ 18-foot boards (No. of sq. ft. _____)

No. _____ 20-foot boards (No. of sq. ft. _____)

FELT: _____ sq. ft. of wall surface = _____ rolls

FLASHING:

_____ linear ft. of flashing above doors

_____ linear ft. of flashing above windows

_____ linear ft. of flashing for wall-roof joints

_____ total linear ft. of flashing

TRIM:

Corners:

_____ No. of boards _____ ft. long

_____ No. of boards _____ ft. long

Gables:

_____ No. of boards _____ ft. long

_____ No. of boards _____ ft. long

Doors:

_____ No. of boards _____ ft. long

_____ No. of boards _____ ft. long

_____ No. of boards _____ ft. long

_____ No. of boards _____ ft. long

Windows:

_____ No. of boards _____ ft. long

_____ No. of boards _____ ft. long

_____ No. of boards _____ ft. long

_____ No. of boards _____ ft. long

NAILS: _____ lbs. of nails.

COST COMPARISON

_____ Store:

Cost per board _____ total $ _____

_____ Store:

Cost per board _____ total $ _____

_____ Store:

Cost per board _____ total $ _____

Daily Checklist

CREW:

_____ Phone _____

_____ Phone_____

_____ Phone _____

_____ Phone_____

_____ Phone_____

_____ Phone _____

TOOLS:

☐ finishing hammer   ☐ pencil

☐ ladders   ☐ nail belt

☐ utility knife   ☐ level

☐ staple gun/hammer tacker ☐ caulk gun

☐ staples   ☐ combination square

☐ measuring tape   ☐ chalkline

☐ hand saw (fine toothed) ☐ sabre saw (if needed)

☐ power saw   ☐ bevel square

☐ finishing blades (for saw) ☐ copycat

☐ extension cords   ☐ nail punch

RENTAL ITEMS:

☐ scaffolding

☐ transit

MATERIALS:

_____ No. of boards _____ ft. long

_____ No. of boards _____ ft. long

_____ No. of boards _____ ft. long

_____ No. of boards _____ ft. long

_____ No. of boards _____ ft. long

_____ rolls of felt

_____ linear ft. of flashing

TRIM:

_____ No. of boards _____ ft. long

_____ No. of boards _____ ft. long

_____ No. of boards _____ ft. long

_____ No. of boards _____ ft. long

_____ No. of boards _____ ft. long

_____ lbs. of nails

DESIGN REVIEW: Make decision on trim design.

INSPECTIONS NEEDED: ☐ YES   ☐ NO

Inspector _____

Phone _____

Time and date of inspection _____

WEATHER FORECAST:

For week _____

For day _____

Do not put boards up wet; they may shrink and crack, or leave gaps.

17 Windows

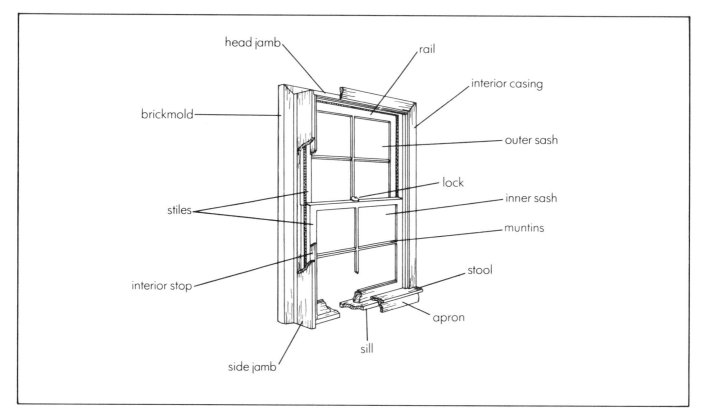

Figure 17-1. Parts of a window.

What You'll Be Doing, What to Expect

WINDOW INSTALLATION is easy, enjoyable and exciting. The exterior appearance of the house will go through final changes as large gaping openings are filled with attractive windows and trimmed. It is one of my favorite days at the job site. In regard to appearance, the home leaps way ahead. It is a job to enjoy, even relish. The only real problems come in two areas: if you ordered the wrong window or if you provided the wrong size opening. This can cause frustration and gnashing of teeth.

Basically, you are placing the window in the opening and levelling it; nailing it in place, and then doing the trim. If the window is pre-trimmed, even the trimming step is eliminated. If you are cutting and installing the exterior trim, precision is required; this will be finished work. Go slow and be careful; dropping a window can be dangerous, and expensive. Use caution installing the windows on high walls, especially on windy days.

Jargon

(Figure 17-1)

APRON. The interior trim piece below the sill.

BRICKMOLD. The exterior trim around a door or window.

CASING. The interior and exterior trim around the top and sides of a window or door.

FINISH FRAME. The stationary pieces surrounding the movable window (side jamb, head jamb, sill, etc.).

JAMB. The tops and sides of a door or window frame that are in contact with the door or sash; referred to as side jamb and head jamb (top).

LIGHT. A window, a section of a divided window, or a pane set in a window.

MULLION. A slender vertical bar (usually wood) separating adjoining windows.

MUNTIN. A narrow vertical or horizontal strip separating panes of a window.

RAIL. The horizontal top or bottom piece of a door or window sash.

SASH. The movable part of a window or the frame that holds the glass.

SILL. The horizontal member at the bottom of the window frame. It is often slanted towards the exterior to shed water.

STILE. A vertical side piece of a door or window sash.

STOOL. A molded, horizontal interior trim piece at the bottom of the frame, resting on the sill and protruding into the interior of the home.

STOP. The narrow strip of trim along the face of a door or window jamb. It prevents the door or window from swinging too far out and retards air flow. It also creates the channels in double hung windows.

Purposes of a Window

1) To provide light
2) To provide view
3) To provide ventilation
4) In solar application, to allow the sun to hit the thermal mass
5) In some instances, to provide an emergency exit

Design Decisions

Design decisions on windows, their size, style, placement, etc., could fill a book or several courses at architecture schools. The decisions you make here are key to creating the proper feeling in each room—the feeling that you really want. Often I have felt that a room simply did not work or fall together properly, due to the window decisions. These decisions are vital to the success of the design and are, all too often, just made at random.

Visualize each decision you are making. If visualizing is difficult for you, visit a room that has a similar window set-up to the one you are contemplating. Then close your eyes and imagine doing what you will be doing in that room and see how it feels. Does the window decision support the activity of the room or does it fight it? If it doesn't support it, then change a few things: the location, the size, the height. Go back into your visualization of the activity in the room and see if this changes the way you feel about the room, better or worse. Your ability to visualize the decisions before you physically enact them, and your ability to honestly get in touch with your innermost feelings about these decisions and their impact (whether they support the activities in the room or not), are keys to successful decisions in this and all areas of the home.

I remember visiting a friend in Florida, who proudly showed me his home, that had just been completed. The livingroom sitting area, an area meant to make people feel relaxed and communicative, was a 20-×-20-foot platform raised 12 inches off the ground that faced a floor-to-ceiling window wall, that looked out onto a ship canal. The couches were arranged in a U-shape with the open section of the "U"

being the window wall. As we sat and talked, a yacht passed through the canal, only 200 feet from the livingroom, people on deck waving and laughing. Our attention, needless to say, was drawn towards the canal, away from our "intimate" conversation. Only after the boat passed could the conversation continue, only to be followed by another boat pulling a couple of skiers, drawing everyone's attention back to the canal again. It was impossible to have a real group interchange in the one area that had been designed just for that purpose. The design was sabotaging the activity.

Before choosing the window design for your house, be aware of what is available. In the following section, the different types of windows are discussed.

Types of Windows

Windows affect the view, privacy, air flow, temperature, safety and security, solar gain, and most important, the feel of a home. The type of window, its size, location, style, and manufacturer all need to be carefully considered. We would all like to put the best windows in our home, but few of us can afford it. Windows for the same house can range from $8,000 to $12,000 for the top of the line, to $3,000 to $4,000 for the same sized windows of a lower quality brand. You have to consider what you will have to give up to put in the more expensive brands (better carpeting or cabinets, a hot tub, the garage, etc.). I usually recommend getting a medium-priced brand, between the best and the cheapest, unless you can afford the best. Avoid, however, installing anything of questionable quality in the home. It is permanent and if it gives you poor service it's very difficult to replace.

Usually you need to have made these decisions about windows by the plan drawing stage, and no later than the wall framing stage, when you need to be providing the proper rough opening sizes specified by the manufacturer (for more on this see "Sizes of Doors and Windows and Rough Openings," p. 227, in Chapter 11, *Wall Framing*). If in your house you are implementing a passive solar energy plan (an essential, unless you live in a very moderate climate—and it is even recommended there), your window size and placement must be co-ordinated with your passive solar design to allow the sun in at the proper times. For these reasons the design decision for your windows would have been made long before the window installation stage discussed in this chapter.

PREHUNG WINDOWS: Most windows today come "prehung." This means that the window, its casing (trim), moulding, sill, weatherstripping, hardware, etc. is assembled at the factory and is installed as one unit. Because of the simplicity and speed of installing "prehung" windows, they come highly recommended. Often old windows, just the window without the frame, can be purchased very cheaply. You can also find classy, decorative windows that add to the appearance of the house. These will take much longer to install. A prehung will take about 1 hour to hang and trim, an old window about 3 to 5 hours. Multiply this times the number of windows you plan to use and it adds up. If you are paying for finish carpenters to install the old window, it might even end up costing you more than a high-quality prehung window.

WOOD OR ALUMINUM: Wood and aluminum are the standard materials used in making windows. Wood windows, usually made from kiln-dried high-quality ponderosa pine, have some real advantages. They are better insulators (wood as opposed to metal), they have a nice appearance and they do not condense water as metal frames do. They may, however, be more expensive and they do require paint periodically.

Aluminum windows, which are often less expensive than their wood counterparts, do not need painting and are more scratch-resistant. They also make use of a smaller frame because of the added strength of metal, thereby allowing more window glass. They do, however, transfer more heat and are not as energy-efficient, and tend to "sweat" in certain climates.

New developments in both wood and metal windows have overcome some of their disadvantages. There is now an aluminum clad wooden window that incorporates an aluminum cladding or covering on the exterior so you do not need to paint it. (Aluminum clad windows are very good in solar greenhouses where moisture content is high.) Aluminum windows are now constructed to have insulation breaks built into the frame so they are more energy-efficient. Both these improvements will cost you more money, but may be worth considering.

The quality of all these windows can vary greatly. The National Woodwork Manufacturer's Association sets minimal standards, but quality varies greatly beyond these. Usually the better brands are known by their reputation. Your local lumberyards will usually carry several different brands to allow the customer a choice in prices and style. If you order the windows with your lumber order, you will get a better price on everything, since the total bill is large.

Styles

Basically there are three different styles of windows: sliding, swinging and fixed. Within these basic styles there are many possibilities. Your choice should not be random, but well thought out, as it will subtly, but profoundly, affect the house. Windows that open out (casement, awning, etc.) allow the full window area to vent and can redirect cooling breezes into the building. However, if they open on to a walkway or porch, they can be dangerous. Fixed windows are inexpensive and adaptable but offer no ventilation. Read this section and choose well. (Be sure to read the Information section in this chapter for the code requirements).

CASEMENT: Casement windows are hinged on the side and open outwards. They are opened by a cranking mechanism or push bar. Because of this opening procedure they are a good choice above sinks or cabinets where the windows are hard to reach. These windows allow 100% opening for ventilation and are weather tight when clasped shut. They can act as a wind scoop to catch the summer breeze. To do this properly, be sure you know the prevailing wind pattern and therefore on which side to hinge the window to redirect these winds into the home. Be careful if they open out into a walkway or porch (especially if children are to live in the home) as they can be dangerous.

AWNING: Awning windows usually have one or more sashes that are hinged at the top and swing at the bottom (usually outwards). Like the casements, they use a push bar or cranking mechanisms. They have the same advantages and disadvantages as the casement windows, but unlike the casement window, they cannot be used as an emergency exit. If the window is hinged at the bottom and opens inward at the top, it is called a "hopper" window. Usually this is just the awning window upside down.

DOUBLE HUNG: Double hung windows are windows with 2 sashes that slide up and down by friction, balancing devices or springs. They are very common and usually very economical. They provide adequate ventilation if opened partially from the top and partially from the bottom, and can be used as an emergency exit. This type of window was one of the first real improvements in windows and it goes back over 100 years. The old, hard-to-repair balancing weights used to open and close the windows have been replaced by springs.

HORIZONTAL SLIDERS: These windows are made up of 2 or more sashes that slide horizontally in a track by pushing on the window. Some are spring assisted. They offer adequate ventilation and can be used as an emergency exit. Though they are not a particularly common type of window, such as their up-and-down sliding double hung cousin, they are available from most window manufacturers.

FIXED: Fixed windows are simply sheets of glass permanently fixed in their frames. Oftentimes these are used in combination with other opening windows to create a certain effect with a series of windows without having to pay the added expense to have all the windows in the series open. They offer the view and lighting qualities of opening windows but offer no ventilation and cannot be used as an exit. A fixed window is generally used in the south wall of a passive solar home where a lot of sunlight, but not as much ventilation, is needed.

These windows can be bought from manufacturers in different sizes, already set in a frame. They can also be built at the job site by ordering the proper size glass and building your own frame. This can be more economical, especially if you are doing the work yourself. Oftentimes you can find double-glazed glass panels very cheap. As seconds used in sliding glass doors, they can run as cheap as 30% of their usual cost (you can then design your openings to fit these panels).

Other than these, there are a few other kinds of windows, but these are the most common. There are jalousies, which are horizontal glass slats in metal frames that operate like venetian blinds. These are used in warm climates or in porches in cold climates, as they are not weather tight. There are multiple use windows that incorporate an outward swinging sash that can be installed either horizontally or vertically. These are usually simple windows with no complicated hardware.

SINGLE, DOUBLE AND TRIPLE PANES: Double or triple paned windows have been around for a long time. At Thomas Jefferson's Monticello, I was surprised to see he had used double paned windows on the entire north side. With energy a great consideration in homes today, double and triple paned windows are becoming increasingly popular. In some states they are even required by code. The reason for the popularity of multiple paned windows is obvious. Glass by itself is a very poor insulator. The R factor (insulation property) of a standard wall is about 14. A single pane window is about .75. If you add another pane of glass with an air space between the two, you have increased the R factor to 2.0. The claim is then made that double paned glass cuts the heat loss through windows by 50%, which is true. But still even a double paned glass has only $\frac{1}{7}$ the insulating property of the wall (insulated).

Fortunately, most heat loss occurs through the roof areas, since the hot air is rising, and through cracks and small openings. Around 10 to 20% of your total heat loss occurs through the windows. Double and triple paned windows also reduce heat loss through radiation to a colder surface (the window) and act somewhat as a noise barrier.

If you are working on a limited budget and are very concerned about heat loss, you may want to use window quilts instead. These are thick (1-inch) insulated window drapes that pull down and fit with velcroe around the edges. They raise the R factor up to 5 or 6, 3 times that of double paned windows. If you can afford it, using both double (or triple) paned windows and window quilts would be good. Check to see the difference in cost between the same windows with double, triple or single panes. Check the prices of window quilts. Evaluate the possible savings in energy in your area with your fuel bills (you may want to consult a professional solar designer to help), and decide which is the best way to go.

What You'll Need

TIME

FOR PREHUNG WINDOWS: About 1 person-hour per window, unless there are problems with the window or openings.

FOR FIXED WINDOWS: About 2 to 4 person-hours per window.

PEOPLE 2 Minimum (1 skilled, 1 unskilled); 2 Optimum (1 skilled, 1 unskilled); 4 Maximum (2 teams).

TOOLS Finish hammer, nail set, utility knife, caulking gun, tape measure, framing square, power saw, hand saw, shim material, pencil, nail belt, chisel, level, safety goggles, nail-pulling tools, saw horses, extension cord.

MATERIALS You will need windows, trim materials (if the windows aren't pre-trimmed), caulking, felt paper and hot-dipped, galvanized finishing nails (16d).

ORDERING WINDOWS: Ordering windows is no simple matter. Aside from choosing the proper size, type and model number, there are several things that must be stated in ordering each window. There always seem to be a lot of mistakes and misunderstandings in this area. Below is a list of

some of the things that must be included when ordering each window:
1) The model number (this tells size, style, etc.).
2) Whether it opens out and if it hinges on the left or the right (usually, facing the window from the outside).
3) What type of exterior trim do you want? What type of interior trim do you want?
4) Can it be used as an exit window in the bedroom if one is required?
5) Do you want blinds or screens?
6) Do you want it painted, stained or primed?
7) Do you want aluminum clad or all wood?
8) Are different types of sills available?
9) Do you want it with mullions, and if so, are there different styles?
10) Do you want single, double or triple glazed?
11) If there are fixed and opening units in a series, which units are fixed, which are open?

THICKNESS OF JAMBS: Depending on the combined thickness of the entire wall panel, special jamb extenders may be needed or the window will have to be ordered with wider jambs. If you are using a 2-×-4 wall, with drywall for the interior surface and plywood for the exterior, your wall will only be about 4⅝ inches thick (stud, 3½ inches; drywall, ½ inch; and plywood siding, ⅝ inch). If you were using, for example, a 2-×-6 wall with ⅝-inch interior drywall surface, ½-inch plywood sheathing outside plus ¾-inch vertical siding on top of that, the wall would be 7⅜ inches thick (stud, 5½ inches; drywall, ⅝ inch; plywood sheathing, ½ inch; and siding, ¾ inch). To accommodate the wider wall, many manufacturers use either wider jamb boards or jamb extenders. Either of these run into an extra cost of $20 to $60 per window. You will need to *tell your manufacturer the wall thickness, from the inside edge of the interior wall to the outside edge of the exterior wall*, so they can order or build your window accordingly.

TRIM: Each manufacturer will have several different types of interior and exterior trim to choose from. Or the window can be ordered with no trim and you can provide this yourself (in this case, some windows will come with an exterior nailing flange). The exterior trim is usually assembled as part of the window, and the interior trim will come precut in a bundle to be installed after the window is in.

Usually the trim is installed around the openings after the siding is in place and the window is installed. A few sidings, such as clapboard, which has a layered rather than a flush surface, may require that the trim be installed *before the siding*. If the trim was installed after the siding and was attached on top of it, gaps would be created between the siding and the trim, since the siding is at different levels. Check before you begin siding to see if the trim should be installed first.

ESTIMATING MATERIALS NEEDED

NAILS: ½ pound per window.
CAULK: ½ tube of caulk per window.
FLASHING: Measure feet of flashing needed.

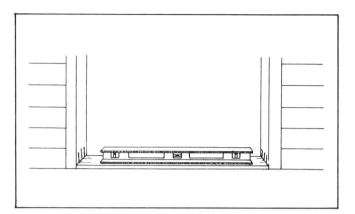

Figure 17-2. Check the level and plumb of the rough opening before installing the window. Use shims to correct it, if needed.

INFORMATION This chapter plus any manufacturer's installation instructions should be all you need, unless you are framing windows in unique shapes or windows that are non-rectangular. In this case, consult a professional or see the Time/Life book on doors and windows.

Be aware of the building codes in your area concerning windows. Most codes require that you have at least one window in each bedroom that fully opens and with its sill not more than 44 inches from the floor. Some codes also require that any windows within 12 to 18 inches of a door or the floor be made of tempered glass.

PEOPLE TO CONTACT Now is the time to contact the insulation contractor, the drywall contractor, and the painter, if you haven't already done so.

INSPECTIONS NEEDED None.

Reading the Plans

As mentioned in the plan reading section of the *Wall Framing* chapter, every window opening will be marked with an identifying number that corresponds, in the window schedule, to a certain manufacturer's model number. The model number will then indicate the exact rough opening that needs to be provided. However, there still may be many decisions to make in regard to that particular window (see "Ordering Windows" under "Materials" in this chapter).

Safety and Awareness

SAFETY: Be very careful working around glass. One clumsy move can cause a serious injury or can break a window worth several hundred dollars.

AWARENESS: Remember, this is all finishing work, so be very particular. Have windows properly flashed and caulked and be sure they are level.

Overview of the Procedure

Step 1: Check opening and window
Step 2: Caulk, set into opening and level the window
Step 3: Nail the window in place

Step 1
Check Opening and Window

Margin of Error: Whatever it takes to get the window into the opening.

Most Common Mistakes: Window will not fit opening; damaged window; improperly ordered window.

Diagram: Figure 17-2.

Photographs: Photos 17-1, 17-2, 17-3, 17-4.

Before the actual installation of the window, uncrate the window and check for the following:
1) For any shipping damage
2) The manufacturer's installation instructions
3) If it is square
4) If it is the correct window
5) If it opens properly
6) If all pieces are there

If you plan to paint the window and it is not primed, you may want to paint it with primer before installing it (be careful not to paint weatherstripping or channels). If diagonal braces are attached to the window to keep it square, leave these on until the window is installed. You can climatize your window to local humidity by leaving it inside the house for a few days before installing it. Manufacturers advise that you wear clean, dry cotton gloves to handle windows. Be careful; frames are made from soft wood.

Be sure to check the thickness of the jambs to see that they were properly constructed to fit your wall panel (see "Thickness of Jambs" under "Materials" in this chapter). If any jamb extenders or interior trim pieces are provided with the windows, be sure they are all there and are the proper length and thickness. (Note: With some windows the sills and side jambs need to be cut flush with the frame before installing the window.)

Check the opening to be sure it is square and the proper dimensions. If there is any siding or sheathing protruding into the opening (thereby reducing the size of your opening), cut this out with a hand saw or power saw, or chisel it out. Some builders wrap the rough frame with felt paper or building paper. The best time to do this, however, is when you install the building paper under the sheathing or siding. Check to be sure the flashing has been included, or that you have flashing at the site for windows that will get wet from water running down the walls above them (see "Step 6" in Chapter 16, *Siding*).

If a window opening is too big or too small for the window, use your best problem-solving abilities to remedy the situation.

Photo 17-1. A rough opening before installation.

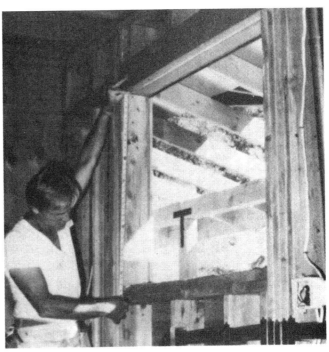

Photo 17-2. Check the measurements of the rough opening before installing the window.

Photo 17-3. Check the measurements and squareness of pre-hung windows.

Photo 17-4. Using a power saw to cut the excess sheathing out of the rough opening.

Photo 17-5. Caulking around the exterior side of the rough opening before installing the window.

Photo 17-6. Press the window flush against the siding.

Step 2
Caulk, Set into Opening and Level the Window

Margin of Error: None.

Most Common Mistakes: Eliminating caulking or flashing; putting window in upside down.

Photographs: Photos 17-5, 17-6, 17-7, 17-8.

Hopefully everything is working fine and the window fits the opening. It will mostly depend on how careful you were back at the wall framing stage. Prepare the window to be installed by running a bead of silicone caulk on the backside of the exterior trim (if the window is pre-trimmed). This adds an extra protection against possible water or air infiltration. If you want, you can caulk the siding where the trim will be, as shown in the photo.

Before installing the window, place some cedar shims every 16 inches on the horizontal rough sill. These will be used in the final levelling of the window. Now lift the window into place and, trying not to disturb the shim stock, set it in the opening. Most windows are installed from the outside, but some can be installed from either direction, especially if they are untrimmed. Installing untrimmed windows from the inside may be the way to go on high walls.

Now, before you nail the window, be sure of four things: (1) that the window is level, (2) that the window is plumb, (3) that the trim is flat against the exterior siding, and (4) that the jamb is the proper thickness for the wall. Check to be sure that the flashing properly overlaps the top of the horizontal trim piece.

Photo 17-7. Check the level of the window after installation.

Photo 17-8. Check the plumb of the window after installation.

Photo 17-9. Nailing through shims. Place shims every 16 inches or so.

Photo 17-10. Toenailing finishing nails into the casing.

Photo 17-11. Applying insulating foam in the cracks between the windows and the trimmer studs of the rough opening.

Photo 17-12. The window installed.

Step 3
Nail the Window in Place

Margin of Error: ⅛ inch off level.

Most Common Mistakes: Trim not flat against siding; no caulking; improper flashing; off level; jamb not wide enough.

Photographs: Photos 17-9, 17-10, 17-11, 17-12.

Once the window is properly positioned, you are ready to nail it in place. Have someone press against the trim (with pre-trimmed windows) as you nail, to be sure it is flat against the siding. Use shims every 16 inches around the window and nail through the jamb into the shims. Be sure there are no bows or waves in the jamb. Use 12d or 16d hot-dipped galvanized finishing nails and countersink them with a nail punch. After you have nailed every 16 inches around the jamb, nail the exterior trim, using the same nails. You may want to insulate around the window with foam or insulation at this time.

Special Situations

FIXED GLASS WINDOWS: Most houses involve some fixed glass as well as the regular opening windows. There are two approaches to fixed glass windows: you can buy windows pre-framed from manufacturers, or frame them yourself at the job site. I highly recommend framing them yourself at the job site; the savings are considerable and it's not very difficult or time-consuming.

You can order the glass from a local glass company. It is wise to have *them* measure the frame openings and install the glass. Then if a piece doesn't fit or gets scratched or broken, they pay the price, not you. Usually the glass is double plated and sometimes it is tempered. Tempered glass is hardened glass that will not break easily. It is twice the price of regular glass, and is used in all opening windows and doors with glass panels. It is not necessary to use in fixed glass areas unless the code requires it. Usually it is required when the fixed glass is within 18 inches of a door. It may also be required if the glass comes within 18 inches of the ground, but even then a railing will satisfy the requirement. If glass panels need to be tempered, see if you can buy seconds, e.g., patio door panels. These are often available at a third the usual cost and come in several sizes.

Some builders build the frame and install it without the glass. Other builders build the frame directly into the opening. I think both processes are about equal. Usually the frame is made of a clear, dry wood such as pine, fir or redwood. Make the frame ½ inch larger, in height and width, than the glass panels. Either 1-inch or 2-inch material can be used (I often use 1-inch for the side and head jamb, with a 2-inch-thick sill). The jambs must be wide enough to extend from the inside edge of the interior wall to the outside edge of the exterior wall. As with prehung windows, be sure the opening is square and has true dimensions; slope the sill outwards to shed water.

After the frame has been nailed level and plumb in the opening, cut the interior stops to go around the entire frame.

Photo 17-13. A bow window after installation.

Spread a small bead of caulk on the underside of the stops, between the stops and the jambs, for added protection.

Once the stops are in, spread a generous bed of silicone caulk on the back stop. Do not use putty.

Next snap pre-made neoprene setting blocks every 12 to 16 inches on the bottom of the glass and spacer strips along the edges and the top. Set the bottom edge in position on the sill. Press the glass in until the spacer strips and setting blocks are flat against the back stop. Nail in the face stop (exterior), again caulking between the face stop and the glass and jambs with silicone caulk. Do not toenail, since you may hit the glass. And there you have it.

BAY AND BOW WINDOWS: (Photo 17-13.) Aside from the windows mentioned, many people use bay or bow windows in their homes. These windows create an appealing interior nook and give the room a larger feeling. Bow and bay windows can be bought pre-made from most manufacturers and are usually made up of several standard casement or double hung windows, along with some fixed glass windows. You can frame these areas yourself at the job site and then install prehung windows into your openings.

Worksheet

ESTIMATED TIME

_____

_____

_____

_____

CREW

Skilled or Semi-Skilled

_____ Phone _____

_____ Phone _____

_____ Phone _____

Unskilled

_____ Phone _____

_____ Phone _____

_____ Phone _____

_____ Phone _____

_____ Phone _____

MATERIALS NEEDED

Windows, flashing, trim, caulking and nails.

ESTIMATING MATERIALS

WINDOWS:

_____ No. of windows _____ size

_____ No. of windows _____ size

_____ No. of windows _____ size

_____ No. of windows _____ size

FLASHING:

_____ ft. of flashing

TRIM:

_____ ft. of trim

CAULKING:

½ tube (per window) × _____ no. of windows

= _____ tubes

NAILS:

1½ lbs. of nails (per window) × _____ no. of windows

= _____ lbs. of nails

COST COMPARISON

_____ Store:

Cost per window _____ total $ _____

_____ Store:

Cost per window _____ total $ _____

_____ Store:

Cost per window _____ total $ _____

_____ Store:

Cost per window _____ total $ _____

PEOPLE TO CONTACT

Insulation contractor _____

Phone _____ Work Date _____

Drywall contractor _____

Phone _____ Work Date _____

Painter _____

Phone _____ Work Date _____

Daily Checklist

CREW:

_____ Phone _____

_____ Phone _____

_____ Phone _____

_____ Phone _____

_____ Phone _____

_____ Phone _____

TOOLS:

☐ finish hammer

☐ nail set

☐ utility knife

☐ caulking gun

☐ tape measure

☐ framing square

☐ power saw

☐ hand saw

☐ shim stock

☐ pencil

☐ nail belt

☐ chisel

☐ level

☐ safety goggles

☐ nail-pulling tools

☐ saw horses

☐ extension cord

MATERIALS:

_____ No. of windows _____ size

_____ No. of windows _____ size

_____ No. of windows _____ size

_____ No. of windows _____ size

_____ ft. of trim

_____ ft. of flashing material

_____ tubes of caulking

_____ lbs. of nails

DESIGN REVIEW: At this point it's a little late to change anything, but if you really don't feel right about some of the window decisions, consider changing them now. Once you move in, you may wish you had.

WEATHER FORECAST:

For week _____

For day _____

Index

Building Your Own House

Part II: Interiors

Contents

How to Use This Book

IN WRITING THIS BOOK, I did not intend a thorough examination of each subject. Rather, my intention was to distill the information into concise, step-by-step construction processes that even the novice builder can follow.

Except for the chapters on plumbing and electricity, each chapter is detailed enough to guide you through the process described. The actual designs you use and the specific challenges you face will of course differ according to many variables, including your taste, the tools you have to work with, and the money you have to spend. What is presented here, in a manner I've not seen elsewhere, is a usable guide to the construction process. Read each chapter and decide whether the process as shown applies to your project. If it does not, you may want to consult other do-it-yourself books, local professionals, or knowledgeable salespeople at your hardware or home center store. Be sure to consult too with your local code enforcement department; you should be familiar with all applicable codes before you begin work.

At the beginning of each project I've listed relevant safety guidelines and the most common mistakes to avoid. These lists are compiled from an accumulation of mistakes made by many people over many years. You can learn from the mistakes of others, but don't be lulled into thinking that therefore your project will proceed without any mistakes or problems. It won't. Most novices—and professionals too—underestimate the time, energy, hassles, and money that building projects demand. Add 30 to 50 percent to your initial estimates in each area.

Be sure to read the entire chapter before you begin work on any project. Throughout the book I've made every effort to tell you what you will be doing, what you will need, and what problems to watch out for; but to really prepare for a project, you need to understand the construction process as a whole before proceeding step by step.

Finally, enjoy yourself. The experience of doing your own work offers not only financial benefits but also a potential for growth and enjoyment rarely found in other endeavors. It will push you to new limits and beyond; and in the end you will have a house that feels more like a home because you took an active part in creating it.

1 Tools

THIS CHAPTER DISCUSSES building your tool inventory, the importance of buying high-quality tools, and the appropriate use and care of your tools and work area.

The types of tools discussed are those for measuring and leveling, those for cutting and drilling, those for attaching and assembling or dismantling, and those for finishing. A section on pneumatic tools—air compressors and accessories—is also included.

Before You Begin

Before you began the job of constructing your new home, you should have acquired a copy of the appropriate building code from the office of your city or county building inspector. Keep this code on hand so you will know what your project requires to comply with the code.

Tools can be fun, and they can open doors to new and interesting projects, but they should never be treated as toys. If you have small children, keep your tools where they won't be accessible to curious young hands.

A workshop is a very personal place, geared to your own specific needs and to the space available. A well-organized workspace makes for an organized, smooth-flowing project, because you always know where to find that special tool when you need it. If you can't designate a separate room for your tools and projects, at least have a closet or locking toolbox that will secure your tools when you're not using them.

SAFETY

Safety should be your primary concern whenever you work with tools. Always follow these general rules.

☐ Wear protective glasses or goggles whenever you are using power tools, and when chiseling, sanding, scraping, or hammering overhead, especially if you wear contact lenses.

☐ Wear ear protectors when using power tools; some operate at noise levels that can damage hearing.

☐ Wear the proper respirator or face mask when sanding or sawing or when using substances that emit toxic fumes.

☐ Be careful that loose hair and clothing don't get caught in tools.

☐ Always use the appropriate tool for the job.

☐ Keep blades sharp. A dull blade requires excessive force and can easily slip.

☐ Don't abuse your tools.

☐ Repair or discard tools with cracks in the wooden handles or chips in the metal parts. Damaged tools can cause injury.

☐ Read the owner's manual for all tools and know the proper use of each.

☐ Keep all tools out of reach of small children.

☐ Unplug all power tools when changing settings or parts.

☐ Don't drill, shape, or saw anything that isn't firmly and properly secured.

☐ Store or dispose of oily rags properly to prevent spontaneous combustion.

☐ Keep a first-aid kit on hand.

☐ Don't work with tools when you are tired. That's when most accidents occur.

PURCHASING YOUR TOOLS

Give plenty of care and thought to the purchase of your tools. Careful investment will carry you through years of enjoyable projects.

☐ Acquire tools as you need them; don't buy unnecessary tools. If you need a tool for a special purpose, you can often rent it.

☐ Always purchase the best tool you can afford. Cheap tools are never a bargain.

☐ Purchase from reputable dealers and manufacturers.

☐ Choose the tool that has the most comfortable fit and weight for your hand.

☐ Examine tools carefully for their sturdiness and smooth finish.

☐ Check all moving parts for smooth action and freedom from play.

CARING FOR YOUR TOOLS

Caring for your tools is also extremely important, if they are to do the jobs for which they are intended.

☐ Keep your tools properly cleaned and lubricated.

☐ Keep your tools out of the weather and store them out of the way when you are not using them.

- ☐ If the storage area is damp (the basement, for example) install a dehumidifier and keep tools covered with a film of rust-inhibiting oil
- ☐ Never throw tools into the toolbox. Handle them carefully to avoid dulling the edges and nicking the surfaces.
- ☐ Hang tools with cutting edges separately to keep them from getting nicked or dulled.
- ☐ Purchase carrying cases for your power tools to protect them and to store their accessories.

Tools for Measuring and Leveling

Most Common Mistake

- ☐ Not measuring accurately. A good carpenter measures twice and cuts once.

Accuracy and care in measuring are all-important. They can mean the difference between a professional project and a sloppy one.

The **carpenter's pencil** is flat, to keep it from rolling away, with a large, soft lead that draws broad, easy-to-see lines.

A **steel tape measure** (Figure 1-1), 20' to 100' in length, is an invaluable tool that belongs in every home. The blade (the tape itself) should be at least 1" wide, with a cushioned bumper to protect the hook of the tape from damage if the

tape retracts back into the case too quickly. The play in the hook allows you to make either inside or outside measurements without having to compensate for the hook. Its flexibility allows it to measure round, contoured, and other odd-shaped objects. When making inside measurements, add the length of the tape case, which is usually marked on the case.

Squares are used for laying out work, checking for squareness during assembly, and marking angles. The **carpenter's square**, also called a **framing square**, is used for marking true perpendicular lines to be cut on boards and for squaring corners, among other things. One leg is 24" long and 2" wide, the other 16" long and 1½" or 2" wide. The better types have a number of tables, conversions, and formulas stamped on the side to simplify many woodworking tasks. (Books are also available that give this information.)

The **combination square** (Figure 1-2) is a most versatile tool. It has a spirit level that locks in place on a 12" steel rule. It is used to square the end of a board, to mark a 45-degree angle for mitering, and to make quick level checks. It can also be used as a scribing tool to mark a constant distance along the length of a board.

Levels are used to make sure your work is true horizontal (level) or true vertical (plumb). To ensure accuracy, always use the longest level possible. The **torpedo level,** used for small pieces of work, is 8" or 9" in length, with vials that read level, plumb, and 45 degrees. A **2' to 4' level** is a must for any home woodworking project, from building shelves to structural carpentry.

A **plumb bob** is a heavy, balanced weight on a string. It is dropped from a specific point to locate another point exactly below it, or to determine true vertical (Figure 1-3).

A **chalkline** is a string or line coated with colored chalk, used to transfer a straight line to a working surface easily and accurately. Pull the line out and hold it tight between the two points of measurement. Then snap it to leave a mark. Some chalklines have a pointed case that doubles as a plumb bob.

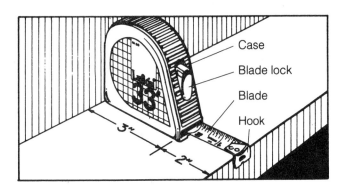

Figure 1-1. To make inside measurements easily, place the tape case against the wall and add its length (in this case, 3") to the measurement.

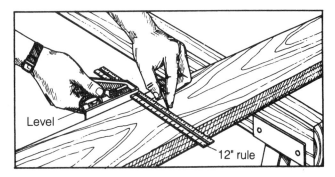

Figure 1-2. Using a combination square to prepare for a crosscut.

Figure 1-3. Using a plumb bob to determine true vertical (plumb).

Helpful Hints on Measuring

☐ When many pieces need to be cut or drilled the same, you can use one accurately cut or drilled piece as a template to mark all the others. However, for greatest accuracy, always rely on your tape measure as well.

☐ Remember the old maxim: "A good carpenter measures twice and cuts once."

Power Tools for Cutting and Drilling

Most Common Mistakes

☐ Setting a board to be cut between two supports. Instead, cantilever the board over the outside of one support to avoid binding the saw blade when the board drops. This method prevents "kick back."

☐ Choosing an inappropriate saw blade for the material being cut and for the job you are doing. Always be sure that your blade is sharp as well as appropriate for a given material.

SAWS

The **circular saw** (Figure 1-4) is one of the most popular tools in the home workshop. It consists of a replaceable blade; a blade guard, part of which is spring loaded to move out of the way as you saw; a sole plate, which may or may not be attached to a ripping fence; a cutting guide; and knobs to adjust the cutting angle and depth of the blade. Set the cutting depth to 1/8" more than the thickness of the board. Keep the sole plate flat on the surface of the wood to avoid binding. There are two basic kinds of circular saws. The **worm drive saw** is used for heavy-duty framing-type work, and many carpenters use this type of saw with a fine-tooth blade for finish work as well. The **sidewinder** is used for lighter jobs.

The **saber saw** or **jigsaw** (Figure 1-5) is used to cut curving lines and to make detailed cuts. When starting your cut from inside a piece of wood to cut an opening in paneling, drywall, or a counter top, drill a starting hole first. Most saber saws can accommodate a circle guide and angle cutting and ripping accessories.

The **reciprocating saw** is generally used to cut openings in existing walls. It is ideal for remodeling or demolition work. It can rough-cut wood, metal, leather, rubber, cloth, linoleum, and plastic.

The **chopsaw,** a motorized miter box saw, makes repeated square and angled cuts with a minimum of effort. This tool is indispensable for making high-grade finish joints, for making trim, for building decks, and for laying hardwood floors—wherever precision end cuts are needed.

The **table saw,** also called a **bench saw** (Figure 1-6), can crosscut, rip, miter, groove, and bevel. With a 7½" to 12"

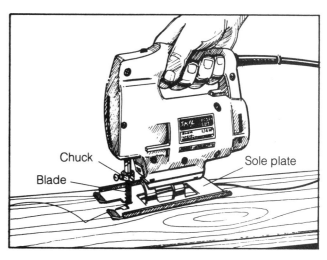

Chuck   Sole plate
Blade

Figure 1-5. A saber saw or jigsaw allows you to make fine cuts in curved and detailed patterns.

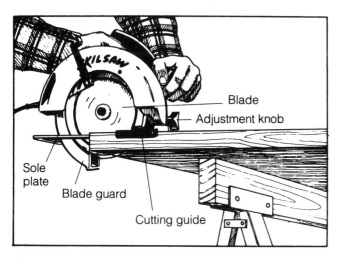

Blade
Adjustment knob
Sole plate
Blade guard
Cutting guide

Figure 1-4. The circular saw is an indispensable tool for all kinds of home building projects.

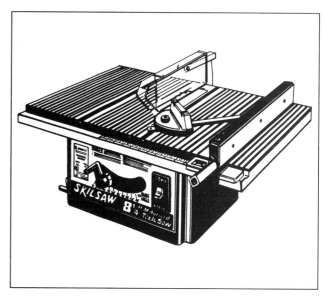

Figure 1-6. A small table saw or bench saw gives you flexibility in cutting trim, making moldings, and cabinetry work.

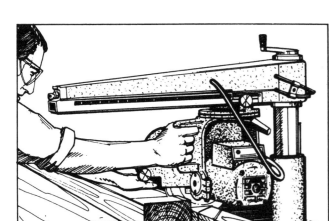

Figure 1-7. A radial arm saw is versatile like a table saw, but is used for heavier stock.

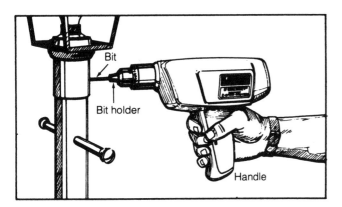

Figure 1-8. A cordless drill can be used where power is not available and where a power cord would be inconvenient.

blade, this saw can cut depths of 1½" to 3⅜". With accessories, it will cut dadoes, make moldings, and sand stock. Smaller table saws are good for general hobby and repair work.

The **radial arm saw** (Figure 1-7) holds the board stationary for crosscutting, angled cuts, notching, and lap joints, with the saw blade bearing down across the board. This saw is also useful for making wide rip cuts, up to about 24" (half the width of a sheet of plywood). The radial arm saw is usually used by professionals.

The **bandsaw** is useful for creative work where precision cutting is needed to produce complex shapes and curves.

SAWBLADES

Sawblades are available to cut tile, stucco, metal, plastic laminates, concrete, and of course wood. Blades for many of these materials are use-specific. Wood-cutting blades differ widely, depending on use. In general, the finer the teeth, the

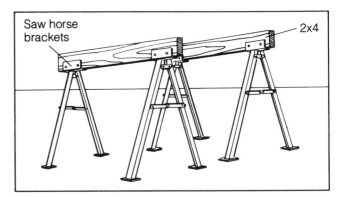

Figure 1-9. Sawhorses are easy to put together with special brackets.

finer the resulting cut. Coarse blades with large teeth tend to cut faster but rougher. If you are cutting into material that could contain hidden nails, use a special carbide-tipped sawblade, which can cut through nails easily.

DRILLS AND ROUTER

A good ⅜" **electric** or **power drill** is one of the most versatile tools in the home workshop. A variable-speed trigger and a reversing switch greatly increase the utility of the drill. A **cordless drill** (Figure 1-8) eliminates the need for cumbersome extension cords and can be used where power is not available. It can speed up many routine jobs, and is a good investment.

You'll need **drill bits** of carbon steel for soft materials like wood and plastic. High-speed steel bits work well on metals, and carbide-tipped bits are used for concrete or masonry.

Other useful **attachments** include hole saws, spade bits, buffing disks and depth stops, screwdriving bits, sanding disks, and power grinders. You can also add a paint mixer, right-angle drive, wire brushes, wood countersinks, and even rotary rasps and files.

The **router** is used to cut contours in wood for edgings and moldings and for more complex relief panels and inlay work, dovetails, and mortises, as well as trimming veneers and plastic laminates.

Helpful Hints on Cutting and Drilling

☐ Always unplug your power tools when adjusting them or changing accessories.

☐ Keep tool adjustment keys taped to the cord near the plug. This will remind you to unplug the tool, as well as keep you from losing the key.

☐ Be sure your tools are properly grounded. Unless a tool is double-insulated, it should be plugged into a three-hole grounded outlet.

☐ Watch cord placement so it does not interfere with the operation of the tool.

☐ Never use a power tool in damp conditions, no matter how well grounded it is. Remember that moisture readily conducts electricity.

□ For sawing and assembling, one of the most useful accessories is a pair of sawhorses (Figure 1-9). They can mean the difference between an easy, comfortable task that can be accomplished standing up, and a difficult one that requires bending repeatedly.

□ When rip cutting, use a rip guide; or tack down a straight-edge to use as a cutting guide.

□ Be sure that both ends of a board are well supported, and don't cut between the supports; cantilever the scrap end instead.

□ When using a table saw, choose the guide that allows the longest edge of the stock to be used against the guide.

□ Always secure your stock before cutting or drilling.

□ Use caution when ripping small, narrow pieces. Use a push stick with a radial arm saw or table saw, and always keep your hands clear of the blade.

□ When drilling into metal, squirt a lightweight oil onto the drill bit and into the hole to cool the bit.

□ Predrill a pilot hole to make nailing into hardwood, as well as screwing into all types of wood, easier. A pilot hole lessens the chance of splitting the wood. The hole should be slightly smaller than the diameter of the nail or screw.

Hand Tools for Sawing and Chiseling

SAWS

The **crosscut saw** is used to cut across the grain of the wood. It has small, offset teeth. An 8-point saw is best for general use.

The **rip saw** cuts with the grain, and has much larger teeth.

The **combination saw** (Figure 1-10) cuts both across and with the grain, but does not perform as well as saws that are designed for specific uses.

The **hacksaw** is used to cut metal, plastic, and electrical conduit.

The **backsaw** is used in conjunction with a **miter box** to cut a perfectly straight line across a piece of wood. The steel backing keeps it aligned (Figure 1-11).

The **keyhole saw** has a blade that is narrower at the tip than at the heel or handle. It is used for cutting openings in drywall or paneling and for curved cuts.

The **drywall saw** is like a keyhole saw, but it has a straight handle and a stiffer blade. Larger drywall saws, shaped more like a wood saw, are available for larger cuts, such as rough openings.

Figure 1-11. A backsaw is used with a miter box to make accurate cuts for trim and molding.

Figure 1-10. Using a combination handsaw with sawhorses.

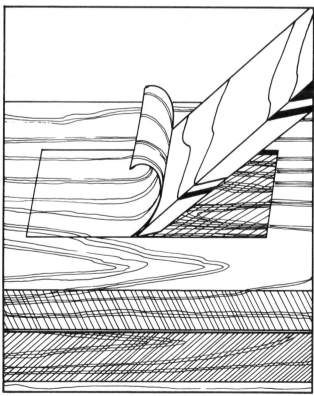

Figure 1-12. Chisels are used to chip and shave wood away.

The **coping saw** is used to follow an irregular, delicate, or intricate cut in wood. The blade is thin, fine toothed, and removable.

Helpful Hints on Sawing

☐ Don't store saws in a toolbox or where the teeth will get damaged. Hang your saws if possible; or buy a plastic tooth guard or make one out of cardboard.

☐ Keep saw blades sharp. Keep them out of contact with metal, concrete, and stone. Remove any nails from wood before sawing.

☐ Protect your saw blades by coating them with paste wax or a light grade of machine oil.

CHISELS

Chisels (Figure 1-12) are wood-cutting tools For general use, purchase a set of chisels ranging from ¼" to 1½" in width. **Socket chisels** are meant to be used with a mallet; don't use a hammer to strike a socket chisel. **Tang chisels** are for working with the weight of the hand only. For more finished work, use a beveled-edge **cabinetmaker's chisel.** The square edge of a **framing chisel** is best for forming work, while the narrow **mortise chisel** is used to break waste away.

Helpful Hints on Chiseling

☐ Approach your work so that the wood's grain lifts the edge as you strike the chisel. If the grain is allowed to direct the edge deeper into the wood, it becomes much harder to control.

☐ First mark with chisel blows the perimeter of the area you want to chisel away. Then start your blade digging into the wood slightly inside your guideline mark.

☐ Be careful not to cut too deeply. Chisels are meant to chip and shave away. The beveled edge constantly directs the chisel out of the wood for better control.

☐ Keep the cutting edge directed away from your body and hands.

Tools for Attaching, Assembling, and Dismantling

Most Common Mistakes

☐ "Choking up" on the hammer handle reduces leverage and does not allow the head to strike flat against the surface.

☐ Not predrilling a pilot hole before nailing or screwing into hardwood or near the end of a board.

☐ Using extra-leverage items to tighten clamps.

At some point in any project you will be attaching, assembling, or dismantling what you have been measuring and sawing. These operations can be done in a variety of ways.

HAMMERS AND NAILS

The **carpenter's hammer** is available in two types. The **claw hammer** is made with a curved claw, well suited to pulling nails. For ripping boards out, the **ripping hammer,** with its straight claw, fits more easily between boards. Both types are used to drive and remove nails. They come with wood (usually hickory), steel, or fiberglass handles and in a variety of face styles and weights. A 12- to 16-ounce **finishing hammer** is recommended for small workshop projects and general use; an 18- to 24-ounce **framing hammer** is used for heavier framing work. Finishing hammers have smooth faces; framing hammers have "waffled" faces.

The **tack hammer** is useful for driving tacks and brads. Some have magnetic heads.

Mallets are used primarily for striking other objects, such as chisels, or to form sheet metal. A soft-faced mallet is used with wood- and plastic-handled chisels (Figure 1-13).

The **sledge hammer** is used for heavy work on concrete, and for adjusting framing.

Figure 1-13. Use a mallet rather than a hammer, which may dent the wood.

Figure 1-14. Place a block of wood under the hammerhead to avoid marring the wood when pulling nails.

The **nailset** is used with a hammer to "countersink" nails by pushing them below the surface of the wood so they don't show.

Nails range from the smallest, thinnest brads to large, weighty spikes. Just be sure you are using the correct nail for the job at hand.

Helpful Hints on Hammering

☐ To withdraw a long nail without marring the wood, place a block of wood under the hammerhead for extra leverage (Figure 1-14).

☐ Use safety goggles when hammering metal. Chips often fly from steel chisels, and nailheads can break off.

☐ Whenever possible, drive the nail through the thinner piece of wood into the thicker one. Use a nail that is at least twice in length the thickness of the thinner piece of wood.

☐ Predrill a pilot hole, slightly smaller than the diameter of the nail shank, to prevent splitting the wood. Pilot holes are recommended for hardwoods, such as oak and maple, and near the ends of boards. A less effective but quicker method is to blunt the point of the nail before driving it to prevent splitting the board. To blunt the point, rest the head of the nail on a solid surface and tap the point gently with your hammer.

SCREWDRIVERS AND SCREWS

Screwdrivers should be used only to drive in and remove screws. Too often they are wrongly used—for chipping, punching holes, scraping, prying, and so forth.

The **conventional screwdriver** has a single blade and is sized to be used with screws of specific sizes.

The **phillips head screwdriver** has a cross-shaped blade that fits into the cross-shaped slot of the phillips head screw. This design reduces blade slippage and lends itself to driving with a power drill.

A **battery-powered cordless screwdriver/drill** is a very helpful tool that can be used to put a hole or a screw almost anywhere without the need to set up an extension cord. Heavy-duty types are recommended for most tasks. Holsters are available.

Screw heads are usually flat, oval, or round, and each has a specific purpose for final seating and appearance. **Flat heads** are always countersunk or rest flush with the surface. **Oval heads** permit countersinking, but the head protrudes somewhat. **Round-headed screws** rest on top of the material and are easiest to remove.

Drywall screws, also called **bugle head screws,** are countersunk phillips head screws, available in very long styles as well as in standard patterns. Originally designed to secure drywall to sheet metal framing, they are very useful for wood work. They are easy to drive compared to conventional tapered wood screws, and are even available in a finish screw pattern, called a **Robertson** screw or a **square-headed** screw.

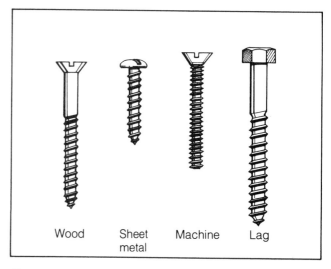

Figure 1-15. *Some common screws.*

Wood screws are used when stronger joining than a nail is needed, and when other materials must be fastened to wood (Figure 1-15). The conventional wood screw is tapered to help draw the wood together as the screw is inserted. Sheet metal screws can also be used to fasten metal to wood, as well as metal to metal, plastic, or other materials. **Sheet metal screws** are threaded completely from the point to the head, and the threads are sharper than those of conventional wood screws. **Machine screws** are for joining metal parts, such as hinges to metal door jambs. Machine screws are inserted into tapped (prethreaded) holes and are sometimes used with washers and nuts. **Lag screws,** or square-headed bolts with screw heads, are for heavy holding and are driven in with a wrench rather than a screwdriver.

Helpful Hints on Screwdriving

☐ When choosing screw length, remember that the screw should penetrate two-thirds of the combined thickness of the materials being joined. To avoid corrosion, consider as well moisture conditions and the makeup of the materials being fastened. If rust could be a problem, use galvanized or other rust-resistant screws.

☐ Lubricate screws with soap or wax for easier installation.

☐ Whenever possible, hold the work in a vise or clamp when inserting a screw. If this is not possible, keep your hands and other parts of the body away from the tip of the driver.

☐ To remove a screw with a damaged slot, cut another slot with a hacksaw blade if the head is exposed enough.

☐ Always make a pilot hole (usually one size smaller than the shank of the screw) before driving a screw into hardwoods and when driving a screw near the end of the board. When working with screws of larger diameter, drill a pilot hole of the same diameter as the shank of the screw into the wood to a depth of one-third the length of the screw.

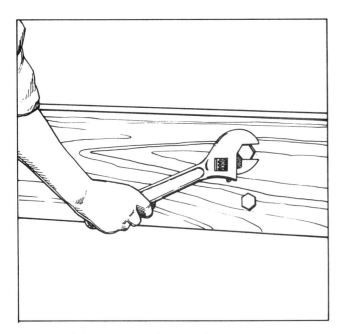

Figure 1-16. Using an adjustable wrench.

☐ To avoid damaging the screw slot and pushing the screw out of line, always keep the screwdriver shank in line with the screw shank.

WRENCHES AND PLIERS

It's not necessary to have a wide variety of wrenches and pliers, but it's handy to have some of the more common ones around the house.

An adjustable **open-end wrench** (Figure 1-16) fits any size nut that is within its opening capacity. The best choice for general use is one that opens to $^{15}/16$".

Box or **socket wrenches** are used for removing nuts and bolts and are fitted to the size of the fastener.

The **allen wrench** is useful for recessed screws and setscrews.

Pipe wrenches are used to tighten or loosen plumbing pipes. Use two wrenches, especially when working on existing pipe—one to hold the pipe in place, the other to turn the pipe or fitting out.

The **locking wrench** works like a clamp for holding pipes and other objects in place.

The **strap wrench** is used to prevent marring chrome-plated finish and to clamp irregularly shaped items.

Slip joint pliers have jaws that lock into normal and wide opening positions.

Lineman's pliers are useful in electrical work. They have side cutters for heavy-duty wire cutting and splicing.

Channel lock pliers, with multiposition pivots, adjust for openings up to 2".

Long-nosed pliers, sometimes called **needle-nosed pliers,** can get into hard-to-reach places. They are used to shape wire and thin metal.

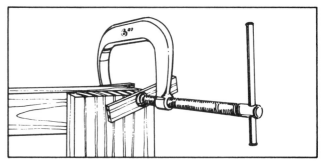

Figure 1-17. C-clamps are available with openings of different sizes.

Figure 1-18. Use wood shims for padding with bar clamps.

Figure 1-19. A handscrew won't mar the wood, and can be adjusted to different angles.

End-cutting nippers are used to snip wire, small nails, and brads.

Helpful Hint on Wrenching and Plying

☐ To avoid slippage and scraped knuckles, make sure the jaws of the wrench or pliers are snug in position before you manipulate the handle.

CLAMPS

Clamps are used to hold objects together while they are being worked on, or while an adhesive is drying.

The **C-clamp** (Figure 1-17) is the most common type of clamp. So named because of its shape, it has a swivel head that makes the clamp self-aligning for odd-shaped pieces.

The **bar clamp** (Figure 1-18) is useful for clamping extra-wide work.

The **vise** is a workbench tool and should be firmly secured before use. It is used for holding work to be sawed, bored, glued, or formed in some way.

The **handscrew** (Figure 1-19) has hardwood jaws that move in opposite directions because of the threading of the screws. The smooth wood and broad jaws protect the surface of the work being clamped. The handscrew is used for cabinet and furniture work.

Spring clamps are for smaller bonding uses and for securing a temporary cutting guide.

Helpful Hints on Clamping

☐ To prevent pressure damage to the surface, use padding or scraps of wood between clamps and your work.

☐ Never pound on the handle or tighten a clamp or vise with a wrench or pipe to obtain extra leverage.

ADHESIVES

The best joining is accomplished with **adhesives,** in conjunction with screws, nails, or other fasteners.

Polyvinyl (carpenter's wood glue) is a white, creamy glue that is available in plastic bottles. Mainly used for furniture, craft, and woodworking projects, polyvinyl sets in an hour, dries clear, and won't stain. However, it is vulnerable to moisture.

Resorcinol and formaldehyde are mixed just before using, and must be used at temperatures over 70 degrees. Both the resin (powdered resorcinol) and the powdered formaldehyde that you mix with water are brown and will stain light wood. Follow the manufacturer's instructions carefully.

Contact cements are used to bond veneers or to bond plastic laminates to wood for table tops and counters. Coat both surfaces thinly and allow to dry until dull before bonding. Align the surfaces perfectly before pressing them together, because once they are joined they cannot be pulled apart. Use in a well-ventilated area.

Epoxy is the only adhesive with a strength greater than most of the materials it bonds. It resists almost everything, from water to solvents. Epoxy can be used to fill cavities that would otherwise be difficult to bond. Read the manufacturer's instructions carefully; drying times vary and the resin and hardener must be mixed with precision.

The resorcinol, contact, and epoxy adhesives should be considered toxic until they cure. Take care not to get them on your skin.

Helpful Hints on Bonding

☐ Except for epoxy, too much adhesive will weaken the hold of the materials you are bonding.

☐ Rough up smooth surfaces slightly before applying adhesives so they will grip more securely.

☐ Apply a thin coat of glue, clamp securely, and allow to dry for the recommended amount of time.

☐ Wipe away excess glue immediately after clamping, except for the resorcinol and contact types. Let them dry before tooling or wiping off the excess.

FASTENERS AND CONNECTORS

You can buy ready-made fasteners and connectors for just about any job you need to do. They can be used to join wood to wood, concrete, or brick, and most meet *Uniform Building Code* requirements.

Safety plates, or **nail guards,** prevent accidental nailing into electrical wires or water and gas pipes that pass through framing.

Nail plates and **plate strips** work well as mending plates and for light-duty wood-to-wood splices.

Fence brackets simplify fence construction and allow easy disassembly when necessary.

Sawhorse brackets turn a 2x4 or 2x6 into an important support tool in just one step.

Stud shoes reinforce joists, studs, or rafters that have been drilled or notched during construction.

Foundation and masonry connectors include foundation anchors, brick wall ties, and floor jacks, among others.

Post anchors are designed to support a post from the ground up. They eliminate the need for deep post holes and prevent wood rot and termite damage.

Post caps and plates are used for a strong support where one or more beams must be connected.

Joist hangers aid in accurate, uniform connections and allow a structure to hold greater loads than do other types of fasteners.

DISMANTLING TOOLS

Everybody makes mistakes, and it helps to know how to fix things the easy way. Some dismantling tools can be a great help in "adjusting" previous work.

Pry bars, which come in various sizes, are useful demolition tools for pulling nails, ripping wood, and prying molding from walls. A pry bar offers more leverage than does a hammer.

A **catspaw** is a small pry bar with a grooved round head that is driven into a board to grab a nailhead that has been driven below the surface.

A **flat bar** is used to remove duplex nails or any nail with the head still accessible. These bars are also used to jack up drywall or door jambs, and to make minor adjustments almost anywhere.

A **reciprocating saw** with a long metal-cutting blade can also be used to cut through nails if necessary.

Finishing Tools

Most Common Mistakes

☐ Allowing a power sander to dig into the wood being finished. To avoid this, always keep the sanding machine moving over the wood.

☐ Putting the belt on a belt sander backward. This tears the seams.

☐ Using an inappropriate grit of sandpaper for the desired effect.

SANDING TOOLS

The **sanding block** or **block sander** (Figure 1-20) is wrapped with the appropriate sandpaper and rubbed across the surface by hand.

A **sanding cloth** is essential for sanding curved or round objects. It's much easier to use than a sanding block, and gives a more even finish.

The **power orbital sander,** or **pad sander** (Figure 1-21), is the ideal all-around sanding tool for finish work on walls, ceilings, floor, furniture, and other woodwork.

The **belt sander** (Figure 1-22) is designed for quick, rough sanding of large areas, or wherever heavy-duty sanding is needed.

The **disk sander** is used for fast removal of wood on uneven surfaces. Disk sanding attachments can be purchased as drill accessories.

The **drum sander** is a powerful machine used to sand floors only.

The **edge sander** is used to sand floors in areas the drum sander can't reach, such as where the floor meets the wall.

Sandpaper runs from very coarse (20 to 40 grits per inch) up to the very fine (600 grits per inch). Materials range from flint and garnet emery to aluminum oxide and silicon carbide. **Flint** is best for hand-sanding painted or pitchy surfaces, which can clog the paper. **Garnet emery** is for hand-sanding clean wood. **Aluminum oxide** is fast and long-lasting for power-sanding wood. It can also be used on plastics and fiberglass and for polishing stainless steel, high carbon steel, or bronze. **Silicon carbide** is harder than aluminum oxide and is best used for hard plastics, glass, and ceramics, or for grinding and finishing brass, copper, and aluminum.

Emery cloth is another option for metal polishing.

Steel wool is available in #3 coarse to #0000, which is very fine.

Figure 1-21. Using an orbital sander to sand a small patch.

Figure 1-20. Using a hand-held sanding block.

Sanding belt

Roller

Tracking control knob

Figure 1-22. Using a belt sander.

Use **pumice** or **rottenstone** (decomposed limestone) for a glassy finish.

Helpful Hints on Sanding

☐ Generally, you should start with a coarse paper and work to a fine paper for the smoothest finish. Whenever possible, sand with the grain of the wood.

☐ When using a power sander, do not press down on the machine. Let its own weight do the sanding. Pressing down inhibits the action of the machine.

☐ When operating any power sander, engage and disengage the machine from the material being sanded while the belt or disk is still in motion to avoid gouging the wood.

☐ To sand inward curves, wrap padding material around a stick or dowel, then wrap sandpaper over it.

☐ When sanding wood, seal heating and air-conditioning ducts and electrical outlets with plastic sheets and/or duct tape. Wood dust is highly flammable.

PLANES

Planes are used for removing very thin layers of wood, for trimming and smoothing, for straightening or beveling edges, and for adding a groove.

The **block plane** is abut 6" long and is used for small smoothing and fitting jobs.

The **trimming plane** is used for more delicate work. This tool is only 3½" in length, with a 1" blade.

The **jack plane** is 12" to 15" in length, with a 2" blade. It is used to smooth rough surfaces.

Fore and **joiner planes,** at 18" to 24" in length, are the best bet for straightening edges.

The **rabbit plane** cuts recessed grooves along an edge.

The **grooving plane** cuts a long slot.

Helpful Hints on Planing

☐ Whenever possible, plane with the grain of the wood to avoid catching and lifting chips of wood.

☐ Prevent splitting the corners of the material you are planing across the grain by approaching each corner from the outside.

☐ Always keep your blades razor sharp; dull blades require excessive force and reduce control.

☐ When it's not in use, rest your plane on its side to avoid dulling the blade.

☐ When starting cuts, apply more pressure to the front of the tool; when completing them, apply more pressure to the rear.

FILES AND RASPS

Files are used for shaping. You will find files that are round, half-round, flat, square, and even triangular. Single-cut file teeth run in one direction. Double-cut teeth, which run in opposing directions, cut more coarsely, but quicker.

Rasps differ from files in that the teeth are formed individually, not connected to one another. In general, a longer file or rasp has somewhat coarser teeth than a shorter one. Files cut smoother than rasps, but when used on wood are much slower and are susceptible to clogging.

Helpful Hints on Filing

☐ Files with an attachable handle are easier to use.

☐ Secure your work with a vise or clamps at elbow height for general filing; lower for heavier filing; and nearer to eye level for delicate work.

☐ File in the cutting direction only; lift the file on the return stroke.

☐ Keep files clean with a file brush (called a **card**) or with a small wire brush.

☐ Store files in a rack or protective sleeves to keep from dulling the teeth.

Pneumatic Tools

Pneumatic tools are fast becoming indispensable to the home builder. Pneumatic tools include an air compressor (pneumatic means "containing air") and a variety of attachments. Although pneumatic tools take some special handling, they actually save a great deal of working time and effort and are relatively easy to use. They also give professional results.

The instruction manual that comes with the compressor should list a toll-free number to call if you have any questions or problems.

Most Common Mistakes

☐ Not reading the instruction manual and safety labels.

☐ Using an improper length or gauge of extension cord. A long extension cord reduces the power and efficiency of the motor. Use extension air hose rather than cord for distance work.

☐ Using an incorrect pressure for the tool and the task.

☐ Not changing the oil every 250 hours of use or every 6 months.

☐ Failing to locate the working air compressor on a clean, dry, level site.

☐ Failing to keep the compressor properly lubricated.

SAFETY

☐ Many tasks for which an air compressor is used require safety glasses or goggles, protective clothing, and a dust- or paint-filtering mask.

☐ Make sure the pressure of the compressor has been completely relieved through the line to the tool before changing to another tool, unless a quick connection is being used.

☐ Make sure the tool has completely stopped before changing or disconnecting it.

- ☐ Never point the blow gun toward your eyes or any other part of your body.
- ☐ Never exceed recommended pressure for the tool being used or the job being done.
- ☐ Never override the pressure switch on a pneumatic nailgun.
- ☐ Never alter the three-pronged plug to fit an outlet other than the type for which it was designed.
- ☐ Always unplug the cord from the outlet before disengaging it from the compressor.
- ☐ Test safety relief valves periodically.
- ☐ Open the tank valve after every use.
- ☐ Never allow children around the compressor, whether it is in operation or stored.
- ☐ Never operate the compressor without the belt guard in place.
- ☐ Read the owner's manual completely and read all safety-oriented labels on the unit before using it.

AIR COMPRESSOR

The **air compressor** is the central power source of the pneumatic tool, and your major investment. Air compressors are available in horsepowers of ¾ to 5, or even larger, with a variety of tank sizes up to 80 gallons. The frequency and duration of use will determine the horsepower and tank size you'll need for your own projects. A 2 HP compressor is adequate for most tasks around the house. Typically, units of this size have tanks from 7½ to 20 gallons. Choose an air compressor for quality and protective features, as well as for capacity.

ATTACHMENTS

If kept properly cleaned and lubricated, air tools are virtually indestructible. They have few moving parts, so maintenance is minimal, and they run cool, because their power source is the compressor.

Most air tools are available at hardware stores and home centers. Specialty air tools can be rented. Instructions for each tool attachment are included with the purchase or rental. Read these instructions carefully before using the tool.

Two of the most obvious and useful tools are an **inflation kit** and **quick-connect couplers.** The inflation kit attachment allows you to inflate everything from beach balls to automobile tires. The quick connect couplers make it fast and easy to change tools.

The **blow gun** attachment is great for blasting away dirt, grease, and dust from hard-to-reach areas. Never point the gun at the eyes or other parts of the body.

Always be sure the **nail gun** is flat against the surface being nailed and know what is on the other side, so you won't cause damage or injury with the high pressure of the gun.

Figure 1-23. Using an air compressor with a spray gun attachment to paint.

Similarly, be sure the **stapler** is flat against the surface being stapled. Large staplers are available for heavy work like attaching roofing shingles.

The dual-action **air sander** should always be touching the surface when it is turned on. This type of sander is frequently used in automotive work and has many other uses around the house, such as rust removal and paint preparation.

The **spray gun** (Figure 1-23) speeds up paint application by 50% or more and gives a smooth finish. There are a variety of spray gun designs on the market for various types of painting.

The **sandblaster** works well for removing rust and old paint and for preparing surfaces for painting. It can also be adapted for use with soap and water for pressure cleaning, such as degreasing auto engines and garden equipment.

The **caulking gun** takes the toil out of caulking by giving a fast, strong, uniform bead. This tool can be used for any tube material, such as adhesive or grease.

The **air ratchet wrench** is great for tightening bolts, whether you are building a deck or working on an automobile engine.

The **air hammer/chisel** is used for jobs like breaking up masonry. It must be up against the surface when started.

The **air drill** makes drilling into any surface an effortless task.

The **impact wrench** is used mostly in automotive and assembly work.

Checklist of Tools for Home Projects

Tools for Measuring and Leveling

Carpenter's pencil

Steel tape measure

Carpenter's square (framing square)

Combination square

Torpedo level

2′ to 4′ level

Plumb bob

Chalkline

Power Tools for Cutting and Drilling

Saws

Circular saw

Saber saw (jigsaw)

Reciprocating saw

Table saw (bench saw)

Radial-arm saw

Chopsaw

Bandsaw

Drills and Router

3/8" variable-speed drill

Cordless drill

Router

Hand Tools for Sawing and Chiseling

Saws

Crosscut saw

Rip saw

Combination saw

Hacksaw

Backsaw and miter box

Keyhole saw

Coping saw

Chisels

Chisel set 1/4" to 1 1/2" in width

Sock

Tang chisel

Cabinetmaker's chisel

Framing chisel

Mortise chisel

Tools for Attaching, Assembling, and Dismantling

Hammers and Nails

Claw hammer

Ripping hammer

Tack hammer

Mallet

Sledge hammer

Nailset

Brads, nails, and spikes

Screwdrivers and Screws

Conventional screwdrivers

Phillips head screwdrivers

Cordless screwdriver

Screws of various kinds

Wrenches and Pliers

Open-end wrench

Box wrench (socket wrench)

Allen wrench

Pipe wrenches (2)

Locking wrench

Strap wrench

Slip joint pliers

Channel lock pliers

Long-nosed pliers

End-cutting nippers

Clamps

C-Clamp

Bar clamp

Vise

Handscrew

Spring clamps

Adhesives

Polyvinyl (carpenter's wood glue)

Resorcinol and formaldehyde

Contact cement

Epoxy

Fasteners and Connectors

Safety plates (nail guards)

Nail plates

Plate strips

Fence brackets

Sawhorse brackets

Stud shoes

Foundation and masonry connectors

Post anchors

Post caps and plates

Joist hangers

Dismantling Tools

Pry bar

Catspaw

Flat bar

Finishing Tools

Sanding Tools

Sanding block (block sander)

Sanding cloth

Orbital sander (pad sander)

Belt sander

Disk sander

Drum sander

Edge sander

Sandpapers

Steel wool

Pumice and rottenstone

Planes

Block plane

Trimming plane

Jack plane

Scrub plane

Fore and joiner planes

Rabbit plane

Grooving plane

Files and Rasps

Single cut file

Double cut file

Rasp

Pneumatic Tools

Air compressor

Blow gun

Nail gun

Air stapler

Air sander

Spray gun

Sandblaster

Caulking gun

Air ratchet wrench

Air hammer/chisel

Air drill

Impact wrench

Miscellaneous Tools

Sawhorses

Vacuum cleaner

2 Electricity

ALONG WITH THE plumbing system, installing the electrical system is among the most demanding tasks for the owner builder. It is not so much that a high degree of skill is involved in running wires, but rather that a great amount of knowledge is required. The electrical code is extensive and varies from area to area. Every electrical system is unique, and every home has its own problems that require special knowledge and experience. In addition, installing the electrical system is physically difficult and can be extremely frustrating.

In the entire process of finishing a house, electrical wiring and plumbing installation are the areas in which you should consider hiring a professional. If you decide to wire the house yourself, you should at least hire a professional consultant to oversee your work. It's also a good idea to read at least a couple of how-to books about electricity, just to determine how much of this work you want to do yourself.

The National Electric Code (NEC), written by the National Fire Protection Association, is designed to provide a basic level of safety in every electrical system in the United States. Within certain limits, each community is free to add or delete provisions in order to address specific local problems or reflect local custom.

Unlike the rest of this book, it is not the intention of this chapter to provide you with all of the information you will need to install the entire electrical system; that would require a book in itself. This chapter does not attempt to walk you through the entire process in a series of detailed steps. Rather, it offers an overview of the electrical system and its components, some of the main code restrictions, and details on installing and connecting cable, boxes, and receptacles. With the information presented here, you should be able to speak knowledgeably with your electrician about your specific needs.

A faulty or improperly installed electrical system can be the source of three major problems. First, inspectors are very demanding that the system be properly installed, and are especially wary of work done by nonprofessionals. Second, wiring mistakes are often discovered after the walls are finished. If the problem is behind the drywall, it will be difficult to repair. Most importantly, a faulty electrical system is a major fire hazard. For all of these reasons, it is a good idea to consult with a professional about all electrical work you plan to do.

Many owner-builders choose to install the branch circuits and subpanels and then have an electrical contractor install the mast and main panel and tie the circuits to the main panel.

Basic Theory

Electricity can be compared to water flowing through pipes, with a small turbine drawing power at each tap, or outlet. **Alternating current** is delivered in pulses to your appliances and lights through a black or red **hot wire.** A white **neutral wire** returns the current to the earth. A bare or green **grounding wire** serves as a safety backup for the neutral wire. The neutral wires and grounding wires are never interrupted by a breaker, as this would compromise the system's safety.

Each circuit in your electrical system is a loop. The power coming in must have a way out to perform its work. For instance, a light fixture is always connected to the neutral wire, and the switch is placed on the hot wire. The current travels through the switch and the bulb and back to the earth. With outlets, the current completes the loop through whatever appliance is plugged in and turned on.

It is vital, and required by the NEC, that you maintain standard color coding throughout your system. The color coding ensures that any electrician or inspector can open any panel, switch, or outlet box in the system and know which wires are doing what. If the neutral wire is used incorrectly to supply power to a polarized appliance or motor, damage to the appliance may occur and the potential for shocks is increased. (**Polarity** refers to the direction the power is traveling, and it must be correct to avoid damage to many electrical devices, computers in particular. If the black wire is always attached to the brass screw, the white wire is always attached to the silver screw, and the bare or green wire is always attached to the green screw, polarity will be correctly maintained throughout your system.)

Before You Begin

SAFETY

Safety is of the utmost importance with working with electricity. Develop safe work habits and stick to them. Electricity may be invisible, but it can be dangerous if it is not understood and respected.

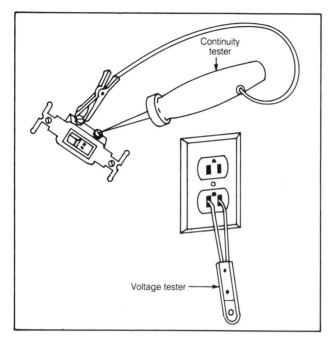

Figure 2-1. Using a continuity tester and a voltage tester.

☐ The first rule of working with electricity is, Never work on any live circuit, fixture, receptacle, or switch.

☐ Make sure the power is off at the breaker box before doing any electrical work.

☐ Avoid electrical shock by mapping and marking your switch and outlet boxes. Put the map on the door of the main power service panel.

☐ Post a warning message at the service panel that you are working on the circuit, and tape the circuit breaker in the off position.

☐ Before working with wires or electrical connections, check them with a voltage tester to be sure they are dead. One prong of the voltage tester is inserted into the hot side of the outlet and one into the neutral or ground side (Figure 2-1). If the circuit is live, the bulb will light up.

☐ Plumbing and gas pipes are often connected to electrical systems. Never touch them while working with or around electricity.

☐ Always work in a clean, dry area that is free from anything wet.

☐ Don't use metal ladders around overhead electricity.

☐ Never attempt to strip wires with a knife. Aside from endangering your fingers, you will nick the wire metal, creating an electrical hazard.

☐ Never change the size of a fuse or breaker in a circuit.

☐ Always correct the problem that caused a fuse or circuit breaker to blow before replacing the fuse or resetting the circuit breaker.

☐ Wires should be connected only at accessible boxes. Never splice wires outside a box.

☐ Ground fault circuit interrupter outlets should be used in damp areas, such as basements, bathrooms, and outdoors. Their use is required by the National Electric Code.

☐ Wear rubber-soled boots when working with electricity.

☐ Wear glasses or goggles whenever you are using power tools, especially if you wear contact lenses.

☐ Always brace the powerful right-angle drill so that it can't spin around and break your knuckles if it gets stuck while you are drilling.

☐ Use the proper protection, take precautions, and plan ahead. Never bypass safety to save money or to rush a project.

USEFUL TERMS

An **ampere** is a measure of the number of electrically charged particles that flow past a given point on a circuit per second.

A **breaker box** or **breaker panel** houses the circuit breakers or fuses and distributes power to various parts of the house.

A **circuit** is all the wiring controlled by one circuit breaker. The NEC requires that lighting circuits be supplied with 15 amps of potential, distributed in size 14 wire (minimum). (See Basic Wiring Information, later in this chapter.) Outlet circuits are always 20 amps unless they supply a large appliance. Size 12 wire is the minimum allowed for 20 amp circuits. Most larger appliances are rated at 220 to 230 volts. These circuits have special outlets and double breakers, and the outlets eliminate the chance of plugging a 110 volt appliance into a 220 volt outlet. The number of light fixtures or outlets per circuit is governed by the NEC and local additions to it. In most cases your designer or electrical contractor will be familiar with these requirements, and your plans will be accepted the first time.

A **circuit breaker** is a protective device for each circuit, which automatically cuts off power from the main breaker in the event of an overload or short. Only a regulated amount of current can pass through the breaker before it will trip.

A **continuity tester** (see Figure 2-1) has its own power supply batteries, and is used to determine if a wire has any hidden breaks, or to locate the other end of a single wire or single circuit.

A **feeder** is a single long run of heavy cable that supplies a distant subpanel or large appliance with power.

Fish tape is a long, flexible metal strip with a hook to which you fasten the cable or wire to pull it through a raceway or conduit.

The **main breaker** turns the power entering your home through the breaker box on or off. It is sometimes found in the breaker box, or it may be in a separate box at a different location.

The above-ground electrical supply from the utility company must terminate at a **mast,** an appropriately sized metal conduit with a meter socket for your utility. The size of the mast is determined by the size of your main breaker and feed wires.

The **neutral bus bar** is the bar to which the neutral wire is connected in the breaker box.

Roughing in is the placement of wires and boxes before the interior walls and insulation are installed, and before the electricity is hooked up.

The power supply from the utility poles or underground cables to the mast is called a **service drop.**

A **subpanel** is a circuit breaker box that draws its power from a main breaker box. It is used to avoid having many long circuit runs.

UL stands for **Underwriters Laboratory.** For safety reasons, each component in your system should bear the UL listed stamp. Underwriters Laboratory is an independent testing facility that tests and then monitors those components to be sure quality is maintained. Each cable, switch, outlet, and panel should bear the initials "UL" or the phrase "UL Listed."

A **volt** is a measure of the current pressure at receptacles and lights. Average household voltage is 120. There are three types of branch circuits—120 volt, 120/240 volt, and 240 volt. Most circuits in the home that are not used for large appliances are **120 volt. 120/240 volt circuits** are used for appliances that may require a lot of electricity for one operation (for example, an electric range) and a smaller amount for another operation (the clock on the range). **240 volt circuits** are used for appliances that require a lot of electricity for all their operations, such as a water heater.

A **watt** is a measure of the rate at which an electrical device, such as a light bulb or appliance, consumes energy. Watts = volts X amperes.

WHAT YOU WILL NEED

Time. The time needed will depend on the scope of the project and on your level of experience. Two inexperienced workers would need at least 5 to 8 days to rough wire a 1500-square-foot house. This isn't difficult work but if you don't follow standard procedure and know how to troubleshoot for electrical problems, it can take much longer.

Tools. Some special (although inexpensive) tools are required for working with electricity.

Long-nose (needlenose) pliers

Wire cutters

Electric drill

Tape measure

Screwdriver

Chalkline

Hammer

Circular saw

Chisel

Hacksaw

Combination square

Utility light

Fish tape

Cable stripper

Wire stripper

Colored tape

Voltage tester

Continuity tester

Safety glasses or goggles

Keyhole saw

Utility knife

Right-angle drill

Pry bar and wood wedge

Materials. After figuring the lengths of cable you will need, it's a good idea to order 10 percent extra for wastage. Depending on the extent of the wiring you are undertaking, your list may include the following materials.

Grounded receptacles

Switches

Junction boxes

Nail guards

Wire nuts

Horseshoe nails (electrical staples)

Push terminals

Screw terminals

Breakers

Track lights and fittings

Dimmer switch

Waterproof junction boxes

Ground fault interrupters

Conduit

Cable

Silicone caulking

PERMITS AND CODES

Most states and municipalities use, and have additions to, the National Electric Code (NEC). Always consult the office of your local building inspector to determine what permits or special provisions must be met. All electrical work, no matter how small the job, must pass local codes. Be sure to get the proper permits, and be certain that you are clear on how to do your work so that it will pass code. Local codes differ, so don't rely on the information given in this book—it may not pass inspection in your area. Obtain a copy of the local building code by contacting the Building Inspectors' Association in your state capital; or check with the building inspector at your county court house or your city building department.

Some minimum design requirements common to most communities are:

1. Illumination: One circuit for each 500 square feet of floor space, and at least one fixture per room. It's a good idea not to have only one lighting circuit per floor in a multilevel building. If that one circuit fails, the whole floor will be in darkness.

2. Kitchens: At least two 20 amp branch circuits for countertop outlets, and one circuit each for a garbage disposal, a dishwasher, and the refrigerator. An electric range will require a separate 220 volt outlet.

3. Laundries: One 20 amp circuit for the washer and a 30 amp 220 volt circuit for an electric drier.

4. Outlets: Two circuits each for the family room, dining room, and breakfast room, and at least one each for the living room and the garage. The NEC requires outlets to be placed no more than 6' apart. Any wall over 2' wide requires an outlet, as does any countertop over 12" wide.

To some extent, meeting these requirements "designs" your installation for you. But don't forget that these are minimum requirements and may not fully address your particular needs.

Besides all these circuits, many people eventually have to add more later, for spas, pools, mood lighting, and so on. Be sure the main panel and subpanels are large enough to add future circuitry without replacing the panels themselves.

Codes specify the size and type of wire that can be used; the type and materials of the boxes; the distance between the supports for the wire; how far from the box the wire must be supported; and the size of the main breaker box, as well as many other details. The code also specifies that you must install lighting and receptacles in basements and unfinished spaces if they appear inhabitable to your inspector or planning department.

Some communities require this work to be done by a licensed electrical contractor. Never are inspectors more fearful of homeowners doing their own work than with electrical systems. The danger of electrocution, or of a house fire resulting from faulty wiring, is significant. Inspectors check electrical work very carefully. And they should. So be sure that all work is done neatly, to code, and in the manner inspectors are used to seeing it done.

DESIGN

A successful wiring project requires a plan so that you know exactly where you want your receptacles (outlets), switches, and fixtures to be placed, and the most efficient way to supply power to them. Since all of this work requires making holes in exposed framing, now is the time to add telephone cable, television cable, alarm wiring, and intercoms or other features. Just be sure to make a detailed plan, to improve your installation efficiency and so you will have a record of where you put these components.

The circuits for lighting, outlets, and heavy appliances are all isolated after the main panel. How many circuits you will need depends on the expected loads designed into the system. It's better to install the largest system you might need, and not the smallest you can get away with.

The **lighting circuits** usually terminate in ceiling fixture boxes. One 15 amp circuit provides current to as many ceiling fixtures as the 15 amp supply will carry, depending on their rated size (wattage). If there are two entrances to a room, plan for a light switch at both doors. Place switches on the unhinged side of the door. Determine the most direct route for cables, and route them accordingly.

The **outlet circuits** are either a long single run to one box, called a **dedicated circuit,** or a series of smaller pieces of cable "leapfrogging" from box to box; these are called **branch circuits.**

A **general purpose circuit** supplies power to both outlet receptacles and lighting fixtures. The code allows the use of a general purpose circuit anywhere but the kitchen or laundry. However, it is common to isolate lighting and outlet circuits in all areas of the home. That way, if you overload an outlet circuit and trip a breaker, you're not left in the dark.

Don't skimp on the receptacles. Aside from the danger of overloading outlets with extension cords and adapters, it can be just plain frustrating to have dark corners where you most need the light. Code usually requires 12' or less between outlets on the same wall. With this maximum distance, a 6' cord on an appliance or lamp can always reach an outlet without an extension cord. It will look better if you plan all of your outlets to be at the same height. This may be determined by local code.

The heavy feeder cable that supplies **subpanels and large appliances** is very expensive. Use the rules for notching and making holes, outlined later in this chapter, to determine the shortest possible route for these large cables.

Draw a rough plan, as in Figure 2-2, to show the general pattern of receptacles, switches, and fixtures and their circuitry. Your plan will assist you in making up your materials list and in calculating the amount of cable you will need.

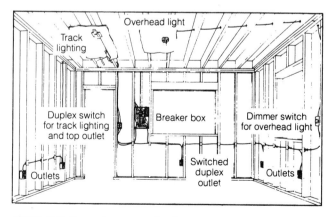

Figure 2-2. General purpose circuits.

BASIC WIRING INFORMATION

Before launching into step-by-step wiring instructions, this section reviews the basic information you need to know about wiring, boxes, receptacles, switches, circuit breakers, and fuses.

Wiring

A **cable** consists of several insulated **wires** wrapped in an outer sheath of insulation. However, the words *cable* and *wire* are often used interchangeably.

The American Wire Gauge (AWG) designation is a system for categorizing sizes and types of wire and cable (Figure 2-3). Wire sizes normally used in homes range from AWG number 10 to 14. The larger the wire, the smaller the gauge number. Wiring that runs inside the walls of a house is called cable. A run is a length of cable between two boxes. The type of cable depends on how the cable is to be used; the size of the cable depends on the amount of current (amperes) to be carried.

The larger the diameter of a wire, the greater the amount of current it can carry without overheating. Therefore, larger wires are used when larger current capacity is required. Every appliance or circuit is designated to carry a certain number of amps, and every size wire is rated as to how many amps it can carry. For instance, a 20 amp circuit requires number 12 cable, while a 50 amp stove requires a much larger cable. Check your local code for required wire sizes. Some current is lost within the wire, as heat generation. This is called **voltage drop,** and it can be significant in long cable runs. You may need to use the next larger size cable.

The insulated wires within the cable are made of either copper or aluminum, although copper is the more common. Aluminum circuit wire is not recommended because of the danger of fire, although it is often used to supply subpanels.

As explained earlier, wires are color coded so that you can tell at a glance which one is which. A black or red hot wire carries power to outlets, fixtures, and appliances. The white neutral wire carries the power back to zero potential by dumping it into the earth. A bare or green grounding wire provides backup in case of appliance failure or lighting surges to your supply lines.

Cable is referred to by the size of the wire and the number of conductors in the cable plus the ground. For example, if you are using a number 14 wire with two conductors plus ground, it is termed *14-2/w ground.* A number 12 wire with three conductors plus ground is termed *12-3/w ground.*

Type NM/NMB cable is often called by a manufacturer's name, Romex. It is sheathed in heavy protective plastic and paper with a thermoplastic covering on each wire. (Thermoplastic is plastic that won't melt from the heat of the electricity.) Type NM (nonmetallic sheathed) cable is used for most indoor wiring projects.

Type UF cable is covered with a plastic jacket that protects the insulated wires from sunlight and moisture. It can be bur-

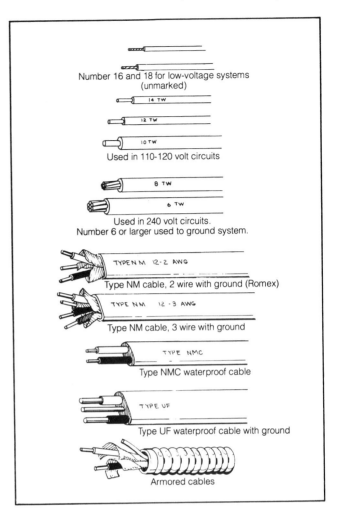

Figure 2-3. Various sizes and types of wires.

ied without any extra protection, and is used indoors and in damp areas. Terminations, spacing, and burial depth of this type is often specified by local code.

Armored cable type BX is a flexible spiral-wound steel casing enclosing two or three insulated wires and a bare, thin tapelike bond, or ground, wire. The steel casing protects against injury by nails or screws. BX cable requires special connectors and small plastic bushings, inserted into the end of the metal jacket to protect the wires' insulation at this sharp edge.

The metal jacket is cut off diagonally about 8" from the end of the cable with a hacksaw or a special BX cable-cutting tool. If you use a hacksaw you must take care not to damage the wires during cutting. This type of cable is subject to corrosion, and should be used only in dry indoor areas.

Metal conduit is rigid or flexible metal pipe or tubing that contains individual wires. It is sometimes required by code. Conduit is installed empty, and two or more wires are pulled into position with a fish tape.

Plastic PVC conduit for electrical systems is similar to the Schedule 40 PVC that is used in plumbing systems, but it is gray and uses different types of fittings. When wiring with

PVC, be sure to use only the long-sweep 90-degree bend fittings to make pulling the wire easier.

Boxes

All splices and joints in electrical wiring must be made inside of boxes, which are commonly made of steel, plastic, or fiberglass. Certain areas do not allow any but metal boxes. Many electricians prefer metal boxes because they can be disassembled if necessary to install a large group of wires. Some metal boxes can also be "ganged," or connected together to accommodate more than two switches at one location. The ground wires can also be attached directly to the metal boxes.

The purpose of the box is to protect the wire ends and to provide an area for any heat coming from the splice to dissipate. Boxes also make it easy to get to wiring connections; every box must be placed so that it is permanently accessible. All runs must be continuous from box to box. No splices may be made outside a box, and the number of wires permitted in each box is governed by code. If a switch box or outlet box is going to be too crowded, a larger junction box is used, with a piece called a **plaster ring** to reduce the visible opening to the original, smaller size. All boxes have **knockouts**—scored areas that can be knocked out to allow cable to enter and exit the box in the direction required. The knockouts are often of two or more different sizes, to accommodate large and small cables.

There are several different types of boxes, with different depths, for different uses.

Junction boxes are octagonal or square, in 3¼" and 4" sizes, with a cover plate that can be removed for access to the wires inside them.

Outlet/switch boxes are rectangular. Steel boxes can be joined together to make larger boxes. Plastic or fiberglass boxes come in different sizes to accommodate the number of switches or outlets needed.

Ceiling boxes are round and are screwed to a joist or header or attached to a crosspiece where a ceiling fixture is to be installed. They are covered by the fixture's trim ring.

Exterior boxes come in the usual varieties, junction, outlet, and switch, and are required in exterior locations. They are much more expensive than interior styles, and must be installed using appropriate weatherproof techniques.

Receptacles

The purpose of a wall receptacle is to tap the circuit to provide electrical power at a given location. The slots in the socket are designed to match the prongs of the plug on the appliance or extension cord; they vary according to the voltage and current (amperage) rating for the receptacle.

Receptacle boxes come in both flush and surface-mounted designs. Flush mounting is more desirable for inside use, except for the large receptacles used for the kitchen range or clothes drier.

Modern installations use duplex receptacles with a grounding terminal on a two-wire cable with ground circuit. These receptacles take either regular two-prong plugs or the three-prong grounding plugs found on many portable power tools and other appliances. The smallest slot is always the hot wire and the wider slot the neutral wire. The round hole connects to the grounding circuit. Some two-prong plugs have a wide tang and a narrow one to ensure that current enters the appliance correctly.

Unless you have nonmetallic cable in plastic boxes, each time a cable enters a box it must be clamped and secured. NMC clamps are available in different sizes to fit different cables, and these clamps must be secured to both the box and the cable to pass a rough electrical inspection. BX clamps are available in different sizes as well, and special 90 percent offset types are useful where clearance is a problem. The clamps and of different sizes and fit in the knockout holes.

These different sizes are intended to fit in the knockout holes in the boxes. Always be sure you are removing the appropriate size knockout plug for your clamp. Many panels have concentrically punched knockouts to accommodate larger supply or feeder cables. To remove the knockout at the right prestamped cut, just insert a flat screwdriver blade into the desired separation and lever or twist the screwdriver to separate the plug slightly from the other knockout rings. With a pair of needlenose pliers, twist or rock the plug until it comes off.

Wire nuts are commonly used to make a connection in a receptacle, switch box, or junction box. Like kinds of wires are spliced and twisted together clockwise with the ends snipped. The twisted wires are capped with a wire nut and turned clockwise to secure the connection. Wire nuts are color coded according to size. You must have the right size wire nuts to cap the number and size of the wire you are working with.

Switches and Switch Loops

Switches take many forms to meet many needs. For most residential purposes, switches come in single-pole, single-throw, three-way, and four-way. Most common is the single-pole, single-throw switch, which usually has three terminals. One terminal connects to the power supply wire, and one carries the power to the fixture. This type of switch controls the light from one location. Three-way switches have three terminals and can control a light from two locations. Four-way switches have four terminals and are used to provide light control from any number of locations between a pair of three-way switches.

A switch can serve a light fixture in two ways. It can be placed on the hot wire as a cable passes through the switch box, or a separate cable, called a **switch loop** or **leg**, can be run to the switch box to control the hot wire to the fixture. A switch loop is used when it is easier to take the cable to the fixture itself rather than through the switch box, such as when a two-story building has its circuit runs in the second floor framing, or when the runs are placed in the attic. Because the supply cable is already where the fixture will be, the switch loops reduce the effort and material that would be required to branch one circuit through three or more switch boxes.

For a single-switch control on a single fixture, the supply cable is routed first to the fixture box itself, and then another piece is routed to the switch box. The white and bare wires are attached directly to the light fixture itself, but the black wire is joined with a wire nut to the black wire of the cable going to the switch. This way, the switch controls the current supplied to the fixture. The white wire in the cable to the switch carries current to the fixture when the switch is turned on, and in this situation standard color coding is improper. You must use black tape, paint, or a felt pen to color both ends of the white wire in the cable to the switch, to indicate that they are hot wires. This is required by the NEC. The marked black end is then attached to the brass screw on the fixture.

Be sure to include this added cable, from the fixture to the switch boxes, in your material estimate.

Two- and three-way switches are basically the same as the single-pole type, but the power supply is routed through more than one conductor. These circuits require special cable, or more than one cable, to accommodate multiple controls to one fixture circuit.

Circuit Breakers and Fuses

A **fuse** is a strip of metal encased in a housing through which current passes into a circuit. The only type used in new construction is the **cartridge fuse.** Used with electric furnaces and air conditioners, cartridge fuses are manufactured to exceed their rated capacity for a short period, to accommodate the sudden peaks those appliances draw on starting. Cartridge fuses show no sign of overload, and must be tested with a continuity tester to see if they have blown. Fuses of different types should never be interchanged.

Circuit breakers are heavy-duty switches that serve the same purpose as fuses. The current is routed through a bimetal spring inside the breaker. When this spring gets too hot, it trips the breaker. When a circuit is carrying more current than is safe, the breaker switches to reset. On most breakers, the toggle-handle switch has to be flipped off and then on after the circuit trips. Always solve the problem before replacing a blown fuse or resetting the breaker.

A special kind of circuit breaker, the **ground fault circuit interrupter** (GFCI), is required by code outdoors and in areas around water, such as the bathroom and kitchen. If there is current leakage or a ground fault, the GFCI opens the circuit almost instantly, cutting off the electricity. This safety feature helps prevent shocks.

A GFCI breaker protects the entire circuit; the GFCI outlet protects only other outlets "downstream" of its location. When a GFCI breaker trips, it is reset in the same way as a regular circuit breaker.

BASIC WIRING

Most Common Mistakes

☐ While it is easy to make mistakes when working with electricity, it is just as easy to avoid them. The single most important mistake to avoid is neglecting to turn off the power before beginning. Other common mistakes include the following.

☐ Not making a plan for the work being done.

☐ Not learning about your community's specific requirements before you start.

☐ Overloading circuits by plugging too many appliances into an outlet, or by using an inadequate extension cord. (See the section on the breaker box, later in this chapter.)

☐ Not labeling circuits at the service panel.

☐ Not using UL-approved materials.

☐ Routing the wiring in an inefficient manner.

☐ Mounting outlets and switches without ensuring that they are flush with the final wall covering.

☐ Not using the correct housing box for the wiring that you plan to install.

☐ Not using weatherproof boxes for outdoor fixtures.

☐ Neglecting to seal around holes drilled through exterior walls.

☐ Not using nail guards where needed.

☐ Not having your work inspected at critical points.

☐ Failing to follow local code.

Step One
Installing the Main Breaker Box

Most Common Mistakes

☐ Using a breaker box that is too small to hold all the necessary circuits, including ones that will be needed in the future.

☐ Locating the breaker box in an inconvenient or unapproved area. (Code often designates where the box can be located.)

Caution: This is the part that even contractors often leave to electricians. If you are going to tackle this yourself, remember that the electrical inspector will scrutinize your circuit boxes and panel closely. This inspector must approve your rough electrical work and panel before your utility company will supply it with power. Work carefully and neatly; this is the heart of your electrical system.

The **breaker box** or **panel box** (Figure 2-4) holds the breakers for all the circuits in the house. It usually contains the main breaker as well. Although it is not required by code, the most logical place for the main breaker (the breaker that controls all the power to the house) is at the main panel. Sometimes, however, the main breaker is located in a separate box, or even at a different location.

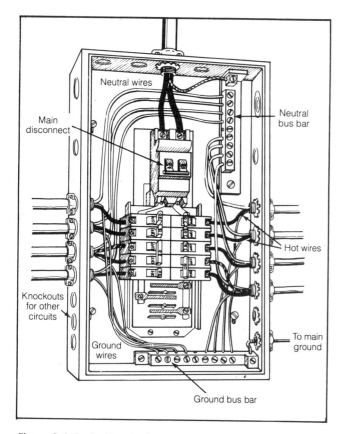

Figure 2-4. Typical breaker box.

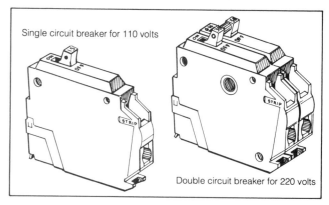

Figure 2-5. Single and double circuit breakers.

The main breaker performs two tasks. It provides a single location at which to turn off all the power, and it can limit the damage of lightning surges through the supply lines.

Circuit breakers (Figure 2-5) control the power going to a particular route of wiring. In case of an overload or a short, the breaker trips and automatically shuts off power to that circuit.

Panel boxes are available in both recessed and surface-mounted models. The recessed box is installed between two studs and securely fastened to them with screws. The box should protrude from the face of the stud by the thickness of the drywall.

After you have run the cables and installed the boxes (see next steps), attaching the wires to the breakers and buses in the panel is not difficult. Bring the neutral and grounding wires into the box and attach them to the neutral bus bar. Do this by sticking the stripped ends of the wires into any hole in the neutral bus bar; secure them by tightening down the screw heads. Some boxes have two bus bars, one for the ground wire and one for the neutral wire. Others have only one bus bar for both neutral and ground

The importance of proper grounding cannot be over-emphasized. Some communities require two grounding rods buried some distance apart. (A **grounding rod** is an 8′ copper-clad steel bar that is driven down into the soil and connected to the grounding bus.) These grounding rods are connected with one large grounding conductor to the grounding bus in the panel. This system provides an escape route for stray current in a faulty appliance so you don't become the conductor.

Code requires joining the neutral bus with the ground bus at *one place* in the system. This is usually accomplished by the metal of the panel box itself, but some panels come with a small strap to connect these buses. In addition, the water supply, gas or oil supply, and all pool and spa electrical equipment is bonded (connected) to eliminate stray current in these components. This is typically done with a bare number 6 copper wire and pipe-to-wire clamps. Larger services, subpanels, and appliances need larger grounding and bonding wires.

Hook up the circuit breaker switch to the black hot wire by tightening the screw in the breaker over the wire, as previously described. Now it's simply a matter of snapping the breaker back into position in the box and turning the main switch back on. There are two hot buses in the main panel. The breakers are installed evenly on these buses to avoid overloading one side. This means that each succeeding 110 volt breaker is staggered from one bus to another, and a 220 volt breaker connects to both buses.

Neatness is particularly important inside your panel. There is little point in labeling your branch circuits if their wires disappear into a jumble in the box. Take your time and lay out all the wires neatly.

Tighten all screws securely; loose terminals lead to strange electrical occurrences and rapid corrosion in the panel buses. Aluminum wire ends are often treated with an antioxidant.

Step Two
Placing the Boxes

Margin of Error: 1/4"

Housing boxes for outlets, switches, and lights are **roughed in** before the cables themselves are installed. In other words, all the outlets, the switch and fixture boxes, and the main panel and subpanels are installed before any wire is installed between them.

Choose the correct boxes for your needs. Check your electrical code for local recommendations. The boxes should be UL-listed and be large enough to hold the wiring, outlets, and connectors you will be using. Use a larger box with a plaster ring if necessary.

Use a ruler or other object of appropriate length, such as a screwdriver, to rest the boxes on when establishing installation points (Figure 2-6). Some boxes have nails attached to them that are hammered into the studs to secure the box to the stud. The octagonal fixture boxes used for light fixtures are attached to studs or to the ceiling joists. Ensure that your boxes will be flush with the final wall or ceiling covering by holding a small piece of the covering (drywall or paneling material) between the box and the wall stud while attaching the box. Switches are commonly located 48" above the floor. Countertop outlets are installed 42" from the floor, and regular outlets are placed so the top of the box is 14" above the floor.

Remove the knockout(s) in each box in the direction of the most efficient route for the cable to run, and install the boxes on the framing.

Step Three
Preparing the Studs for Cable

Margin of Error: 1/4". Always wire according to code. There is no room for error!

Drilling with a right-angle drill and an auger bit is the most common way to run the cable through interior, uninsulated walls. Some larger drills have right-angle attachments; or you can rent a right-angle drill.

Drill your holes in the center of the studs. Make the holes large enough to hold all the cables you will run through them. Brace yourself and the cumbersome drill so that it does not spin around if it binds.

Notching rather than drilling the studs works best when you have insulation you don't want to compress with cable. To notch a stud, follow this procedure.

1. Snap two horizontal chalklines 1" apart across the studs. These lines indicate the path the wire will follow.

2. Use a circular saw set to cut notches 1/2" deep at those lines on each stud.

3. Use a chisel to carefully remove the blocks from the notches.

4. After placing the cable, cover the notches with nail guards to prevent accidental nailing into the wire.

To avoid weakening structural members, the NEC states specific rules about the notches and holes used to route circuits. Notches in exterior (bearing) walls and plates may not exceed 25 percent of the stud width; in interior (nonbearing) walls and plates, up to 40 percent of the width is permitted. Holes in exterior walls and plates are limited to 40 percent of the stud width, and in interior walls to 60 percent.

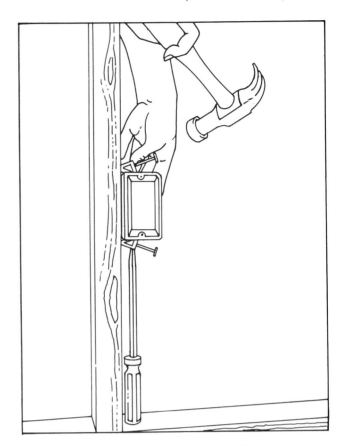

Figure 2-6. Use a ruler or screwdriver to set outlet boxes at a uniform height.

Avoid notching in all girders, headers, and beams. No top or bottom notching is allowed in the middle third of any joist. Notches within a third of each end cannot exceed one sixth the depth of the joist. No holes exceeding one third of the joist depth are ever allowed, and no holes in a joist should be any closer to the edge than 2".

Step Four
Installing the Wiring

Margin of Error: 1/4"

The NEC requires wire that is insulated and is of appropriate size for the application. The size depends on how much current the cable is expected to carry, and how far it must be carried. The most commonly used interior wiring is number 12 or 14 gauge NM sheathed cable, sometimes called Romex. Many electricians use only number 12 gauge wire even for the lighting circuits. When used behind 15 amp breakers, this provides an extra margin of safety. Of course, larger wire is required where demands will exceed 20 amps.

Within the cable are plastic-coated copper wires, colored for each function. (As explained earlier, hot wires are always black or red; neutral wires are always white; and grounding wires are bare or green.) In most cases your circuits will have to make sharp bends somewhere in their runs. Enough cable for the whole run is pulled through the straight portions of the framing and piled up just before the turn. The cable is then fed through the next straight series of holes until the cable(s) terminate. This is called **pulling cable.** Enough excess cable is provided at each end and intermediate box to be sure that components can be installed easily. Try not to let the cable twist in the corner bends or in the runs.

Beginning at the breaker box, or the hole where it will be, expose enough wire to reach the breaker switch and neutral bus bar. Use a cable stripper to prevent cutting the plastic coating on the wires (Figure 2-7). Knock out a tab in the breaker box that will provide a direct route to the switch for

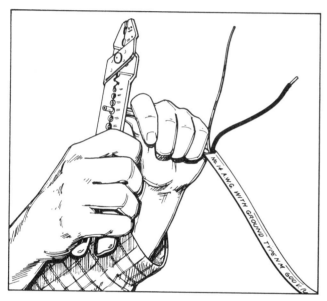

Figure 2-7. Use a cable stripper to remove insulation from wire ends.

the wiring, and knock out the tab in the housing box that provides the closest connection for each separate cable.

Push the wire through each hole or notch in the studs, keeping the wire smooth and free from kinks. After it is in place, secure it with horseshoe nails or staples at the notches. With holes, the staples are applied near the boxes, on the studs and joists, and at turns. The cable should never be pulled tight; some slack is necessary to allow for expansion and contraction of the building. Local code will tell you how close to the box the first horseshoe nail or staple must be (usually 8"), and how often the wire must be supported (usually every 4½').

Pull the cable to the box(es) and secure it with a staple. Put on the connector clamp and make sure that only the uncut sheathing is clamped at the box opening. Peel the jacket back far enough to allow at least 6" of wire to stick out of the face of the box.

The wire must reach from one junction box to another; do not tape two wires together to make them longer. If they must be joined, wires should be connected only at junction boxes. Code requires that a solid coverplate be placed over every junction that will not make use of a receptacle (Figure 2-8). This should not occur in a well-planned system except in basements, attics, and garages. If it does happen, consider putting outlets in those locations.

Provide at least 6" of wire to spare at each end of a run (Figure 2-9). Local code may vary on this length, so be sure to check. It's a good idea to label both ends of each wire with colored tape to make it easy to determine where it leads.

Where there is less than 1¼" between the face of the stud and the wiring, nail guards should be placed on the studs to

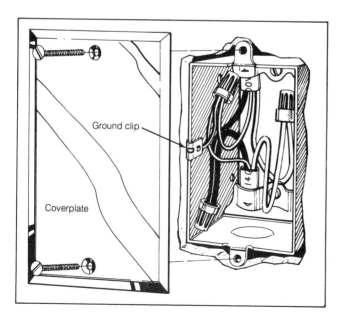

Figure 2-8. A junction box that will not make use of a receptacle must be covered with a solid coverplate.

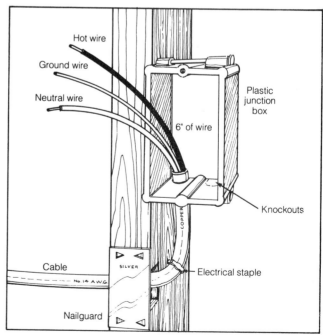

Figure 2-9. Typical box installation. Note staple, nailguard, and at least 6" of stripped wire.

protect the wire from the nails or screws that will attach the wall covering.

Now it is time to call in the inspector to check your work, before you complete the connections. This is called the **rough electrical inspection,** and it must be signed off before any of your circuitry can be covered with finish wall surfaces.

Step Five
Installing the Receptacles and Switches

Margin of Error: Exact; good tight connections, correctly done

Switches and outlets with push terminals or connections (Figure 2-10) make connecting the wiring easy, but they should be used only for copper or copper-clad wire—never for aluminum. Strip away insulation to the length indicated on the strip gauge when you push the color-coded wires into the correct push terminal. The terminal automatically clamps down when the wire goes in, so the fit is nice and tight. To release it, just insert a small screwdriver into the release slot.

Some communities, however, do not allow the use of push terminals. Check with your designer, electrician, or local authorities before installing push terminal devices.

Many electricians use screw-type terminals (Figure 2-11) because they feel that the screws hold the wires more securely. If you choose to use a screw-type terminal, strip only enough insulation for the bare end to be wrapped three-quarters of the way around the screw. With long-nose pliers, make a loop on the bare end wire to hook clockwise around the terminal screw, then tighten the screw with a screwdriver.

The black wire always is attached to the brass screw or the push terminal marked +. The white wire is always attached to the silver screw or the terminal marked -. The bare copper or green wire is always attached to the green screw.

The loop of wire is installed clockwise so that turning the screw to the right to secure it also tightens the loop itself.

Step Six
Connecting the Wires

Margin of Error: Exact; good tight connections correctly done

Wire nuts are used to make connections between wires of the same color. Wire nuts come in various sizes, distinguished by color, and may differ from manufacturer to manufacturer. (Yellow, red, or gray will cover most household uses.) Always select the proper-size connector for the wires being used; wire nuts should be used only where the connection won't be pulled or strained in any way.

When using wire nuts to make connections between solid wires, strip 1½" of insulation from the wires to be joined and hold them parallel. Twist these together securely with pliers in a clockwise direction. Then cut off the ends of the wires to fit the twist-on connector. Slip the wire nut over the bare ends of the wires and twist the nut clockwise around the wires, pushing them hard into the nut. Tug gently on each wire to be sure that it is secure, then wrap electrical tape around the wire nut and the wires connected.

Pigtailing

Simple wiring connections of two cables can be handled easily. More complex types of wiring, such as lights that have switches and junctions where the wiring is continued on to

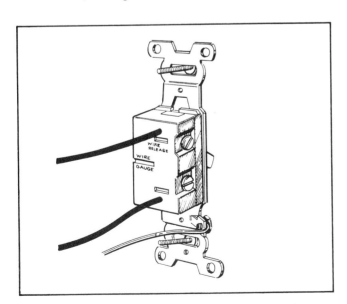

Figure 2-10. Push terminals are quicker and easier to use than screw terminals, but are not allowed in some areas.

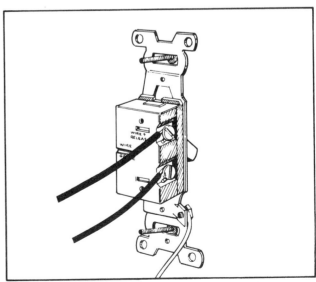

Figure 2-11. Screw terminals hold wire more securely than do push terminals.

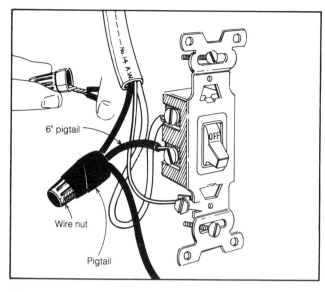

Figure 2-12. When pigtailing, always twist the wires together before attaching the wire nut.

other receptacles, involves more wires. To simplify such arrangements, wires are joined by a method called **pigtailing** (Figure 2-12).

Pigtailing connects two or more wires together with another 6" pigtail wire that has been stripped 3⁄4" on each end. The pigtail wire is the wire you connect to the outlet or switch. This reduces the number of wires to be connected at the receptacle. When you are using pigtailing, always twist the wires together securely before twisting on the wire nut. Following are some of the common uses of pigtailing.

Connecting Wires at a Duplex Receptacle

Strip all wires 3⁄4" and then bind all the wires of like color together with another 6" wire of the same color. Twist the ends of the wires being connected with the pigtail wire tightly together clockwise (see Figure 2-12). Then screw on a wire nut of the appropriate size. Check the security of your connection by holding the wire nut and giving a good tug to each wire. The bare grounding wires are also ganged together and joined with a wire nut, or with metal boxes, with a crimp ring. Crimp rings come in different sizes, and are required in some areas because of their superior holding power. All the ground wires are ganged and twisted, the ring is slipped over the ganged wires, and special pliers are used to lock this assembly together with the crimp.

Now it is a simple matter to connect the pigtail portion of the connection to the terminal—black to brass, white to silver, and the bare grounding wire to the grounding screw. Once pigtailed, it is easier to fold all of the wires together to fit them into the box. Then you can simply screw the duplex receptacle (outlet) onto the electrical box.

Splitting a Receptacle on a Push Terminal

Three cables (nine separate wires) come into the outlet box—one cable supplies power directly from the breaker box, another cable carries that power on to other receptacles or outlets, and the third cable comes from a wall-mounted switch so that half of that receptacle will be controlled by that switch. This type is often used in bedrooms, where you may want to control a lamp with a wall switch while still being able to plug in an alarm clock. Either the upper or the lower

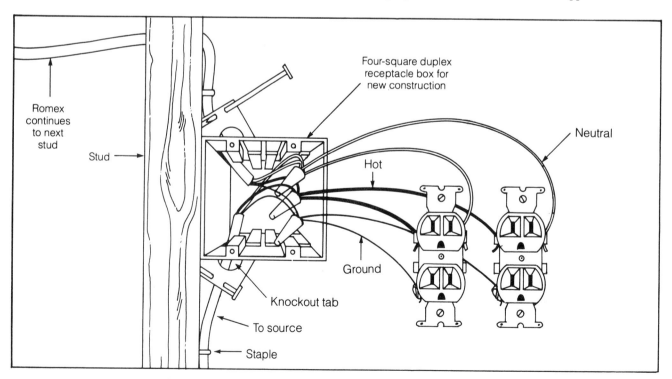

Figure 2-13. Wiring to split a receptacle so that one outlet is controlled by a wall switch.

socket always remains hot, while the other is controlled by a switch.

The white wire within the cable that goes to the switch will be hot, so it must be marked with black electrical tape at the outlet box, to distinguish it from the other neutral (white) wires. This same "hot white" wire should also be marked with black tape up at the switch; this is a switch loop, as described earlier.

Pigtail all of the same-colored wires together, as previously described, omitting the white wire that is marked with black tape.

Hook up the pigtail ground wire to the ground screw on the receptacle, and the pigtail neutral wire to a silver screw. (If it is a metal box, use two ground pigtail wires—one grounded to the receptacle and the other to connect onto the box grounding screw.) The hot pigtail wire goes into the permanently hot (lower) side. The white wire marked with black tape goes into the hot outlet side, upper or lower, depending on which socket you wish to control with the switch.

Breaking off the knockout tab on the hot side linking the two brass terminals of the outlet will allow half of the receptacle to work off of a switch, while the other half receives continuous power (Figure 2-13).

Wiring Switches

If your box and your switch are plastic, connect all grounding wires with a wire nut and push them into the box. If you are using metal boxes, pigtail all of the bare ground wires together with two separate pigtail wires. Then connect one pigtail wire to the grounding screw in the box, and the other one to the separate grounding screw on the switch. A switch is never connected to neutral wires; so join all of the white wires together with a wire nut and push them back into the box. Then pigtail the hot wires together.

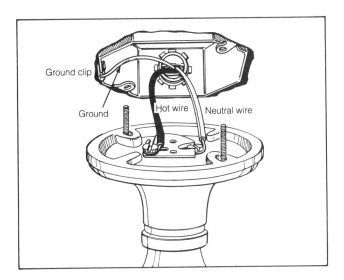

Figure 2-14. Wiring an end run ceiling fixture.

Because a switch interrupts the flow of electrical current, it should only be connected to hot (black) wires; or, in this case, a white wire marked with black electrical tape designating it to be hot. On the first switch, push the hot pigtail wire into either push terminal, and push the color-coded hot wire that goes to the overhead light into the push terminal of the same switch. On the second switch, push the black hot wire, which was pigtailed to the other black wires at the outlet, and the white wire wrapped in black tape into the back of the switch. This white wire will be made hot when the switch is turned on and will take the electrical power to the controlled outlet.

Step Seven
Hanging Light Fixtures and Receptacles

Margin of Error: Exact; well secured, properly connected, and UL approved

When you did the roughing in, the electrical boxes were attached to the studs or to the attic or second floor joist so that the plate is flush with the finished ceiling material. At this point you are ready to complete your wiring (Figure 2-14).

Pull the electrical wire through the cutout and box and strip it, just as you would for the outlets. Attach a surface-mounted hanging light with the attachment strap that is screwed to the electrical box. Follow the manufacturer's instructions. Connect the wires together, black to black, white to white, and ground to ground, using the wire nuts as previously described. Push the wires up inside the electrical box. Then it is a simple matter to attach the coverplate of the hanging light or the plate of the receptacle with the screws provided.

Track Lighting

Track lighting is connected to a ceiling box on one end.

Screw the mounting bracket into the electrical box. Push the wires of the electrical connector through the slot in the track connector, then through the slot in the box adapter. Now attach the wires to those in the box, matching colored wires and using the appropriate wire nuts. Push the wires up into the electrical box. Attach the two-piece mounting assembly to the box tabs (Figure 2-15).

Hold a ruler flat against the ceiling (or wall) so that one edge of the ruler is lined up with the center slot on the track connector. Using the ruler as your guide, draw a line along the ceiling (or wall) from this slot, as long as necessary, to where the track will end. Make marks at even intervals on the penciled line to indicate the attachment positions of the clips that will hold the track. Hold the clips in position, mark and drill pilot holes in the ceiling (or wall) at these points, and screw the clips in place. Secure the clips to the studs, rafters, or joists, or use toggle bolts if this is not possible. Partially

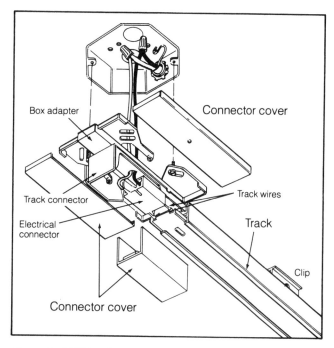

Figure 2-15. Wiring for a track light.

insert the side screws into the clips. Then push the track wires firmly into the electrical connector and slide the track into the track connector. Tighten the side screws of the clips to hold the track in place. The connector cover will then snap in place to cover the wires.

Recessed Light Boxes

Recessed lights are set into ceiling joists and secured so they will be flush with the finished ceiling surface. They come prewired from the socket to a small junction box built into the unit. This is due to heat buildup in these lights, and conventional cable joins high temperature rated wire in the box.

Outdoor Wiring

Outdoor wiring is essential for many uses outside the home—for security lights, porch and yard lamps, and tool operation, to name a few.

Basically, wiring for exterior use is the same as wiring indoor fixtures. However, exterior moisture-proof cables and boxes must be used. To simplify installation, choose an outdoor outlet location that is easy to get to, and close to an indoor receptacle.

Remember that the power supply can return to zero through the grounding wires, or through wet soil or concrete. You also have bonded your water pipes, gas pipe, and so on to the grounding bus in the panel. Since you can become the conductor if you are touching a faulty appliance and a water faucet, or if you are standing in a wet location, ground fault circuit interrupters (GFCIs) are required for safety.

Outdoor outlets, and those in the kitchen, bathroom, and garage, must include ground fault circuit interrupters. GFCIs measure the amount of power the hot wire brings in and the neutral wire returns. If there is a small difference, the outlet automatically shuts off, reducing your chances of receiving a fatal shock. A GFCI is a circuit breaker that is installed in the breaker box or in a special receptacle. The circuit of your exterior light is then attached into the GFCI breaker. The NEC requires GFCI protection on all outlets within 6' of all sinks or exposed plumbing, and within 12' of a pool or spa. Some local codes require GFCI protection for all pool and spa electric equipment.

Drill a hole through the exterior wall to take the power outside. Even though you want to be near the indoor receptacle, do not position the receptacles back to back. Run the cable or the conduit and install the appropriate junction box. Connect the wires as previously described, matching the color-coded wires to the correct terminal. Be sure to seal around the holes with silicone sealant, once the conduit and/or wires are in place.

Alternative Technologies

Code compliance is a matter of minimum requirements. As previously mentioned, you should install the whole system with future expansion or changes in mind. For instance, if you intend to use a personal computer in your den, consider running a dedicated circuit to that location to reduce current drops and peaks as other appliances cycle off and on.

The NEC also addresses solar-powered low-voltage systems in buildings, although the chances of fatal shocks and fires from these systems are currently quite low. As the costs of these technologies come down and efficiency rates go up, however, larger systems will become possible and concerns about safety will rise.

Nevertheless, photovoltaic and battery backup systems are easy to install and maintain. And they can be very reassuring to people who live in remote areas and to those who live where hurricanes or earthquakes occur.

3 Plumbing

EVEN MORE THAN installing an electrical system, installing the plumbing system is the most demanding task for the owner builder. A high degree of knowledge is required for this task. The *Uniform Plumbing Code* is extensive, and local codes, which vary from area to area, may take precedence. Every plumbing system is unique, and every home has its own problems that require special knowledge and experience. In addition, installing piping systems is physically demanding and can become extremely frustrating.

In the entire process of finishing a house, electrical wiring and plumbing installations are the areas in which you should consider hiring a professional. If you decide to do the plumbing yourself, you should at least have a professional consultant to oversee your work. It is also a good idea to read at least a couple of how-to books about plumbing and to keep them handy for reference.

Unlike the rest of this book, it is not the intention of this chapter to provide you with all the information you will need to install the plumbing system; that would require an entire book in itself. This chapter does not attempt to walk you through the plumbing process in a series of detailed steps. Rather, it offers an overview of the plumbing system and its components, some of the main code restrictions, and details of the actual cutting and assembling of the different types of pipe involved. With the information presented here, you should be able to speak more comfortably with your plumber about your specific needs.

Before You Begin

SAFETY

Special caution: Natural gas and propane pipe runs can look very similar to a water supply system because all three products are often piped in the same Schedule #40 galvanized steel piping. The gas piping begins at the gas meter and runs directly to the appliances and/or heating system. These lines should be clearly marked from the meter to the furnace or appliance. Some plumbing suppliers sell bright, easy to read identification tags that you can attach to the pipes. Hot water and steam heating systems require equal caution. All of these systems should be installed and repaired by an expert.

☐ *Have a fire extinguisher available at all times.*

☐ When using a torch to solder copper, wear a cotton or wool long-sleeved shirt and high-top shoes or boots.

☐ Use one of the several kinds of insulation materials available at the plumbing supply store to insulate the torch flame from the structure.

☐ Be careful about storing and discarding oily rags, which can cause fires through spontaneous combustion.

☐ When using a ½" drill to bore the large diameter holes needed for the piping, be sure that the drill is properly braced, according to the manufacturer's recommendations, so that if the boring bit binds up, the drill handles won't slam your hands into the wood.

☐ Wear safety glasses or goggles whenever you are using power tools or when soldering, chiseling, sanding, scraping, or hammering overhead, especially if you wear contact lenses.

☐ Be careful that loose hair and clothing do not get caught in tools.

☐ Keep blades sharp. A dull blade requires excessive force and can easily slip.

☐ Always use the right tool for the job.

☐ Don't try to drill, shape, or saw anything that isn't firmly secured.

☐ Don't work with tools when you are tired. That is when most accidents occur.

☐ Read the owner's manual for all tools and know the proper use of each.

☐ Unplug all power tools when changing settings or parts, and be sure that they are properly grounded.

☐ Keep all tools out of reach of children.

USEFUL TERMS

An **adapter fitting** is used to connect pipe of different types.

Air chambers or **air cushions** are piping configurations or devices that are used on the supply lines (usually at the fixtures) to reduce "water hammer," or water banging in the pipes.

The **building drain** is the part of the lowest piping of a drainage system that receives the discharge from soil, waste, and other drainage pipes inside the walls of the building and conveys it to the building sewer, which begins about 2' outside the wall.

A **cleanout** is a fitting that allows access into a drain, waste, soil, or sewer line so that an obstruction can be cleared.

A **closet bend** is a 90-degree fitting that carries waste from the closet flange, under the toilet, over to the closest drain.

An **escutcheon** is a doughnut-shaped ornamental plate that slides over a pipe to cover the hole in the wall where the pipe has penetrated.

A **female-threaded fitting** is any fitting in which the threads are located internally.

Finish plumbing is all the materials (and their installation) required to complete a plumbing system after the rough plumbing has been installed.

Fixtures, such as sinks, tubs, toilets, and bidets, are devices that are attached to the waste system and that require a fresh water supply in order to function.

A **male-threaded fitting** is any fitting in which threads are located externally.

A **reducer fitting** allows pipes of different diameters to fit together.

A **riser** is a water supply pipe that extends vertically one full story or more to convey water to branches or fixtures.

Rough plumbing is all parts of the plumbing system that can be completed before installation of the fixtures. This includes drainage, water supply, gas and vent piping, and the necessary fixture supports.

A **soil pipe** is any pipe that carries the discharge of water closets, urinals, or fixtures with similar functions, with or without the discharge from other fixtures, to the building drain or building sewer.

Sweating is a method of soldering copper pipe and fittings together.

A **trap** is a fitting or device designed to provide, when properly vented, a liquid seal that prevents the back passage of air without affecting the flow of sewage or waste water through the trap.

A **wet vent** is a vent that also serves as a drain.

WHAT YOU WILL NEED

Tools. Many of the tools that you will need for working with copper and plastic pipe and fittings are already in your tool chest or can be purchased or rented inexpensively.

ABS hand saw

Hacksaw

Reciprocating saw

Saber saw

Hole saw kit (sized for plumbing pipe)

Large boring bits (sized for plumbing pipe)

Adjustable wrenches

Basin wrench (telescoping)

No-hub ratchet wrench

Pipe wrenches

Chain cutter (soil pipe snap-cutter)

Hammer

Chisels

Screwdrivers

Tinsnips

Putty knife

Tape measure

Torpedo level

Chalkline

Tubing cutter

Plumber's sandcloth

Copper fitting brushes

Acid brushes

Propane torch

Striker (torch)

Goggles

Uniform Plumbing Code book

Materials. Communities differ in what piping materials you are allowed to use. A licensed plumber will know what is permitted in your area. If you are going to do the work yourself, check with the local building inspector and find out what is required. Some communities even have lists of fixture model numbers and manufacturers that you may or may not install. The inspector can also show you how to size your piping system from the tables in the *Uniform Plumbing Code (UPC)* book.

Guesstimating the amount of pipe and fittings you will need for a large job is close to a science. A good job schematic will help you estimate closer to the mark, with fewer trips to the supplier. Do some investigating to find out which suppliers in your area cater to the owner builder, and make sure that they will allow you to return any full-length pipe and extra fittings you don't use.

Some hardware stores have facilities for you to cut and thread pipe, and may even lend out tools like the soil pipe snap-cutter to customers who purchase supplies from them. These loans can be worth more than a 5 to 10 percent savings in materials from a supplier who does not offer such perks.

Of course, the materials you need will depend on the nature of the plumbing you plan to do. The following list should cover most of your needs.

ABS (or PVC) rigid plastic pipe (1½", 2", or 3")

ABS (or PVC) fittings

ABS solvent cement, cleaner, and primer

Traps

Pipe hangers

Stop valves

Closet flange and set of closet bolts

Wax ring

½" copper tubing

¾" copper tubing

Copper fittings

Faucet assemblies

Shower head

Tub spout

Plumber's putty

Roof vent

Water supply tubes

Teflon tape

Flexible tubing

Silicone caulk

Pipe joint compound

Standpipe

Filter washers

Vacuum breakers

Rags

Electrician's tape

Duct tape

Fixtures (bathtub, shower, toilet, sink)

PLANNING

Any project is only as good as the planning that goes into it. Before beginning any work, it's a good idea to map out your system in detail on graph paper, as shown in Figure 3-1. If you are not a licensed contractor, the building inspector may require a detailed plumbing schematic before issuing a permit. A detailed plan also makes it easy to compile your materials list.

Begin with the basement or crawlspace and sketch the building drain. Then sketch the branch drains, vents, and cleanouts. Trace the planned network of hot and cold water supply piping as well as the fuel gas piping, even if you are hiring a professional for this aspect. Locate your toilets, sinks, lavatories, tub, shower, hose bibbs, ice maker, wet bar, sprinkler system, and anything else you can think of that might need supply and/or waste lines.

What size water meter will you need? If your urban job is an entirely new structure, not a remodel of an existing home, then you can request a water meter of a particular size from the utility company to meet the requirements of the code. You may find that anything other than the basic 5/8" (inside diameter) water meter will be an expensive one-time proposition. However, if you have high water requirements, or if there is a long distance from the meter to the structure, you will have to provide a larger meter.

Other issues you need to consider include the municipal water pressure in your neighborhood, and whether or not the water pressure is adequate for your demands without a booster system.

In designing your plumbing system, it's important to keep the work simple, grouping fixtures so that they are as close to the piping runs as possible. It's a good idea to have a professional plumber go over your plans to be sure that you haven't forgotten anything or created extra work for yourself, and that all obvious problems are anticipated and worked out as much as possible *before the work begins.*

All rough plumbing should be installed before the electricians begin their work, and needless to say, before insulation

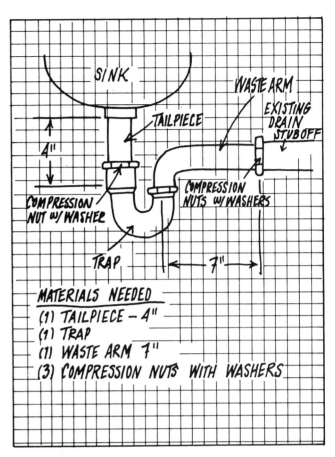

Figure 3-1. Sketch your proposed system in detail on graph paper, and use your drawings to make your materials list.

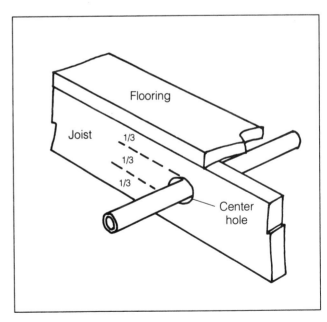

Figure 3-2. Running pipe through the center of a floor joist.

and wall coverings are in place. Your local code may require that all underfloor pipes (especially with slab construction), wires, and ducts be installed and inspected before the subfloor is put down. Some communities charge an additional

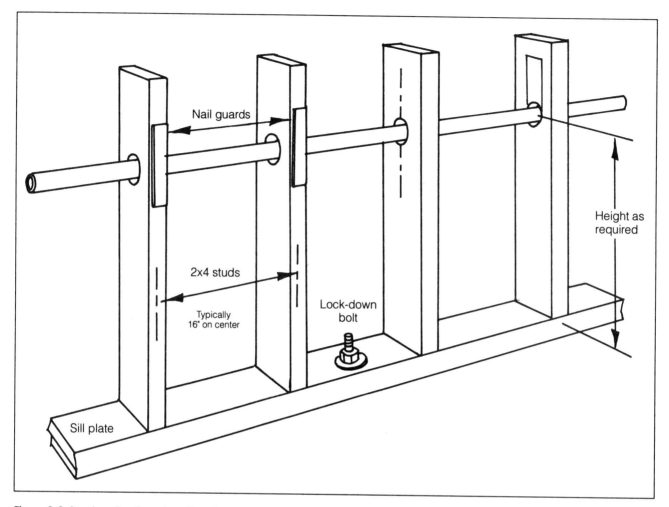

Figure 3-3. Running pipe through wall studs.

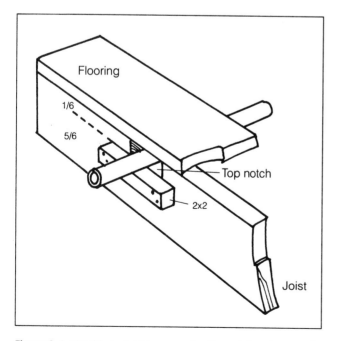

Figure 3-4. Notching a joist to run a pipe through it at the top of the joist.

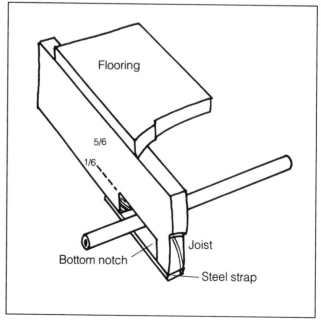

Figure 3-5. Notching a joist to run a pipe through it at the bottom.

fee for more than one rough inspection. Even if the code does not require it, it is much easier to install any underfloor piping before the subfloor is laid, especially if you have a low crawlspace.

For the structural integrity of your building, it is always best to run pipes *parallel* to framing members, especially floor joists. However, it is sometimes necessary to cut into or through the joists at some point. Always check your local building code before doing this. The code may include such regulations as the following:

☐ If a pipe must run through a floor joist near its center, you can usually drill a hole, as long as the hole is less than one-third the depth of the joist (Figure 3-2). Studwall may have holes drilled up to 40 percent of the stud depth in load-bearing walls and up to 60 percent in non-load-bearing walls (Figure 3-3).

☐ If a pipe runs through a joist near the top (Figure 3-4) or bottom (Figure 3-5), you can notch the joist. The notch should be no greater than one-sixth of the depth of the framing member. Nor should the notch be located in the middle third of the span. In studs, you may notch up to 40 percent in non-load-bearing walls.

☐ If you are notching the framing member on the top, nail a length of 2x2 under the notch on both sides of the member to give added strength.

☐ Bottom notches should be covered with a steel strap for added support.

☐ The larger DWV pipes may necessitate the removal of an entire joist or stud. Check your local building code for specifications in this situation.

CODES AND PERMITS

The plumbing industry has established standards to protect the health and safety of the community. Plumbing codes and the permit process exist to make sure that those standards are adhered to. Faulty plumbing can result in serious health and safety hazards such as fuel gas leaks, sewer gas leaks, raw sewage leaks, flooding, and electrical shorts. However, regulations governing design, methods, and materials differ from community to community. Nationwide, most communities adhere to the *UPC*, with local codes superceding it in various areas.

The main areas of code enforcement include the following issues:

☐ Type of piping materials.

☐ The proper size (diameter for length) of water supply, drain/waste/vent, and fuel gas piping.

☐ Slope of drains (inches per foot of fall).

☐ Distance from fixture traps to vent pipe.

☐ Use of certain fittings, such as sanitary T's versus Y's.

☐ Placement and number of cleanouts.

☐ Distance between pipe supports.

☐ Height of laundry standpipe traps off the floor.

☐ Height and slope of horizontal vent piping off the floor.

Contact your building inspector about any local codes that may supercede the *UPC* and about the permits that you will need. Some communities have a separate inspector for the building sewer and the outside of the structure.

In some areas plastic pipe is prohibited entirely. In other areas only ABS, not PVC, is permitted inside the structure for the drain/waste/vent system. Some codes may require you to insulate your hot water supply piping or to install vacuum breakers on your outside hose bibbs. Before beginning any work, be sure your plans conform to all local codes and ordinances. Discuss your plans with the local building inspector and find out what work you may do yourself and what, if any, must be done by a licensed plumber. Be sure to follow the code to the letter, or you run the risk of having to rip out all your hard work.

INSPECTIONS

Your plumbing work will require inspections at at least two and maybe even several stages, depending on the overall construction sequence of your particular project. Usually there are two basic plumbing inspections: rough and finish. **Rough plumbing** is all the work that goes into the project before the fixtures are installed, including the water supply systems, drain/waste/vent system, any fuel gas lines, and gas appliance flue venting. Do not cover up any of your rough plumbing work until the inspector has seen it, or you may have to uncover it for inspection.

The inspector will require a pressure test on the water supply, DWV, and fuel gas systems. For the water supply and DWV, you can usually choose either an air pressure test or a water test. Fuel gas systems can only be tested with air pressure.

After signing off on the rough plumbing, the inspector will look at the **finish plumbing.** It is at this point that the hook-up to the municipal sewer main or septic system is usually performed. Some communities require that a licensed sewer contractor make the connection to the municipal sewer main.

THE PLUMBING SYSTEM

The water-related plumbing system in your home, shown in Figure 3-6, has three main functions:

1. To carry fresh water through the supply system.

2. To use this water in the fixtures.

3. To dispose of sewage and liquid waste through the drain/waste/vent system.

To make it easier to see how each part functions, the hot and cold water supply system (Figure 3-7) and the drain/waste/vent system (Figure 3-8) are separated out from Figure 3-6.

The Supply System

The supply system carries fresh water from a well or municipal water main into your house and distributes it to your fixtures. A main shut-off valve is usually installed in the main water supply line to stop the flow of water into your home. If your water is supplied by a private well, the shut-off valve will probably be at the wellhead, or at the house where the main supply enters the building, or at both places. If your water is brought to your site by a municipal main, then you will have a water meter with a key-operated shut-off valve installed directly in front of the meter. If this is the only shut-off valve, you will need a water meter key to operate the valve. However, you will probably also have a hand-operated shut-off valve where the main supply line enters the house. Depending upon the weather conditions where you live, this hand-operated valve might be outside, at or near ground level, or inside the basement or crawlspace.

Since the 1960s, copper has become the most common material for fresh-water supply lines, due largely to ease of installation. Don't attempt to install cast iron or galvanized pipe yourself. These require the skill of a professional. As

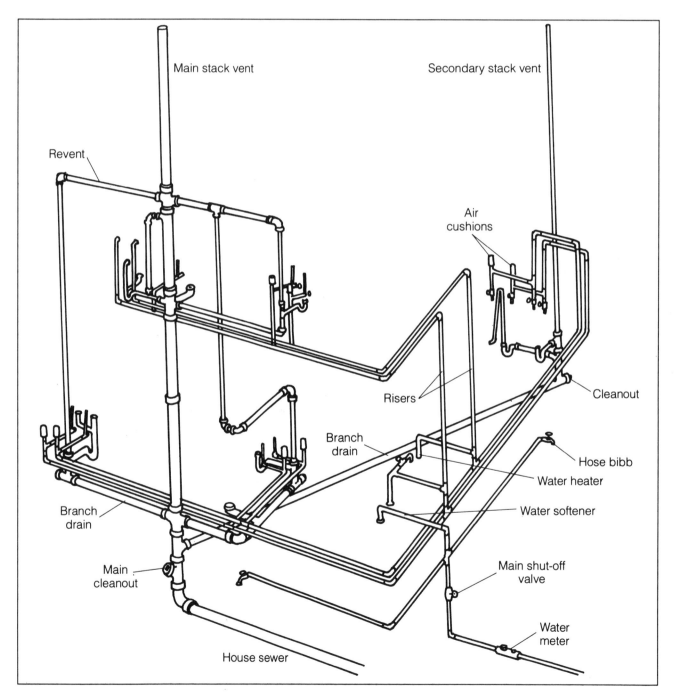

Figure 3-6. Hot and cold water supply and drain/waste vent system.

shown in Figure 3-7, the supply line enters the average-sized modern house in a ¾" pipe (inside diameter). This ¾" line continues, full size, to the cold inlet to the water heater. On the way to the water heater, branch lines lead off from the main, taking *cold water* to the various fixtures. Depending upon the distance of each fixture from the main, the branch lines may start out as ¾" and then reduce to ½", closer to each fixture. A ¾" *hot water* line leaves the water heater and threads its way through the building, with branch lines taking hot water to the fixtures. The diameter of the sections making up these branch lines depends upon the "developed length" between main and fixture.

If your home has a water softener, it is customary to soften only the hot water, and perhaps the cold water for the kitchen sink. These systems can require a fair number of gate or ball valves and check valves to isolate and direct flows. You should produce a good schematic of the finished design so that you can understand your system years from now.

Supply pipes that rise vertically from one floor to another are called **risers**. All supply lines are pressurized—although sometimes not as high as you'd like. The water supply is pressurized at its source, and this pressure drives the water throughout the system. In very cold climates, it is customary to install supply lines on a slight pitch, sloping back to a low point where valves may be installed so that the entire system can be drained down to make repairs on pipes damaged by freezing, or to drain the system for times of winter vacancy. It is very difficult to solder copper pipe that has water in it. When joining copper pipe to galvanized pipe, a **dielectric union** must be installed to prevent corrosion. Even the pipe

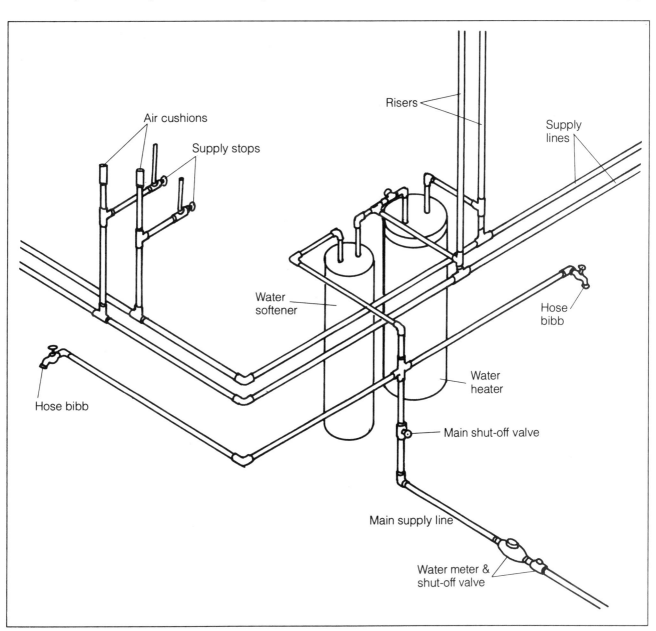

Figure 3-7. Hot and cold water supply lines.

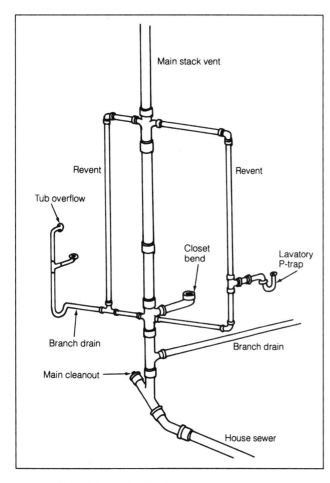

Figure 3-8. Drain/waste/vent system.

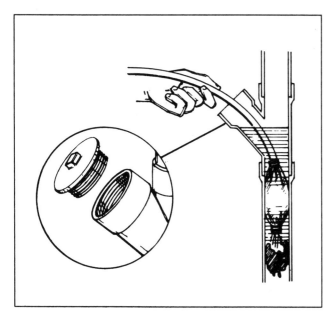

Figure 3-9. Cleanouts at key positions in drain pipes allow obstructions to be cleared away.

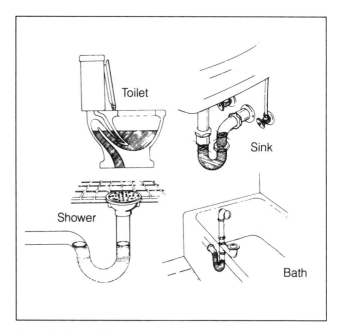

Figure 3-10. Each fixture must have its own trap to prevent noxious gases from coming up the pipes into the house.

hangers and pipe straps used to support copper pipe should be solid copper, or at least plastic coated or felt lined.

The Drain/Waste/Vent (D/W/V) System

The drain/waste/vent system (Figure 3-8) moves sewage and liquid wastes out of the house to the building sewer, from where it runs to a septic system or municipal sewer. Because the drain/waste system is driven by gravity rather than by pressure, it is harder to understand than is the water supply system.

ABS plastic is the most common material for residential DWV systems. The *Uniform Plumbing Code* establishes a minimum slope of ¼" per foot for drain/waste piping. If the slope is much greater, over a long distance the liquid waste may travel faster than the solid waste, leaving it behind to block the pipe. If the pipe does not slope enough, drainage is not sufficient and *all* wastes may back up. The drain/waste piping connects to the main drain, or **building drain**, which becomes the **building sewer**, once it is outside the structure.

Occasionally a waste line, drain line, or sewer line will clog, for any of several reasons: improper fall (slope); improper number of type of fittings; a clogged vent; objects such as toys being flushed down the toilet; or that common sewer stopper, roots. For this reason, **cleanouts** are placed at designated locations in the drain/waste system. Depending upon the type of piping and method used, a cleanout can be a hubbed female fitting with a male screw plug to seal it, attached at the end of a pipe run; or it can be a blind cap held to the end of a piece of pipe by a removable coupling. It can also be a branched fitting with a threaded branch and plug,

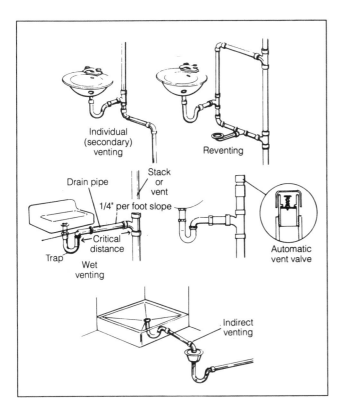

Figure 3-11. Vent types.

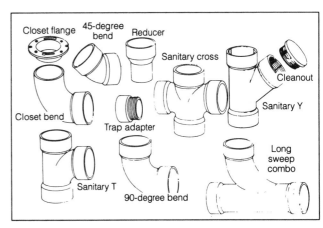

Figure 3-12. Plastic fittings.

these fixtures are attached to the waste system and to either the cold water supply line or to both the hot and cold lines.

Fittings and Valves

Figure 3-12 shows a sampling of various kinds of **fittings**—the connecting links that join pipes. They are used to connect lengths of pipe in a straight run, to change the direction of pipes, and to connect two intersecting pipe runs together. Common categories of fittings include the following:

A **coupling** joins two pipes in a straight run.

An **elbow** or **bend** changes the direction of a pipe run.

A **T fitting** joins two perpendicular pipes.

A **Y fitting** joins two intersecting pipes at 45 degrees.

A **union** is a straight, three-piece fitting with female threads at opposite ends and both male and female union threads in the middle, which allow the fitting to be broken in two.

A **reducing coupling** joins a larger diameter pipe to a smaller diameter one.

A **bushing adapter** allows smaller diameter pipe to be joined with larger diameter fittings.

A **cap** covers the end of the pipe.

A **plug** fits inside a threaded fitting, sealing off internal access to the pipe run.

Valves are used to control the flow of water and gas in the plumbing system. Common types of valves found in residential plumbing are the shower valve, sink faucets, angle stop (stop valve), gate valve, ball valve, and globe valve.

Air Chambers

Water supply pipes will sometimes "hammer" when a valve is shut off suddenly. This noisy, irritating condition occurs most often when water pressure is high. Because water does not compress when its flow is stopped suddenly, unsupported piping may bang against structural members; or worn valve parts may begin to "chatter." Instead of a pressure-re-

as shown in Figure 3-9. The criteria for the placement of cleanouts is complex. Section 406 of the *UPC* explains the many variables involved.

The **vent system** is part of the drain/waste system. Vents are designed to let atmospheric pressure into the drain/waste system so that waste can flow out of the house. (This is much like opening a can of juice. You punch two holes in the can, one for the juice to flow out and the other to let atmospheric pressure in.) To ensure a safe vent system, each fixture is protected by a water seal in the trap. Without the trap and water seal, sewer gas could enter the building through the fixture (Figure 3-10). The vents terminate above the roof, sending any gases safely skyward.

All plumbing fixtures must be vented and trapped (Figure 3-11). The minimum size of the vent pipe is specified by code. Generally, a toilet vent pipe must have a minimum diameter of 2". Kitchen sinks, tubs, showers, and laundry standpipes require 1½" vents, while a single lavatory basin needs only a 1¼" vent. The section of the *Uniform Building Code* called, simply enough, "Useful Tables" will help you to determine the size vent pipe you are likely to need for your fixtures. Once again, however, your local code takes precedence over the *UPC*.

The Fixture System

The fixture system includes all fixtures in the home that use water, including sinks, toilets, showers, and bathtubs. All of

ducing valve, you may try installing air chambers (water-hammer arresters) at fixtures. These manufactured devices have a factory-charged air bladder that absorbs the line shock in the pipe. Some designs have an air valve that allows you to increase the internal pressure with a tire pump.

INSTALLING THE DRAIN/WASTE/VENT SYSTEM

Most Common Mistakes

☐ Violating code restrictions.

☐ Not installing the DWV system with horizontal slope of at least ¼" per foot.

☐ Not properly venting or trapping all fixtures.

☐ Exceeding the fixture unit capacity of a drain or vent.

☐ Not providing enough cleanouts, or not providing cleanouts at the prescribed places.

☐ Venting the fixture too far from the fixture's trap.

☐ Reducing pipe size downstream.

☐ Not allowing for the proper insertion distance in the fittings when cutting pipe.

☐ Misunderstanding code requirements.

The type of pipe you use in the DWV system will depend on your local code. Plastic pipe is the easiest to work with, but it is not acceptable in some parts of the country.

Two types of plastic pipe are commonly used in DWV systems. PVC (polyvinyl chloride) is white to off-white in color and has the words "PVC" and "Schedule #40," along with other industry standards, printed in a continuous stripe down its entire length. ABS (acrylonitrile-butadiene-styrene) is the preferred type; it too has the material and schedule printed down its entire length.

Plastic pipe is less expensive, lighter weight, and easier to work with than cast iron or DWV-weight copper pipe and fittings. A transition cement is available to join ABS to PVC, but most codes demand that you stay with one type of plastic pipe. Most communities require Schedule #40 ABS or PVC pipe and fittings for residential DWV use. Both materials are manufactured in 20' lengths; wholesale suppliers generally sell it at this length. Shorter, cut lengths are often available at home improvement centers and hardware stores.

One of the most difficult things about installing any residential DWV piping is boring the large 1½" to 4⅝" diameter holes that are necessary to run pipe through plates, studs, and joists. You will need special heavy duty plumber's bits and hole saws and an extremely powerful drill motor to accomplish this feat. You can rent the drill motor (Milwaukee Hole Haug), but the large-diameter bits are so expensive and easy to break that you will probably have to buy your own.

Cutting Plastic Pipe

Cutting plastic pipe by hand is best accomplished with a handsaw made expressly for this material, commonly referred to as an **ABS saw**. Modern "chopsaws" also accommodate a blade made expressly for cutting plastic. When cutting the pipe (Figure 3-13A), remember to take into account the distance that the pipe is to be inserted into the fitting on each end. Trim the rough edges with a small knife and finish with rough sandpaper to remove the burr from the freshly cut pipe (Figure 3-13B).

With a rag and cleaning solvent, clean the ends of the PVC pipe and the inside of the fittings where the pipe will join (Figure 3-13C). (You do not need to use solvent with ABS pipe; it can be wiped clean with a clean rag.)

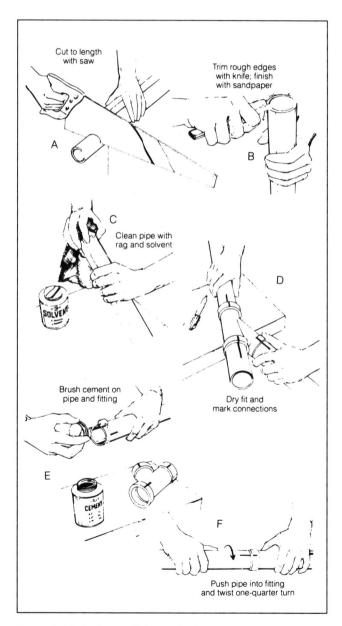

Figure 3-13. Cutting and joining plastic pipe.

When you glue the pipe and fitting together, the glue will dry almost immediately and you can never get them apart. To change the connection, you must cut off the outer fitting and start over. Therefore, you want to get it right the first time. This is easy in some cases, but often you will have a number of pipes coming into one fitting from different angles. To ensure that all of these elements meet correctly once they are permanently glued, you should "dry fit" the fittings and pipe by cutting and assembling the pieces without glue to be sure that everything fits correctly.

Make marks across fittings and pipes so that you will know exactly how to realign everything when the glue is spread and the pipes are inserted permanently into the fitting. Be sure to make the marks long enough so that they will not be covered by the glue you will spread on the pipe (Figure 3-13D).

Spread a generous amount of pipe cement around the end of the pipe and on the inside of the fitting (Figure 3-13E). Finally, insert the pipe into the fitting until it "bottoms out" and give it a little twist—about a quarter turn—to be sure that the cement is spread evenly (Figure 3-13F).

INSTALLING THE WATER SUPPLY SYSTEM

Most Common Mistakes

☐ Violating or ignoring code restrictions.

☐ Using undersized supply pipes.

☐ Moving or knocking copper pipe while the newly sweated fittings are still hot.

☐ Leaving materials smoldering after sweating fittings, thereby creating a fire hazard.

☐ Forgetting lines for such things as outside hose bibbs, ice makers, and wet bars.

☐ Placing pipe supports too far apart.

☐ Trying to solder a pipe joint when water has not been completely drained or contained.

☐ Failing to run the outside hose bibbs to bleed dirt and air from the lines after turning the water on.

The water supply lines are generally made of copper tubing and fittings. Cooper is easily joined by soldering, or sweating. It resists corrosion when not contacting or mated to ferrous metals, and its smooth surface minimizes resistance to water flow and small particles that could become lodged in the pipes. Although supply piping is not required to slope the ¼" per foot that DWV piping is, a slight slope makes it easier to drain and repair when necessary.

Hard copper supply pipe is sold in lengths of 20'. This kind of pipe is most commonly used for supply systems. Bending hard pipe results in crimping; therefore it must be cut and joined with fittings whenever a change of direction is made. It comes in three weights: K, L, and M. K is the

heaviest and M is the lightest. M can be used in most areas, above the ground and within the structure.

Soft copper is sold in 60' coils. Although it is more expensive, it can be bent around corners, which makes it easy to lay in trenches. It is used primarily for the main building water supply from the meter or wellhead to the house. It is not allowed for use within the structure in most communities. The rolled copper is available in K and L weights. Check with your building inspector to find out which type is specified in your area.

PVC and CPVC schedule #40 plastic pipe is manufactured in 20' lengths. Most communities only allow plastic main supply lines, in the ground, up to the structure. However, a few areas do allow it for fresh water supply within the building.

Cutting and Soldering Copper Pipe

Figure 3-14 shows several kinds of copper fittings, which are joined to the pipes by soldering. The supply lines are usually installed after the DWV piping. The supply system has more flexibility, and the smaller-diameter supply lines can be run over, under, and around the larger DWV piping. It is generally easiest to begin where the water source enters the house, usually with a ¾" pipe from the source to and through the house. However, don't take any chances; design your system according to the sizing Tables 10-1 and 10-2 in the *UPC*. To install the pipe, you simply begin at one end and work your way to the other end. As the pipe is run, supports are installed according to code, at intervals of about 6' to 10'.

Insertion distances vary for different types of joints, although ½" pipe usually inserts ½" and ¾" pipe inserts ¾". Cut your pipe with a tubing cutter. Place the pipe in the opening of the tubing cutter and twist the knob until the cutting wheel just contacts the pipe. Then rotate the cutter around the pipe, tightening the knob after each revolution until the pipe separates in two. Whether you are using hard

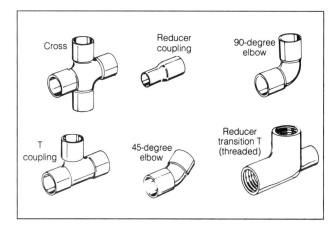

Figure 3-14. Copper fittings.

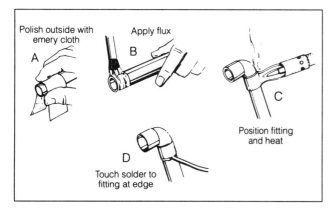

Polish outside with
emery cloth

A

Apply flux

B

C

D

Position fitting
and heat

Touch solder to
fitting at edge

Figure 3-15. Soldering copper pipe.

or soft copper tubing, be careful not to damage it as you work.

Use the special blade on the back of the cutter to ream out the burr on the inside of the newly cut pipe. Then use plumber's sand cloth to polish the last inch of the outside of the pipe (Figure 3-15A). Use a wire fitting brush to clean the inside socket of the fitting. It is important to clean both the pipe and fitting thoroughly. The time you spend on this step now will save you a lot of time later fixing leaking joints. It's much easier to do it right the first time.

To solder your joints you will need an acetylene or propane torch, a striker, flux and acid brush, and some lead-free solid wire solder. Apply the flux with the acid brush on the inside of the fitting and around the outside of the pipe end (Figure 3-15B). Place the fitting on the pipe, twisting it back and forth a couple of times to ensure even distribution of the flux.

Heat the bottom of the pipe and fitting with the torch flame. Slowly pass the torch back and forth across the fitting to distribute the heat evenly (Figure 3-15C). Be careful not to get the fitting too hot, or the flux will burn away to nothing. To avoid overheating, touch the solder to the joint occasionally as you heat it. The moment the wire melts on contact and doesn't stick, the joint is ready.

Remove the torch and touch the soldering wire to the edge of the fitting (Figure 3-15D). Capillary action will cause the solder to pull in between the fitting and the pipe. Continue to solder until a line of molten solder shows all the way around the fitting. Be sure there are no air gaps between the solder and the fitting.

Wipe the excess solder off the surface with a clean soft rag before it solidifies, leaving a trace of solder showing in the crevice between fitting and pipe. Keep your hands well away from the heated joint, and take care not to bump or move the newly soldered joint until it has cooled.

Completing and Testing the System

Run hot and cold lines to all of the planned fixtures, and at the fixture locations screw galvanized caps onto brass nipples that are threaded into special fittings called "½" FIP by ½" copper winged 90s" that are soldered to the end of each line. Now the system can be charged with water and tested for leaks.

The tub and shower valves are usually installed in the rough phase of the plumbing. You may hook the water up to an existing main line from the meter or borrow a neighbor's water by using an adapter to hook a garden hose to your cold water supply line at the edge of the building. Be sure to leave the tub and shower valves on the "warm" setting so that cold test water can get into the hot side also and test those lines and connections. If the tub and shower valves are not installed yet, you can use a loop at the water heater's roughed-in connections to connect the hot and cold lines together here.

You'll have to drain the system of water before you can repair any leaks you may find. You can try reheating the joints, applying more flux, and then resoldering. If that fails to do the trick, you may have to cut out the offending fitting and sweat in a new fitting, using repair couplings that slide down the pipe. (It can be very difficult, or even impossible, for an inexperienced plumber to get a previously soldered fitting off the pipe by trying to reheat it and pull it free.) If you cut open the line for repairs and water continues to show up, making soldering impossible, shove a chunk of white bread into the pipe upstream of your joint. Work quickly while the bread is absorbing the water. When you turn the water on again, the bread will disintegrate and flow through the pipe and out into a bucket before you install the finish plumbing materials.

INSTALLING THE FIXTURES (FINISH PLUMBING)

This section discusses general instructions and precautions for roughing in your fixtures, as well as installation procedures for tying into the DWV and supply systems.

Bathroom Sink

The most common sink styles are cabinet-mounted, pedestal, and wall-hung. Some sinks have no holes for the valve; others have holes 4", 6", 8", or more apart for 4"-center sets, and 6" or 8" for kitchen faucets. Lavatory bowls and counter tops may have holes spaced up to 14" apart for faucets. Some sinks also have holes for spray attachments and lotion dispensers. Standard height from floor to sink rim for lavatory sinks is 31" to 32", and 36" for kitchen sinks. Common ma-

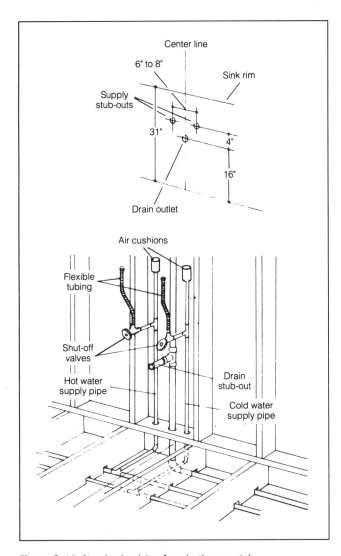

Figure 3-16. Rough plumbing for a bathroom sink.

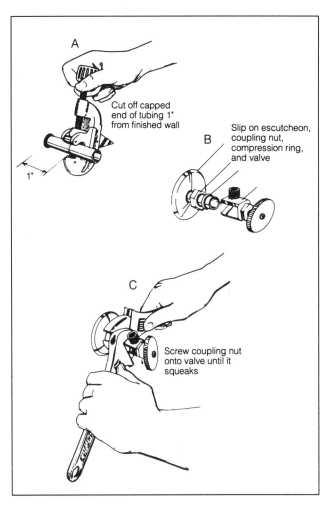

Figure 3-17. Installing a shut-off valve.

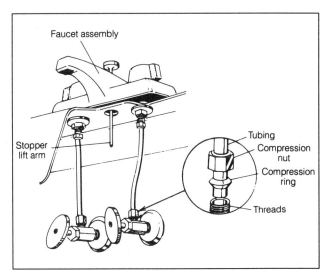

Figure 3-18. Faucet assembly.

terials include vitreous china, enameled cast iron, enameled steel, synthetic marble, and various forms of plastic.

Pipes required for roughing-in the bathroom sink include hot and cold supply stub-outs, shut-off valves, transition fittings, and possibly flexible tubing to go above the shut-off valves. Air chambers may also be required. When the roughing-in has been completed and you are ready to install your sink, your rough plumbing should resemble that shown in Figure 3-16.

Depending on the local code, you may or may not be able to wet vent the lavatory basin. Clearance from the side of a bathroom sink to a toilet tank or finished wall should be at least 4", while the distance to a tub may be as little as 2". There must also be a minimum of 21" from the front edge of the sink to a wall or fixture.

When cutting the capped lines to install your compression shut-off valves, cut the ½" copper supply pipe at least 1½" from the finished wall to allow for an escutcheon and shut-off valve compression nut and ferrule, and still have the supply pipe bottom out in the valve.

Use plumber's sand cloth to clean any drywall mud or paint off of the pipe. Cut slowly and carefully so that you don't deform the pipe (Figure 3-17A). You can also use a

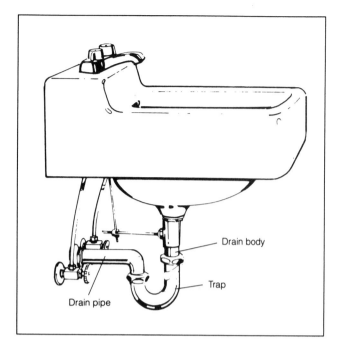

Figure 3-19. Trap connected to the drain body and drain pipe.

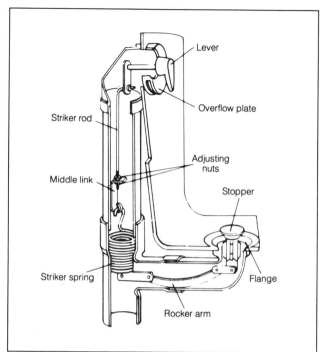

Figure 3-21. Drain overflow assembly.

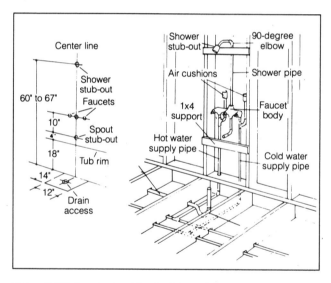

Figure 3-20. Rough plumbing for tub and shower.

plumber's minihacksaw. The compression ferrule will make a water-tight seal only if the pipe is perfectly round and clean. Then slip on the escutcheon, the compression nut and compression ferrule, and finally the valve (Figure 3-17B).

Use two adjustable wrenches to tighten down the compression nut onto the valve body (Figure 3-17C). The nut will usually squeak when it is sufficiently tightened.

Assemble the faucet according to the manufacturer's instructions (Figure 3-18).

Finally, connect the trap to the waste's tailpiece and the drain pipe at the wall, as shown in Figure 3-19. Align and tighten the compression fittings carefully to prevent leaks.

Shower and Bathtub

The highest quality bathtubs are made from porcelain-covered cast iron. Other choices are enamel on steel, fiberglass, and acrylic. Cast iron tubs are the heaviest, steel tubs the lightest. Prefabricated shower units are made of fiberglass, ABS, or acrylic. The standard bathtub is 60" long and almost 30" wide . Plastic shower units are usually between 32" and 48" wide and from 78" to 84" high. The outside dimensions of a plastic tub/shower combination are usually a little wider and longer than a cast iron tub, and the actual usable space of the plastic fixture is less than that of the iron or steel tub. Installing a tub/shower or shower is one of the most difficult plumbing projects.

Pipes required to plumb a shower or bathtub include the hot and cold supply lines; pipe from valve to the tub spout, to the shower head, or to both; a trap; and drain and vent line. When roughing in has been completed and you are ready to install your tub or shower, your rough plumbing should resemble that shown in Figure 3-20.

Bathtubs and shower stalls usually require support framing. A bathtub filled with water is extremely heavy. At 80° Fahrenheit, a cubic foot of water weighs 62.19 pounds. At the same temperature, a gallon of water weighs 8.314 pounds.

Take into account the capacity of your tub, and be certain that your floor framing is up to code. The minimum floor area required for a shower stall is 1,024 square inches (32"x32"), and you should allow 24" from the stall to any other fixture or to the wall.

Install as much of the piping as possible before installing the fixture itself. Lower the tub into place so that the continuous flange fits against the wall studs and the back edge is properly supported for its full length. If you have a plastic fixture, nail or screw the fixture in place through the flange.

Assemble the drain connections by connecting the tub overflow (Figure 3-21) with the tub drain. The trap has a slip joint that connects to the tailpiece out the bottom of the overflow T.

Hot and cold water lines are run to the tub or shower mixing valve, where they are attached by sweating them directly into the valve, in the case of valves with copper connections. A valve with threaded connections can be used with copper pipe by installing copper adapters into the threaded ports of the valve and then sweating the copper into the socket.

For the shower riser, it is best to sweat a special fitting called a "brass ½" copper by FIP winged 90" to the top of the riser to accept the threaded end of the shower arm. *Also* screw (don't nail) one or two copper pipe straps on the riser right below the "winged 90." Do the same thing for the tub spout. Use a brass nipple of the appropriate length to secure the spout snug against the waterproof finish wall of the enclosure.

The shower valve trim (handles, cover plates, tub spout, shower arm and head) are the "icing on the cake," to be installed only after the drywall has been painted and the enclosure fully cleaned. Use both teflon tape and pipe joint compound on your threaded connections, and you shouldn't have any leaks.

Toilets

Toilets come in many styles and several sizes. The degree of efficiency with which the toilet flushes depends upon many factors, price not always being one of them. *Consumer Reports* is a good place to comparison shop for performance.

Pipes required include a cold water supply stub-out with a shut-off valve, flexible tubing for above the valve, and possibly one air chamber. When the roughing-in has been com-

pleted and you are ready to install your toilet, your rough plumbing should resemble that shown in Figure 3-22.

The minimum side distance allowed from the center of the toilet bowl to a wall is 15" or 12" to a bathtub. Clearance from the front of the bowl to a wall or other fixture should be at least 21".

It is best to install the finished floor first so that this water-resistant material runs underneath the bowl. Install the angle stop (shut-off valve) now, before going any further. Closet bolts come in two lengths. *Buy the longer length.* Use only brass or stainless steel washers and nuts with the closet bolts. First bolt the closet bolts to the closet flange. Be sure to use a stainless steel washer and nut. Align the bolts so that they line up with the center of the drainpipe in the floor (Figure 3-23A).

Now place the plastic-sleeved bowl wax gasket on the closet flange, with the sleeve protruding into the drain pipe (Figure 3-23B). Using the longer closet bolts as guides, lower the bowl down onto the flange and wax. Next, drop another *two* stainless steel or brass washers down each closet bolt, and thread a brass or stainless nut as far down as you can, using your fingers. Use a ¹⁄₁₆" box ratchet wrench or 4" adjustable wrench to tighten each nut, going back and forth between the two nuts half a dozen revolutions each until the toilet is firmly on the floor. Go slowly after the toilet begins to compress the wax. You usually hear the wax being compressed and squishing out sideways. You may or may not

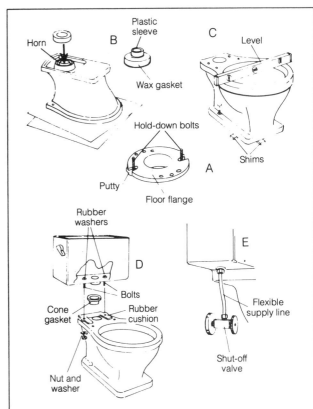

Figure 3-23. Installing a toilet.

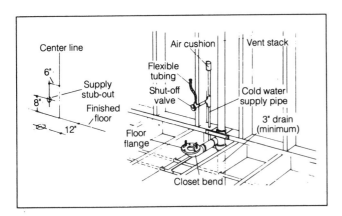

Figure 3-22. Rough plumbing for toilet.

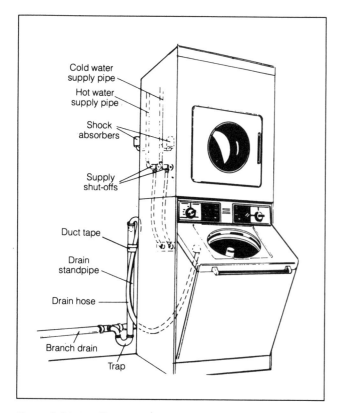

Cold water
supply pipe

Hot water
supply pipe

Shock
absorbers

Supply
shut-offs

Duct tape

Drain
standpipe

Drain hose

Branch drain

Trap

Figure 3-24. Installing a washing machine and drier.

want to level the bowl, as shown in Figure 3-23C. Shimming the bowl increases your chance of a leaking seal.

Place the rubber tank cushion (if one is needed) into position on the rear of the bowl and fit the rubber gasket onto the tank's flush valve threads on the underside of the tank (if it hasn't come from the factory already in place). Position the tank over the bowl; then install the tank bolts as shown Figure 3-23D. Now install your supply tube between the angle stop and the fill valve connection and turn the valve on, filling the tank (Figure 3-23E). Flush the toilet at least a dozen times, checking for leaks. If you have no leaks, use a minihacksaw to cut the closet bolts down to three threads above the top of the closet nuts. Then put a lump of plumber's putty inside the closet bolt covers and shove them over the tops of the bolts. The final step is to install the toilet seat.

Washing Machine

Pipes required for the clothes washer and drier installation (Figure 3-24) include hot and cold supply lines, drain line and vent line, and trap and standpipe or trap for the wash tray.

Thread brass nipples into the special fittings inside the wall. Then slip escutcheons over the nipples and thread female washing machine bibbs onto the nipples. Washing machine bibbs differ from hose bibbs in that the male hose threads for the washing machine hoses point straight down so that the machine can fit closer to the wall.

Install a 2" drain standpipe with a trap above the floor for the wasteline. Read *UPC* Section 604, "Indirect Waste Receptors," so that you understand how to rough in the standpipe and trap.

Install the hot and cold hose bibb valves and standpipe so that they can be reached when the machine is in place. The drain standpipe should always be taller than the highest water level in the machine, to protect against the possibility back-siphoning.

Hook up the washing machine supply hoses. Make sure that you have filter washers (screen cone pointing out) on the hose ends that attach to the appliance. Hand-tighten the hoses, then use pliers for a final quarter turn. Make sure to connect hot to hot and cold to cold. You may also want to use a filter washer, with a screen cone facing out, at the ends of the appliance supply hoses that connect to the washing machine bibbs.

Set the drain hose into the standpipe and secure the hose to the standpipe with duct tape. Finally, level the machine by adjusting the legs under the machine.

Clothes Drier

Allow at least 4" of clearance at the back of the drier for the venting elbow. Most driers have a vent stub-out at the bottom of the machine to attach the vent run to. Many brands also have knockouts at other locations, usually near the top and side, so that you can begin the vent run at more than one location. If you are installing a gas drier, you may want to get a professional to hook up to the gas supply.

The method that you choose to vent your drier depends on several factors. Always choose the most efficient route from the appliance to the outside. Use a hinged damper weather hood placed at least 12" off the ground to prevent backdraft. Be sure that the weather hood will not exhaust to a window well, gas vent, chimney, or any unventilated area such as an attic or crawlspace. The accumulation of lint can be a fire hazard.

Create the vent system with rigid aluminum or galvanized steel conduction pipe and wrap all joints with duct tape. Do not use sheet metal screws. Hang the rigid vent pipe from hangers on a horizontal line at the distance prescribed by your local building inspector.

4 Insulation and Ventilation

HEAT NATURALLY FLOWS from a warmer area to a cooler one. It does this in only three ways: **conduction,** the transfer of heat directly from mass to mass (Figure 4-1); **convection,** the movement of heated air from one space to another (hot air rises, heavier cool air sinks); and **radiation,** which simply means that any warm body gives off heat toward a cooler one (Figure 4-2).

The function of insulation is to minimize the transfer of heat so that the house stays warmer in cool weather and cooler in warm weather.

This chapter discusses the merits and uses of various types of insulation and explains how to evaluate R-values. "R" stands for "resistance to heat flow." The greater the R-value, the greater the insulating power. R-value requirements depend on such factors as local climate and the surface you are insulating (walls, ceiling, floor), and are regulated by local building code. Each region of the country has different requirements for adequate amounts of insulation.

This chapter also describes how to install vents in your attic. Ventilation is extremely important in an insulated attic for two reasons. First, of course, it allows hot air to escape during the summer months, so that the attic does not become superheated. And second, it prevents condensation from being trapped and causing the insulation and even the wood of the house itself to deteriorate.

In most areas, local utility companies will offer suggestions on insulating your home to reduce your energy bills. Often there is no charge for this service. You may be eligible for low-interest or interest-free loan programs. And don't forget that you may receive state or federal tax credits for energy-saving home improvements.

Before You Begin

SAFETY

☐ Wear a mask and goggles when working with any type of insulation or when sawing wood.

☐ Cover your body as fully as possible, with long sleeves, a hood, long pants, and gloves. Insulating materials can irritate your skin.

☐ Wear a hard hat when working in the attic, and watch your back; roofing nails may be sticking through the sheathing.

☐ To reduce fire hazards, keep the insulation at least 3" away from objects that transfer heat. Install sheet metal baffles (dams) around recessed light fixtures, chimneys, and flues.

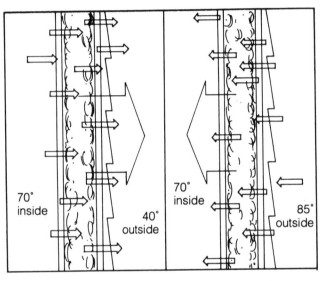

Figure 4-1. Heat is transferred from mass to mass through conduction.

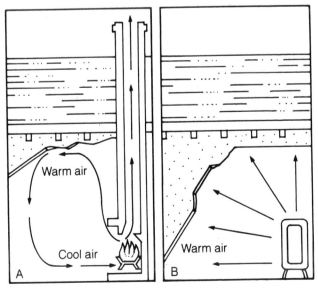

Figure 4-2. (A) Convection is the movement of heated air from one space to another (warm air rises, cool air sinks). (B) A warm body radiates heat toward a cooler one (radiant heat).

☐ When working outside on the roof, wear shoes or boots with rubber soles, stay clear of power lines, and secure extension ladders with safety hooks that clamp over the ridge.

☐ Delay your work until the roof is free of dampness from rain, frost, snow, or dew.

☐ When working high on the outside of the house, consider renting a scaffold to provide a balanced, level working surface.

☐ Do not step between attic floor joists onto the ceiling of the room below. It will give way.

☐ Some types of insulation are flammable and may give off toxic gases when they burn. Check with your local building department and fire department for special precautions or restrictions.

USEFUL TERMS

Caulking is available in several types, depending on what it is supposed to adhere to. It is a pliable material that is usually forced into a gap or crack with a caulking gun to inhibit moisture and air from entering the crack.

Cellulose is a shredded paper product, used for insulation, that has been treated with borate salts to resist burning.

Blown-in or **loose insulation** (Figure 4-3) can be cellulose, rock wool, glass fiber, vermiculite, perlite, or a combination of these materials. It is used in floors, walls, and hard-to-reach places. This type of insulation is poured between joists or blown in with special equipment. It is best suited for use in irregular-shaped areas and is the preferred option for blowing into existing finished walls.

Fiberglass blankets or **batts** (Figure 4-3) are widely used for insulating walls, floors, ceilings, roofs, and attics. They are easily fitted and stapled between studs, joists, and beams.

Flexi-vent is a strip of corrugated styrofoam designed to allow air circulation to carry away moisture that could build up under insulation.

Foam insulation (Figure 4-3) can be extruded polystyrene, isocyanurate board, or sprayed in place. The rigid panels are used on unfinished walls, on basement and masonry walls, and on exterior surfaces. The panels are glued or cut to friction fit between studs, joists, or furring strips, and must be covered with drywall or paneling for fire safety. They offer a high insulating value for a relatively thin material, but are highly flammable, and some chemically based foams may discharge poisonous fumes over a period of time. For exterior applications and in high-moisture areas, be sure to use a closed-cell, waterproof rigid panel. Sprayed in place foam insulation is usually shot directly on the roof sheathing with a foam mixer and gun. This type of insulation is most commonly used on exposed beam ceilings.

Furring is strips of wood that are used to level out a surface before finishing.

A **scab** is a piece of wood nailed into place to extend or "shim out" the surface of a rafter or stud.

Shims are thin wedges of wood used to bring furring strips level with each other when used on an uneven wall.

Silicate compound is an insulating material made of glass and sand. It does not burn or release toxic fumes. It comes in lightweight, easy-to-handle bags, and is used in the same manner as loose fill or cellulose.

Vapor barriers are used to control moisture build-up in closed areas. The most common vapor barrier is sheets of 6

Figure 4-3. The most common types of insulation.

mil plastic attached over the insulation. In the eastern United States, vapor barriers control condensation in northern regions and humidity in the coastal and southern regions. In the West, they are usually intended to limit the accumulation of vapor to an amount that will be dispersed by normal interior heating. Elaborate vapor protection is vital in some areas and undesirable in others. Consult your local code about whether or not you should install a vapor barrier.

WHAT YOU WILL NEED

Time. The time required depends on the type of job to be done. Allow 3 to 4 person-hours per 100 square feet when installing fiberglass batts and a vapor barrier in an attic. Allow 4 to 6 person-hours per 100 square feet when installing furring, insulation, and a vapor barrier in a basement.

Tools. Most of the tools required for installation of insulation are found in the home toolbox. Others can be rented from your home building center.

Dust mask

Goggles

Gloves

Pencil and paper

Trouble light

Extension cord

String

Steel tape measure

Level

Circular saw

Hole saw

Utility knife

Spackling knife

Hammer

Drill

Shovel (for exterior application)

Staple gun and heavy-duty staples (or an air compressor with a stapling attachment)

Ramset (rental)

Caulking gun

Blowing machine (rental, if you choose to blow in your insulation)

Sump pump

Shovel (for exterior application)

Materials. As with any home project, the materials you need depend on the type of insulation used and the extent of the work to be done. Your list will include many of the following items.

For Interior Application

2x4 boards

16-penny nails

Flexi-vent material

Duct tape

2x2 furring strips

Adhesive

Shims

Spackling compound

Long straight board

Fiberglass insulation

Vapor barrier (6 mil visquine)

Cellulose

Rigid foam panels (regular or closed cell)

For Exterior Application

2" extruded foam panels

Construction adhesive

Tar paper

Sheet metal flashing

Drain pipe

Gravel

Tape

Plastic

Caulk

Waterproofing sealant

Closable vents

Drain tiles

Pipes

Soffit ventilation plugs

Sheet metal, louvred and screened vent

Continuous ridge vent

Wind turbine

PERMITS AND CODES

Codes for insulation requirements vary in different parts of the country. Check with your local building inspector. Codes also indicate regional R-factors required.

Most Common Mistakes

The most common error in insulating projects is neglecting to find out the locally required R-value and insulating accordingly. Other common mistakes include the following.

☐ Not providing for good air circulation between the roof and the insulation.

☐ Installing fiberglass batts or blankets with the paper side (vapor barrier) facing toward the outside instead of toward the heated area.

☐ Distorting, compressing, or squeezing the batt or blanket out of shape.

☐ Using paper-faced batts or blankets against a heat source like a chimney or a heating duct.

☐ Neglecting to insulate all of the small spaces and corners.

☐ Covering the vents in the eaves with insulation, thereby cutting off ventilation.

☐ Making unnecessary trips up and down the attic stairs during installation. Assemble all tools and equipment in your work area before beginning the job.

☐ Not using closed-cell (waterproof), rigid foam insulation panels on below-grade installations.

THE ATTIC

The attic is usually the single greatest heat loser in the home. The lighter heated air rises, while the heavier cool air drops (convection). That's why your feet are usually cold or why your much taller mate says it's already too hot when you want to turn the heat up.

Adequate insulation in attics is imperative to reduce heat loss. The most widely used insulation in the attic is fiberglass, which commonly comes in rolls 3½" to 6" thick. It is available in widths to fit between 16" and 24" on center framing members. It comes in two forms, batt and blanket. Batt insulation is available in 4' or 8' lengths. Blanket insulation in cut-to-fit rolls is available in lengths from 30' to 70'. Blanket insulation is generally preferable because there are fewer gaps when it is installed.

Fiberglass is available in foil-backed, paper-backed, and unfaced batts and blankets. Both the foil and the paper act as vapor barriers. Unfaced fiberglass is used in conditions of potential fire hazards and as the top layer of a two-layer application. Otherwise, paper-backed fiberglass is usually recommended.

Step One
Preparation

Margin of Error: Leave no gaps

Before you begin installing the fiberglass, you need to do some preparatory work. The first thing to do is to decide how you want to use your attic space.

Take a pencil and paper, a clipboard, and a trouble light with an extension cord with you up to the attic. Examine your attic space carefully to determine whether you will ever want to use the space for living and therefore want to heat it, or whether you prefer to insulate the main part of the house below the attic. You want to place your insulation so that it encloses the heated areas only (Figure 4-4).

If you want to finish and heat the attic space, look closely at the rafters, checking them for depth and uniformity. They must be deep enough to accommodate the depth of batting needed for your region., or you will have to use foam insulation. The rafters will also give you a point of attachment for the vapor barrier and a structure capable of supporting the finished walls of drywall or paneling.

Step Two
Ventilating the Attic

Margin of Error: Not applicable

Ventilation in the attic is extremely important. If the attic isn't properly ventilated, condensation can be trapped, causing the insulation as well as the wood structure to deteriorate. Insulation loses its R-value as it takes on moisture. Also, if heat is not vented in the summer, the attic becomes superheated. Types of ventilation include soffit plugs, gable vents, continuous ridge vents, and wind turbines (Figure 4-5). When insulating between the rafters, a 1½" space is commonly left below the roof sheathing. With soffit vents and a ridge vent, a very efficient positive airflow is obtained.

Soffit Plugs

Soffit ventilation allows air to travel from outside the house into the attic space rafters and out the gable wall vents, or out the ridge at the peak of the roof. Soffit ventilation plugs, available in various sizes, are screened and louvred so that air can pass through but insects can't. To install these vents, first drill a hole with a hole saw between each pair of rafters

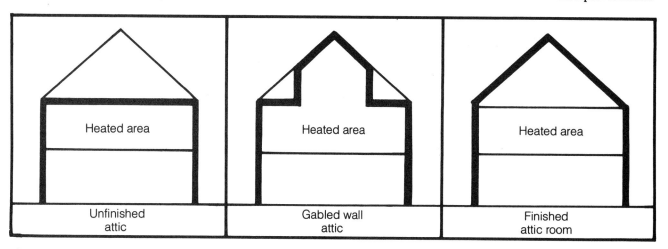

Figure 4-4. Insulation is placed to contain the heat in the heated areas only.

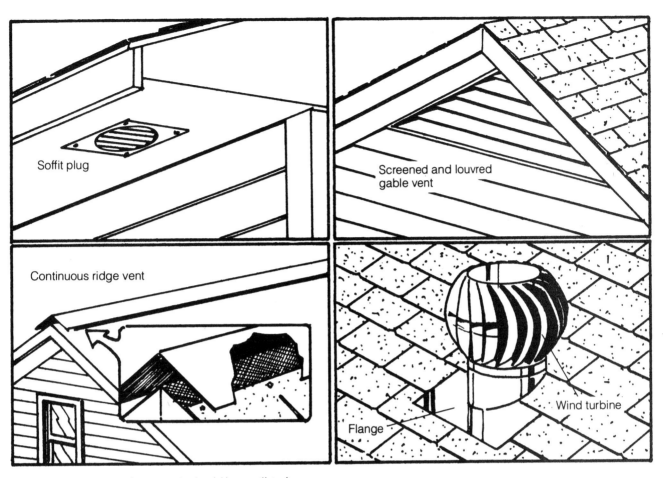

Soffit plug

Screened and louvred gable vent

Continuous ridge vent

Wind turbine

Flange

Figure 4-5. Key places where an attic should be ventilated.

all the way around the soffit (the outside roof overhang of the house). Then push in the soffit ventilation plugs.

Gable Vent

On the gable wall, near the peak, you may want to install a louvred sheet metal vent to let the air out as it moves toward the top of the attic. These gable vents are triangular, square, or round screened vents that are installed in gable walls in the same way you would install a window.

Continuous Ridge Vent

Screened continuous ridge vent is available in 10′ lengths. This kind of vent is designed to be installed along the entire ridge of the house, letting air out but keeping rain from getting in. These ridge vents are difficult to retrofit in existing homes, but they are recommended in new construction.

Wind Turbine

The wind turbine is a most effective means of ventilation. It is installed near the roof peak; wind or air rising through the turbine turns the vanes and gets the air moving near the top of the house, drawing heat and moisture out. They can be noisy, however. To install a wind turbine, follow these steps.

1. Find a good location near the peak and between rafters. Remove the roof shingles, tiles, gravel, or other roofing material down to the tar paper to expose an area slightly larger than the flange of the wind turbine. Place the flange in position and mark the flange opening on the tar paper with chalk or colored pencil.

2. Drill a starter hole, using a hole saw or a 1″ or larger drill bit. Once the hole is started, use a reciprocating saw to finish cutting out the entire flange opening area. Be sure the blade is long enough and suitable for cutting tar paper and sheathing. Keep a few extra blades on hand; the long, narrow reciprocating blades tend to break easily. Wear safety glasses or goggles when sawing.

3. Use a caulking gun to apply a generous amount of roofing asphalt cement to the outer perimeter to seal the opening, then nail the flange of the base of the turbine to the roof with roofing nails. Roll some tar paper over the flange to create a double seal, and carefully replace the shingles to cover the tar-papered flange. Be sure the flange is properly interwoven with the shingles. Like the scales on a fish, each component must overlap the last piece.

4. Install the wind turbine into the base, and you're in business with a very fine ventilator. Be sure to install the turbine properly to prevent leaks. It is essential that the flange and

Figure 4-6. Staple flexi-vent material between the rafters against the roof sheathing for ventilation behind the insulation.

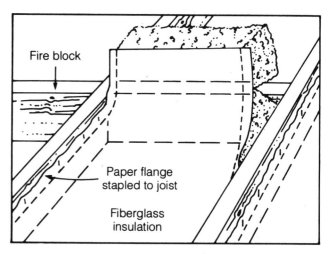

Figure 4-7. Back-cut insulation to fit around fireblocks and other obstacles.

shingles are installed in the proper overlapping fashion. You may want to call in a professional to install a turbine in a flat or nearly flat roof, because leaks are more likely to occur in these roofs.

Flexi-Vent

Complete your preparations by stapling flexi-vent material (a corrugated strip of styrofoam) between the rafters and against the roof sheathing (Figure 4-6). This material is designed to allow air circulation to carry away moisture that could build up between the roof and the new insulation. The sheets are made to fit between 16" or 24" on center rafters. Butt these strips right up next to each other and against the roof itself. Then use a staple gun or heavy duty stapler to staple them to the inside surface of the sheathing.

Step Three
Insulating the Rafters
with Fiberglass

Margin of Error: Fill entire cavity and plug holes exactly

Working on your knees in an attic for long periods of time can be very hot and tiring. You will be covered from head to toe in protective clothing and gear. Plan a few breaks into your work schedule, and try to work early in the morning before it gets too hot. Add a spray container of cold water to your toolbox to spray yourself and your fogged-up goggles.

Begin by measuring the length of each space between the rafters. Cut each piece of insulation an inch or two longer than needed, to ensure a snug fit. Use your straightedge to compress the insulation while cutting it, and use a very sharp utility knife to ensure a good straight cut across the fiberglass blanket or batt. Staple the insulation to the rafters or scabbed-on boards with a staple gun or a hammer-tacker. An air compressor with a staple gun attachment can also save you time and effort in this task.

The paper backing must be toward the interior of the room, to act as a vapor barrier. Use ⅜" heavy-duty staples every 6". Attach the insulation to the rafter scab board by the paper flange, being careful not to compress the fiberglass. Make sure the paper flange and the staples lie flat against the board, to create an even surface for attaching the finished wall material.

Where the spacing of the rafters is uneven, odd-angled, or not standard, you will need to cut the insulation to fit. Cut the fiberglass 1" wider than the necessary width, and tuck it in to create a flange for stapling.

Once the insulation is in place between all the rafters, it's time to staple up a vapor barrier of 6 mil visquine (a painter's drop cloth) over the insulation and across the face of the rafters. Although the paper or foil backing of the insulation is a vapor barrier, the visquine gives added protection against moisture forming in the cavity because it covers completely, with no breaks. As you staple the vapor barrier to the rafters, draw it as tight as possible, being careful not to puncture the plastic unnecessarily. Repair any punctures with duct tape.

When installing insulation over wires, pipes, bridging, or fireblocking, it is necessary to back-cut the fiberglass to fit over the obstacle, leaving the paper intact and the insulation uncompressed (Figure 4-7).

Step Four
Insulating the Attic Floor

Margin of Error: Fill all cavities and plugholes exactly

Fiberglass Insulation

Any floor is most easily insulated before the subfloor is applied. Just staple the flanges to the top of the joists and cover with the subfloor material. Working overhead to insulate the top ceiling is much more difficult than installing this insulation after the ceiling is in place.

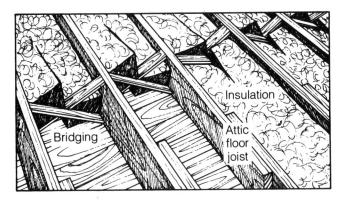

Figure 4-8. Cut insulation to fit around floor bridging.

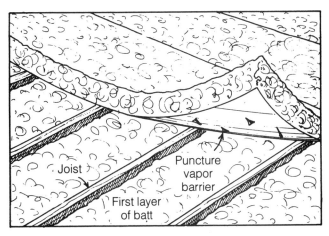

Figure 4-9. When installing two layers of batt insulation, puncture the vapor barrier of the upper layer all over so that moisture is not trapped between the layers.

Placing the fiberglass insulation between the attic floor joists effectively insulates the lower portion of your home so that you don't use expensive energy to heat unused attic space. Use this method if you don't plan to convert your attic into living space (see Figure 4-4).

Begin by unrolling the insulation between the joists, paper face down (toward the heated area of the house). Start in a corner, using a stick to tuck the insulation into the corner. Be careful not to compress the fiberglass. If there are soffit vents, leave a space at the eaves for air circulation.

Your aim in installing attic insulation is to slow the infiltration of heat from the rooms below through any cracks that may occur around the insulation. Therefore, be very careful in installing the insulation around any obstacles in the joist space, such as plumbing, heating ducts, chimney stacks, or bridging. Cut the insulation to fit snugly around the object (Figure 4-8).

The paper facing on most insulation is flammable. Therefore you must use unfaced fiberglass when working around heat sources like a chimney, flue, or heating duct. A 3" air space between the chimney and the insulation is recommended. With prefabricated flues and chimneys, check the manufacturer's recommendation.

You can insulate closely around electrical junction boxes (but not electrical fixtures boxes) because they do not give off heat. Don't cover the junction boxes, however; the code requires you to have permanent access to them. Again, be careful not to distort or compress the fiberglass. Leave about 3" around recessed lighting fixtures for air to circulate and to keep the fixture cool. Wrap pipes separately to cut off all air passage around them, and stuff scraps of fiberglass into small, hard-to-cover areas.

If one layer of fiberglass insulation between floor joists does not meet the R-value you need, a second layer can be added on top of, and at right angles to, the joists. There is less thermal loss with this method because the joists are covered, but you will no longer be able to see them. With this method, you must be careful to avoid trapping moisture between the two layers. It is best to use unfaced insulation for this layer. Or install the second layer with the paper side down and puncture the paper barrier on this second layer all over (Figure 4-9). Your main concern with this layer is that the fiber-

glass fits snug and tight, side by side and end to end. Begin this second layer in a corner, butted against the bottom of the rafters. Continue to install it end to end until you get to the center of the floor or near the stairwell. Then begin again at the opposite side and install to the center again, to avoid walking on and compressing the insulation over the joists. Finally, don't forget to insulate the hatch door to the attic.

Blown-In Insulation

Another method of insulating the attic is to pour or blow in loose insulation up to the top of the joists or beyond to provide an even surface. The loose insulation comes in large bags, and is easy to pour between the attic floor joists. Or you can rent a blower to blow the insulation between the joists. Then unfaced or punctured fiberglass insulation can be installed perpendicular to the joists. A trouble light will help you see that hard-to-reach places are adequately filled with the cellulose. To achieve a higher R-level, blow in the cellulose to fill the joist spaces past the top of the wood framing.

As you work back into the corners and around the vents in the eaves, be careful not to cover any ventilating areas. Use a long straight board to help you even out the cellulose (Figure 4-10). Drag the board along the joists to push loose piles of insulation into the spaces between the joists.

By the end of the day, your skin may itch from the small particles of glass that have found their way beneath your protective clothing. Vinegar makes an effective rinse when you bathe or shower after working with fiberglass. It almost eliminates the itching.

THE WALLS

Your choice of insulation in your walls depends on personal preference, the intended use of the space, and the R-value necessary for your area. Fiberglass (blankets or batts) is the most common material, although in some areas people install rigid foam board on the outside of the frame underneath

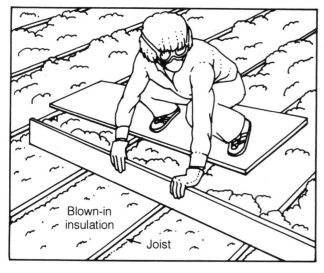

Figure 4-10. Use a long board to level blown-in or poured-in insulation, especially if you plan to cover it with batts or a blanket.

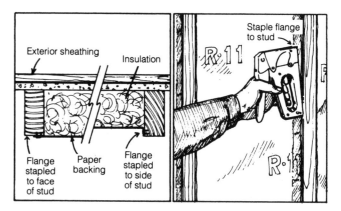

Figure 4-11. Staple the flanges of the insulation to the inside or front surface of the wall studs.

the siding, in addition to the fiberglass in the walls. The principles for installing fiberglass in the walls are the same as for installing it between the rafters. You will need a uniform surface to staple the insulation to and to hold the drywall in place (Figure 4-11).

Blown-in cellulose insulation is not recommended in new walls because it tends to settle toward the bottom of the wall, leaving uninsulated areas at the top, where insulation is most needed.

THE BASEMENT

Step One
Preparation

Margin of Error: 1/4"

Before adding insulation to the basement walls or to the crawlspace foundation, it is essential that you repair any leaks and solve any problems with dampness. Wait until the basement is thoroughly dry before you install the insulation, to be sure that all problems have been eliminated. If you are not sure whether you have a moisture problem, tape a square-foot piece of plastic to the basement wall or floor and leave it in place for a week. (Or use one of the methods described in Chapter 9, Vinyl Floors.) If condensation builds up under the plastic, you have a problem that needs to be solved.

There are several possible causes of such condensation. The most common causes are seepage, condensation, and drainage problems around the foundation, and leaks and cracks in the concrete. Taking care of these potential trouble spots will save you a great deal of time, trouble, and money later on.

Caulk any visible cracks, and control seepage by painting waterproofing sealant on the interior walls. In cold areas, it's a good idea to install closable vents in the crawlspace at the time the foundation is built. Local code will specify the location and number of these foundation vents. Open them in warm weather to air out the crawlspace and close them in winter to prevent heat loss. Some vents are automatic, and will open and close at set temperatures. You can minimize crawlspace moisture and basement seepage by diverting the most obvious sources of moisture (usually the roof downspouts) well away from the foundation.

In hilly or low-ground areas where drainage is a problem, install a "French drain" or pipes around the perimeter of the foundation to carry the water away from the foundation. You may even need a sump pump to pump out excess water. If you have a clothes drier in the basement, make sure it is properly vented to the outside; the exhaust from clothes driers is very humid.

For more information on caulking and sealing, see Chapter 14, Weatherizing. Weatherizing goes hand in hand with insulation, and many of the solutions overlap.

Insulating the basement and crawlspace sometimes calls for a different type of insulation than that used in the attic and walls. In the first place, the basement and crawlspace are more susceptible to moisture seepage, which can lead to problems like damp surfaces, stained finishes, and mildew. Water vapor moves easily through most materials used in construction, including brick and concrete blocks. A basement wall that is not adequately insulated with a moisture-resistant material will conduct warm moist air from the living space through to the cooler outer wall, where it is likely to condense. If you are not heating the basement or crawlspace, you will want to insulate beneath the first floor. This is best done before installing the subfloor; if it must be done later, it cannot be stapled correctly.

In this case, the insulation is held in place either with chicken wire or with support wires called "tiger tails," as shown in Figure 4-12. These wires are flexed, inserted between the joists over the insulation, and then released so that

Figure 4-12. You can also buy wire insulation supports, called "tiger tails," that snap between the joists to support the insulation.

Figure 4-13. Use furring strips and shims to furr out before installing rigid insulation over masonry walls.

the sharpened end digs into the joists and supports the insulation.

If you plan to heat the space below the first floor, you may want to use a closed-cell, rigid foam panel or the reflective layered type of insulation. In this case, you will insulate the basement walls, not the basement ceiling.

Rigid foam panels can be used to insulate both interior and exterior walls. The closed-cell type is not as susceptible to moisture as are other types of insulation. Use only closed-cell insulation in below-grade applications. It usually comes in 2'x8' sheets, and should be covered with a fire-resistant material, such as drywall, when exposed to the interior.

Some types of foam make an ideal habitat for certain kinds of ants. Consult your local supplier about insect problems in your area.

Step Two
Installing Rigid Foam Insulation in the Basement

Margin of Error: ¼"

Interior Application

After correcting any moisture problems, you are ready to begin installing rigid insulation. First, you will need to install 2x2 furring strips to the wall 16" on center (Figure 4-13). Attach the furring strips with the appropriate adhesive, or with a rented ramset if you are attaching them to masonry walls. The furring gives you a firm, level surface to which you can attach drywall or paneling. Chapter 6 gives complete instructions for attaching furring strips.

To even out the slight irregularities in masonry, you will need some shim material to get the furring surfaces flush with each other. Hold a level on the face of the furring strips and shim out as needed.

Measure and cut the foam to size with a utility knife or saw. Then press the foam between the strips. It should fit snugly.

Staple a sheet of 6 mil plastic as a vapor barrier over and at the top of the insulation. Do not staple it to the furring strips or puncture it unnecessarily. Gravity and the wall covering will hold it in place.

Exterior Application

On the exterior side of the basement walls, you will need to dig a trench 2' deep all around the foundation (deeper in cold weather areas with a deeper frost line). Check with your local building inspector and follow the local code.

Use at least 2" extruded foam panels on the exterior; they are denser than regular foam and will hold up longer. Be sure they are the moisture-resistant type, made to be installed below ground, or they will become waterlogged and lose their insulating value. Place the panels right up against the concrete, butted tightly against one another. Use exterior adhesive to glue the panels to the foundation wall.

To keep water from getting in between the foam panels and the exterior foundation walls, push sheet metal flashing up under the siding and a few inches over the insulation, and hammer it in place over the foam panels with galvanized nails.

If you have a drainage problem, this is a good time to install a French drain (Figure 4-14). Install it at a slope and connect it to the municipal storm drain or other drainage outlet. Check your local building code to see how foundation drains and storm drains are installed in your area. The system, which is usually installed 1' to 3' below the ground, consists of a plastic pipe surrounded by gravel. The top half

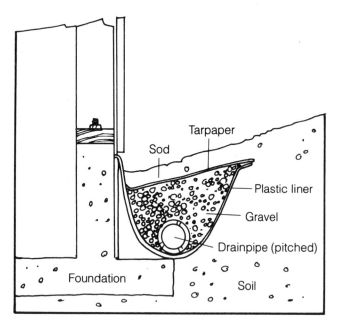

Figure 4-14. Typical French drain.

of the pipe is perforated. Water runs into the holes and down the solid lower half of the pipe, away from the foundation. A piece of tar paper on top of the gravel prevents dirt from getting in the gravel and clogging the holes in the pipe. New plastic fabric filtration materials, called **geotextiles**, can simplify these drains and reduce costs.

Backfill the excavated dirt to hold the foam panels in place around the foundation walls. Then plant grass seed or plants to prevent erosion. If you are installing a drain pipe, backfill with gravel, cover the gravel with tar paper, and cover the tar paper with the excavated dirt.

5 Drywall

BEFORE THE INVENTION of drywall, walls were usually made of hardened liquid plaster spread over wood slats (lath). Drywall is basically that same plaster, encased in large paper-covered sheets.

Drywall, often called **sheetrock, gypsum, gyp board,** or **wallboard,** is made of a crumbly fire-resistant mineral substance called gypsum, wrapped in a thick paper coating. It is durable and is easily cut, trimmed, and repaired. In this chapter you will learn how drywall can be used to cover bare stud walls.

Because of drywall's resistance to flames, most model codes require that it be used to protect wood frames. Even if the final surface will be wood paneling, drywall is often needed to protect the frame and slow the spread of fire.

Because of its unique construction, drywall can be cut, sawed, drilled, bent, nailed, glued, screwed, and painted or papered over. In addition to being easy to work with, drywall is inexpensive. Although a novice may find it difficult to finish to a smooth surface, if you practice your finishing in a closet or other area of low visibility, you can master the techniques presented here. A poorly finished job can be quite visible, especially in a well-lit room. You may find a textured finish easier to master, because it does not require the perfection of a smooth finish. Once in place, the drywall can be painted or papered (unless it is textured), which makes it ideal for designing new interiors.

Before You Begin

SAFETY

Developing and practicing safe work habits is especially important when you are working with heavy, unwieldy materials like drywall. Use the proper protection, take precautions, and plan ahead. Never bypass safety to save money or to rush a project.

☐ Drywall is heavy and difficult to lift and maneuver. It is best to work in pairs, especially when working on ceilings and high areas.

☐ Be careful not to strain your back when lifting the heavy drywall.

☐ Wear the proper respirator or face mask when sanding or sawing drywall.

☐ Use the appropriate tool for the job.

☐ Keep your blades sharp, and change the blade in your utility knife frequently.

☐ Use stepladders properly. Never climb higher than the second step from the top—use a taller ladder instead. Be certain that both pairs of legs are fully open and that the spreader bars are locked in place. When leaning the ladder against a wall, the distance between the wall and the feet of the ladder should be about one quarter the height of the ladder.

☐ Do not use an aluminum ladder near electrical wires.

☐ When setting a plank between ladders as a scaffold, be sure it extends a foot on each side and is clamped or nailed to its support.

USEFUL TERMS

A **corner bead** or **corner strip** is a protective metal cover for outside corners to prevent damage to the drywall and to allow a smoother finish.

Drywall nails have a concave head and a barbed shank or annular ring shank, or a powdered adhesive coating.

A **drywall saw** is similar to a keyhole saw. It has a stiff, pointed blade to punch through drywall and to cut curves easily.

Drywall screws are tapered number 6 screws.

Greenrock is moisture-resistant (MR) drywall that is specially treated for use in bathrooms and other damp areas. One side of these panels is usually covered with green paper.

Mud, drywall compound, or **vinyl spackling** is the substance used to cover tape and nailheads and to smooth the wall to an even, unblemished surface

A **pole sander** is a sanding block that is pivot-mounted on a long pole to let you sand the ceiling and high areas of the walls without having to stand on a ladder. It uses special precut, prepunched sandpaper.

Tape, available in **paper** or **fiberglass,** is used to cover the seams between drywall panels.

A **taping knife** is a wide, flat-bladed tool from 6" to 10" wide, used to apply drywall compound to taped seams and nail heads.

WHAT YOU WILL NEED

Time. The time you will need depends on the type and extent of job to be done.

Tools. Although you may already have a number of the tools you will need, many specialized tools, like drywall saws, pole sanders, drywall knives, and corner knives, are available to make the job easier and the finished results more professional looking.

If you plan to use adhesives to attach your drywall, a rented air compressor with a caulking attachment will make your job a great deal easier.

Ladder
Sawhorses
Steel tape measure
Level
Plane
Drywall saw
Drywall T-square
Carpenter's pencil
Drywall hammer
3/8" drill with phillips head screwdriver attachment
Chalkline and chalk
Scraper (plane)
Mud knives (4", 6", and 10" widths)
Corner mud knife
Mud trays
Drywall square
Utility knife and blades
Keyhole saw
Dust masks
Goggles
Buckets and sponges
Flashlight
Drywall sander (pole sander)
Orbital sander
Sanding block
Sandpaper in various grades
Tin snips

Materials. Standard drywall comes in several thicknesses—1/4", 3/8", 1/2", and 5/8". Thin drywall offers the advantages of being lightweight and easy to manage. Thick drywall is stiffer and tends to go up flatter. The most common thicknesses are 1/2" and 5/8". A so-called "one-hour fire-resistant wall" requires 5/8" drywall. A single layer of drywall that is thinner than 1/2" can result in a "spongy" wall. The standard panel is 4'x8', although 4'x10' and 4'x12' panels are available.

The long edges of the panels are tapered to compensate for the thickness of mud and tape used to finish the seams, where two panels butt together.

To determine the amount of drywall you will need, find the area to be covered by multiplying the length of each wall by its height. Add these numbers together and divide the result by 32, the area in square feet of a 4'x8' panel. (Of course,

if you are using 4'x10' panels, you will divide by 40.) It's a good idea to make a sketch of each wall to help you decide on the most efficient layout. The idea is to minimize the number of joints and to be sure that all panels meet over joists, and never at the corner of a window or door. Add 10 to 20 percent to cover waste.

Drywall
Joint compound
Drywall nails
Drywall screws
Drywall adhesive
Shim material
Butcher paper or cardboard for templates
Metal corner beads
Paper or fiberglass tape

PERMITS AND CODES

The thickness and type of drywall used, as well as the nailing pattern employed, are governed by local building codes in most areas. Contact the office of your local building inspector before beginning work.

DESIGN

Drywall can be installed either vertically or horizontally. You should plan your installation to create the smallest possible number of seams. Use this criterion to choose the size of drywall for your project and to plan its application. Remember that where two boards butt up against each other at their long edges, both must have a tapered factory edge where they meet. Of course, when you are using small pieces in odd-shaped areas, this will not always be possible; but do your best.

Most Common Mistake

☐ The single most common mistake in working with drywall is not practicing the finishing steps before you do the actual work. Other common mistakes are listed with each step in the process.

Step One
Prepping the Walls

Margin of Error: 1/8"

Most Common Mistakes

☐ Neglecting to install insulation, ventilation, vapor barriers, plumbing, phone lines, wiring, and ductwork before installing the drywall.

☐ Neglecting to have the rough electrical and rough plumbing inspections done before applying the drywall. These elements *must* be inspected before they are concealed.

☐ Neglecting to install nailguards (safety plates) where wires or pipes run within the studs.

☐ Joining pieces so that the seam occurs at the edge, not the center, of a window or door opening.

All electrical and plumbing work, including phone wires, cable TV lines, and alarm systems, and all insulation must be complete before the drywall is installed. You should also take care of any leaks, poor ventilation, or other repairs. It won't make your new walls look any better if a faulty pipe soaks them a few days after you put them up. The codes require you to place metal nailguards over studs to protect wires and pipes (Figure 5-1).

Once you are sure all potential problems within the walls have been addressed, you can begin your layout work for drywall application. First, observe your framework carefully to see if any studs or rafters are badly bowed and would cause the drywall to protrude or bow inward. This is especially important around doors and windows where trim will be applied later. Correct these places with shims or by chiseling, planing down, or even replacing the faulty studs if necessary before proceeding. Next, mark the centers of all wall studs on the ceiling and the floor. These marks will show you where to snap chalklines on the drywall to act as nailing guides.

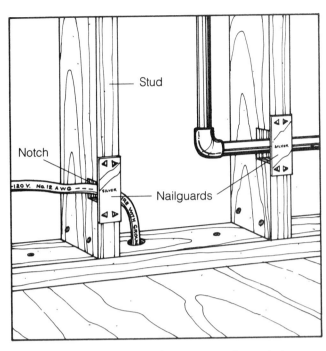

Figure 5-1. Nailguards (safety plates) protect wires and pipes from drywall nails or screws.

Step Two
Applying the Drywall

Margin of Error: 1/4"

Most Common Mistakes

☐ Not getting insulation and utilities inspected before covering with drywall.

☐ Not getting nail pattern inspected if necessary (check local code) before covering nails with compound and tape.

☐ Driving nails or screws so deep that they break the paper on the panels.

☐ Not using drywall nails or screws.

☐ Not butting two panels of drywall at the beveled factory edge (although this is unavoidable at the ends).

☐ Not butting panels at the center of a stud or rafter.

☐ Applying the drywall sheets with the wrong side exposed.

☐ Creating more seams than are necessary—for example, by using small scraps.

Final taping and mudding will go much more smoothly if the joints where sheets and odd shapes of drywall meet are smooth and fit closely together. Use a scraping plane or rasp on cut edges to smooth any rough places.

When you are positioning a drywall panel, align the top of each panel with the ceiling edge or the angle break to ensure a clean edge. (Always install the ceiling panels first. The wall panels will help hold up the edges of the ceiling panels.)

All joints between boards should be positioned to meet over the center of a stud or rafter. The exception, of course,

Figure 5-2. Use a foot fulcrum to support drywall while nailing. Place the nails directly across from each other in adjacent panels.

is where the long edges meet on a horizontal application. Any gaps should fall close to the floor, where a baseboard will cover them. To raise the panels, you can make a foot fulcrum with two pieces of wood (Figure 5-2).

Start a couple of drywall nails at the corners and across the top of the drywall panel before you lift it into place. Once the panel is positioned, these nails will make it easier to attach while another person holds it in place. Be sure to align each panel to meet over a stud.

There are several kinds of drywall nails. Some have cupped heads, which make them easier to cover when mudding and taping. Those with barbed shanks have greater

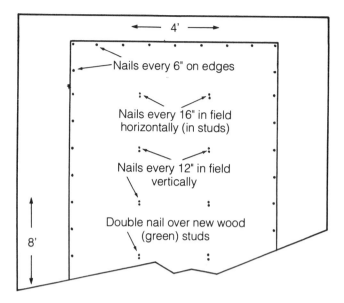

Figure 5-3. Sample nailing pattern for sheetrock. Check your local code to find nailing requirements in your area.

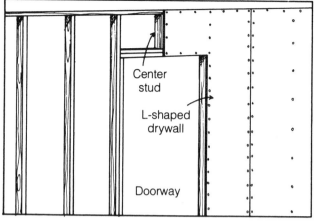

Figure 5-4. Seams should fall at the center of the window or door opening, not the edge.

holding power and reduce nail popping. Nail along the edge of the panel about every 6" or 7" on ceilings and every 8" on walls. In the middle of the panel, nail about every 12". Check your local code on this point for variance. If the studs are new wood, it is advisable to double-nail in the field, using two nails every 12" instead of one (Figure 5-3).

Hammer the nail firmly into the stud until it is forced slightly below the surface of the panel—this is called "dimpling." But don't drive it so deep that you break the paper. Be careful not to ding the edge of the panel when nailing or handling. Dings require extra mudding and finishing work. If you are using drywall screws, be sure to screw them to just below the surface of the panel.

Whether you are using drywall nails or screws, be careful not to break the paper when you are forcing the head below the surface. If the paper is broken, drive another nail or screw directly below to ensure a good hold. A special drywall hammer or a cordless electric drywall screw gun speeds the job and makes it easier.

Place nails in adjoining sheets directly across from each other where they meet at a stud. This makes mudding easier. If you miss the stud, pull out the nail or screw and dimple the hole so you can mud and tape over it properly.

In areas where the code allows it, drywall adhesive is an alternative to all these nail or screw holes. Apply the adhesive with a caulk gun onto the studs in a continuous 3⁄8" zigzag bead. A caulking gun attachment to an air compressor can make this method almost effortless. Follow the manufacturer's instructions when using adhesive. The adhesive minimizes the number of nail holes that need to be finished with mud and reduces nail pops. When using adhesive, however, at least one nail is needed in each corner to hold the panel in place.

When using adhesive, lay your stack of drywall panels over a stud so that the stud is in the center of the stack of panels, with the panels' back sides facing up. This creates a slight bow in the panel, which will force it into the stud and into the adhesive once the panel is applied.

As you apply the drywall, be careful not to leave a gap between boards greater than 1⁄4", and less if possible. Larger gaps make taping more difficult. Before embedding the tape over a seam, fill this void with joint compound. Never jam the sections of drywall together; they should be lightly butted but never forced.

Step Three
Cutting Openings

Margin of Error: 1⁄4"

Most Common Mistakes

☐ Dinging or damaging the edges of the panels.

☐ Joining pieces so that the seam occurs at the edge, not the center, of a window or door opening.

The easiest and most foolproof way to cut door and window openings in drywall is to cover a portion of the rough opening with a drywall panel, then cut along the edge of the opening with your drywall saw. By cutting out half of a doorway or window opening at a time, you will always have the framing visible as a reference.

Applying drywall around openings like doors and windows calls for a little extra care and very accurate cutting. Never try to fit around a large opening with just one panel. You'll end up with an unstable panel of thin "arms" that are likely to break off from the main piece. It's best to work with two L-shaped pieces of about the same size, with a joint that meets in the middle over the opening, as described earlier. (Figure 5-4)

Joints must always meet at a stud, and they should never occur at the edge of a door or window. The joint will even-

Figure 5-5. Snap the drywall at the cut line over the edge of your worktable.

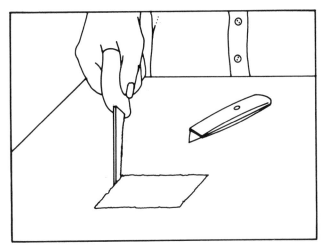

Figure 5-6. Use a drywall saw to cut out holes in your drywall.

tually crack from the stress of the movement of the door or window.

Be particularly careful not to damage the board when cutting a notch or corner. If your cuts are not clean and precise, the entire job will end up looking sloppy.

When applying drywall around right-angle openings, use a drywall T-square or a chalkline to mark the board for cutting. Cut the shorter length first with a drywall saw or keyhole saw. Then use a utility knife to score the longer cut. Use several light strokes of the knife to cut into the core. Position the cut over the edge of your worktable and snap the panel (Figure 5-5). Finish by cleanly undercutting the paper on the back side with the utility knife. Always cut with the tapered-edge side up, not the flat-edged side with the coarse paper.

Most stud walls are designed to allow for a ½" drywall ceiling, 8' of drywall (two sheets horizontally or one sheet vertically), and a small gap above the floor. This gap keeps the drywall away from any moisture on the floor and compensates for any unevenness. The gap is hidden by the baseboard.

When cutting an opening to fit around a window, measure the opening carefully and mark it on the back of the panel, using a drywall square, a straightedge, or a chalkline. Raise the panel until it hits the ceiling, so that any gaps will fall at the floor where they will be covered by the baseboard. Cut the panel to fit the edge of the stud at the rough opening. Don't let the drywall overlap onto the jamb or into the rough opening.

For smaller openings like outlets, you can coat the rim of the opening with lipstick or colored chalk. Then fit the panel into place and apply hand pressure over the outlet area. The lipstick will transfer to the back of the panel for a cutting pattern. Cut this patch out with your drywall saw (Figure 5-6). When cutting from the back side of a panel, be careful not to tear the paper beyond the patch hole area.

Step Four
Angles and Ceilings

Margin of Error: ¼"

Most Common Mistakes

☐ Not putting up the ceiling drywall before doing the walls.

☐ Creating more seams than are necessary—for example, by using small scraps.

Cutting a complex piece of drywall is a precise job that can best be done with a paper or cardboard template. Tape the paper or cardboard over the space to be covered and mark the perimeter. Then use that template to transfer a precise pattern to the drywall. Your fit will be more exact, and your job will look more professional.

Ceilings and slanted walls, such as attic ceiling-walls, pose a serious problem in the form of gravity. More than one person is needed to support each drywall panel as it is secured in place. If an assistant is not available, you could use a 2x4 T instead. Actually, it is best to use both a brace and an assistant; as you know by now, drywall is heavy, cumbersome, and not easily aligned overhead by one person, even with the use of supports. Make a couple of supporting T's by cutting 2x4s to the height of your ceiling, less 2" to accommodate the crosspiece and the drywall thickness. Nail a 3' section of 1x4 to the 2x4 as a brace (Figure 5-7). You can also rent a drywall jack to help you lift and hold the panels.

Drywall screws used with a screw gun (Figure 5-8) are generally more efficient and convenient than are nails for

Figure 5-7. You can make T's out of 2 x 4s to support ceiling drywall until it is nailed in place, although it is better to have an assistant as well.

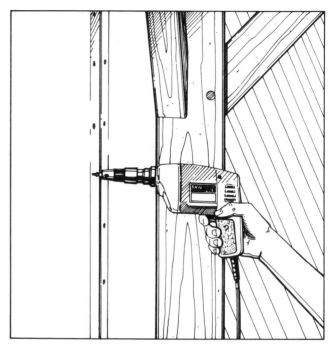

Figure 5-8. An electric screwdriver or drywall screw gun is quicker and easier than nailing.

attaching drywall, especially overhead. The pattern for fastening with screws is a maximum of 12" on center (o.c.) for ceilings and a maximum of 16" o.c. in walls with studs placed 16" o.c. When framing members are spaced 24" o.c., the pattern for using screws is a maximum of 12" o.c. for both ceilings and walls. Local codes may vary on this point, however, so be sure to determine what is acceptable before you begin.

Again, remember that ceiling pieces should always be installed before the wall boards, and the wall boards held tight against them to serve as added support for the ceiling drywall. Angled ceilings, however, may be installed after the walls.

Step Five
Curves and Odd Spaces

Margin of Error: 1/8" to exact

Most Common Mistakes

☐ Contaminating the compound with debris or dried chips of compound.

☐ Not completely covering the tape with compound.

☐ Creating more seams than are necessary—for example, by using small scraps.

As with angles, the easiest way to transfer the exact measurement from an odd-shaped space to the drywall is with a paper or cardboard template.

Tape or staple the paper to the studs and use a chalkline or pencil to mark the proper perimeter on the paper. Tape the

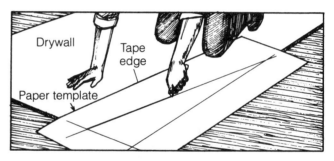

Figure 5-9. Use paper templates to cut odd-shaped pieces of drywall.

template to the drywall. Then cut right through both the pattern and the drywall (Figure 5-9).

Cut the template pieces slightly smaller than the actual space defined by the studs by cutting to the inside of the chalkline. As with other cuts of this type, use your utility knife to make the edge cuts, and a drywall or keyhole saw for the penetrations.

Fasten the drywall to the studs by the means you have chosen—nails, screws, or adhesive. With these pieces, tapered factory edges are not needed between adjoining pieces. Do smooth the cut edges for a more uniform joint, however.

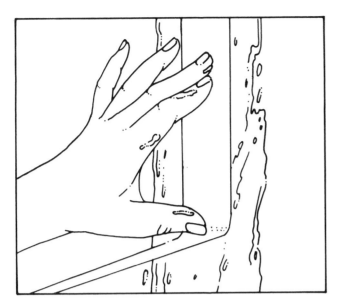

Figure 5-10. Paper tape is embedded into a layer of mud.

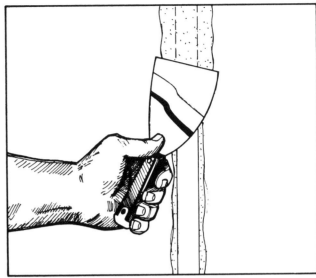

Figure 5-11. Use a 4" knife to apply a second layer of mud over the wet tape.

Step Six
Tape and Mud

Margin of Error: Exact

Most Common Mistakes

☐ Not sanding after applying each layer of drywall compound.

☐ Not sanding the final coat of drywall compound to a smooth finish.

Now is the time to muster all of your patience and diligence, because faulty workmanship will show in the taping and mudding.

Many home owners will choose to carry the project to this point, then call in a professional to finish with finesse. If you are one of the more confident, have faith that a little practice in a less noticeable area, such as inside a closet, can go a long way. If you're still not confident, select a textured application that will hide any flaws.

Materials

To achieve a smooth finish, all screw and nail heads must be covered with joint compound (mud) and the seams taped over. Joint compound has a drying time of 8 to 24 hours, depending on the thickness of application, temperature, and humidity. Allow longer for deeper cracks and crevices.

Cleanliness is very important to a smooth finish. Working from a mud tray keeps dried pieces and bits of debris out of the can. Keep the lid on the can airtight when you're not using it to keep it from drying out. Smooth the top of the mud flat and add a small amount of water before sealing the can if it is to be stored for some time.

Compound can be purchased in 1- or 5-gallon buckets, or in powdered form, which you mix yourself. Unless you are doing a very small job, the 5-gallon bucket is a much better buy. As you are using the mud, scrape the inside wall of the container often to keep residue from hardening and dropping pieces into the compound.

Drywall tape is available in paper tape and in fiberglass tape. Paper tape is precreased and can be used on straight seams as well as for corner taping. Self-adhesive fiberglass tape has the advantage of not needing a coating of mud underneath. This tape is recommended for the novice, except for use on the inside corners, where it is necessary to use the paper tape.

Application

Taping joints is a lengthy project, requiring three applications of mud to each joint. Begin by covering all nail dimples, applying the mud flush with the panel.

If you have chosen to use paper tape at the joints, apply the first layer of mud to the full length of the seam with a 4" mud knife. Apply enough mud to the seams for the drywall tape to adhere and to cover the entire seam. Then soak the paper tape in water for a moment.

Apply the wet tape to the joint (Figure 5-10), smoothing it with a 1½" putty knife. Embed the tape into the mud slightly with the knife. While the mud is still wet, apply a second layer to cover the tape completely, using a 4" taping knife (Figure 5-11). At the same time, draw your blade tightly over the surface to squeeze the mud out from underneath the tape so it is good and flat, being careful not to create any bubbles.

If you are using self-adhesive fiberglass, an undercoat of mud is not necessary. Simply apply the tape over the seam (Figure 5-12), then add a layer of mud over the tape, as shown in Figure 5-13.

Inside corners must be covered with paper tape, because the fiberglass tape will not fold. Again, apply a layer of mud first, covering each side of the corner, then embed the wet folded tape. Use a special corner knife to finish smoothing the mud over and around the corner (Figure 5-14).

Outside corners are easily damaged and require special metal corner beads to provide stability and to protect the drywall (Figure 5-15). Use tin snips to cut the bead to fit the full height of the corner. Fasten it to the framing with screws or nails every 6". Mud over it like any other seam, feathering the mud out from the corner.

Curved corner beads are now available for both inside and outside corners. These new types soften the edges and shadows, and are best suited to smooth wall applications.

Let the first coat of mud dry completely, usually overnight, before you apply the next. If you want to speed the process, a heater with a fan will cut down the drying time considerably. If the first coat is to dry overnight, clean your tools thoroughly and throw away any unused mud. You will need to make up a new batch for the second coat.

To prepare the wall before the second application, use a mud knife to scrape off any dried chips at the seams and nailhead areas. Then sand the first coat smooth with a drywall sander or a sanding block, using 100 grit sandpaper. A pole sander (Figure 5-16) is invaluable for this task. Sand only the mud—be careful not to scuff the surface of the drywall itself. When sanding drywall, always use a paper face mask to avoid inhaling the dust.

Instead of sanding, you can wipe the first coat down with a large wet sponge. Just be careful not to soak the paper or wash away the mud.

Thin the second coat of mud with a little water and apply it with a 6" mud knife. (Each coat you apply should be thinner than the last.) Again, let this coat dry properly, then scrape it and sand or sponge it smooth. Be sure to feather the edges of the layers of mud at the joints so that raised areas don't build up. After the second coat, the tape should not be visible under the mud.

Apply the third and final coat with a 10" mud knife for a smooth, even finish. When it is thoroughly dry, sand or sponge it smooth.

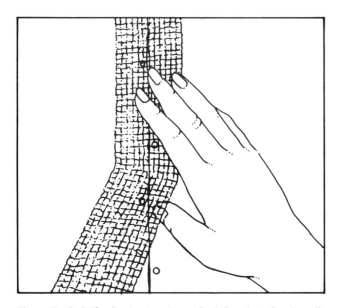

Figure 5-12. Self-adhesive tape is applied directly to the drywall.

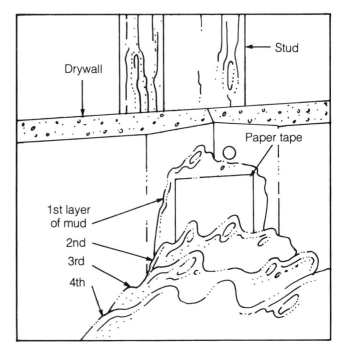

Figure 5-13. Self-adhesive fiber tape is applied directly to the drywall (left); paper tape is applied over a thin layer of mud (right). Note how the edges of the panel are tapered to avoid a bulge at the seam.

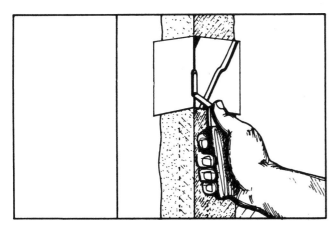

Figure 5-14. Use a special corner knife to smooth the mud over the tape in the inside corners.

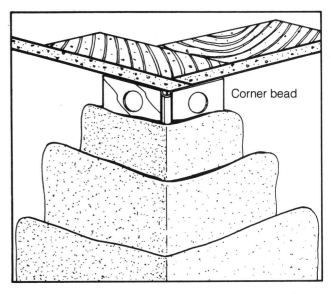

Figure 5-15. A metal corner bead protects outside corners from damage.

If you have decided on a textured finish, only two coats of mud may be needed. Obviously, sanding and smoothing are not as critical as with a smooth wall finish.

Once the final coat has dried and been sanded, use a flashlight to check for blemishes. Hold the light at an angle so that any irregularities cast a shadow. A good smooth finish is essential before applying paint or wallpaper. Always do a final sanding or sponging just before painting or papering the new drywall.

You will need to seal the drywall with a primer-sealer undercoat before painting it. Many primer-sealers are also sizings, a prerequisite to wallpapering. An oil-based primer is recommended for bare drywall, but be sure to check with your dealer for special considerations with the wall covering you plan to use.

Repairs

Small dents and dings in drywall can usually be covered with an application of wallboard compound and smoothed with sandpaper.

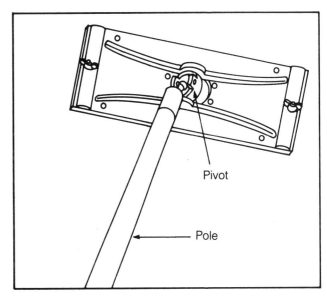

Figure 5-16. With a pole sander, you don't have to stand on a ladder to sand the ceiling and the upper part of the walls.

6 Wall Paneling

PANELING IS A dramatic and easy way to finish your walls. Paneling comes in two basic forms—solid-wood boards in various widths (many of which are tongue-and-groove) and wood-faced 4′x8′ panels. In this chapter you will learn how to install both kinds of panels on several different surfaces, with special techniques to fit panels around openings and to custom-cut uneven intersections. You will also learn some shortcuts and useful tips on trimming your new walls, as well as how to construct shelves to complement them.

Before You Begin

SAFETY

Paneling your new walls is not a difficult task. However, it is always a good idea to develop safe work habits and stick to them.

☐ Use the appropriate tool for the job.

☐ Keep your blades sharp. A dull blade requires excessive force and can easily slip.

☐ Be sure that electric tools are properly grounded.

☐ Use proper precautions when using power tools.

☐ Wear safety glasses or goggles when hammering or sawing, especially if you wear contact lenses.

☐ Wear the proper respirator when using adhesives that emit toxic fumes.

☐ Wear rubber gloves when using solvents.

☐ Be careful to avoid unnecessary strain when lifting heavy panels.

☐ Some adhesives are highly flammable. Don't smoke, and extinguish pilot lights and open flames when working with adhesives.

USEFUL TERMS

Furring is strips of wood, usually 1x2 or 2x2, that are attached to masonry or uneven walls to create cavities in which to install insulation or to provide a suitable surface over which to apply paneling or drywall.

A **ramset** is an explosive-powder-activated nailgun.

Scribing is a method of using a compass to transfer the line of an uneven surface to a panel to be cut.

A **shim** is a small wooden wedge used to even out furring strips, or to hold panels in place while attaching them to the wall.

A **vapor barrier** is a sheet of material, usually 6 mil plastic, that is installed between insulation and paneling to prevent the buildup of moisture.

Wood-faced panels are 4′x8′ sections of plywood, one side of which is laminated with hardwood veneer.

WHAT YOU WILL NEED

Time. The installation of paneling and trim over bare studs will take 2 to 4 person-hours per 100 square feet. If you must install furring first, allow 3 to 5 person-hours per 100 square feet. When using adhesives, allow 2 to 4 person hours per 100 square feet.

Tools. Paneling a new room can be a very satisfying project, resulting in a warm and inviting new living space. The process requires only tools that are usually found in any home toolbox.

If you are paneling over masonry, you may want to rent a ramset or nail gun, which will make the job of attaching the furring strips much easier. These guns use a .22-caliber powder cartridge to fire a nail into a concrete block or masonry. Though intimidating, they are safe when used properly. As an alternative, there are heavy-duty construction adhesives on the market that will adhere furring strips to masonry. Check with your home center or hardware store.

If you plan to use adhesive as your primary means of attachment, you will also need a caulking gun. You can save yourself time and effort by renting an air compressor with a caulking gun attachment. If you are not familiar with the use of pneumatic tools, read over the information in Chapter 1 before beginning work.

Tools for All Projects
Hammer
Tape measure
Electric drill with various drill bits
Nails
Chalkline
Level
Circular saw
Saber saw
Handsaw

Assorted screwdrivers

Carpenter's pencil

Tools for Paneling

Chalkline

Caulking gun or air compressor with caulking gun attachment

Nail gun or ramset

Compass

Finishing hammer

Finishing or plywood blades

Tools for Trim

Miter box with backsaw

Coping saw

Caulking gun

Tools for Shelves

Orbital sander

Clamps

Chisels and mallet

Materials. To determine the amount of sheet paneling you will need, multiply the length of your wall by its height and then divide by 32 (the area of a 4'x8' panel). For tongue-and-groove paneling, determine the area and consult with your dealer, as widths of this type of paneling vary. For molding, measure the linear feet. To avoid splicing short pieces, purchase extra lengths of paneling from which to cut.

Materials for All Projects

Shims

Screws

Nails

Adhesive

Materials for Wood-Faced Paneling

Wood-faced paneling

Color-coordinated paneling nails

Furring strips

Rigid foam insulation

Vapor barrier

Materials for Tongue-and-Groove Paneling

Tongue-and-groove planks

Adhesive

Color-coordinated paneling nails

Materials for Trim

Quarter-round molding

Caulk

Baseboard moldings

Finishing nails

Carpenter's wood glue

Materials for Shelves

1x2 board

1x4 board

Pole

Wood or plastic pole supports

Supporting pole bracket

Plywood

Quarter-round molding

Adhesive-backed mirror squares

Posts

Tracks and brackets

Shelf boards

Finishing screws

PERMITS AND CODES

In some areas, the installation of wall paneling is governed by local building codes. Contact your local building inspector before beginning work.

DESIGN

Paneling is a good choice of wall covering for casual playrooms or dens, and it can lend an air of elegance to a library or study. Use paneling to pull together spaces that are separate but adjoining, such as a family room/kitchen, or in rooms with cathedral ceilings that expose portions of upper hallways or other rooms. It is important to choose your paneling during the design stage for two reasons. First, it's easy to hide cuts around doors and windows under the trim; and second, the thickness that the paneling adds to the walls will require thicker door jambs.

Properly chosen, paneling and shelves can enhance a room and make it a warm, inviting place. Before you decide on a type of paneling, take the time to become familiar with the variety of colors and textures available and to consider your choice in the light of your particular needs. For example, a pastel panel may be well suited to a dark attic room, while a deeper wood tone will work better in a sunny family room.

The choice between sheet paneling and tongue-and-groove paneling is a matter of personal taste and expense. Solid wood tongue-and-groove is more expensive and more difficult to install. However, many people find the richness of the natural wood to be well worth the extra expense.

WOOD-FACED PANELING

Wood-faced panels, usually 4'x8' sheets with a thin veneer surface, are the most popular wall paneling material on the market. They come in many styles, patterns, and wood types.

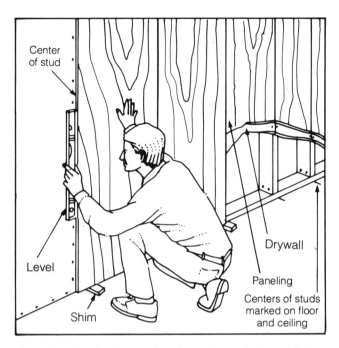

Figure 6-1. Check to be sure that all panels are plumb and that adjacent panels meet over the centers of studs.

Step One
Applying the Paneling

Margin of Error: 1/4" where covered by trim. Exact where exposed.

Most Common Mistakes

☐ Installing windows and door jambs before paneling.

☐ Not taking the time to plan carefully before purchasing the materials for your project.

☐ Failing to fasten the electrical boxes to the frame so that they are flush with the final paneled surface.

☐ Neglecting to furr out an existing wall, if necessary, before installing the paneling.

☐ Not adding insulation or a vapor barrier over an outside or basement wall, where appropriate.

☐ Splintering the veneer panel by cutting panels face up with a saber or circular saw.

☐ Failing to check that each panel is plumb on the wall before applying the next.

☐ Transferring measurements to the panel incorrectly or to the wrong side of the panel.

☐ Not using a finishing hammer and finishing saw blades when working with paneling.

Paneling Over Drywall

If you are starting with exposed stud walls, you will need to cover them with drywall before applying the wall paneling to provide adequate fire safety. Follow the instructions in Chapter 5, Drywall. Your mudding and taping can be rough, because the drywall will be covered with paneling. However, you must cover the nail or screw heads and fill the joints as part of the flame barrier. Don't forget to get a drywall nailing inspection, if required, before you cover the fasteners.

Like drywall, plywood paneling comes in 4'x8' sheets, and you can special-order it in 4'x10' or 4'x12' sheets. The panels are usually installed vertically, and the edges always nail to a stud or plate. The standard 4'x8' sheet will fit flush against the ceiling of a common stud wall and leave a small gap at the bottom. This gap is hidden under the baseboard. Trim cuts for taller walls should be made at the bottom of the panel, not the top, so the baseboard will cover them. If you must cut the tops to accommodate a sloping ceiling, you can hide the cut edge with molding.

Once your drywall is in place, mark the stud centers on the floor and ceiling. To avoid marring the surfaces, make your marks on small pieces of tape. Begin applying your paneling in an inside corner. Start in the corner (you should have one per wall) where the full sheet of drywall was applied, and work out from there (Figure 6-1). Trim the corner edge so the uncut edge will be plumb and on a stud center when installed. A technique called **scribing** allows you to accommodate walls or ceilings that are not plumb and level, as well as other irregularities such as a fireplace or built-in furniture. For information on how to use this technique, see "Scribing," later in this chapter. When cut, the scribed contour will fit the irregular wall closely, and the uncut edge will still be plumb and centered on a stud.

After the first sheet of paneling is scribed, trimmed as necessary, and installed, the other sheets are applied in sequence to cover the rest of the wall. The last sheet will probably require trimming and/or scribing. Be sure you trim the corner edge and butt the factory edge in the field. For example, if the factory edge of the next-to-last sheet is 30" away from the corner at the bottom and 30½" away at the top, both scribing and trimming will be required. If the wall is straight, this procedure is fairly simple; just trim the panel so it is just shy of 30" at the bottom and 30½" at the top. If the wall is not straight, or if you are terminating the panel at a chimney or a piece of built-in furniture, careful scribing is necessary. In this case you may find it worth the effort to make a paper or cardboard template.

Manufactured paneling usually has flush or recessed lines, simulating separate boards. When the panels are installed vertically, two of these lines in the field and one on each side should line up with the stud center marks on the floor and ceiling. Nail in these darker lines, using nails with colored heads, typically every 8" in the field and 6" on the perimeter. If you are using hardwood plywood paneling, just countersink the finish nails and putty the holes before applying your finish. Thinner panels are more likely to reflect any unevenness in the wall than are thicker panels.

Figure 6-2. Use a ramset to attach 2x2 furring strips to concrete block or masonry walls.

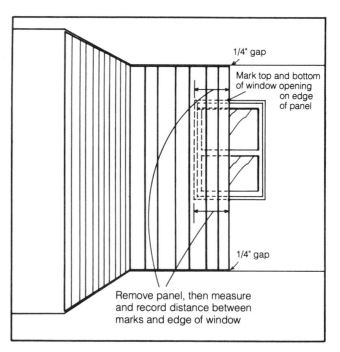

Figure 6-4. Measure paneling to fit around doors and windows.

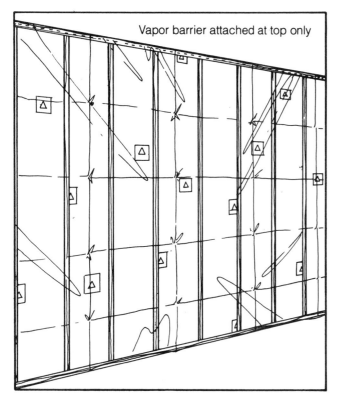

Figure 6-3. Rigid insulation and vapor barrier installed over an uninsulated wall.

Paneling Over Masonry Walls

Because a masonry surface is so hard to penetrate with paneling nails, and because the surface is often uneven, 2x2 furring strips must be attached to such walls (Figure 6-2). It's a good idea to use decay-resistant lumber here.

To attach the furring strips, use adhesive, special concrete shields and lag screws, or a ramset, which can be rented from your local building center. Wear safety glasses and ear protectors when using a ramset, and be sure to use the right size nails and "bullets." Never use a ramset on hollow concrete blocks. Instead, drive the fasteners into the mortar at the joints.

Install horizontal strips at the top and bottom of the wall and vertical strips 16" on center, simulating a stud wall. Use a level on the furring strips and even out the irregularities in the masonry with shims to get the furring surfaces level and even with each other. You are building a sturdy wooden structure that will hold the paneling solidly. If you are working on an uninsulated exterior wall, you should cut rigid foam-core insulation to friction-fit in the spaces between the furring strips. Measure the space between each furring strip separately, at both top and bottom, before you cut the insulation to fit. For full information about installing insulation, see Chapter 4.

Next, add a vapor barrier of 6 mil visquine (a painter's plastic drop cloth) to prevent the buildup of moisture between the insulation and the new paneling. Attach the vapor barrier only at the top (Figure 6-3); gravity and the paneling will keep it in place without puncturing it in too many places. If too many holes are punched in the plastic, its ability to keep moisture out will be destroyed. Now you can attach your paneling over the vapor barrier with nails. Adhesive cannot be used in this situation. Insulated walls that already have a vapor barrier do not require one directly behind the paneling.

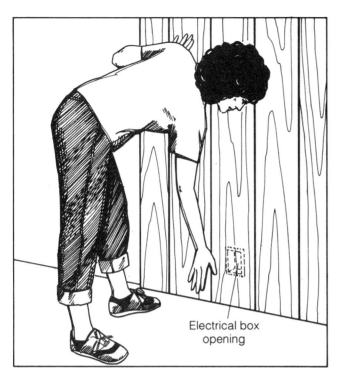

Figure 6-5. Outline edge of electrical box opening with lipstick or colored chalk. Whack the paneling over the box to imprint the outline on the back of the paneling.

Step Two
Fitting Panels
Around Openings

Margin of Error: ¼" where covered by trim. Exact where exposed.

Most Common Mistakes

☐ Cutting window, door, and electrical openings too large.

☐ Failing to complete all electrical wiring and plumbing before installing the paneling.

The easiest way to get an accurate door or window opening measurement is to have one person hold the sheet of paneling in place over the opening while another traces the line of the opening on the back of the panel. This works best for doors. If you are unable to get to the back of the panel, you will need to measure, as described next.

Place the panel in position over the window, butting it tightly up against the previously installed panel. Be sure that the ¼" gap between the floor and the panel is maintained with shims, and use a level to check that the panel is plumb. Mark the edge of the panel to indicate the top and bottom of the window or opening. Remove the panel, then measure and record the distance from each mark to the side of the window (Figure 6-4). Transfer these measurements to the back of the panel, and use a straightedge to connect the points.

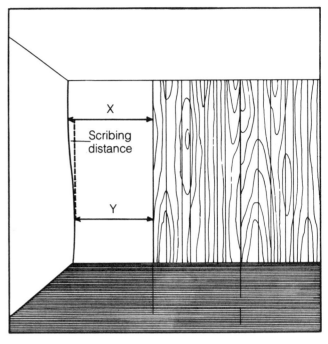

Figure 6-6. Overlap the paneling by the scribing distance. (X - Y = scribing distance.)

For smaller openings like electrical outlets, outline the edge of the outlet with chalk or lipstick. Then position the panel over it (Figure 6-5). Give the panel a whack over the opening so that the imprint appears clearly on the back of the panel. Then make your pencil lines and cut from the back of the panel to avoid splintering the veneer with your circular saw.

Scribing

Now and then a panel must butt up against an irregular surface, such as a fireplace. In these situations, you will find scribing to be the easiest and most effective way of fitting the panel tightly in place.

Begin by measuring from the edge of the last full sheet to the farthest point of the irregularity. Make a note of that distance, and then cut your panel to that width. Next, measure the distance from the edge of the last full sheet of paneling to the nearest point of the irregularity. Make a note of that distance. Subtract the second measurement from the first to get the "scribing distance." Then position the panel you need to trim and fit so that it overlaps the last full sheet of paneling by the scribing distance (Figure 6-6). Use your level to position it plumb and tack it temporarily in place.

Set a compass to the scribing distance. Hold the compass with the tip and the pencil horizontal and ride the tip along the irregular surface while the pencil runs along the panel you wish to trim (Figure 6-7). Trace the irregular line carefully from the top of the panel to the bottom. The result will be a line on the panel that is the exact profile of the irregularity. Cut along this line with a jigsaw and smooth the cut edge with a rasp or sanding block. Now glue or nail the panel

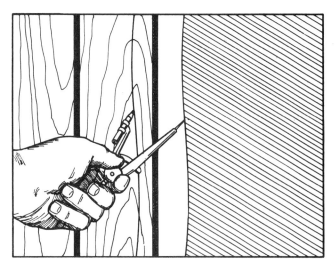

Figure 6-7. Use a compass to scribe a panel to fit an irregular corner.

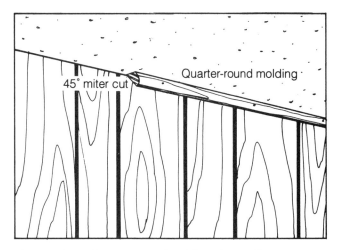

Figure 6-8. Install quarter-round molding where ceiling meets wall.

Figure 6-9. Use the shank of a screwdriver to smooth the outside edge of the baseboard.

into place. (A contour gauge can also be used to transfer complex details accurately.)

If you are starting at a corner that is irregular, measure out from the corner 48" or to the center of the farthest stud *under* 48". Snap a plumb chalkline down the center of the stud to establish the position of the edge of the panel. Mark the center on the ceiling and floor for a nailing reference. (Use tape to avoid marring the finished surface.) Place the panel as snug into the corner as possible, using a level on the opposite vertical edge to plumb the panel. Temporarily tack the panel in place at the top and bottom. Remove the panel, then measure from the panel edge to the center line of the stud marked on the floor to determine your scribing distance. This measurement should be the same at the ceiling and at the floor. Scribe and cut with a jigsaw, smooth the cut edge, and install the panel as described above.

Step Three
Applying the Trim

Margin of Error: Exact

Most Common Mistake

☐ Leaving a gap of more than ¼" at the corners.

Trim comes in a huge variety of styles and many materials, including wood, metal, and plastic. Trim is used to cover gaps at inside corners, floor and ceiling joints, and around door and window openings. Trim is the key to an attractive, professional-looking job. And with trim you cover any minor mistakes you may have made in cutting and installing your paneling.

Floor and Ceiling Joints

The basic trim piece for a floor-to-wall or ceiling-to-wall connection is quarter-round molding or baseboard (Figure 6-8). It offers a clean, neat appearance and is easy to install. Simply nail it in place with finishing nails, countersink the nails, fill the holes with matching wood putty, and sand smooth. Or use color-coordinated paneling nails.

Outside Corners

Place your paneling so that the edges meet as tightly as possible at an outside corner. If the intersection of the two pieces is neat enough, it can be left uncovered. Special outside corner molding can also be applied.

Inside Corners

If your walls are straight enough, and your paneling job neat enough, you can sometimes run one sheet tightly into the corner and butt the mate right up to it for a snug fit and a good-looking job. If this is not the case, however, quarter-round molding will work nicely.

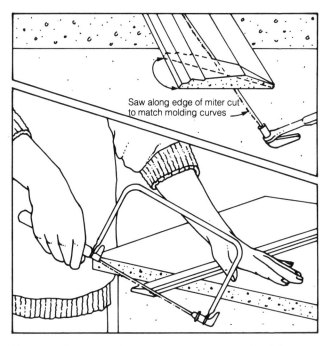

Figure 6-10. Use a coping saw to cut quarter-round molding.

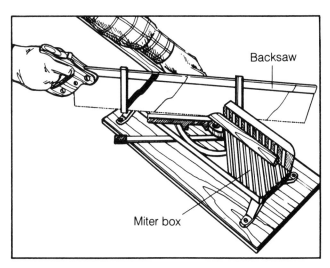

Figure 6-11. Use a miter box and a backsaw to cut the ends of quarter-round molding at a 45-degree angle.

Outside Baseboard Corners

These pieces should first be cut to a 45-degree angle with your miter box and back saw. Nail them into place so they are snug. Then run the round part of your hammer handle or screwdriver shank tightly up the joint to seal any gap that may be left (Figure 6-9).

Inside Baseboard Corners

These corners can be tricky when you are using the types of baseboards that have curves and shapes to them. If you use a miter cut here, the nails will pull the intersecting pieces away from each other, creating a gap. If you are painting, this crack can be filled with caulk and painted over. If you prefer the appearance of the natural wood trim, a process called "coping" will give you a beautifully fitted corner every time.

Begin by butting the first piece of molding up to the wall and nailing it in. Place the second piece of molding in your miter box for a 45-degree cut with the inside wood grain showing on the front of the molding. Then use a coping saw at a 90-degree angle to cut away the wedged portion of the molding, following the outside line exactly (Figure 6-10). This method transfers the perfect profile to the molding, which will fit snugly over the mate at the wall.

Joints Along the Length of the Baseboard

These joints can be ugly and tend to pull apart if they are merely butted up against each other. To avoid these unsightly gaps, miter cut the ends at a 45-degree angle (Figure 6-11) so that the two pieces of molding overlap each other. To keep the joint tight in case the molding shrinks, apply some glue at the miter joint between the two pieces. Allow

the glue to ooze out of the joint and get tacky. Then lightly sand the joint; the glue will mix with the sawdust and conceal the joint.

TONGUE-AND-GROOVE PANELING

Paneling with solid wood tongue-and-groove boards is more difficult than with hardboard panels, but their weight and stiffness mean they require less support. They also offer a real wood appearance. Either special boards for paneling or tongue-and-groove hardwood flooring can be used.

Step One
Applying the Boards

Margin of Error: Exact

Most Common Mistakes

- [ ] Not adding insulation or a vapor barrier over an outside or basement wall, where appropriate.
- [ ] Splintering boards by cutting them face up with a saber or circular saw.
- [ ] Neglecting to check that each piece is plumb on the wall before applying the next.
- [ ] Transferring measurements to the board incorrectly or to the wrong side.

If you want to install the boards vertically, you need to put up horizontal furring strips so that the boards have something to hold onto. Use adhesive or nails to attach the furring strips horizontally across the studs or masonry wall every 24" to create a solid backing.

If you choose to install this type of paneling horizontally or on the diagonal, furring is not necessary if your drywall is in good shape. Always start the boards in a corner with the groove edge facing the direction in which you are paneling. (The tongue edge of the first board should be in the corner.) The boards can be glued or nailed to the walls, or both glued and nailed. The easiest method is to use a paneling adhesive and toothed trowel, or a paneling adhesive applied with a caulking gun. Follow the adhesive instructions for application. If you are using the cartridge method to apply the adhesive, an air compressor with a caulking gun attachment will save you a great deal of time and effort.

For vertical paneling, apply the boards in a stairstep pattern, being sure that each one is plumb. All butt joints should occur on a furring strip. Use a small piece of paneling board as a pounding board to force two adjacent pieces together. The last piece, next to the intersecting wall, should be cut to size and its tongue slipped into the groove of the adjacent board. It should snap right into place. You may need to make a scribe cut, as explained earlier. However, corner trim often eliminates the need for scribing.

Step Two
Fitting Boards
Around Openings

To panel around large openings, first apply full-size boards as close to door and window jambs as possible. Then measure the distance from the edge of the last board to the edge of the opening. (You will need to notch some boards, and cut others to specific length, above doors and windows and below windows.) Then transfer your measurements very carefully to your board, and just as carefully make your cuts. If you are notching, cut only to within 1/4" of the intersection of two cuts with your circular saw. Then finish the cut with a handsaw if the board must butt up against existing trim, since the trim will not cover the cut. Otherwise, use a circular saw with the board face down so the circular blade doesn't cause the board to splinter. This step takes patience and care because you are working with such narrow pieces. Each board must be measured and trimmed separately. To apply trim, see Step Three under Wood-Faced Paneling.

SHELVES

The shelves described here can be applied to both wood-faced and tongue-and-groove wall paneling. Properly planned shelves will enhance the appearance of your new walls, as well as providing you with storage space.

Design

To minimize design and fitting problems, use graph paper to draw plans of your proposed shelves. Take into account the visual effect you want from your shelves, what they will hold, and the structural elements needed. Your plan will be

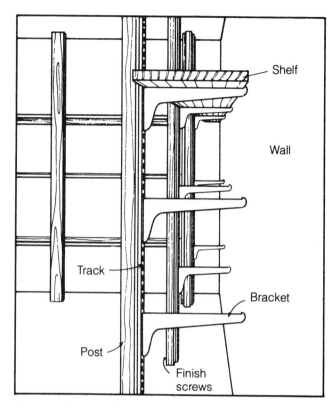

Figure 6-12. Suspended shelving is attached to floor-to-ceiling posts.

helpful in making a materials list. Be sure that you have all necessary materials on hand before you begin building.

Suspended Shelves

Probably the most common shelves are those in which metal tracks are attached to a wall for hanging movable metal brackets. Wood or plastic shelves rest on the brackets. These shelves are very simple to construct. The main installation concern is that the tracks must be attached to the studs behind the wall.

There are several ways to find the studs in a finished wall. Studs are usually located every 16" or 24" on center. Tap along the wall with a hammer until you hear a solid sound; look for seams in the drywall; or use a magnetic stud finder to locate the nails in the studs. Mark the position of the stud on floor and ceiling and snap a chalkline to mark the center of the stud on the wall.

If you don't want to detract from the visual effect of your paneling, or you want more wood and less metal to show, suspended shelves are an attractive alternative to mounting conventional shelving brackets directly to the wall.

Create front supports for the shelves with posts running from floor to ceiling. Space them out from the wall to the depth you want your shelves to be (Figure 6-12). Attach the supports to the floor by drilling holes at the base and securing them with finishing screws. With these supports you can still use the inexpensive track-and-bracket construction. Attach the shelving brackets to the insides of the posts, facing the wall. This hides the tracks and minimizes the amount of metal that is visible.

Closet Shelves

Before building your closet shelves from scratch, you may wish to investigate the many ready-to-assemble closet storage systems that are now available.

To build closet shelves, attach 1x4 ledger boards to the closet wall at the height and length you wish your shelf to be. These ledgers should be level and attached directly to the wall studs for maximum support. The shelf board will rest on the lip of the ledger.

To install a clothes-hanging pole, you will need plastic or wood pole supports that screw into the frame or ledger at ei-ther end of the closet. If the pole is over 5′ long, add a supporting pole bracket in the middle, attached to the ledger.

Recessed Shelves

Many rooms have recessed cubbyhole walls or even closets that will accept shelving. It is a simple matter to measure and cut boards to fit vertically on either side of the alcove. Inexpensive hardware tracks can then be screwed plumb into the boards and adjustable clips installed in the tracks for multi-position shelving.

First measure and cut the shelving from plywood or 1" stock. Then face the plywood with a 1x2 to cover up the rough edges. If you are using wood boards, sand the shelves with an orbital sander for a fine finish, then paint or stain them as desired.

7 Ceilings

BECAUSE CEILINGS are more "out of the way" than other parts of the interior of the home, such as walls and floors, their impact on the general feel or mood of a house is often underestimated. The floor and ceiling are the largest surfaces in any room, and different ceiling heights, materials, colors, slopes, and angles have a definite effect on the way a room feels to its occupants.

In addition to being an important design feature, ceilings hide the framing of the floor or roof above, plumbing pipes, ductwork, electrical wire, insulation, and so on. Suspended ceilings (ceilings that are hung a set distance from the ceiling joists) work very well for this kind of cover-up because they create a cavity that allows easy access for repairs or further work.

In the past, the most common types of ceilings were lath and plaster, and wood planks. Metal ceilings were also popular because of their light weight and ease of installation. In the past few decades drywall has replaced lath and plaster as a wall and ceiling material, largely because of the ability of drywall to slow the spread of flame. The techniques of working with drywall are presented in Chapter 5. This chapter describes the installation of acoustical ceilings and suspended ceilings.

Before You Begin

SAFETY

☐ Always use the appropriate tool for the job.

☐ Wear the proper dust mask or respirator when sanding or sawing.

☐ Wear safety glasses or goggles whenever you are using power tools, or when hammering overhead, especially if you wear contact lenses.

☐ Watch power cord placement so that it does not interfere with the operation of the tool.

☐ Keep blades sharp. A dull blade requires excessive force and can easily slip.

☐ When using a stepladder, have both pairs of legs fully open and the spread bars locked in place. Never climb higher than the second step from the top. When bracing a ladder against the wall, a safe distance between the feet and the wall is one quarter of the height of the ladder. Do not use an aluminum ladder when working near electrical wires. Consider using scaffolding.

USEFUL TERMS

An **acoustic ceiling** has tiny noise-trapping holes to improve the quality of sound in a room.

Ceiling joists are the horizontal framing elements that support the ceiling drywall.

Ceiling tiles are 12"-square tongue-and-groove tiles that are attached to wood or metal furring strips or tracks.

Cross tees are the elements of a gridwork that connect at right angles to runners.

Furring strips are strips of metal or wood attached directly to an old ceiling (perpendicular to the ceiling joists), onto which ceiling tiles are clipped or stapled.

The **runners** form the main support grid for suspended ceilings. They are installed perpendicular to the joists.

A **suspended ceiling** is a ceiling hung from the original ceiling or from the joists by a grid system, often used to hide exposed joists, rafters, and ductwork.

Tegular panels are two-level panels in which the face is lower than the flange, which rests on the grid.

Tracks are metal furring strips that support the ceiling tiles.

A **valance** is a canopy or covering above a window that creates a decorative effect.

WHAT YOU WILL NEED

Time. Installing either a tile ceiling or a suspended ceiling in a 9'x12' room will require 14 to 20 person-hours, longer if unusual situations are encountered. This job is best undertaken by two people.

Tools and Materials. Suspended ceilings are often hung from the ceiling joists with a metal grid. This creates the cavity between the joists and the ceiling where pipes, wires, and ductwork can be installed and worked on. A tile ceiling is glued directly to a drywall ceiling or onto furring strips that are glued or nailed to the ceiling joists. This type of ceiling works well where height is a consideration and a suspended ceiling would drop too low.

A suspended ceiling isn't hard to install and requires no unusual tools. However, the metal gridwork, especially the 12' runners, requires two people for installation.

Tile ceilings are even easier to install than suspended ceilings. Wood furring strips are nailed to the old ceiling or to the ceiling joists and the new tiles are stapled to these strips. An even simpler system uses 4′ metal tracks instead of long furring strips. Instead of staples, clips snap onto the tracks to lock the tiles in place. Everything you need for this method of installation comes in one kit.

Tools for Suspended or Tile Ceilings

20′ to 25′ metal tape

Straightedge

Framing or combination square

2′ to 4′ level

Utility knife

Ladder

Hammer

Chalkline

Pencil

String

Putty knife

Nails

Drywall pan

Handsaw

Nail belt

Face mask

Safety goggles

Drill

Miter box

Coping saw

Additional Tools for Suspended Ceilings

Pliers

Aviation snips

Additional Tools for Tile Ceilings

Fine-toothed hacksaw

Screwdriver

Materials for Suspended Ceilings

Eyehooks

Hanger wire

Wall molding

Molding nails

Runners

Cross tees

Panels

Lighting fixtures

Materials for Tile Ceilings

Furring strips

Adhesive

Shims

Tiles

PERMITS AND CODES

Check with your building department to find out about necessary permits.

DESIGN

Ceilings can do many things besides looking nice and hiding structural elements. They can also muffle noise, support lights, and retard flames. An acoustical ceiling with tiny noise-trapping holes or fissures is a wise choice for noisy rooms like entertainment centers.

If you want to put lights in your new ceiling and to be able to relocate them easily, a suspended ceiling is your best bet. You can buy fluorescent fixtures that fit into the grid system in place of a standard-size ceiling panel.

The acoustical ceiling uses space-age technology to create a ceiling that is both lightweight and durable, in addition to absorbing sound. These ceilings come in many different styles, sizes, and colors. You can get tiles or larger panels that look like marble, oak, and other natural materials, with the designs authentically hued, shaded, veined, and striated. If you prefer traditional white, pattern choices include reproductions of bleached wood, sculptured plaster, and rough-troweled stucco.

Most Common Mistakes

☐ Not planning the ceiling layout on paper before beginning work.

☐ Not checking local code for minimum ceiling height and clearance.

☐ Failing to plan grids to avoid running into columns or posts.

☐ Measuring the ceiling height line on the wall from a sloping floor, which creates a sloping ceiling.

☐ Failing to lay out runners so that border tiles will be more than half a tile.

☐ Not installing the runners level.

☐ Soiling the tiles during installation.

☐ Neglecting to have the rough electrical work inspected before installing a suspended ceiling.

☐ Failing to correct any leaks in the roof before installing the ceiling.

☐ Applying loose-filled or roll insulation directly above the ceiling panels rather than in ceiling joist cavities.

THE SUSPENDED CEILING

A suspended ceiling is used where enough height is available to hang it from the ceiling joists and still have enough height between the floor and the new ceiling. A minimum of 7′6″ is usually required. The advantage of a suspended ceiling over a tile ceiling is that it allows easy access to pipes, wires, and ductwork. This is very useful if you need to repair or add on to any of these systems. It is also useful in rooms where you want to reduce the ceiling height, either for es-

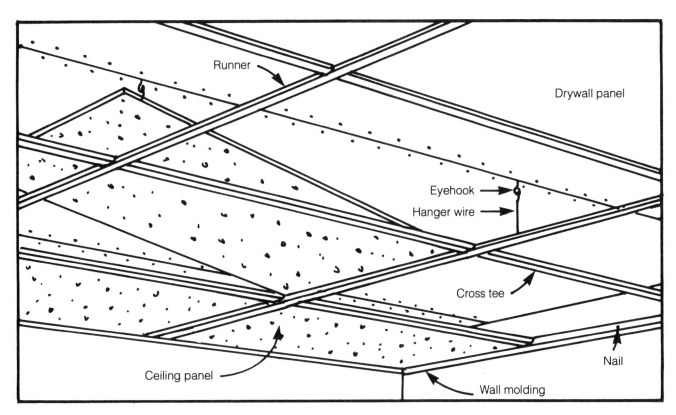

Figure 7-1. The anatomy of a typical suspended ceiling.

thetic reasons or to save energy in rooms where high ceilings allow hot air to gather at the top.

Suspended ceiling systems vary from manufacturer to manufacturer, but the installation techniques are similar. The typical suspended ceiling is formed of a grid and panels. A few simple pieces hook together to form the grid, and the panels are placed in the cells of the grid. The angle molding or wall molding is attached to the wall and supports the runners, which run the long way in the room. These runners are also supported by wires that are attached to the ceiling joists by eyehooks. Cross tees, running perpendicular to the runners, hold the runners together and create spaces or cells into which the panels are inserted. The panels are usually 2'x2' squares or 2'x4' rectangles. Figure 7-1 shows how the entire system goes together.

Here are a few important things to remember about how the grid goes together.

1. The runners, not the cross tees, run perpendicular to the ceiling joists.

2. The runners are hung from the joists with wire and eyehooks.

3. The cross tees run parallel with the joists and perpendicular to the runners.

4. The wires supporting the runners are often at an angle, since the holes in the runners into which the wires are threaded occur periodically and are not always directly below the wire.

5. Because the wires are often at an angle, each wire must be bent individually around a reference string to be sure their

bends all occur at the same level. Communities in areas of seismic activity often require diagonal braces, also made of wire, to inhibit movement in these ceilings.

6. The slots in the runners, into which the cross tees interlock, occur at intervals along the runners. These slots must all line up perfectly in the rows of runners so that all of the cross tees will be on line and perpendicular to the runners. To achieve this, you will need to cut the first runner installed in each row to be sure the cross tee slots in each row of runners are aligned.

Step One
Planning Your Ceiling

Margin of Error: Not applicable

Careful planning is essential for a successful ceiling installation. Neglecting or shortchanging this step will cause you great frustration later in the project. Plan well, and the rest of the job will go easily. This section outlines the important issues in planning and ceiling installation.

Border Tiles

Border tiles are the tiles that run next to the walls. Because room dimensions are rarely multiples of exactly 2' or 4', these tiles must usually be cut to fit. For appearance' sake, you want the border tiles on opposite sides of the room to be equal and all border tiles to be more than half a tile. You will

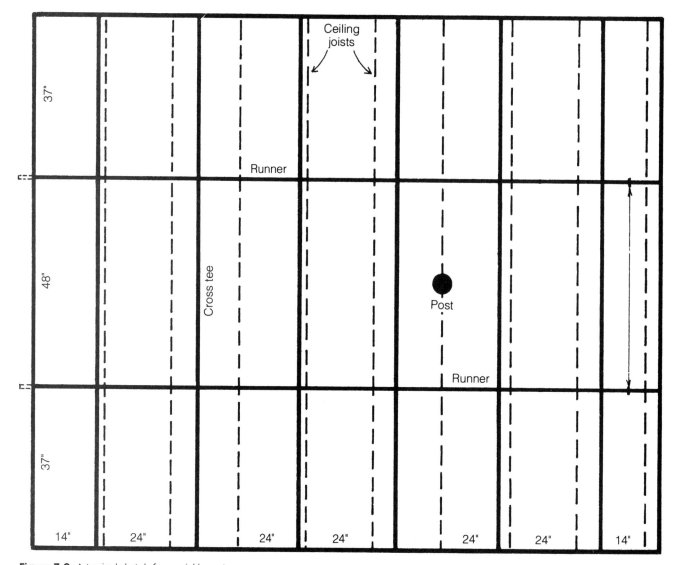

Figure 7-2. A typical sketch for a grid layout.

need to place the cross tees and runners in such a way that you can accomplish this goal.

Using graph paper, make a diagram of where all your runners and cross tees will go. First you must determine the size of all border tiles so that they are more than half a tile and equal in size. To do this, convert the length of the room's short wall into inches. If you are using 2′x4′ panels, divide this measurement by 48 (inches) if the panel length is to run parallel to the short wall. If you are using 2′x2′ panels, or if the panel length will run parallel to the long wall, divide by 24″. Take the remainder (in inches) of this division and add 48″ if the panel length will run parallel to the short wall or if you are using 2′x4′ panels. Add 24″ if the panel length will run parallel to the long wall or if you are using 2′x2′ panels. Half of this final figure equals the border dimensions at each side of the room.

Here's an example. For a room that is 10′2″ wide with the panel length running parallel to the short wall and using 2′x4′ panels:

1. Convert 10′2″ to inches (10′2″ = 122″).

2. Divide 122″ by 4′ = two full panel lengths and a remainder of 26″.

3. Add 48″ to 26″ = 74″; then divide by 2 = 37″.

Thus the border panels at each side will be 37″. There will be one full-size panel and two 37″ border panels (37+48+37 = 122). This dimension of 37″ also equals the distance of the first runner from the sidewall.

Repeat these calculations using the length of the room to find the end border panel size. In our example, the length of the room is 12′4″, or 148″.

Laying Out the Grid

Now that you have determined the size of the border tiles, you can draw a full grid on the graph paper, as shown in Figure 7-2. Indicate the runners on your graph paper by drawing the first and last runners at a distance of one border tile from the sidewalls and perpendicular to the ceiling joists. Add the in-between runners at intervals of 4'. Using a different color pencil, mark the cross tees on the layout sheet. Start at the border tile distance from the end walls (14" in our example); then add the in-between cross tees every 2'. If you are using 2'x2' panels, additional cross tees will be needed, locking into the perpendicular cross tees halfway between the runners.

Columns or Posts. You should take into account any columns or posts in the room that support the floor above. This is a common situation in basements. The grid must be planned so that *no* runners or cross tees run into a column or post. If this does happen, you will need to make slight adjustments, lengthwise or crosswise or both, so that the column falls in the open area of the grid. This adjustment may require unequal border tiles.

Light Fixtures. The location of light fixtures will not require any changes in your grid, but they should be marked on your sketch or grid plan. Remember to do all your rough wiring before installing the grid. Then cut out for the fixtures as you place the ceiling tile.

Estimating the Materials

Wall moldings are available in 10' lengths that are butted together. Measure the perimeter of the room and divide by 10 to find the number of wall molding pieces you will need.

Runners are available in 12' lengths. Tabs at each end of the runners make it possible to join them for lengths longer than 12'. However, no more than two sections can be cut from each 12' length of runner.

Cross tees come in 2' and 4' lengths with connecting tabs at each end. These connecting tabs are inserted into indentations or slots that occur at intervals in the runners and cross tees.

Wire fasteners or eyehooks are necessary at each support point to attach the hanger wire to the ceiling joist.

Hanger wire of 12 to 16 gauge is needed to hang the runners every 4'. The wire should be 6" longer than the distance between the eyehook and the ceiling.

Your graph paper layout will tell you how many ceiling tiles you need to buy. Don't forget to allow for waste on the cut border tiles. Also purchase any lighting fixtures now.

Step Two
Establishing Level Lines and Installing the Wall Molding

Margin of Error: 1/4"

The following installation method may vary from manufacturer to manufacturer, but most methods are similar. Also, the method is usually the same whether you are installing 2'x2' panels or 2'x4' panels.

The first thing to do is to establish level lines on the wall at the height of the new ceiling. Remember that the code may

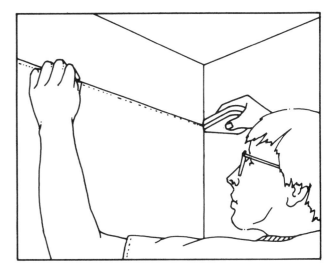

Figure 7-3. Use a chalkline to snap level lines for wall molding.

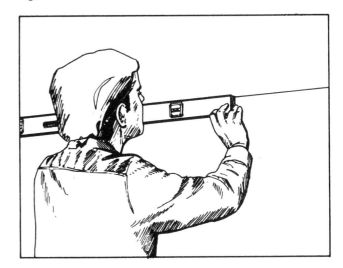

Figure 7-4. Use an 8' level to mark the placement of the wall molding.

require a minimum floor-to-ceiling height, usually 7'6". Try to leave as large a cavity as possible, a minimum of 2" if no lighting fixtures are involved.

Marking the level lines can be simple if your floor is level. In this case, all you need to do is measure up in each corner the height of the ceiling to be installed (in our example, 8'), mark the walls at this point, and snap chalklines connecting these points (Figure 7-3). You now have a level line running around the wall at ceiling height. You will nail your wall molding at this level.

However, if you use this method with a sloping floor, your ceiling will slope to match the floor. If the floor is off level to any great degree (1/4" or more), you will need to determine the difference at each corner and adjust your measurements accordingly. Use an 8' level to determine the dip or rise of your floor. Or you can measure down from the ceiling or joists above, if they are level. You can also simply use an 8' level to draw a level line on all the walls, starting from any point (Figure 7-4).

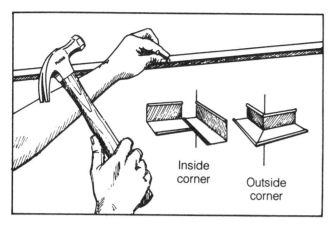

Figure 7-5. Nail the wall molding to the wall, lining up the top edge of the molding with your chalkline. Butt the inside corners and miter cut the outside corners.

Figure 7-6. Use a scrap of cardboard to hold masonry nails being driven between mortar joints and the edge of the block.

Next you will need to prepunch holes in the wall molding to correspond to the wall studs. You can locate these studs by tapping along the wall with a hammer until you hear a solid sound; look for seams in the drywall; or use a magnetic stud finder to locate the nails in the studs.

Nail the wall molding to the wall around the entire perimeter of the room, lining up the top edge of the molding with your chalkline (Figure 7-5). Butt the molding at the inside corners and miter cut it at the outside corners. If you are fastening the molding to a concrete block wall, use short masonry nails and direct the nail between the mortar joints and the edge of the block. Use a piece of scrap cardboard with a notch cut in it to hold the nail before driving it in (Figure 7-6). If the wall is solid concrete or otherwise unable to accommodate wall molding, hang a section of runner directly next to the wall as a substitute.

Step Three
Snapping Chalklines
for the Runners

Margin of Error: 1/4"

Now you are ready to snap chalklines across the bottom of the ceiling joists to mark where the runners will be. The object of this runner layout is to be certain that no border tile is less than half a tile (24" with 4' tiles). This process was explained in Step One, Planning Your Ceiling. Using your tape measure, snap chalklines perpendicular to the joists. In the example given in Step One, these lines are 37" from either wall, leaving a 48" space in the middle (see Figure 7-2).

Step Four
Installing the Wire Fasteners
and Hanger Wire

Margin of Error: 1/4"

Now you are ready to attach the wire fasteners (eyehooks) and hanger wire that will support the runners. Screw the eyehooks into the ceiling joists where the joists intersect the runner chalklines. Your eyehooks can be placed at 4' intervals; you do not need a fastener at every intersection. Then thread your wire through the eyehooks and wrap it securely around itself three times. The wire should be long enough to extend at least 6" below the level of the new ceiling. Add extra hangers and wire at light fixtures, one for each corner or as instructed by the manufacturer.

Step Five
Stretching Your First
Reference Strings

Margin of Error: Exact

Now you need to stretch some reference strings, which must run exactly at the level of the bottom of the wall moldings. These strings show you precisely where the runners and cross tees will be and serve as guides for cutting the runners and cross tees.

To stretch these strings, choose the corner of the room where you plan to start your installation. As shown in Figure 7-7, you will stretch only two strings: string AB, which will be the border tile distance from one wall (14" in our example); and string CD, which will be the border tile distance from the other wall (37" in our example). Be sure the strings are at the bottom of the wall molding so they will be out of the way while you install the grid. Check your strings with a framing square to be sure they are square. Adjust them until they are square, even if it means that the border tiles must be irregular.

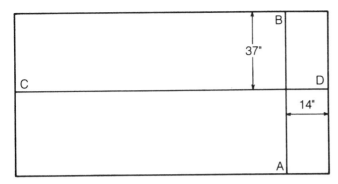

Figure 7-7. Stretch reference strings AB and CD in place at the border tile distances from the walls.

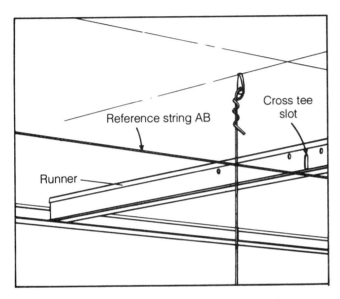

Figure 7-8. Reference string AB aligned with cross tee slots.

Step Six
Measuring and Cutting the Runners and Stretching New Reference Strings

Margin of Error: 1/8"

You are now ready to install the runners. The most important thing to remember is that the runners have to be cut and installed in such a way that the cross tee slots in the runners occur exactly where each cross tee is planned to intersect the runners. The cross tees occur at line AB and then every 24" from line AB (Figure 7-8). You will need to cut the runners' length so that a slot appears exactly along line AB. Slots will automatically be aligned every 24" from there, since they are spaced on the runners every 24". Measure and cut each runner so that the slots are aligned with line AB. Measure each runner; do not use the first runner as a pattern for the rest.

Now hold the cut runner at the level of the strings and next to string CD (Figure 7-9). Locate the wire support hole

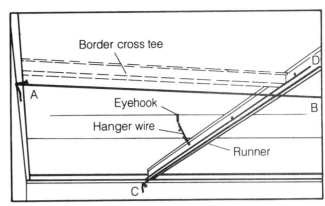

Figure 7-9. Cross tee location in grid. Note that wire hanger is at an angle.

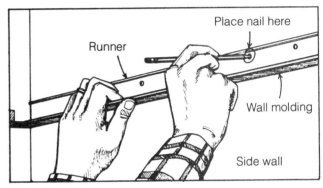

Figure 7-10. Mark location of first wire hanger hole on wall. Drive nail at this point.

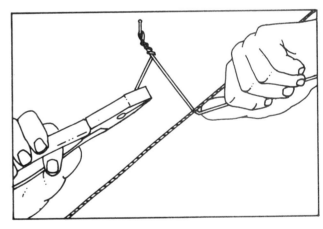

Figure 7-11. Bend wire at right angle around reference string.

(holes on tops of runners that hanging wires are attached to) that is closest to the first wire hanger from this wall. Circle this hole on the runner.

Carry this runner to the long side wall and rest it on the wall molding with the cut end butted against the short end wall. Mark the wall with a pencil or felt-tip pen through the circled hole (Figure 7-10). Remove the runner and drive a nail at this mark to attach another string to. Repeat this procedure on the opposite wall. Then stretch a string tightly from nail to nail. This string is used as a guide to bend your hanger wires at the proper place. Note that the string is aligned with the wire support holes for the first set of wire hangers (although not necessarily with the hanger wires themselves). It is also running level with these holes and therefore indicates where the wire must be bent at an angle of 90 degrees to ensure that the runner is at its proper level.

Align each of the hanger wires to intersect with the string and make a 90-degree bend in the wire where it intersects the string (Figure 7-11). This usually results in support wires that are at an angle in their final position, but the runners are hung perfectly level with the surrounding wall molding.

Step Seven
Hanging the Grid
(Cross Tees and Runners)

Margin of Error: ⅛"

You are now ready to hang the runners and clip in the cross tees. You can run additional strings to show where to bend each wire. Run these strings aligned with each series of wire support holes, into which wires will be inserted.

An easier method is to place a level on the runners and bend the wires as you go. To do this, rest the cut end of the runner on the wall molding. Thread the first prebent wire hanger through the support hole. (This wire hanger has been prebent so that the runner will hang at just the right level; the rest of the wire hangers have not been prebent.) Place a 4′ level on the runner and have someone hold it exactly level as you thread the unbent wire hangers through their support holes and bend them so that they keep the runner level. Wrap all wires around themselves three times after threading through the runners.

As you work toward the opposite wall, connect the runner sections together as needed. Be very careful with your measurements. You will need to cut the last piece to fit. (You can sometimes use the excess piece to start the next row.) As you connect and install the pieces, check to be sure that you have cut each runner so that the cross tee slots align exactly with reference string AB. You will have better support if you position an additional wire hanger close to the point where two runner sections are connected. Be sure you are hanging the runners at the right level.

After all the runners are in place, install the cross tees (Figure 7-12). For 24x48 tiles, these 48" pieces are installed every 24" perpendicular to the runners. For 24x24 panels, add a 24" cross tee to the midpoint of the a 48" tee. This process is easy if you have done everything correctly up to here. First install the full-size 4′ cross tees in the interior of the room. Once these are in place, lay in a few full ceiling panels to stabilize the system.

Now you must cut the cross tees in the border areas. Most tiles can be cut with a straightedge and a utility knife. Score the tile deeply on the finished surface and flex it on a hard edge, just as you would drywall. Rasp away any unevenness.

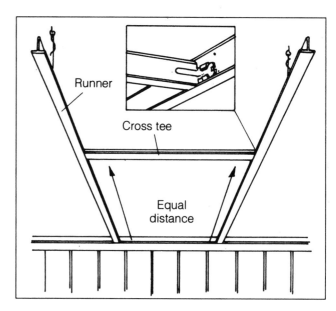

Figure 7-12. Place cross tees into slots on runners. (Slots must be directly across from each other on adjacent runners.)

Measure each border cross tee individually, using string CD, not the runners, as a measuring guide. The runners still have a lot of play in them. To measure, align the edge of the first runner with reference string CD. Then make your measurement from the side wall to the near edge of the runner. Measure and cut accurately. Repeat this process on the opposite wall.

Step Eight
Installing the Panels

Margin of Error: Not applicable

Installing the full panels is as easy as just slipping them into the grid from below and dropping them into place. Make sure your hands are clean, or wear clean gloves, so you don't soil the white panels.

The border panels will need to be cut to size. The larger 24x48 flat panels usually require only a straight cut with a sharp knife or saw to fit snugly between the grids and the wall molding.

If you are using tegular 24x24 panels, you will have to cut in a flange on the freshly cut edge of the border tiles so they will fit the grid (Figure 7-13). (Tegular means two levels; the face of the tile is lower than the grid level.) To cut in the flange, measure the size of the opening plus the flange. Place the cut panel in position in the grid. The cut side without the flange should rest on the wall molding. Draw a pencil line where the wall molding meets the border panel. Be sure the opposite end of the panel is centered in the grid.

Remove the panel and lay it face up on a flat surface. Cut along the pencil line so that your utility knife only penetrates half the thickness of the panel. Then, laying your utility knife flat on the table, cut into the side of the panel, halfway

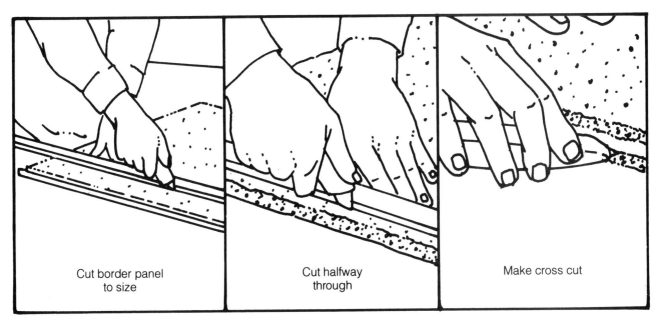

| Cut border panel to size | Cut halfway through | Make cross cut |

Figure 7-13. Three-step process for cutting border tiles to size and cutting flange.

through its thickness, to meet your original cut, thereby creating the flange. Use special paint, available where you bought the tile, to restore the surface that you cut. Now install the panel in your grid.

Step Nine
Special Situations

Margin of Error: Not applicable

Boxing Around Basement Windows

To install a suspended ceiling around basement windows, build a three-sided valance around each window, as shown in Figure 7-14. Use ¼" plywood for the top and 1x6 pine for the three sides. Be sure to build the valance wide enough to allow the window to open and long enough to allow open drapery. In most cases, 9" on either side of the window, for a total of 18", is sufficient for open drapes.

Attach the top of the valance to the bottom of the ceiling joist and install the wall molding for the ceiling panels at the desired level. Curtain rods can be attached to the inside of the valance.

Boxing Around Iron Support Beams and Ductwork

Many basements have horizontal iron support beams that support the first floor. To box around them, follow these steps (Figure 7-15).

Construct a wooden lattice to attach to both sides of the support beam. Use 1x1½ wood strips (1x3 ripped in half) and 1x3 center supports spaced every 16" to build the lattice. Nail the lattice to the floor joists that run on top of the iron beam.

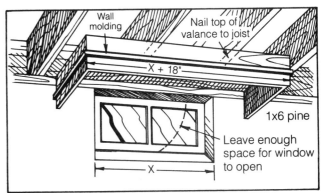

Figure 7-14. Boxing around a basement window.

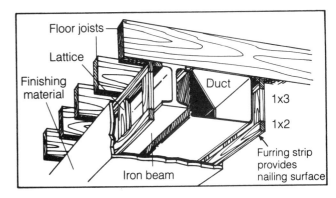

Figure 7-15. Boxing around a beam and duct.

Enclose the support beam by nailing a finishing material, such as drywall or paneling, to the lattice. Attach the same material to the bottom of the lattice to cover the bottom of the beam. Attach the outside corner molding. Now snap a chalkline onto this finished box at the height of the new ceiling and nail the wall molding along this line. This method can also be used to box around ductwork.

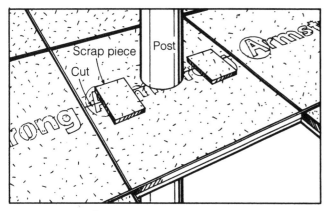

Figure 7-16. Cut tiles to fit around a post or column.

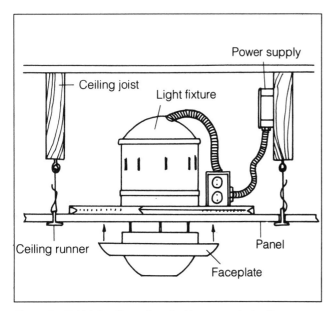

Figure 7-17. Lighting fixture installed in suspended ceiling.

Fitting Around Columns and Posts

Cut the panel in two at the midpoint of the column. Then cut semicircles to the size required so that the rejoined panel will fit snugly around the post, as shown in Figure 7-16. Make all cuts with a very sharp utility or fiberboard knife.

Installing Electrical Lighting Fixtures

Lighting fixtures are easy to install in suspended ceilings (Figure 7-17). It is best to install all your rough wiring before you begin the project. There is often a cavity between the panels and the joists, and the panels can be removed, so it is possible to install fixtures later. However, it takes considerably more effort after the ceiling has been put up.

You can use either incandescent or fluorescent fixtures. Fluorescent fixtures often come with a translucent panel that fits directly into a panel cell. Incandescent fixtures are either flush-mounted or recessed. You need to be more careful with incandescent fixtures, because they get much hotter than fluorescent. Incandescent fixtures come with adjustable arms that are attached to the ceiling joists to carry the weight. A hole is cut in the panel for the fixture, and a finishing collar fits around the fixture to hide the rough-cut hole. (For more information on wiring, see Chapter 2, Electricity.)

Care of the Ceiling

Caring for a suspended ceiling is an easy matter. If a panel becomes water damaged, marred, or very dirty, it can simply be replaced. Also, the panels are washable and can be painted.

THE TILE CEILING

Standard ceiling tiles are usually 12" square and come in many patterns and colors. They attach directly to wood or metal furring strips, and leave a large cavity as do suspended ceilings. Unless you use the clip method described in Steps One through Seven, following, the tiles are permanently attached; unlike suspended panels, they cannot be easily removed. However, they are quick and easy to install and work well in rooms where you do not want to lower the ceiling height.

The method of installation described here is a new method that uses metal furring strips and clips. These strips, or tracks, are attached to the ceiling joists and support the tiles. The clips make it possible to remove ceiling tiles without damaging them in order to correct minor mistakes, insert light fixtures, or reach wiring and pipes between the joists. You merely slide the clip back along the track to release the tile. This system reduces nailing by two thirds. Since the metal tracks don't have to be spaced precisely 12" apart as do wood furring strips, this system gives you a greater margin of error. And it doesn't require that you saw the tracks; they simply overlap at the end wall.

This simple system comes in a kit with all the materials and instructions you will need, except for the tiles. Steps Seven through Ten, following, discuss installation using conventional wood furring strips.

Step One
Preparation

Margin of Error: Not applicable

Leave the tiles in open boxes in the room where they will be installed for at least 24 hours, so they can become acclimated to the temperature and humidity. Also, repair any leaks or moisture problems before you begin.

Step Two
Laying Out the Ceiling

Margin of Error: 1/4"

The layout for ceiling tiles is somewhat simpler than that for a suspended ceiling. The main thing you are planning for is the border tiles, the rows of tiles that run along the walls of a room. You want to plan these tiles so that they are never less than 6" (half a tile). Rarely do room dimensions work out exactly in multiples of 12".

For example, assume that the width of the room between two of the walls is 9'8". If you made no adjustments, there would be nine full-size tiles and two border rows of 4" each. To correct for these small border rows, covert 9'8" (116") into 8'20" (116"). Now you can have eight full-size courses and two border rows of 10" each. Work out a similar layout in the other direction (in our example, 12'4" coverts to eleven full-size tiles and two 8" border tiles). In this way, all four border rows of a rectangular or square room are more than 6", and the two opposite border rows are equal.

Remember to make cuts for light fixtures and parts.

Step Three
Installing the First Row
of Metal Furring Strips

Margin of Error: 1/2"

The metal furring strips or tracks need to be nailed to something solid in order to hold the ceiling and prevent it from sagging. The ceiling joists will work fine when the furring strips are nailed perpendicularly across them. These strips usually come in 4' sections. There are a few things to remember about installing the strips. They must be well secured to the ceiling joists, level, and parallel.

Nail the first row of metal furring strips 1" away from the wall. Nail at each joist with the nails provided in the kit. Give the strip a tug to be sure it is well secured. Use two nails for every 4' of track nailed into the joists to prevent it from sagging after the tiles are up. Continue placing the 4' strips 1" from the wall. It is not essential that the ends of the strips butt right up against each other; there can be a gap of as much as 1/4" between the 4' strips. Also, the joints do not need to meet over a joist. Just be sure that the tracks are in line, since you may need to attach a clip over the junction of two tracks.

When you reach the other end of the wall you may need to cut a section of the 4' strips to fit. This can be done with a fine-tooth hacksaw.

Step Four
Installing the Remaining Rows

Margin of Error: 1/2"

After the first row of furring strips is in place, proceed with the second row. This second row of furring strips should be placed a distance from the wall that is the size of the border tile less 2". In our example, where we have 10" border tiles, this second row would be placed 8" (10"-2"=8") from the wall.

Measure as you go, being sure to keep a constant 8" from the wall. Nail this row up in the same manner as you did the row next to the wall, being sure to stagger the joints. (The joints between two 4' sections of strips should not occur at the same place in two adjacent rows, but rather should be staggered. A series of joints in a row across the furring strip could cause a noticeable dip in the ceiling.)

After the second row is installed, go on to the remaining rows, which are all placed 12" apart. Again, the next to the last row should be at a distance from the wall equal to the size of a border tile less 2" (8", in our example). The last row should be 1" from the wall.

Step Five
Squaring the Ceiling
with Strings

Margin of Error: Exact

If your walls meet at true right angles so that the room is a perfect square or rectangle, you can omit this step.

Unfortunately, this is often not the case. In this instance you must set up strings that are at true right angles and lay the tiles according to these strings. The strings will serve as guides for installing and cutting the border tiles. If you omit this step, the tile ceiling will look out of line with the walls.

You need to set up two strings that run at the level of the new ceiling and out from each wall the width of the border tiles at that wall, similar to those shown in Figure 7-7. The two strings are attached to nails driven temporarily into the walls, and intersect near a corner. The strings outline the edges of the two intersection border rows.

For example, let's say that one row is a 10" row and the intersecting row is 8". Set up one line so that it is at the level of the ceiling and running 10" from the wall. Do the same with the intersecting string so that it is running 8" from the wall. Once the strings are in place, use a framing square to be sure that the two strings are square. Remember, if they are just a little off square over the 2' width of the tile, the walls could be a few inches off over the distance of the whole wall. Therefore, you need to be sure that both lines are running exactly along the arms of the square. If they are not, adjust one of the strings (preferably the one that outlines the border row that will be least noticeable from below) until the lines are

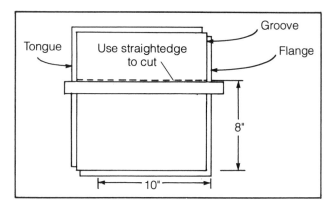

Figure 7-18. Cutting border tiles.

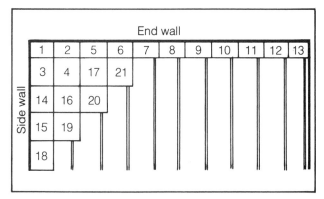

Figure 7-20. Install ceiling tiles in a pyramid pattern.

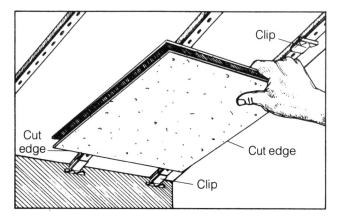

Figure 7-19. Installing the first border tile. Note that the cut edges are against the wall.

exactly at right angles. You now have a square reference from which to start installing the tiles.

Step Six
Installing the Tiles

Margin of Error: 1/8"

Begin installing your tiles in the corner of the room where the two strings intersect.

You will need to cut border tiles to fit. Cut the tile face up with a sharp utility knife. Cut the tile so that the edges next to the wall are the cut edges with no tongues or grooves. All the edges pointing toward the interior of the room should be factory edges (Figure 7-18).

After cutting the first few tiles, snap a clip into each of the first two tracks and push the clips flush against the wall, as shown in Figure 7-19. Place the first tile in the corner and push the cut edge of the tile into the clips. Secure the other edge of the tile by snapping clips into the tracks and pushing them onto the flange. Be sure your hands are clean, or wear clean gloves, so that you don't soil the tiles.

Install the next three tiles that surround your first corner tile. Make sure your tiles fit snugly into the corner and line up with the border reference string lines. This is very impor-

tant, so take your time. Adjust them until they are perfectly aligned.

Once these first four tiles are in place, continue with the installation pattern shown in Figure 7-20. When you reach the opposite side wall, cut the border tiles 1/2" short of the wall to leave a gap for expansion due to moisture. Install the remaining tiles diagonally across the room, in the pattern shown shown in Figure 7-20.

To install the last tile in a row, snap a clip on the end of each track and push it flush up against the wall. Cut the last tile to fit so that it is 1/2" short of the wall. Position the tile in place, then slide the clip into the edge of the tile with a screwdriver, as shown in Figure 7-21.

Finally, nail on the molding. (Paint or stain the molding before installing it.) Be sure to nail the molding every 24" into the wall studs, not into the ceiling (Figure 7-22).

Step Seven
Installing a Tile Ceiling with Wood Furring Strips

Margin of Error: 1/4"

Tile ceilings can also be installed using wood furring strips; in fact, some manufacturers require this type of system. Preparing for installation and calculating the border tiles are the same as for the clip method (Steps One and Two).

The furring strips are nailed perpendicular to the joists. If you are installing 12x24 tiles, furring strips must be installed on 12" centers. When stapling 24" tiles to furring strips, make sure that the tiles run lengthwise along the furring strips, not perpendicular across them.

Use two 8-penny nails at each joist to keep the strips from warping. You need to penetrate into the joists at least 1"; don't forget the drywall thickness. Use a 4' to 8' level to be sure the strips are level, shimming as needed.

The first furring strip is flush up against the wall, as shown in Figure 7-23. The second strip is positioned so that its center is the same distance from the wall as the width of the border tile that butts against the wall plus 1/2" to allow for the

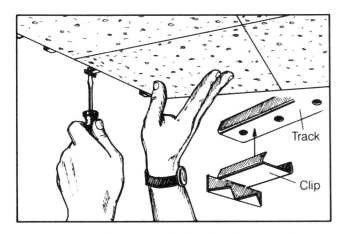

Figure 7-21. Slip the clips onto the last tile with a screwdriver.

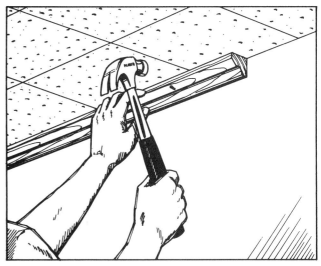

Figure 7-22. Be sure to nail molding into the wall studs, not the ceiling.

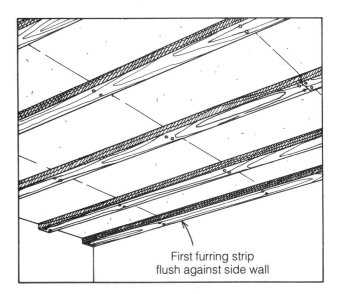

First furring strip
flush against side wall

Figure 7-23. First wood furring strip is flush against the wall. The center of the second is the width of the border tile plus ½".

stapling flange of the tile. For example, if the width of the border tile that butts against the wall is 10", add ½" to the 10" width and position the second strip so that its center is 10½" from the side wall.

After installing the second strip, work across the ceiling, nailing furring strips spaced 12" on center. Be exact. Nail the final strip flush against the wall.

When two furring strips are joined, the butt joint should always be over the center of a joist. Never allow these joints to fall in between joists where they cannot be nailed. Also, stagger the joints so that joints in adjacent rows do not fall on the same joist.

Step Eight
Squaring the Ceiling
with Strings

Margin of Error: Exact

Same as explained in Step Five for the clip method.

Step Nine
Installing the First Row of Tiles

Margin of Error: ⅛"

After your lines are in place, you can cut the border tiles. Cut them face up with a sharp utility knife. Be sure to include the face and flange of your tile in your measurement. Each border tile should be cut and measured individually. Cut them ¼" shy of the actual measurement to make them fit easier. The gap between the wall and the tile will be covered by molding.

Cut the tiles so that the cut edge is against the wall and the factory edge points toward the interior of the room. The out-

side edges of the flanges should line up exactly with your reference strings. Fasten each tile with four ½" or 9⁄16" staples, two in each flange (use six staples for 12x24 tiles).

Work across the ceiling, installing two or three border tiles at a time, filling in between with full-size tiles. When you reach the last row, measure and cut each tile individually. If the stapling flange isn't large enough for stapling, face-nail the tile in place near the wall where the nailhead will be hidden by the molding.

Finally, nail on the molding as described in Step Six for the clip method.

8 Doors and Trim

DOORS ARE DIVIDED into two classes, interior and exterior. Their installation differs slightly, because exterior doors must be windproof and watertight. **Prehung doors,** for either interior or exterior application, are already installed in their jambs when you purchase them. Some exterior doors have a threshold in place, some do not. Interior doors also come in two types. Some come fully installed in their jambs and some come partly assembled, their hinge jambs attached to the door and the head and strike jambs attached to the hinge jamb temporarily. (It is also possible to buy just the door and build your own jambs. However, this chapter describes installing prehung doors.)

All of these prehung varieties are a vast improvement over the old labor-intensive types, which required a great deal of fitting work. Prehung doors are prehinged, with holes for locksets and deadbolts already drilled, and with their jambs cut to exact dimensions. Installing a prehung door is a fairly simple task that requires little physical effort. However, accuracy and precision are essential. If the door is not hung properly, it will be too tight or too loose, and it may open or close by itself.

This chapter also discusses trimming windows and doors. As with hanging doors, accuracy is crucial. A sloppy trim job is highly visible. If you plan to stain your trim or finish with varnish or polyurethane, your work needs to be exact, because every crack will be visible. If you plan to paint the trim, you don't have to be as exact, because any cracks will be filled with painter's caulk. For information on applying trim to wall, ceiling, and floor joints, see Chapter 6, Wall Paneling. For information on installing thresholds, see Chapter 14, Weatherizing. And for information on putting locks in your doors, see Chapter 15, Home Security.

Before You Begin

SAFETY

☐ Use the appropriate tool for the job.

☐ Keep blades sharp. A dull blade requires excessive force and can slip and cause accidents.

☐ Be careful when lifting, to avoid unnecessary strain.

☐ Be sure that electric tools are properly grounded.

☐ Use safety glasses or goggles when hammering, sawing, or drilling, especially if you wear contact lenses.

USEFUL TERMS

The **casing** is the outer trim of the door that covers the gap between the door jamb and the drywall or paneling.

The **head jamb** is the upper portion of the door jamb.

A **reveal** is a small portion of jamb left as a visual detail or to provide clearance for hinge barrels when applying the casing.

The **rough opening** is the hole framed into the wall where the door will be installed.

A **shim** is a small, usually tapered piece of wood used to fill in the space between the door jambs and the surrounding framing studs.

The **side jamb** is the inner portion of the door trim on either side of the door. One side is called the **hinge jamb,** the other the **strike jamb.**

The **stop molding** is the portion of the door trim that stops the door's swing when it is being closed.

WHAT YOU WILL NEED

Time. It takes about two hours to hang a prehung door. The amount of time required for trim work depends on the difficulty of the project and on your familiarity with this work.

Tools

Level

Hammer

Drill and bits

Nailset

Tape

Square

Circular saw

Handsaw

Miter box

Pencil

Electric screwdriver or drill

Materials

Prehung door

Screws

Finishing nails

Trim material

PERMITS AND CODES

The fire code requires that exterior doors in dwellings must be at least 32" wide, and no interior door may be under 24" wide. The code also requires a solid core "one hour" door between the living space and the garage for fire safety.

DESIGN

These days, doors come in a huge array of styles and materials. Modern manufacturing techniques have brought costs down, and modern materials have greatly improved security and reduced maintenance.

Doors make a highly visible design statement; it's a good idea to purchase the best you can afford. Your main design decision is the type and material of prehung door to install. Solid core doors are recommended over hollow core because they offer greater sound reduction and security. Besides wood, doors are made of materials ranging from fiberglass to metal. Foam-filled doors with fiberglass or steel skins look like painted wooden panel doors but are much tougher. These doors also deaden sound and provide some insulative value. In fact, many communities now require insulated exterior doors.

When you purchase prehung doors, you will need to specify the direction in which you want them to swing and on which side you want them to be hinged. This subject deserves a lot of consideration, because the way a door swings can have a large effect on the way a room feels. Most prehung doors are built to fit walls made with 2x4 studs and drywall on each side. If your walls are thicker than that, you will have to order wider jambs and trim them to the proper width.

Trim also makes a statement about your house. Carefully chosen, it can highlight the best features of your home and draw attention away from the less attractive ones.

If you plan to paint your door jambs and trim, they can be of inexpensive "finger jointed" wood—several small pieces of wood joined together to make a long piece of trim or casing. If you plan to stain the wood or simply seal it with varnish or polyurethane, you should order solid jambs and trim. When you order your doors, you will also need to indicate the size and style of your door trim.

HANGING AND TRIMMING A PREHUNG INTERIOR DOOR

Most Common Mistakes

☐ Installing the door off plumb.

☐ Installing the door in an opening that is too small and lacks shim space.

☐ Dinging the casing, trim, or jamb while hammering.

☐ Shimming improperly or insufficiently.

☐ Ordering the door with the wrong direction of swing or hinged on the wrong side.

Step One Inspecting the Door and Opening

Margin of Error: 1/4"

Hanging doors can be frustrating because there are so many things that can go wrong. The trick is to approach the process one step at a time, in the proper sequence.

All doors are made in standard sizes. The dimensions of your rough opening are determined by the size of the door to be hung. Door sizes are designated in feet and inches; a 3'0" door is 36" wide. Figure 8-1 shows the anatomy of a typical prehung interior door.

First, remeasure your rough opening to be sure that the prehung door will fit. The opening should be 1" to 1½" wider and ½" to 1" taller than the prehung door and its jambs. Also measure diagonally in both directions across the opening. If the diagonal measurements are equal (within approximately 1/4"), then the opening is square. However, this

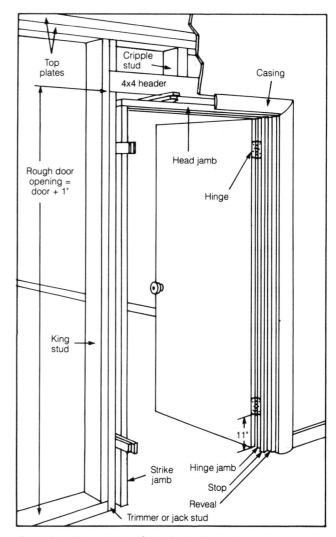

Figure 8-1. The anatomy of a typical prehung interior door.

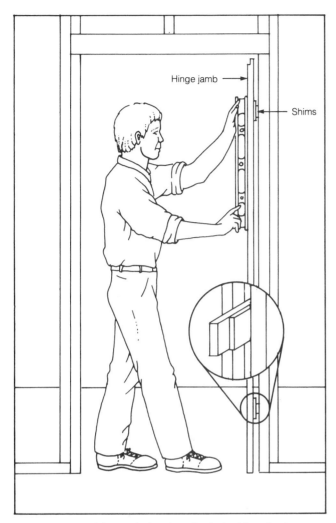

Figure 8-2. Use shims and a level to plumb the hinge jamb.

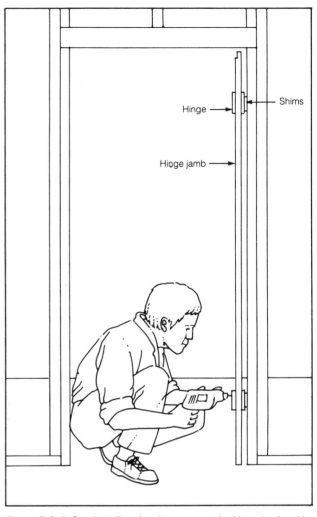

Figure 8-3. Before installing the door, secure the hinge jamb with screws or nails through the jamb and shims into the trimmer stud.

is seldom the case. If the opening is not square, or if it is too large or too small, it must be corrected before you proceed. If the opening is too small, it will have to be reframed; or you can use a smaller door. If it is too large, 1x4s or 2x4s can be added to make it smaller. With wide casing this should not present a problem. With narrow casing, however, there may be gaps that need to be filled in with drywall. Also check the area around the door opening on both sides where the trim will be attached to be sure the area is flat.

As you hang the door, bear in mind that the clearance between the door and the jambs may need to be increased if you are going to paint the jambs and the door itself. One coat of primer and two coats of enamel are about 1/32" thick. A total of 12 coats are applied to the two jambs and two door edges, so the layers of paint approach 1/8" in thickness. On the other hand, if your gaps are too large, weather-stripping becomes a problem on exterior doors, and interior doors don't look right. Ideally, when all varnishing or priming and painting are completed, a gap of 1/8" is left on both sides and at the top of the door.

Step Two
Installing the Hinge Jamb

Margin of Error: Within 1/16" of plumb and flush with the wall surfaces on each side

Remove the blocks that temporarily attach the jambs and casing to the door. Mark the door bottom on the hinge jamb. Take off the hinge jamb by removing the hinge pins, *not the hinges themselves.*

The height of the door off the floor is determined by the length of the hinge jamb. Therefore you should check the floor for level, not only where the door will be placed, but where it will swing as well. The distance the hinge jamb protrudes below the door is the distance between the bottom of the door and the floor. If you plan to install thick carpet or wooden finish flooring later, now is the time to take the thickness of the material into account. If you have a forced air furnace, leave a 3/8" gap at the bottom of the interior doors, to allow warm air into closed rooms and to allow cold air to cycle back to the furnace.

The hinge jamb, cut or shimmed to the correct length, is the first piece to be installed. Traditionally, the hinge jamb was secured with finish nails through small shims. These days, bugle-head drywall screws are often used instead of finish nails because of their superior strength. Shims are necessary because all buildings settle over time, and the space provided by the shims allows adjustment to accommodate this settling.

If your king stud and trimmer on the hinge side of the door are perfectly plumb, you can use small pieces of plywood as shims. Otherwise, use two tapered shims, slid past one another, to adjust the hinge jamb to plumb, as shown in Figure 8-2. The shim thickness should be roughly half of the difference between the rough opening measurement and the total width of the door plus its jambs and gaps. Always shim under the hinges, since this is where the weight is borne.

The edges of the hinge jamb must line up with the wall surface on each side. Hold the jamb in place with your knee and screw or nail through the center of the jamb into the trimmer stud, where the screws will eventually be covered by the door stop trim piece (Figure 8-3). Install the screws with an electric screw gun or a reversible drill.

For increased durability, place additional screws under or through the hinge plate. Tighten the screws enough to compress the shims. If the jamb is not perfectly plumb after the shims are compressed, just back the screws out a little, adjust the shims, and resecure the screws. To put a screw under a hinge, simply remove the hinge jamb plate and screw into the mortised part (but not in the hinge screw holes), and then replace the hinge plate.

Once the hinge jamb is plumb and secure, install the door by mating the door's hinge plates to the jamb's hinge plates and inserting the hinge pins. Make sure that the door doesn't swing on its own.

Step Three
Installing the Head Jamb

Margin of Error: Within 1/16" of level and flush with the wall surfaces on each side

Some head jambs have grooves, called **dadoes;** some have notches, called **rabbets;** and some are simply butt jointed, depending on the manufacturer. The distance above the hung door is determined by the manufacturer with the grooved and notched models, and by you with the butt-jointed models. Some flexibility is possible, even with the notched types. Shims are used to provide appropriate gaps on the butt-jointed types.

With the door in its closed position, stand on the side opposite the swing. Set the head jamb in place and shim the gap from the top of the door as necessary. Use additional shims on top of the head jamb in two places, near the ends, and line up the edges with the wall surfaces. Holding the head jamb in place with one hand, push the door open and screw or nail

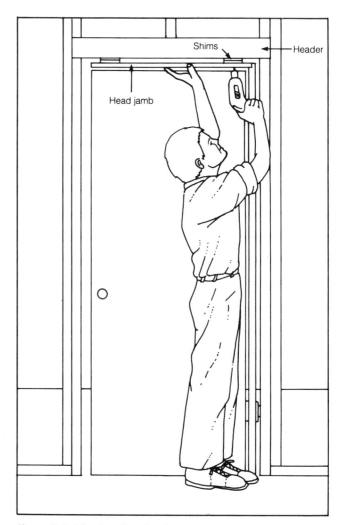

Figure 8-4. After installing the door, secure the head jamb with screws or nails through the jamb and shims into the header.

through the jamb and shims where the doorstop will cover the screw heads, as shown in Figure 8-4. Now close the door and check the gap at the top. It should be even, and about the same size as the gap on the hinge side. Don't use a level or square to set the head jamb; instead, make the jamb conform to the door itself.

Step Four
Installing the Strike Jamb

Margin of Error: Within 1/16" of plumb and flush with the wall surfaces on each side

Like the head jamb, the strike jamb is installed to conform to the door and not to plumb, although if your hinge jamb is plumb and your gaps are even, the strike jamb will automatically be plumb. Be sure to install any premortised strike plate notches on the correct side, as described in Chapter 15, Home Security, and line up the jamb edges with the finished

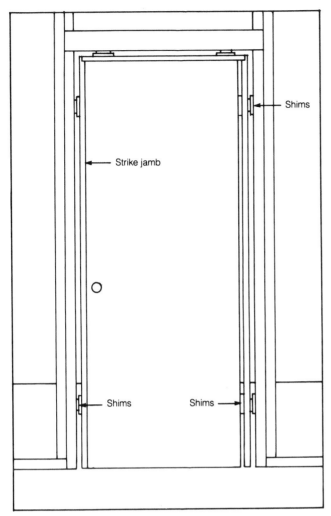

Figure 8-5. Finally, install the strike jamb as you did the hinge jamb.

wall surfaces. By opening and closing the door, you can check the gaps and adjust as necessary (Figure 8-5).

Use shims and screws to straighten all the jambs and perfect the gap widths. For more durable and better-sounding operation of the lockset, it's a good idea to put at least one screw near the strike plate location.

Step Five
Installing the Lockset

Margin of Error: 1/16"

Prebored and premortised doors and jambs make the locksets easy to install. Simply follow the manufacturer's directions, which are included with the hardware.

If you don't have this prebored and premortised type, then you can follow the detailed instructions given in Chapter 15, Home Security.

Step Six
Installing the Door Stops

Margin of Error: 1/16"

After installing the lockset, close the door and mark the location of the door's edge in pencil on the jamb. Now open the door and use small finish nails to install the top piece of stop, lining up the edge of the stop with your pencil mark. If you intend to paint the door and jamb, remember to leave a suitable gap to allow for the thickness of the paint. Be sure to maintain this gap as you install the stops on the side jambs.

Step Seven
Trimming Out the Door Jamb
with Beveled Casing

Margin of Error: 1/16"

Before trimming out the door jamb, you will need to cut off all the shim ends that are sticking out. Try to cut them even with or below the drywall surface. A handsaw works best for this, but you may have to use a chisel for shims that are close to the floor.

If your casing is premitered, you should adjust the length by trimming the bottom, not the mitered end. Start at the hinge side, remembering that the hinge barrels and pins require a certain amount of clearance from the jamb edge. This clearance is known as a **reveal,** because it exposes the jamb

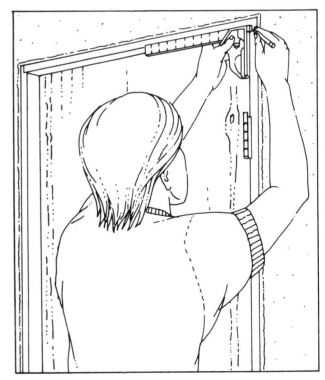

Figure 8-6. Use a combination square to mark the reveal.

Figure 8-7. Secure the head casing with finish nails.

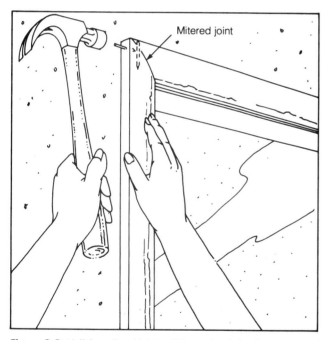

Mitered joint

Figure 8-8. Nail the mitred joints of the casing in both directions.

slightly. The reveal is needed so that you can get the hinge pins out later, and it is sometimes widened for esthetic reasons. To determine the length of this piece of casing, mark the reveal, using a combination square, as shown in Figure 8-6. Add the height of the inside of the hinge jamb to the width of the reveal, and measure this distance from the *inside* of the mitered corner to determine the cut at the bottom. Install this first piece of casing with finish nails so that the beveled end of the mitered casing lines up with the inside corner of the jambs, at the top. Now hold up the head casing in place and check the mitered joint. Plane as required for a tight fit.

Make a mark on the opposite inside edge, again including the reveal distance of the side jambs, and miter this end cut. Check to be sure that the new mitered edge of the head casing lines up with the interior corner of the jamb and secure it with finish nails, as shown in Figure 8-7.

Now check the mitered corner of the strike jamb casing for fit with the head jamb trim and adjust the length if necessary. Maintaining the reveal, secure the last piece of casing. Nail each mitered corner in both directions, as shown in Figure 8-8, to prevent the joints from opening up. Finally, go to the other side and repeat this process. It's a good idea to predrill before you nail the mitered corners to avoid splitting the wood.

Casing that is not premitered is installed in the same way, except that you must make all the miter cuts. Butt-jointed casings are much easier to cut and install, although they are less attractive.

HANGING AND TRIMMING A PREHUNG EXTERIOR DOOR

Hanging an exterior prehung door is generally easier than hanging an interior door, because most exterior doors come completely assembled in their jambs, with thresholds already in place. The whole unit is set into position, leveled and plumbed, and secured through shims to the adjacent framing in one operation.

Exterior doors have their stops built into their jambs. Although this makes them more weatherproof, it means that you can't conceal the screws under the door stop trim. Screws are still a better choice than nails, however; just use small wood plugs or putty to cover the visible ones.

The interior casing of exterior doors is the same as for interior doors, but the outside trim work must be weatherproof. Some manufacturers ship exterior doors with the outside casing already installed. All you need to do is caulk the back of this casing, and under the threshold, before pushing the whole unit into position and securing it with screws.

Many communities now require that the space between the jamb and the frame be insulated. If the code requires this in your area, be sure to insulate this cavity before you trim out the interior.

INSTALLING WINDOW TRIM

As with doors, casing for windows is available in a wide variety of styles, sizes, and shapes.

Beveled casing, the same kind that is commonly used on doors, must be mitered to fit together on the ends (see Figure 8-8). The plainness of beveled or thin flat casing focuses attention on the view, not on the window itself.

Butt-jointed styles closely resemble exterior window casing but are usually smaller in size and are often embellished

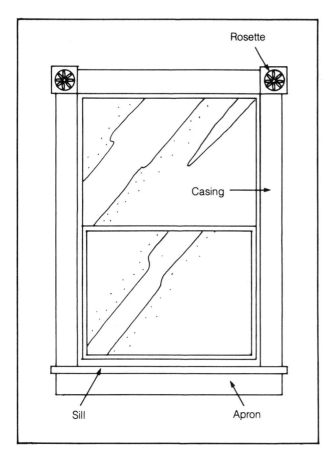

Figure 8-9. Window casing with butt-jointed corners, decorated with rosettes.

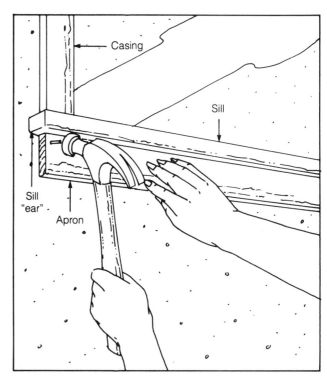

Figure 8-10. The apron is installed last, after the window casing and sill.

with routed edges or decorative corner blocks, called *rosettes* (Figure 8-9).

Both types of casing usually butt against the interior window sill and a small piece of trim, called an *apron,* is nailed below the sill to cover any gap between the drywall and the bottom of the sill.

All casing is installed with finish nails to minimize puttying work later. Number 4 finish nails are fine for attaching the inside casing to the jamb, but at the outside edge the nail must go through the drywall and penetrate about 1" into the trimmer stud. The length of nail needed depends on the thickness of the casing and the wall-covering material. Pilot holes are often required for larger nails.

The first piece of trim to be installed is always the sill. An "ear" extends out from the sill on either side so the casing on the sides can butt to the sill. The depth of the sill is up to you, but these ears must be at least as wide as the side casing.

Level the sill in the rough opening with shims and secure it with finish nails. Next install the side casing, as described in Step Seven for doors, with the sill acting like a floor. Adjust the miter joints and reveals as you work. The last piece to be installed is the apron, below the sill. The apron is usually the same length as the width of the casings and window together; the apron's edges line up with the casing edges, as shown in Figure 8-10.

For a simple, modern appearance, some people prefer no casing or sills at all on their windows. In this case, drywall is applied where the jamb would be. Metal corners are installed around the opening and finished with drywall compound. However, the drywall "sill" lacks impact resistance, and may be damaged if you leave the window open on a rainy day.

INSTALLING SLIDING GLASS DOORS

Sliding glass doors require careful placement of the rough opening during the framing process. Most are designed to be installed over the exterior siding with a small nailing flange, as shown in Figure 8-11. These doors come in precut kit form, and the jamb is assembled on the floor with screws provided in the kit. Some manufacturers recommend installing the jamb first, and some recommend installing the fixed glass panel in the jamb before placing the jamb in the rough opening.

To install a sliding glass door in your rough opening, you must first flash (waterproof) the opening according to the manufacturer's directions. Then run a bead of caulk along the outside edge of the bottom of the opening and, with a helper, tilt the jamb into position. Split the shim space equally at each end so you can plumb the sides later. Using your level, shim at the bottom to adjust the bottom to level. These jambs are flexible, so sight down the exterior edge to

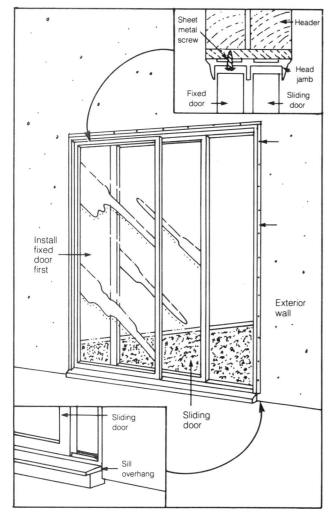

Figure 8-11. Anatomy of a sliding glass door.

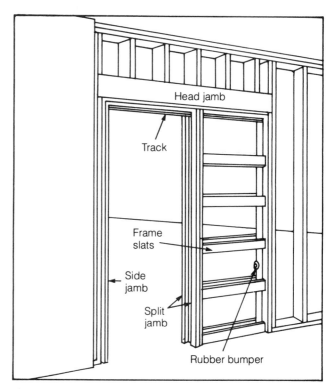

Figure 8-12. Anatomy of a pocket door.

be sure it is also straight. Now nail or screw the flange to the siding about every 12". Use galvanized nails or screws to secure the flange. Use your level and shims to plumb the side jambs, and secure these flanges to the siding.

If the fixed panel is not already in place, install it now. This will help to make the head jamb straight. Be sure to sight down the edge and verify that the head jamb is straight before you secure it. Now install the movable panel and its frame. Two holes, one on each end of the bottom of the door panel, provide adjustment access for the door rollers. Insert a #2 phillips head screwdriver into the hole to engage the adjustment screw. Turn the screw clockwise to raise the door or counterclockwise to lower the door, as necessary. Adjust the rollers so the door clears the bottom jamb and slide the door almost closed. Adjust the rollers to make the strike jamb and door edge parallel, then close the door and check the latch operation.

Follow the manufacturer's directions for exterior trim, caulking and flashing as necessary to waterproof the door. Interior jambs are usually provided with sliding glass doors, so only casing is usually required to finish off the interior side.

INSTALLING POCKET DOORS

Like sliding glass doors, pocket door openings require careful placement during framing. Pocket doors, which come in various sizes, are usually hollow core, to reduce the weight borne by the overhead track. They are prehung in a carefully sized lightly built "pocket," which is basically a box, as shown in Figure 8-12. The box is 3⅝" wide, the same width as standard framing. The rough opening is much larger than the door itself—twice as large. Unlike other doors, pocket doors are always installed *before the drywall is applied*.

To install a pocket door in a suitable rough opening, first remove all the packaging, but *not* the wood braces and blocks. The wood cleat near the bottom is left in place to hold the strike jamb, and the block across the door keeps the assembly square. Place the closed door unit in the rough opening. Make sure the jambs and sides of the box are flush to the framing and adjust the door opening and jambs to the desired location laterally.

Level the bottom if necessary and drive nails through the bottom of the pocket portion into the subfloor. Plumb the pocket side jamb and shim between the frame and the back of the pocket portion. A nail or screw here will keep the pocket side jamb plumb. Shim and screw or nail the head jamb to the header above it. Be sure that the nails are long enough to penetrate the frame. Remove the block holding the door shut, and remove the cleat on the strike jamb. Pull out the door and shim and secure the jamb to provide a good fit with the door edge. When the door is fully pulled out, the door is slightly narrower than the opening itself. These small gaps will be covered with the doorstop that comes with the door, which prevents the door from swinging in the opening. The doorstop is nailed in on both sides of the door. Use only as many nails as needed to hold the stops in place. To renew the door rollers, or to paint a pocket door, remove one set of stops, swing the door out of plumb, and disengage the rollers and door from the track.

Finally, drywall is applied over the pocket "box" and brought flush with the jambs. Use drywall screws, but be sure they don't go through the box sides and into the door. Casing is applied just as with a prehung door.

9 Vinyl Floors

This chapter describes the kinds of vinyl flooring materials on the market today and the techniques used to install them. Vinyl is an easy, affordable way to customize your floors, offering a huge variety of patterns, colors, styles, and installation choices.

This chapter also discusses floor preparation, template use, cutting and installing sheet vinyl flooring, the application of perimeter bond materials and full-spread adhesive floors, installing a self-adhesive tile floor, and the application of vinyl flooring over concrete.

Before You Begin

SAFETY

☐ Some adhesives are toxic. Be sure to provide adequate ventilation with window fans when using them.

☐ Some adhesives are flammable. Turn off gas valves on appliances in your work area.

☐ Wear the proper respirator when using substances that give off toxic fumes.

☐ Wear rubber gloves when using solvents.

☐ Keep blades sharp. Dull blades require excessive force and can easily slip.

☐ Wear safety glasses or goggles whenever you are hammering, prying, or cutting.

☐ Use the proper protection, take precautions, and plan ahead. Never bypass safety to save money or to rush a project.

USEFUL TERMS

Full-spread adhesives require the application of the adhesive with a trowel under the entire vinyl floor.

Ledging is a situation where one side of a seam rises up higher than the other.

A **perimeter bond** sheet vinyl floor is attached only at walls and seams, where flooring shrinks after installation.

A **subfloor** is a layer of plywood or boards that covers the floor joists. An underlayment is sometimes added over the subfloor if a perfectly smooth surface is needed over which to install a vinyl floor.

A **template** is a paper pattern used to ensure accurate, error-free installation.

An **underlayment** is a layer of thin particle board installed over the subfloor that gives a smooth, level surface over which the vinyl floor is applied.

WHAT YOU WILL NEED

Time. Most vinyl floor installations can be completed in one or two days by a single worker. For a 9′x12′ room, using perimeter bond sheet vinyl or self-adhesive tiles, plan on 7 to 9 person-hours. Full-spread adhesive will require 9 to 11 person-hours.

Tools. Vinyl floor installation requires very few tools that are not in every home owner's toolbox. In addition to the tools listed below, a hair blow drier may come in handy to heat up tiles before making complicated cuts.

Tools for Perimeter Bond Sheet Vinyl or Full-Spread Adhesive

Scissors

Pencil or ballpoint pen

1"-wide ruler

Notched trowel

Notched blade or utility knife

Rolling pin or seam roller

Staple gun (for perimeter bond sheet vinyl)

Tools for Self-Adhesive Vinyl Tiles

Scissors

Pencil or ballpoint pen

Utility knife

Chalkline

Carpenter's square

Materials for Perimeter Bond Sheet Vinyl or Full-Spread Adhesive

Craft or butcher paper

Do-it-yourself installation kit (for perimeter bond sheet vinyl)

Masking tape

Sheet vinyl

Adhesive

Seam sealer kit

Staples (for perimeter bond sheet vinyl)

Materials for Self-Adhesive Vinyl Tile

Craft or butcher paper

Tiles

Vinyl wall base

PERMITS AND CODES

Laying a new floor may be regulated by local building codes, and a permit may be necessary. Check your local building code before beginning work.

DESIGN

The color you choose for your new floor is likely to be determined by the predominant color of the room. Neutral tones and single-color rooms will make your decision easy. For a multicolored space you will need to decide which color you prefer to emphasize.

In selecting a pattern, keep in mind that the most effective way to combine patterns in a room is to use one large pattern, one medium pattern, and one small pattern distributed among walls, fabrics, and flooring. Of course, these patterns should be color-coordinated.

If your floor offers a wide expanse of uninterrupted space, you may wish to use a large pattern. However, if the space is broken up, has alcoves, or is interrupted by counters or appliances, a more pleasing effect will be accomplished with a smaller pattern.

You can use the pattern to create a visual span from one space to another by repeating like shapes in regular sections both vertically and horizontally.

Most Common Mistake

☐ The single most common mistake in any vinyl flooring project is applying the flooring over an improperly cleaned or prepared surface, such as a basement floor that has a moisture problem.

SHEET VINYL FLOORING

Step One
Preparing the Subfloor
or Underlayment

Margin of Error: The subfloor or underlayment should be level within ⅛", with no gaps larger than ¼"

Most Common Mistake

☐ Not leveling the floor or applying an underlayment if needed.

If your subfloor is level and structurally sound, with an even finish, simply fill any holes or cracks with floor filler and be sure no splinters or nailheads are protruding. Then sweep and vacuum thoroughly and mop the floor with a mild cleaner.

If any areas slope or dip badly, fill them with leveling compound, allow it to dry thoroughly, then install an underlayment of ¼" paraticle board to create a smooth, level surface. Plan the seams of the underlayment so they do not match those of the subflooring beneath. Over an open joist system, first apply a layer of ¾" tongue-and-groove plywood as a subfloor. Check local code for specific recommendations for your area.

Nail the underlayment with 6-penny ring-shank nails every 4" to 6" around the edge and every 4" in the middle. Be sure all the heads are slightly below the surface. Leave a ½" gap between the plywood and the wall all the way around, and ¹⁄₁₆" between sheets, to allow for expansion and contraction of the wood in cold, wet, or humid regions. In temperate regions, it is common to fill these joints with joint filler and then sand them smooth.

Step Two
Making the Template

Margin of Error: Exact

Most Common Mistake

☐ Neglecting to make a template when working with sheet vinyl.

A template or pattern is essential for accurately cutting your vinyl floor (Figure 9-1). Craft paper, butcher paper, and the paper that comes in the do-it-yourself installation kits all work equally well. The template will enable you to transfer

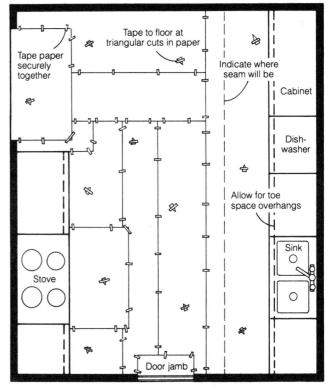

Figure 9-1. A typical paper template.

accurate measurements to the vinyl flooring without making unnecessary or awkward cuts during installation.

To keep the pattern from shifting while you work, cut little triangles in various areas of the paper. You can then tape the pattern to the floor with masking tape pressed over the cutouts (Figure 9-2).

If you need a bigger piece, tape overlapping edges of the paper together, keeping the pattern smooth and flat as you progress. Flooring usually comes in 12′ widths. If you are working with a floor large enough to require two pieces of flooring (over 12′), you will need to follow special steps for seam fitting, pattern matching, and seam sealing. If at all possible, it is best to lay out the flooring so that the seam will fall in a low-traffic area.

Many floors have irregular, odd-shaped, or just plain hard-to-fit objects like pipes or built-in cabinets. Use smaller pieces of paper to make templates for irregular sections of the floor, and attach them to the main template with tape. Finally, check to be certain that the template is an exact fit, and make any necessary corrections.

If you are using an installation kit, the accompanying roller disk will aid you in marking your pattern. When using this roller, leave a gap of about ½″ between your template and the wall. The roller disk is designed to transfer the wall line to the pattern paper, leaving a space of exactly 1″ between the pencil line and the wall (Figure 9-3). In other words, this line on the template is actually one full inch shy of the wall everywhere you use the roller disk. With this method, when you transfer your template to the vinyl, you must use a 1″-wide straightedge placed along your outline mark to put that inch back onto the flooring when you mark it. Whether you use an installation kit with a disk roller, a makeshift marker, or just press the paper up to the wall, creasing it for your outline, it is best to work a short distance at a time. Don't try to mark a wall in one continuous line.

Note on the pattern the position of any object you had to fit around, so that you can check its fit and its relative location one last time before you complete the installation. When you have finished your template, you will have a paper floor

with all of its "landmarks" clearly indicated. If there are any holes for registers for the heating system, just ignore them when making your template. These holes can be cut after the flooring is laid. Now you are ready to get an accurate transfer of your paper pattern onto the vinyl material.

Step Three
Cutting the Vinyl

Margin of Error: Exact

Most Common Mistakes

☐ Laying out the template on the wrong side, causing you to cut the floor backward.

☐ Not lining up the template seamline with a pattern (grout) line on the flooring material.

Unroll the new flooring face up on a clean, smooth surface. (Otherwise, small stones or dirt can become embedded in the back, eventually wearing through or tearing the new floor.) The basement, garage, attic, or driveway is often the best place for this. However, if you are working outside, don't expose the vinyl to direct sunlight.

Overlap the vinyl pieces where the seams will fall. Check the two sections for pattern match all along the overlap and at each corner of the pattern. if you are working with a pronounced design that calls for a perfect match, keep both pieces running in the direction they came off the roll.

Tape the two vinyl sections securely together after they are matched so they won't move when you cut them. Then cut straight through the overlapping edges of the two pieces so that they fit together perfectly (Figure 9-4). (It's a good idea to practice on a piece of scrap flooring before making the actual cut.) Getting a good-looking seam is not difficult if you make your cut in a simulated grout line or other pattern feature that can serve as camouflage. Be sure to keep a sharp blade in your utility knife.

After you have made your cut, double check the pattern match before continuing. Then tape the seam together.

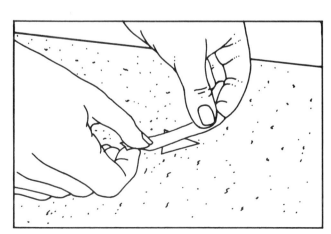

Figure 9-2. Tape paper template to the floor with masking tape over triangular cutouts.

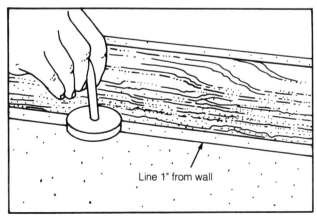

Figure 9-3. A roller disk enables you to draw a line exactly 1″ from the edge of the wall onto your paper template.

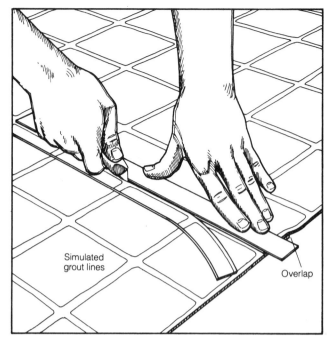

Figure 9-4. To match the pattern at a seam, overlap the two pieces until the pattern matches. Then tape the two pieces together and cut through both.

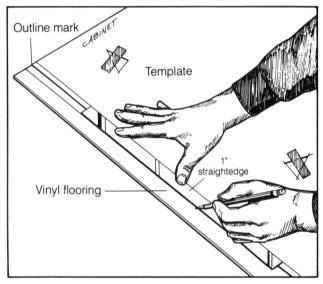

Figure 9-5. If you used a disk roller, you will need to transfer the "lost" 1" back onto the vinyl.

Now lay out your paper pattern or template over the flooring material, making sure that the seam falls in a low-traffic or low-visibility part of the room. Be sure that you are not positioning the template upside down and that you are cutting the flooring as you want it. Compensate on both sides for any out-of-line walls by shifting the template in a direction that will split the difference of the error. If possible, don't position a line in the vinyl pattern too close to the out-of-line wall, where it will be more obvious.

Once you have the template situated where you want it, tape it to the vinyl through the triangular slits, just as you taped it to the subfloor.

As mentioned earlier, if you used a roller or a 1" marking guide, remember to transfer the "lost" 1" back onto the flooring material, using a pencil or ballpoint pen (Figure 9-5). After you have transferred the dimensions to the flooring material, remove the template.

Use a notched blade knife or a utility knife to trim the vinyl. Many kits include a notched blade knife. Always make sure the blade is sharp. Cut very carefully and true along your line for a precise fit.

Once the floor has been cut out, roll it up with the pattern showing on the outside and any narrow protruding areas on the outside end of the roll.

Step Four
Installing the Flooring

Margin of Error: Within ¼" at edges, exact at seams

Most Common Mistakes

☐ When estimating the amount of sheet vinyl or perimeter bond material, forgetting to allow for pattern matching at seams.

☐ Unrolling perimeter bond sheet vinyl too early, or waiting too long to lay it, so that it shrinks before it is permanently laid in place.

☐ Failing to use flooring materials with the compatible adhesive and appropriate trowel at seams.

PERIMETER BOND
SHEET VINYL FLOORING

Carry your roll of cut vinyl flooring into the room in which it will be installed. Carefully unroll and position it over the clean, dry subfloor, matching up the landmarks you indicated on your template.

Carefully assess your cutting job. If any additional trimming needs to be done, now is the time to do it—before any adhesives have been applied.

The first part of the new floor to be secured is the seam. This is done by applying the adhesive along the floor between the two sections of flooring. First, gently fold one section back and tape it out of the way. Draw a pencil line along the edge of the other section to mark the seamline. Gently fold the second section back and tape it out of the way.

Apply a band of adhesive to the subfloor along the seamline, using the recommended notched-tooth metal trowel (Figure 9-6). Check the manufacturer's recommendations about the width of the adhesive. Some require only a 3" band (1 ½" on either side of the pencil line); others may require as much as 6" of adhesive (3" on either side of the seam).

Apply the adhesive all along the pencil line to about 1 ½" away from any cabinets. You want to stop the adhesive here so that, after the seam is pressed together and rolled, you will be able to fold back the flooring under the cabinets to apply adhesive there. You can't get a staple gun under the cabinet overhang, so you have to glue the areas under the cabinets.

Lay one piece of the flooring into the adhesive, and then the other. Make sure the edges of the vinyl are tight against each other. If you don't, you'll get a condition called "ledging," where one side rides up higher than the other. Dirt can build up here and draw attention to the seam.

Now go over the seam with a rolling pin or seam roller, to press the vinyl into the adhesive and eliminate ledging.

To prevent moisture from getting under the vinyl along this seam, use a special seam sealer kit. Read and follow the instructions carefully. When applying solvent, hold the bottle at the proper angle and don't wipe up any of the excess. It will evaporate, and you won't see it after a short time. Give the seam a few hours to set up before walking on it.

Use the adhesive, as instructed, on the perimeter areas that will not be covered by molding and in areas where you are unable to use your staple gun. Roll the edge of the flooring back, apply the adhesive in the proper amount with the notched trowel or manufacturer's suggested applicator, and press the flooring into place with the roller.

For the edges that will be covered by quarter-round trim or baseboards, staples applied with a staple gun work best. They are fast and easy to apply and they provide great holding power. In addition, if the flooring material shrinks or expands with temperature and humidity, staples will cause fewer problems than will adhesives. Staple close to the wall, so that the molding will cover the staples (Figure 9-7).

Perimeter-bonded vinyl flooring will contract slightly, tightening like a drumhead, over the 24 to 48 hours after installation. You should wait until the floor has contracted to its final tension before moving the furniture and appliances onto it.

Most vinyl floors today are "no-wax." Once the floor has contracted, all you have to do is damp mop. As always, follow the manufacturer's suggestions for cleaning and care.

FULL-SPREAD ADHESIVE SHEET VINYL FLOORING

Full-spread adhesive is fast becoming the least popular type of vinyl flooring. It is more difficult to install than perimenter bond, and it cannot be removed as easily as other floorings if you want to install another floor in the future. One advantage of full-spread adhesive flooring is that coved (curved) edges can be obtained around the perimeter of the room, eliminating the need for baseboards. Properly sealed, these coved edges can greatly improve water-proofing in bathrooms and kitchens. However, this type of flooring is difficult to install. If you choose coved edges, consider calling in a professional. Check around to see if your pattern is available in a perimeter bond application before you opt for

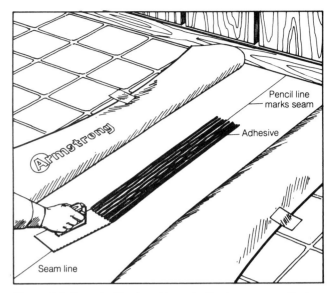

Figure 9-6. Tape vinyl back and trowel adhesive over penciled seamline.

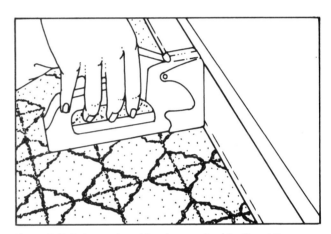

Figure 9-7. Perimeter bond flooring requires stapling at the edges only. Staples will be covered by the quarter-round trim.

the full-spread. If your pattern is available only in a full-spread application, try to limit its installation to small rooms. It is much too cumbersome to try to align on larger floors that are already spread with adhesive.

1. Full-spread adhesive must be applied on a surface that is smooth and free of grease, dirt, and any irregularities. Clean the subfloor well. See Step One, Preparing the Subfloor, earlier in this chapter.

2. Follow the template procedure described in Step Two.

3. Roll up the material and carry it into the room. Place the roll against a wall and unroll it as you work toward the opposite wall (Figure 9-8).

4. One of the most important things about working with adhesives is to use the correct type of applicator or trowel and adhesive. This information is included in the manufacturer's instructions. The adhesive should be applied to enough of

Figure 9-8. For full-spread adhesive application, spread adhesive with a toothed trowel and unroll the vinyl as you go.

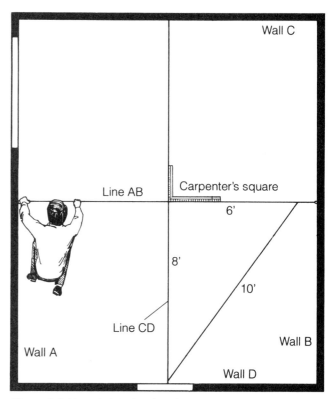

Figure 9-9. Use a 3-4-5 triangle to be sure your layout lines are square.

the floor at a time to allow you to place the sheet vinyl and give you some working and adjusting room.

5. If you are applying a seamed floor, lay the smaller section first. Follow the instructions for seams as described for perimeter bond vinyl flooring.

6. If you have done a careful job of outlining the odd and irregular shapes of the floor, and of transferring the template to the right side of the flooring, getting the sheet to fit should present no problems.

7. Once the flooring is in place, you should go over it with a rolling pin or a 100-pound roller to ensure a good bond between the floor and the adhesive. Roll from the center of the floor out toward the edges to get rid of all the air bubbles and waves.

SELF-ADHESIVE VINYL TILES

Self-adhesive vinyl tiles are the easiest type of floor to put down. Because of the sticky backing, there is no need to mess with gooey adhesives and seam sealers. You need only acclimate the tiles by storing them there overnight in open boxes in the room in which they will be installed. This is especially important when installing floors over concrete. You will need to buy about 10 percent more tile than the total area you plan to cover, to allow for border tiles.

Step One
Preparation

Margin of Error: Subfloor should be level within ⅛" and with gaps no larger than ¼"

Most Common Mistake

☐ Not purchasing about 10 percent more tile than the total area you plan to cover, to allow for border tiles.

When laying a tile floor, you will have a more balanced floor if you work from the center of the room outward. You will need to establish layout lines and find the center of the floor. You must also be sure that the seams in the tiles do not line up directly over seams in the subfloor.

First, measure to find the midpoints of two opposite walls (wall A and wall B in Figure 9-9). Snap a chalkline between the midpoints of these walls (line AB). Measure to find the midpoints of the other two opposite walls (in this case, wall C and wall D).

Before snapping the second line (line CD), place a carpenter's square in the center where the two lines intersect. If necessary, adjust the second line (line CD) so that the two lines intersect at a 90-degree angle. These lines must be perfectly square. You can easily check for square with the 3-4-5 triangle technique, as follows.

Start at the intersection of the two layout lines and measure 6′ along line AB. Then measure 8′ along line CD. If the lines are perfectly square, the measurement between the two

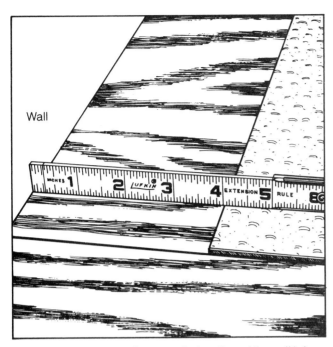

Figure 9-10. If the space between the last tile and the wall is less than half a tile (6" with 12" tiles), move the center line over half a tile space.

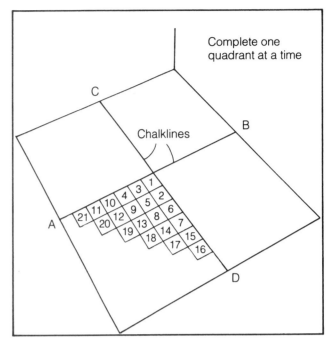

Figure 9-11. Pattern for laying tiles.

points will be exactly 10′. (Of course, you can use any multiples of 3-4-5, but it is best to use the longest possible measurements to ensure the greatest accuracy.)

If the lines are not perfectly square, adjust line CD until they are square. Do not adjust line AB! Lay loose tile from the center point along line AB to the wall. If the space between the last tile and the wall is less than half a tile, move line CD half a tile (6" with 12" tiles) closer to one wall and snap a new line CD (Figure 9-10). Repeat this process by laying tiles along line CD and adjusting line AB. This will ensure equal borders on opposite walls, and all borders will be more than half a tile.

Step Two
Laying the Tiles

Margin of Error: Exact

Most Common Mistakes

☐ Not laying out and squaring your working lines when placing square tiles.

☐ Placing a pattern line too close to an irregular wall, thereby accentuating the out-of-line wall.

☐ Neglecting to lay the floor tiles in the direction indicated by the arrows on the backs of the tiles.

After making sure that your subfloor is clean and free of debris and dirt, begin laying the tiles. Vinyl floor tiles should be installed in a certain direction, indicated by an arrow on the back of the tile. The tiles have a paper backing that is peeled off to expose the adhesive.

Beginning at the intersection of the working lines, the first tile goes down along the first working line (line AB), with the second along line CD, as shown in Figure 9-11. Work out toward the borders, completing one quadrant at a time. When each tile is in place, press it firmly onto the subfloor. Be sure the tiles are snug so that they will end up properly at the wall with a good finished look.

Vinyl tiles are easily cut with a utility knife or scissors, and are even easier to cut if they are first warmed with a blow drier. This technique is especially helpful in making a cut to fit around an intricate shape. In the case of a complex cut, a template of butcher paper or craft paper is needed as well. Cut the paper to fit around the obstacle, then trace the pattern onto the tile.

A consistently accurate method of cutting a border tile for a perfect fit at an outside corner, even though the wall may be slightly out of line, is shown in Figure 9-12. Lay the tile you want to cut squarely on top of the tile that is already in place on the row adjacent to the border row. Take another loose tile and butt it up against the wall so it overlaps the tile you want to cut. The second loose tile thus becomes a straightedge for marking the border tile. Repeat this procedure on the other side of the corner. Then score the border tile with a sharp utility knife to outline the portion of the tile to be removed. Make a diagonal cut across the back from the inside corner. The tile should break cleanly along the scored lines, leaving you with an L-shaped piece that will fit perfectly into place.

When fitting border tiles along irregular or curving walls, you will need a paper template. Place the template in position and push it into the corner or curve as tightly as you can,

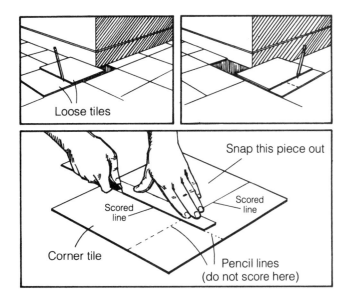

Loose tiles

Snap this piece out

Scored line

Scored line

Corner tile

Pencil lines (do not score here)

Figure 9-12. Cutting a tile to fit around an outside corner.

creasing the paper as you go. Mark the crease with a pencil as accurately as you can manage. Transfer the pattern to the tile (making sure the tile is facing in the proper direction) and make your cut very carefully along the line with a utility knife.

DRY TILES

Dry-backed flooring tiles that need an adhesive can be laid the same as sticky-backed tiles, but they are messier and more difficult to lay. Follow the manufacturer's instructions for specific preparation, adhesive and trowel use, and drying time. This type of flooring, like full-spread adhesive sheet vinyl, can be difficult to remove later.

APPLYING VINYL OVER CONCRETE

If you plan to apply sheet vinyl or tiles over concrete, note the following special considerations.

The concrete must be dry. New concrete should be cured to a hard, dry, nonpowdery finish. New concrete should also have at least a 4 mil moisture barrier between the ground and the concrete slab.

To be on the safe side, test all concrete subfloors for moisture, preferably during the rainy season. This test can be done in three different ways. The easiest is to completely tape down $2'x2'$ polyfilm squares in several places on the floor. Leave the squares for 24 to 48 hours, then check for condensation under the plastic.

A second test involves chipping small sections of concrete from the floor in several areas. To each chipped area, apply a solution of 3 percent penophalene in alcohol. This solution can be purchased at most drugstores. A red color reaction indicates the presence of moisture in your floor.

The third test involves the use of calcium chloride crystals, also available from a druggist. Make a 3"-diameter putty ring on the slab, place $\frac{1}{2}$ teaspoon of the calcium chloride in the circle, then cover the crystals with a water glass to seal them off from the air. If the crystals dissolve within 12 hours, the slab is too wet to use adhesives.

To carry an applied floor, a concrete subfloor should have a density of 90 pounds per cubic foot or more. A lighter slab tends to hold moisture longer, and to retain a scaly or chalky surface. Either moisture retention or concrete dust can lead to problems.

10 Hardwood Floors

A HARDWOOD FLOOR is perhaps the most beautiful and durable floor you can install. Earlier in this country's construction history, hardwoods were the most common of all flooring materials. With the advent of vinyl and linoleum flooring, and later with the widespread use of carpeting, they fell out of favor. In recent years, however, they have regained some of their lost popularity.

Part of the reason for their comeback has been the introduction of new types of hardwood flooring. These new types are as beautiful and as durable as the earlier tongue-and-groove plank flooring, and they are so easy to install that their application is a fairly simple project. It may surprise many people to learn that a parquet or a plank floor can easily be installed in a weekend.

This chapter discusses installing both parquet floors and prefinished plank (strip) floors, and finishing unfinished hardwood floors. For the most part, this chapter deals with installing prefinished floors. These floors are already stained and have a protective top coat. However, you can also install unfinished hardwood floors that will need to be finished by sanding, staining (if you choose), and application of a protective finish.

Before You Begin

SAFETY

☐ Wear safety glasses or goggles whenever you are using power tools, especially if you wear contact lenses.

☐ Always unplug your power tools when making adjustments or changing blades, drill bits, or sandpaper.

☐ Be sure your tools are properly grounded.

☐ Watch power cord placement so that it doesn't interfere with the tool's operation.

☐ Wear ear protectors when using power tools, because some operate at noise levels that can damage hearing.

☐ Be careful not to let loose hair and clothing get caught in power tools.

☐ Wear the proper respirator or face mask when sanding, sawing, or using substances that give off toxic fumes.

☐ Some adhesives are toxic. Be sure to provide adequate ventilation with window fans when using them.

☐ Some adhesives are flammable. Turn off gas appliances in the vicinity.

☐ Wear rubber gloves when using solvents.

☐ Use the appropriate tool for the job.

☐ Keep blades sharp. A dull blade requires excessive force and can easily slip.

☐ Seal all heating and air conditioning ducts and electrical outlets when sanding.

☐ To prevent spontaneous combustion, dispose of oily rags correctly.

☐ Use the proper protection, take precautions, and plan ahead. Never bypass safety to save money or to rush a project.

USEFUL TERMS

Baseboard is the trim that is used where walls and floors meet. Also called shoe molding.

A **floor register** is an opening in the floor, usually covered by a grate, that brings heated or cooled air into a room.

Glazier's points are small metal triangles that are used to hold glass panes in their frames.

A **penetrating sealant** is a finish that soaks into the wood as well as providing a hard finished surface.

Polyurethane is a synthetic rubber polymer sealant for wood.

Reducer strips are prefabricated door thresholds for use where two rooms with different floor levels come together.

A **surface finish** is a finish that provides a hard surface coat without penetrating the wood.

A **subfloor** is a layer of plywood or boards that covers the floor joists. An underlayment is sometimes added over the subfloor if a perfectly smooth surface is needed over which to install a finished floor.

An **underlayment** is a layer of thin particle board installed over the subfloor that gives a smooth, level surface over which the finished floor is applied.

PERMITS AND CODES

Laying a new floor may be regulated by local building codes, and a permit may be necessary. Check your local building code before beginning work.

WHAT YOU WILL NEED

Because the tools and materials lists for the various projects are so extensive, they appear at the beginning of each section in this chapter: Installing Parquet Floors, Installing Strip Flooring, and Sanding and Finishing Hardwood Floors.

INSTALLING PARQUET FLOORS

Some people look at parquet floors, with all their small beautiful inlaid pieces, and think, "What patience and skill that person must have had!" Actually, patience and great skill are no longer required to install this beautiful flooring. In fact, this is now a fairly simple project because all of those little pieces have been prefabricated into larger tiles. These days, a prefinished parquet floor is no more difficult to install than vinyl floor tiles.

WHAT YOU WILL NEED

Time. Plan on approximately 10 to 15 person-hours to complete a 9′x12′ room. This can be a one-person task.

Tools for Installing Parquet Floors

Ear protectors

Respirator/face mask

Eye protection

Carpenter's pencil

Hammer

Utility knife

Heavy-duty shop vacuum

Fan(s)

Tape measure

Carpenter's square

Pry bar and wood wedge

Extension cords (heavy duty)

Chalkline and chalk

Jigsaw

Notched trowel

150-lb. roller

Handsaw

Radial arm saw or table saw

Materials for Installing Parquet Floors

Mastic or adhesive

Parquet tiles

Plywood

Reducer strip

Adhesive cleaning solvent

Rags

Step One
Preparing the Subfloor

Margin of Error: Within 3/16" of level

Most Common Mistake

☐ Not leveling the floor or applying an underlayment if needed.

If your subfloor is level and structurally sound, with an even finish, simply fill any holes or cracks with floor filler and be sure no splinters or nail heads are protruding. Then sweep and vacuum thoroughly and mop the floor with a mild cleaner.

If any areas slope or dip badly, fill them with leveling compound, allow it to dry thoroughly, then install an underlayment of 1/4" plywood to create a smooth, level surface. Plan the seams of the underlayment so they do not match those of the subflooring beneath. Over an open joist system, first apply a layer of 3/4" tongue-and-groove plywood as a subfloor. Check local code for specific recommendations.

Nail the underlayment with 6-penny ring-shank nails every 4" to 6" around the edge and every 4" in the middle. Leave a 1/4" gap between the plywood and the wall all the way around, and 1/16" between sheets, to allow for expansion and contraction of the wood.

Step Two
Locating Your Layout Lines

Margin of Error: 1/4"

Most Common Mistakes

☐ Not using layout lines.

☐ Not adjusting the border tiles correctly.

As with vinyl floor tiles, you need to mark your floor with layout lines that will guide you in installing the parquet floor tiles. The two layout lines cross approximately in the center of the floor and divide your room into four quadrants that are roughly equal in size. The tiles are then laid out from the center, where the lines cross, toward the corners. This process needs to be done precisely, because it affects the final outcome of the project considerably.

First, find the center of the room and establish your layout lines. In Figure 10-1, walls A and B are 20′6" and walls C and D are 14′8". Wall D is the most visible wall, having a wide opening into another room, so you will want a full tile at this opening for appearance' sake. Using increments of a full tile (usually 12"), measure along walls A and B from wall D approximately half the room—in this case, about 10′. Snap a chalkline AB between these midpoints. Now find the midpoints of walls C and D and snap your second line (in Figure 10-1, the dotted line).

You may need to make some adjustments to these lines. You want the border tiles along walls A and B to be equal in

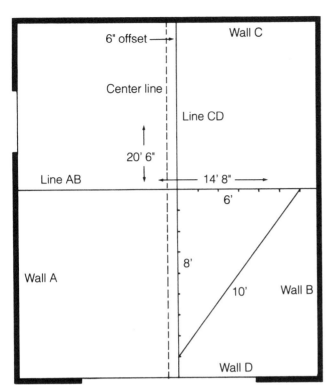

Figure 10-1. Layout lines for a parquet floor.

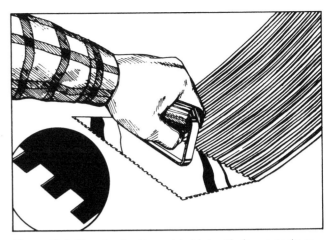

Figure 10-2. Use a toothed trowel, held at a 45-degree angle, to apply the adhesive.

Step Three
Applying the Adhesives

Margin of Error: Not applicable

Most Common Mistakes

☐ Using the wrong adhesive.

☐ Applying the adhesive when it or the room is too cold.

☐ Using the wrong trowel.

Clean and vacuum the subfloor thoroughly and be sure there are no protruding nails, screws, or splinters. Be sure to provide good ventilation, such as a window fan, because adhesive fumes can be toxic. Turn off nearby gas appliances; some adhesives are highly combustible.

Be sure you have chosen the proper adhesive for the flooring, and follow the manufacturer's instructions closely. Most adhesives should be stored in a room heated to 70 degrees for 24 hours before applying, and the wood tiles should be stored loosely in the room for at least 24 hours so they can acclimate to its temperature and humidity. If you are not able to store the adhesive in a heated room before applying it, you can place the unopened can in warm water to heat it up.

Use a toothed trowel to spread the adhesive in the area you will be working in. Hold the trowel at a 45-degree angle to get even ridges (Figure 10-2). Spread up to, but do not cover, your layout lines. Be careful not to spread a larger area than you can cover in 2 to 3 hours. A little experimentation will show you the proper amount of adhesive. Too much adhesive will squeeze up between the tiles; too little will not allow the tiles to adhere properly. Ridges ⅛" high are about right. Let the adhesive thicken and become tacky before laying the tiles. Most adhesives take about an hour to set up properly, but there are some variations. Follow the manufacturer's instructions.

width and more than half a tile. To do this, calculate the length of line AB in inches (14'8" = 176"), divide that number by 2, then divide the result (88") by the size of the tile (12"). If more than half a tile remains along walls A and B, you will not need to adjust your chalkline. In our example, however, the remainder is only 4", less than half a tile (88" divided by 12" = 7'4"). In this case, you would end up with two rows of 4" border tiles and 14 rows of full-size tiles. Remember, you want border tiles that are 6" or more in size. To accomplish this, move the center chalkline (line C/D) by half a tile (6"). This becomes your second layout line CD (in Figure 10-1, the solid line), which will leave border tiles of 10" along walls A and B, and 13 rows of tiles in between. (It's a good idea to check your calculations by dry-laying your tiles to see how they work out.)

Before proceeding, you must check to be sure that your two layout lines, AB and CD, are perfectly perpendicular to each other. If all the walls were perfectly parallel, the lines would automatically be at right angles. However, this is rarely the case; the walls are usually a little off parallel.

Use a 3-4-5 triangle to check your right angle. Start at the intersection of the two layout lines and measure 6' along line AB and then 8' along line CD. Mark these points and measure the distance between them. If the lines are perfectly square, the distance between the two points will be exactly 10'. (You can use any multiple of 3-4-5, but it is best to use the longest possible measurements to ensure the greatest accuracy.) If the lines are not perfectly square, adjust line CD until they are. Do not move line AB or you will have odd-cut tiles at the highly visible doorway.

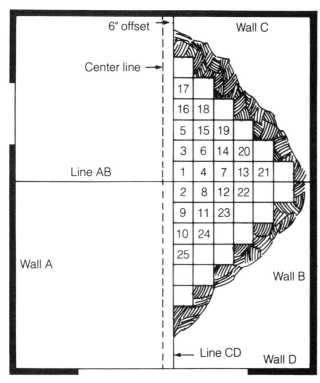

Figure 10-3. Installation pattern for parquet floor tiles.

Figure 10-4. Use a plywood board to distribute your weight over freshly laid tiles.

Step Four
Laying the Tiles

Margin of Error: ⅛"

Most Common Mistakes

☐ Not leaving a ¼" expansion gap between tiles and the wall.

☐ Not allowing the adhesive to set up properly.

After your adhesive has set up, you can start laying the tiles. This is finish work, and you'll be living with it for a long time, so go slow and do a good job. Lay the tiles in the pattern

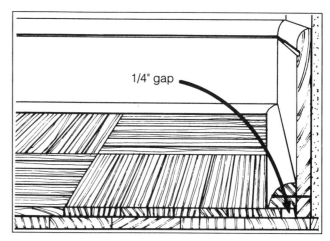

Figure 10-5. Leave an expansion gap of ¼" at the wall. This gap will be covered by quarter-round molding. Be sure to nail the molding to the wall, not to the floor.

shown in Figure 10-3, or as instructed by the manufacturer. Do one side of the room and then the other in the same pyramid fashion. Place the tiles exactly up to your layout lines. Here is where your earlier accuracy will pay off.

Take special care in laying your first 10 or 12 tiles, because they will determine the appearance of the entire floor. Check to be sure there is no debris lifting the tiles. Engage the tongues and grooves of the tiles as you lay them next to each other, but don't slide them into place. Occasionally you may need to use a mallet or hammer and a block of wood to force them together.

Every few tiles, stop and tap the tiles with your mallet or hammer and wood block to be sure they are properly seated. If any adhesive gets on the tiles, clean them immediately with a rag soaked in solvent. Never apply the solvent directly to the tiles; it could mar the finish.

Don't step on the tiles for 24 hours after they are laid. To avoid putting your weight directly on any one tile, place a sheet of plywood as a kneeling board over the newly laid tile as you continue to work (Figure 10-4). Be sure there is no adhesive between the board and the tiles.

You will need to cut the border tiles to fit. This is best done with a fine-tooth handsaw, although a table or radial arm saw is easier to use. Don't forget to leave a ¼" expansion gap between these border tiles and the walls, as shown in Figure 10-5. Many manufacturers recommend placing a cork expansion strip in this gap.

Step Five
Rolling the Floor

Margin of Error: Not applicable

Most Common Mistake

☐ Omitting this step.

Figure 10-6. After the floor is completed, roll it with a 150-pound roller to ensure proper adhesion.

After you have finished laying the floor, or a section of the floor, go over it with a rented 100- to 150-pound floor roller (Figure 10-6). If you can't get a roller, use a kitchen rolling pin, putting all your weight on it. To ensure proper adhering, this step must be done within 4 hours of when the adhesive was originally spread.

After you have installed and finished the entire floor, and the adhesive has set, nail your molding in place. Be sure to nail the molding into the wall, not the floor, to allow the flooring to expand. Reducer strips may be needed if your new floor meets another floor that is either higher or lower. These reducer strips are the same color and material as your flooring, and are nailed to the subfloor where the two floors meet.

INSTALLING STRIP FLOORING

Laying a hardwood strip or plank floor used to be a good deal more difficult than it is today. Strip flooring came only in rather thick (¾") boards that had to be nailed to the subflooring. This process, which requires a special nailing gun, takes considerably more work than gluing. Although this flooring is still available, the new, thinner (⅜" or 5⁄16"), prefinished floorings are much easier to install. They are glued instead of nailed, and they come prefinished, so no sanding and finishing are required. Like parquet flooring, strip flooring comes in many colors and styles.

WHAT YOU WILL NEED

Time. Plan on 10 to 15 person-hours to complete a 9'x12' room. This can be a one-person task.

Tools for Installing Strip Flooring

Ear protectors

Respirator/face mask

Eye protection

Carpenter's pencil

Hammer

Circular saw

Jigsaw

Handsaw

Utility knife

Heavy-duty shop vacuum

Fan(s)

Tape measure

Carpenter's square

Straightedge board

Pry bar and wood wedge

Extension cords (heavy duty)

Chalkline and chalk

Notched trowel

150-pound roller

Materials for Installing Strip Flooring

Mastic or adhesive

Tongue-and-groove planking (glue-down type)

Plywood

Reducer strip

Adhesive cleaning solvent

Rags

Step One Prepping and Repairing the Floor

Margin of Error: Within 3⁄16" of level

Most Common Mistakes

☐ Not leveling the floor or applying an underlayment if needed.

This step is the same as Step One under parquet flooring.

Step Two Planning the Floor

1. Strip flooring installation traditionally begins at the most visible wall of the room.

2. Strip flooring is usually installed parallel with the long wall of the room, perpendicular to the joists.

3. The piece of flooring next to the most visible wall should be a full-size piece. If you are using random-width boards,

Figure 10-7. Nail down a 2x4, at a distance of 30" to 32" from the wall, to use as a straightedge.

this should be a wide board. The piece next to the wall opposite this most visible wall can be a special-cut piece.

4. When installing flooring in adjacent rooms and halls perpendicular to the main room, you may want to turn the direction of the plank. However, don't lay the flooring in the short direction of a long hallway; this creates an unpleasing "ladder" effect.

5. To ensure a good blending of color, work out of several cartons at once.

6. Don't use boards that have a substantial bow.

To ensure that the most visible wall has a full-width board, and that the planks are installed straight, use a layout line. This is a chalkline snapped on the floor approximately 30" to 32" from the most visible wall and running in the direction in which the flooring will be laid. Its exact distance from the wall depends on the width of the boards you are laying. Again, your object is to have a full-size plank against the most visible wall.

The following example uses both 3" and 5" planks, so eight planks will equal 32" (four 3" planks and four 5" planks). You therefore measure out from the wall 32 ¼" at each end, mark these spots, and snap a chalkline between them. (The additional ¼" is for the expansion gap between the flooring and the wall.) Then install a straight 2x4 as a straightedge to act as a guide when laying the floor (Figure 10-7). Be sure the straightedge is temporarily nailed to the subfloor perfectly straight and exactly on the layout line. This straightedge will determine the appearance of the entire floor.

Step Three
Applying the Adhesive

Margin of Error: Not applicable

Most Common Mistakes

☐ Using the wrong adhesive.

☐ Applying the adhesive when it or the room is too cold.

☐ Using the wrong trowel.

Apply the adhesive between the straightedge and the most visible wall, 30" to 32" away. Never cover an area with adhesive that cannot be laid within 4 hours. Other than that, the adhesive is applied as discussed in Step Three under parquet flooring.

Step Four
Laying the Planks

Margin of Error: Exact

Most Common Mistakes

☐ Not wiping off excess glue.

☐ Not laying a full-size board next to the most visible wall.

☐ Not achieving a tight fit between boards.

☐ Using bowed boards.

☐ Not leaving a ¼" expansion gap between all walls and flooring.

Your first row of flooring will be laid tight up against the straightedge. Install it so that the tongue is against the straightedge. If you are using random-width boards, the piece next to the wall should be the widest piece.

Make sure that this first board is good and straight, and install it tight up against the straightedge. If you are using flooring that has fake pegs, cut the end of the first plank in each row so that all the pegs don't line up next to a wall. Also be sure that there is at least 5" between the ends of the planks in adjacent rows (see Figure 10-7). Leave a ¼" expansion gap between all walls and the flooring.

As you lay the planks, use a hammer or mallet and a scrap piece of flooring to force the planks tightly together and ensure a snug fit. Remember to let the adhesive set up to a sticky feel before you apply the flooring. Don't slide the planks into place, because this will cause the adhesive to ooze up between the boards. Rather, insert the tongue into the groove and adjust it into final position. If any adhesive gets on the planks, wipe it off immediately with a rag wet with solvent. Never apply the solvent directly to the plank; it could mar the finish.

Work toward the most visible wall until that section is completed. You can use a pry bar against the wall to force the last piece snug up against its adjacent course.

Don't kneel or walk on the newly laid planks. Use a clean piece of plywood as a kneeling board to spread your weight out. Be sure there is no excess glue on the planks before placing the plywood on top. As you cover each section, go over it with a 150-pound floor roller rented from your supplier.

After you have done the first area, remove the straightedge and continue laying the floor, using the first plank as a straightedge. Your final piece, across from the most visible wall, may need to be cut to fit.

After you have installed and finished the entire floor, install the molding or baseboard, nailing it into the wall, not the floor, to allow the floor to expand.

SANDING AND FINISHING HARDWOOD FLOORS

Unfinished hardwood floors must be sanded to a smooth, even finish, free of ridges, dips, and irregularities, before they are stained and/or finished. Sanding involves several passes with a large drum sander and a smaller edge sander with progressively finer grades of sandpaper. Sand with the grain wherever possible.

WHAT YOU WILL NEED

Time. Finishing a 9′x12′ hardwood floor may take anywhere from 8 to 18 person-hours. Remember, however, that more time will be needed between steps to allow coats of finish to dry.

Tools. Except for a few specialty tools, most of the tools you will need to finish hardwood floors are common ones. Specialty tools that you will probably need to rent include a drum sander, an edge sander, a floor buffer, and a heavy-duty vacuum cleaner. The drum sander is a large sander used on the main body of the floor. The edge sander is designed to sand the floor where it meets the wall, where the drum sander will not reach. The floor buffer is used for very fine sanding and for polishing. The shop vacuum is needed to remove as much dust as possible before you apply the stain and protective finish.

Tools for Repairs

Variable-speed drill
Drill bit stop
Drill bits
Nailset
Chisel
Putty knife
Pry bar
Caulking gun
Screwdriver

Tools for Finishing

Putty knife
Drum sander
Edge or disk sander
Block sander
Orbital sander

Coarse-, medium-, and fine-grit sandpaper
Nailset
Paint tray and roller
Lamb's wool mop
Tack cloth
Cheesecloth
Floor buffer

Materials for Repairs

Graphite (tube)
Wood dough
8-penny finishing nails
3/16" round-head wood screws
3/4" hardwood flooring strips
Subflooring adhesive
2x4, 2x6, or 2x8 for bridging
Rags

Materials for Finishing

Duct tape
Plastic sheets
Wood stain
Wood sealer
Polyurethane
Paint stick or coat hanger for stirring paint
#2 steel wood machine disks and pads

Step One Repairing Squeaky or Cupped Boards

Margin of Error: Not applicable

Most Common Mistake

☐ Not fixing any problems before sanding.

Before you sand your newly laid floor, some prep work will probably be in order. Make a careful survey of the floor. You will need to repair any squeaky areas or ridged or cupped boards before sanding.

Squeaky boards are annoying, but they can often be easily fixed. The first thing to try is tapping the squeaky area with a hammer and a 2x4 wrapped in a towel. Another simple technique is to insert a shim from below the floor, between the floor joist and the area where the floor is squeaking. If that doesn't work, try squirting some lubricant such as graphite, talcum powder, floor oil, or mineral oil between the boards. Next, you can try forcing metal glazier's points between the boards every 6" to separate the boards.

If these simple techniques don't work, try drilling a pilot hole through the board, nailing from above with a finishing nail, and then countersinking the nail and filling the hole

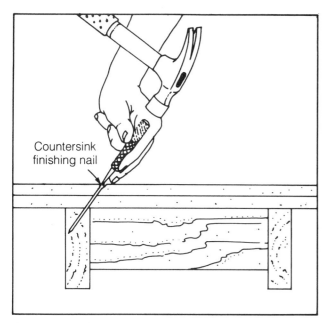

Figure 10-8. To fix a squeaky board, nail it from above into a joist.

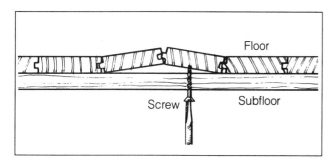

Figure 10-9. Use a screw to pull down a warped floorboard.

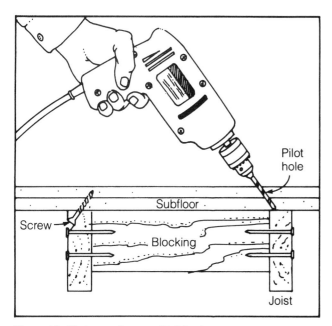

Figure 10-10. A wood or metal bridge between joists may stiffen the floor enough to eliminate a squeak.

Figure 10-11. Use a putty knife and wood dough to fill cracks.

with wood dough (Figure 10-8). This technique also works for repairing cupped or warped boards.

If the floor joists are exposed from below, you can drill a pilot hole up through the floor joist and/or subflooring and 1/4" into the squeaky board (Figure 10-9). Use a drill bit stop to prevent drilling through the floor surface. Wrap masking tape around the bit as a drill guide. Then you can grab the board from below with a 3/16" round-head wood screw with a large washer. This technique of screwing the board from below also works for repairing cupped or warped boards. Another possible solution is to add metal joist bridging or wood blocking between the joists near the squeak (Figure 10-10). This will often stiffen the floor and eliminate the squeak.

If there are any small holes or cracks in your new floor, now is the time to fill them. Use a putty knife to fill cracks thoroughly with a wood dough that matches the color of your unfinished floor (Figure 10-11). Allow the wood dough to dry thoroughly before sanding.

Step Two
Preparing the Room
for Sanding

Margin of Error: Not applicable

Most Common Mistake

☐ Neglecting to prepare the room.

After you have prepared the floor, you will need to take a few additional steps before sanding. Remove the floor registers, if they have been installed, and seal the openings with plastic. Also seal any heating or air conditioning ducts. Search out any protruding nailheads and countersink them with a nailset. Sanding produces a very flammable dust, so turn off all gas and electrical appliances and seal off all electrical outlets. Seal the room off tightly from the rest of the house to avoid dust problems. However, you should open a window and use a window fan to provide adequate ventilation while sanding. After all this is done, sweep the floor well and vacuum with a heavy-duty shop vacuum.

Step Three
Sanding the Floor

Margin of Error: Everything level within ⅛"

Most Common Mistakes

☐ Allowing sander to gouge floor.

☐ Leaving high spots or ridges.

☐ Not using an edge sander.

☐ Not sanding with fine paper.

Most oak floorings are ¾" thick and can be sanded a number of times. Thinner floors—½", ⅜", or 5/16"—must be refinished with caution to avoid sanding through to the subfloor. If your floor is thinner than ¾", consult a professional floor refinisher.

You are now ready to begin sanding. This is the only difficult part of the process; you need to be very careful or you can gouge the floor past repair. This happened to a friend of mine who had not bothered to learn to use the drum sander properly. Fortunately, his gouge was in an area covered by a sofa; but you don't want to have to arrange your furniture according to your gouges.

When you rent the drum sander, be sure to get a manufacturer's instructional manual and some hints and a demonstration from the store where you rent it. Be sure that the sander is in good shape and functioning well, and check to be sure you have all the dust bags, special wrenches, and attachments. The drum sander is a powerful machine. Although it does not require any great strength to handle, you should practice on a piece of plywood, or with fine sandpaper, until you get the hang of it. Always use a dust mask and ear protection when you are working with a drum sander.

Purchase several grits of "open face" sandpaper. You will use the coarse grits for rough sanding and the finer grits at the end to provide a smooth finish. If you need to sand cupped boards, start with a 20-grit paper. For the second sanding, use a medium, 80-grit paper. The final finish sanding requires a fine, 100-grit paper, or even finer for certain woods. The number of sanding passes, two or more, depends on the condition of your floor.

Be sure to buy enough sandpaper. The average room requires about ten sheets of each grade for the drum sander and ten sheets of each grade for the edge sander. Get a surplus and return what you don't use. Be sure the paper is properly installed in the sander, and remember to change it regularly.

In your first sanding, with rough-grit sandpaper, you will be removing any stains or discolorations and leveling the floor to a smooth surface. If there are warped boards or ridges that you were not able to repair earlier, you will need to sand diagonally across the floor in those areas with a rough-grade sandpaper until the floor is smooth. Then sand

Figure 10-12. Use a professional drum sander to sand the floor.

with the grain of the floor to get out the sanding marks left by the diagonal sanding. Except for these badly cupped areas, always sand with the grain of the floor.

Never turn the sander on while the sandpaper and drum are touching the floor. Tilt the sander back by the handle until it is out of contact, start the sander, and when it reaches full speed slowly lower it until the sandpaper touches the floor (Figure 10-12). Begin to move the moment the drum touches the floor. Let the sander pull you forward at a slow, steady speed. You can sand both forward and backward, but always keep the sander in motion. Never allow it to stop while it is in contact with the floor. Sand in straight lines with the grain of the floor. As you approach the end of your run, lift the sander while it is still moving forward.

You should sand about two-thirds of the floor in one direction and one-third in the other, as shown in Figure 10-13. Whenever you need to reposition the sander, make certain the drum is off the floor. Overlap your passes to be sure you are sanding all areas thoroughly, and to ensure an even finish with no sanding marks. Go forward and then return over the same area as you go backward. Move sideways in 3" to 4" increments to overlap each pass.

After you have done your first pass on the main body of the floor, use an edge sander, with the same coarse-grit paper, where the floor meets the wall and in other areas missed by the drum sander (Figure 10-14). Follow the manufacturer's instructions with this machine. It can also gouge the wood, although not as easily as the drum sander.

After the first sanding, fill any dents, gouges, or cracks with wood dough and allow it to dry. Then repeat the sanding process with both the drum and edge sanders with medium-grit paper.

Before you begin the final sanding, use a hand sanding block to get to any areas the power tools could not reach, such as under radiators and in corners. Then repeat the sanding process a final time, using a fine-grit paper.

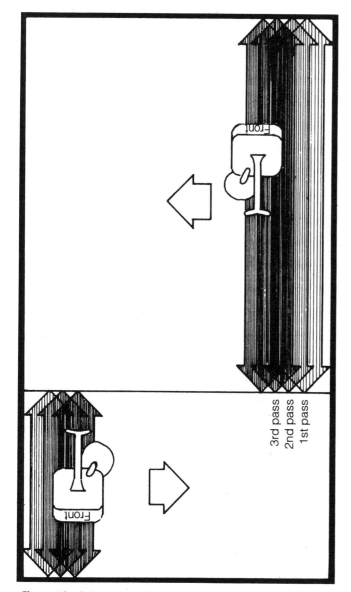

Figure 10-13. Recommended pattern of sanding with a drum sander.

1st pass
2nd pass
3rd pass

Step Four
Applying the Stain

Margin of Error: Not applicable

Most Common Mistakes

☐ Not removing all the sanding dust before staining.

☐ Staining unevenly.

It is best to finish the floor and apply a protective sealant as soon as possible after sanding the floor, preferably the same day. This protects the floor from moisture and other problems that could cause the wood grain to rise and create a rough surface. Before you begin finishing the floor, you need to be sure that it is perfectly clean and free of dust or debris. Also check carefully for any flaws or imperfections.

Fill any remaining cracks or holes with wood dough, allow it to dry, and then sand it smooth. Sand off any swirls or sanding marks. Vacuum the floors well and rub them with a tack cloth to pick up all the fine dust. Finally, rub your entire floor well with cheesecloth.

Staining the floor is optional. You may want to just put on a protective finish and let the natural color of the wood show through. If you decide to stain the floor, it's a good idea to test the color on an area that will be hidden before applying it to the entire floor. The color may appear quite different from the small sample you saw at the store.

Before staining your floor, you may want to apply a wood sealer, especially if you have installed a softwood floor. This is not the heavy-duty protective finish that you will apply last, but rather a light-weight sealer that seals the open pores of the wood, making it easier to apply the stain evenly. Some

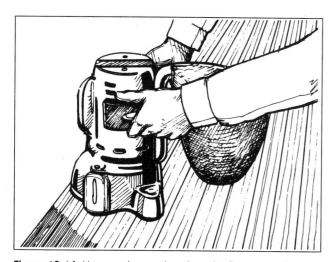

Figure 10-14. Use an edge sander where the floor meets the wall.

Figure 10-15. Use a professional buffer with a steel wool pad to sand between coats of stain and between coats of polyurethane.

wood sealers are colored and stain and seal the floor in one application.

Applying stain can be tricky; follow the manufacturer's instructions carefully. It takes concentration and some skill to get a good even finish with no blotchy areas where the stain is unevenly applied. The sealer will help, because the pores will be sealed and the stain will not penetrate as deep or as quickly.

Pour all the stain you will need into one container and mix it thoroughly to ensure even color. You can use rags, brushes, or a roller to apply the stain. Go carefully and be sure that the stain is penetrating to give an even color. The pigments of the stain are in suspension, not in solution, so the stain must be stirred regularly during application. Apply a generous coat; after 5 to 10 minutes, wipe vigorously with a rag to remove the excess. The amount of time you allow the stain to set on the floor will determine the darkness of the color. After wiping off the excess stain, buff the floor with a professional buffer and a #1 steel wool disk (Figure 10-15). As always, follow the manufacturer's instructions. Vacuum the floor thoroughly before applying the final coat.

Keep rags handy to wipe up any excess stain. To avoid the danger of spontaneous combustion, never store oily rags together or in a closed container.

Step Five
Applying the Protective Finish

Margin of Error: Not applicable

Most Common Mistakes

☐ Not sanding after every coat.

☐ Not stirring the finish properly.

There are several types of protective finishes. The two most common types are penetrating sealers and surface finishes, including polyurethane, varnish, and shellac. Penetrating sealers (mentioned under staining) may be clear or tinted. They penetrate the pores of the wood, so the finish wears as the wood wears and can be retouched with wax in heavy-traffic areas. These penetrating sealers are often used as an undercoat with surface finishes. However, make sure that the two finishes are compatible before you begin.

Surface finishes provide a tough, clear coating over stained or sealed wood. Polyurethane has largely replaced varnish, shellac, and lacquer. A heavy-traffic wax is often applied over the polyurethane.

Finishes come in a clear gloss finish (high gloss) and a satin finish (low gloss). The satin finish shows dust less.

Applying Polyurethane

Be sure to buy the slower-drying air-drying polyurethanes, rather than the faster-drying moisture-cured types that professionals use. Make sure that whatever type you use is compatible with any undercoat you may have applied.

Always stir polyurethane well before you apply it. The hardeners settle to the bottom, and if the polyurethane is not well stirred the floor will not dry evenly. Use a paint stick or install a bent coat hanger in a variable speed drill. Stir at a low speed, being careful not to create bubbles in the polyurethane. Ventilate the room well and wear a mask made for use with toxic fumes. In ventilating, however, be careful not to create a situation that will allow dust particles to settle on the wet floor.

It is best to apply the finish with a lamb's wool applicator and a paint tray. Use a brush at the walls and in hard-to-reach areas. Apply the polyurethane evenly, moving the applicator in the direction of the grain.

Allow the first coat of polyurethane to dry completely (Drying time will vary with the temperature and humidity.) When the floor is completely dry, with no tacky feel, sand it with a buffer equipped with a #2 steel wool disk. This is much easier than using a hand sander, although you will still have to sand hard-to-reach areas by hand. Vacuum after sanding, and then go over the floor with a damp mop to remove all the dust.

The second coat of polyurethane should be applied across the grain. When the second coat is completely dry, repeat the sanding process. Three or four coats are usually applied. the final coat, which is applied with the grain of the wood, does not need to be buffed. For one final layer of protection, you may apply a coat of heavy-traffic wax.

You can move in as soon as the final coat is dry. the protective finish may give off unpleasant fumes, but the odor should not be noticeable after a week or so.

11 Ceramic Tile

INSTALLING CERAMIC TILE in your bathroom or kitchen is one of the most satisfying projects you can undertake around your home. That's because the result is so beautiful—no other surface combines the colorful appeal of tile with its practicality and durability. Properly installed, a tile floor can retain its easy-to-clean good looks for a lifetime.

Tile comes in a bewildering array of sizes, shapes, colors, and finishes. Larger, thicker tiles are usually used for floors; smaller, thinner ones are used for walls, countertops, and shower stalls.

Vitreous tile and **glazed tile** both resist water absorption through their finished surface, and are a good choice for areas where water will be present. Unglazed tiles absorb water, and are a poor choice for wet areas. Unglazed tile has a dull surface; glazed tile is satin or shiny; and the vitreous type is available in dull, satin, and shiny finishes. All tiles have dull backs so the adhesive has something to stick to.

Tiles are available in shapes like squares, rectangles, hexagons, thin strips, and thick strips, and in many sizes. Standard wall and counter tiles are about 4" by 4" and of uniform thickness. Smaller sizes, called **mosaic tiles,** are often glued to a fiber backing. Mosaic sheets are available in solid colors and in patterns.

You will need color-matched trim pieces to complete your installation. Trim for countertop edges, backsplashes, and wall accents can be quite elaborate; trim for the larger format floor tiles is usually limited to bullnoses for doorways and baseboard pieces.

A trip to a tile showroom can be overwhelming with the variety of shapes, sizes, colors, and glazes of tile available. Unless you have a firm idea of what you want, plan to visit the showroom at least twice before you make up your mind; you'll be living with your selection for a long time.

Until about 30 years ago, tile was set into mortar—a difficult job that was best left to a professional. These days, tile is set with a strong adhesive on top of backer board, a much easier task.

This chapter describes the process of installing ceramic tile on kitchen and bathroom walls, floors, and counters, from putting up the backer board through grouting and sealing the tile. The job requires patience and precision, and it can be messy, but it's not really difficult once your layout lines are in exactly the right place.

Before You Begin

SAFETY

☐ Wear safety glasses or goggles when cutting tile to protect against flying chips.

☐ Cut or broken tile edges are very sharp. Use a tile sander to smooth these edges.

☐ Wear ear protectors when using power tools. Some tools operate at noise levels that can damage hearing.

☐ Be careful not to let loose hair and clothing get caught in tools. Roll your up sleeves and remove jewelry.

☐ Wear the proper respirator or face mask when sanding, sawing, or using substances that give off toxic fumes.

☐ Keep blades sharp. A dull blade requires excessive force and can easily slip.

☐ Always use the right tool for the job.

☐ Don't drill, shape, saw, or cut anything that isn't firmly secured.

☐ Don't work with tools when you are tired. That's when most accidents occur.

☐ Read the owner's manual for all tools and understand their proper use.

☐ Keep tools out of the reach of small children.

☐ Unplug all power tools when changing settings or parts.

WHAT YOU WILL NEED

Time. The time you will need depends on the size and complexity of your tiling job. Allow yourself plenty of time— laying tile is a painstaking process.

Tools. Most specialty tile tools can be purchased inexpensively. You can probably rent or borrow an electric tile saw from your tile dealer.

Tape measure

Tile cutter

Tile nippers

Tile sander

Combination square

Carbide rod sawblade

Framing square

Notched trowel

Level

Hammer

Rubber mallet

Floor scraper

Caulking gun and caulk

Chalk line

Screwdriver or can opener

Putty knife

Utility knife

Large sponge

Grout mixing tray

Grout trowel

Margin trowel

Materials. Field tiles, including mosaic sheets, are purchased by the square foot. Trim tiles, such as counter edges, backsplashes, and edge detail pieces, are figured in linear feet. Before you order your tile, you should make accurate measurements of the space you want to cover with tile and draw your plan out on graph paper. To allow for cuts and damage, order about 5 percent more tile than you need, or even more for complicated jobs. The color of the glaze will vary from shipment to shipment, so if you have to go back to the store for more tile, you may find that the color does not match very well.

Self-spacing tiles have small tabs cast into their back edges, in the corners. These tabs typically produce a very narrow grout line. If you are not using self-spacing tiles, you will need to buy small plastic spacers to maintain proper alignment and grout spacing. Plastic spacers come in the form of X's, Y's, and T's, with legs in widths varying from 1/16" to 1/2". The width of the leg determines the width of the grout line. Spacers are made of a flexible material that can be compressed slightly so that you can make minor adjustments.

Grout comes in many colors. If you decide to use a colored grout, be sure that it won't stain the tile you've selected. In choosing kitchen tile, keep in mind that a dark grout does not show dirt like a white grout.

Tile

Backer board or underlayment (for floors)

Grout

Roofing nails for backer board

Nails for underlayment (for floors)

Mastic

Spacers

Wood for battens

Grout sealer

Grout fortifier

USEFUL TERMS

Backer board is a thin sheet of concrete sandwiched between pieces of thin fiberglass mesh and made into sheets like plywood. It is used as a backing for tile.

A **batten** is a straight piece of wood nailed to the wall or floor as a guide.

Mastic is a generic name for cement mortar *or* tile adhesive.

A **rod saw** is a small wire encrusted with carbide chunks that fits into a hacksaw. Small holes are best cut in tiles with a rod saw.

A **spacer** is a piece of plastic or wood that is used to keep uniform spaces between the tiles.

Thinset is a cement-based adhesive.

Tile adhesive is a solvent-based adhesive that is used to stick tiles to backer board or drywall.

An **underlayment** is a piece of heel-proof plywood installed over the subfloor to provide a base for installing tile.

Most Common Mistakes

☐ Cutting tile with equipment and blades not designed for that purpose.

☐ Not using the proper backing or underlayment as a base for the tile.

☐ Not having the tile and underlayment at the proper temperature. Follow the manufacturer's directions.

☐ Not sealing the joints of the backing well.

☐ Not laying out the tiles correctly, and therefore ending up with very narrow tiles on the ends of the rows.

☐ Misaligning the tiles, so the job looks sloppy and out of level.

☐ Not using waterproof mastic when applying tile where it will get wet.

☐ Poor adhesion of tiles to the mastic, allowing the tiles to pull away from the wall.

☐ Not applying silicone caulk around the lip of the tub or shower pan.

☐ Not allowing the mastic to dry long enough before applying the grout.

☐ Not wiping the grout off before it sets up.

☐ Not sealing the grout with a silicone sealer a few days after installation, when the grout has had time to cure.

TILE WALLS

Step One
Installing the Backer Board

Margin of Error: 1/4"

Before you begin, cover any drains with tape so that debris won't fall in and clog them. Line the tub, shower, and/or sink with cardboard to avoid scratching them. In bathrooms, kitchens, and other areas that are exposed to moisture, the basic wall surface should be greenrock (moisture-resistant drywall).

For the best tiling surface, a mortar-based backer board is installed over greenrock. Backer board is easy to apply and compares in quality to the traditional but difficult method of doing a mortar bed.

Backer board is made of a concrete core that is coated on both sides with a thin fiberglass mesh. It comes in plywood-like sheets that are 7/16" thick, and is applied in much the same way as drywall. It is installed with the smooth side out if you plan to use adhesive and the rough side out if you plan to use epoxy or acrylic mortar.

The height of the backer board determines the height of the tile job. Once the tile has been laid, a line of quarter-round or other trim tiles will cover up the rough edge of the backer board.

If the tile is not going all the way to the ceiling, you will need to make some level layout lines at the correct height so that you have a line to run the backer board to. In laying out these lines, make sure that there is at least one row of tiles above the shower head. Use a level to establish an accurate line all the way around the surface to be tiled.

Determine the height of the backer board by measuring carefully so that when the tiles are installed up the wall you won't have to cut the tile for the top row. Also check to make sure that the tub is level. If you are working around a tub or shower pan that is not level and cannot be adjusted, cut your backer board so that the cut edge is along the lip of the tub or shower pan and is at the same angle as the tub, before cutting the top.

If your tub can be adjusted to be level, do so by placing shims under the tub before you put up the backer board. If you are tiling around a shower, be sure that the shower pan is correctly seated so that it is level and will drain well.

Start with the back wall, because the backer board that goes on there requires the fewest cuts. Mark the locations of the studs on small pieces of tape on the floor and ceiling. Later, when you are putting the backer board up, you can snap a chalkline from floor to ceiling on the backer board along the center of the stud to show you where to nail. (There are several ways to find the studs in a finished wall. Studs are usually located every 16" on center. Tap along the wall with a hammer until you hear a solid sound; look for seams in the drywall or mud over fasteners; or use a magnetic stud finder to locate the nails in the studs. Drive a finish nail into the wall in an area that will be covered by tile to verify the location of the stud behind the drywall.)

If you plan to use a floor and wall tile adhesive, install the backer board with the smooth side out. If you plan to use an epoxy or acrylic mortar, install the textured side out. If you are using epoxy or acrylic, make sure that the room is very well ventilated.

Cutting the backer board is easy. Make careful measurements, and mark them with chalk on the front of the backer board. Use a straightedge and a utility knife to score along your chalkline. The backer board will crack along that scored line when it is bent, just like drywall. Then turn the piece over and score the back to cut through the mesh.

Use galvanized roofing nails to hang the backer board. The nails must be long enough to go through the backer board and the drywall and penetrate 1" into the studs. Nail at 6" intervals around the edges and in the center over the studs. Nailheads should be flush with the surface but not countersunk. Countersinking the nails can cause the backer board to crack or break. Joints should be close together but not tight. Some backer board manufacturers recommend the use of a nail and a large washer at the edge for better holding. Check the manufacturer's instructions.

Position 1/4" spacers along the rim of the tub to hold the backer board up slightly. After all the backer board is up, but before you lay the tile, fill the gap with silicone caulk to form a water-tight seal.

When making the holes for the faucets and shower head, measure very accurately. Remember the cardinal rule of building: Measure twice, cut once.

Cutouts and holes for plumbing pipes and fixtures can be made by breaking through the fiberglass mesh with the edge of a hammerhead. For neater cuts, use a saber saw with a carbide blade or a masonry hole saw attachment for a drill.

Step Two
Taping the Joints
of the Backer Board

Margin of Error: 1/4"

After the backer board is nailed in position, all the joints—including the gaps between dissimilar materials, such as backer board and greenrock—must be filled, much like the procedure for taping drywall (see Chapter 5, Drywall).

For the seams and corner, a 2"-wide coated fiberglass mesh tape is embedded in a dry-set or latex portland cement mortar. The gaps and cracks, including the one along the top of the tub, are filled with a thin coat of the same mortar.

When this mortar is dry, it needs to be waterproofed. Coat the entire backer board surface with a waterproofing sealer.

Step Three
Laying Out the Tile

Margin of Error: 1/8"

You have more leeway in laying out the tile for walls than for floors or countertops, because the finished tile surface doesn't have to go all the way to the ceiling. For appearance' sake, it's best to use full-sized tiles at the top and bottom of the walls and avoid having to cut tiles for top and bottom. This means you must add up the dimensions of all the field tiles, accent and trim strips, and grout lines to determine the height of the finished wall. Around the shower, the height of the finished surface will determine the height of the backer board, so measure and add carefully.

The following explanation of a tile wall installation assumes that you are using standard square tiles. If you are using prealigned mosaic sheets, adjust your layout lines to

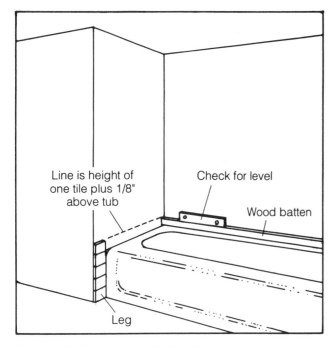

Figure 11-1. Horizontal working lines for tile layout.

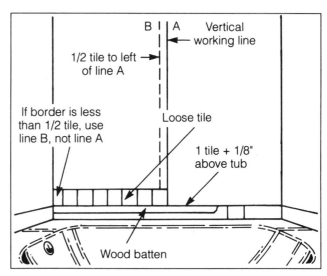

Figure 11-2. Vertical working lines for tile layout.

reflect the sheet format. Although these sheets are uniform squares, their sizes vary by manufacturer and style.

Before you roll up your sleeves and go to work, you want to make sure that you have everything you're going to need. Double check your supplies carefully.

Do the colors of the tile match? There is often some discrepancy in tile color from box to box. It's a good idea to pull tiles out of different boxes as you work so that any slight color difference is blended in.

Horizontal Working Lines

It is crucial that you make accurate layout lines. These lines are the map you follow in installing your tile, so be sure they are bold and easy to see.

The horizontal lines come first (Figure 11-1). The way you lay them out depends on whether or not your tub or shower pan is level.

For a Level Tub or Shower Pan. If the tub or shower pan is level to ⅛", measure and mark your horizontal line from the high point of the tub. A difference of ⅛" is easy to disguise at the bottom. Measure up from the lip of the tub the width of the tile plus ⅛". If you are using standard 4 ⅛" square tile, measure up 4 ¼" and make a level line, using a level, along the back wall and the two end walls.

For an Out of Level Tub or Shower Pan. If the tub or shower pan is not level to ⅛", mark your horizontal line from the low point. If you don't do this, the gap along the edge of the tub and tile will show. Determine a level line in the same way as for a level tub, and then run a batten along the bottom of the line so you have a level line to work off of. To do this, nail a straight wooden batten so that the top of the

batten is set to the horizontal line. This provides an exactly level surface on which to begin laying the tile. After all the tile is laid, remove the batten and install the bottom row of tile. In this case, you will need to custom-cut the bottom row of tiles to fit.

Vertical Working Lines

When you lay the tiles out, adjust them so that the border tiles (the tiles on the vertical edges) are more than half a tile in width, and so that the the tiles on each edge are the same size.

To do this, first locate the midpoint of the back wall and mark it on the horizontal line (Figure 11-2). Then line up a row of loose tiles along the back of the tub, making sure that a joint matches up with the center mark. If your tiles are not self-spacing, use plastic spacers. The distance that is left at either end gives you the dimension of your border tiles.

If the end tiles turn out to be larger than half a tile, mark the vertical center line all the way up the wall, using a level and straightedge. The edge of the tile will be set to this line (line A in Figure 11-2).

If it turns out that the end tiles are smaller than half the width of a tile, move the center line exactly one half the width of a tile to the left or right (line B in Figure 11-2). By making this adjustment you avoid having narrow tiles on the borders. Very narrow tiles are unattractive and hard to cut.

The vertical layout lines for the end walls are usually marked after the back wall has been tiled. Just position the vertical working lines to minimize the number of tiles to be cut, and locate any cut tiles in the corners. Also decide whether you want to align the joints between the tiles or to stagger them so that they fall in the middle of the tile in the rows directly above and below them. In this case you will need to develop vertical layout lines so that no row has edge tiles that are less than half a tile.

Determine where you are going to put the towel rods and the soap dish or any other special accessory tiles, and mark

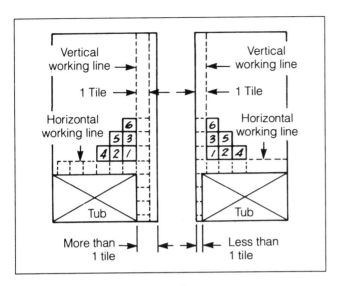

Figure 11-3. Tile layout on side walls.

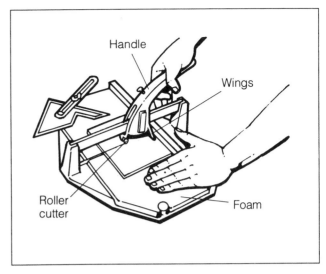

Figure 11-4. Use a tile cutter for straight cuts.

their locations. These will be installed last, and usually take up the room of ½, 1, or 1 ½ tiles.

If your soap dish fits into the wall, cut your hole before you spread the adhesive, and position it to minimize or even eliminate the need to cut any tiles to go around it.

Step Four
Spreading the Adhesive

Margin of Error: Not applicable

For a tub or shower enclosure, you should start with the back wall, unless you are also tiling the ceiling, in which case you should start there.

It is very important to use waterproof mastic in areas that are exposed to moisture. Before you apply the adhesive, read the manufacturer's instructions and note the drying time. Don't spread any more adhesive than you can work with before it sets up. Generally it's best to spread enough for 30 to 40 minutes work.

When applying the adhesive, be careful not to obscure your working lines. Also, be sure to leave blank spaces where you plan to install accessories.

Use the flat side of your trowel to spread the adhesive over a wide area of the wall. Then run the notched side through the adhesive at a 45-degree angle to create grooves. The peaks of the grooves should be nearly as thick as the tiles; the valleys should be only a thin film.

Step Five
Laying the Tile

Margin of Error: ⅛"

When laying the tile you can either align the joints in the tile of each row or stagger the joints. If you stagger the joints, the joints of one row will fall in the middle of the tiles in the row directly above and below. This is an esthetic decision; staggering the joints slightly is easier, but aligning the joints is not difficult. The tabs on self-spacing tiles do not align if you stagger the joints. Use plastic T's to space staggered joints.

Set the first tile along one side of the vertical and horizontal working lines (Figure 11-3). When setting the tile, use a gentle twisting motion. Don't slide the tile into place or you will squeeze the adhesive to one side.

If you are working with a batten, make sure that the tiles are firmly seated on it. If you are not working with a batten, be sure to line the top edge of the tile up along the horizontal line. If necessary, use two spacer legs underneath the tiles along the lip of the tub to hold the tile up along the line.

Lay the tiles row by row in a triangular pattern, using the spacers at the intersections, as shown in Figure 11-3. Keep a watchful eye out for correct alignment along the working lines. Tap the tiles with a rubber mallet and a block of padded wood as you go.

On inside corners, the tiles are laid up to the corner, leaving a small gap for grout. Outside corners may be treated with a bullnose tile; or you can use a special outside corner tile that has a curved lip. As you work, keep checking with your level to be sure that you are staying plumb and level, and adjust your tiles as needed. Also check to be sure the tiles are flush and are not pushed out from the wall.

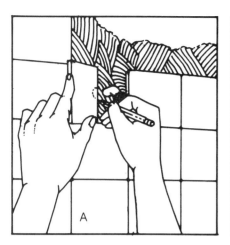
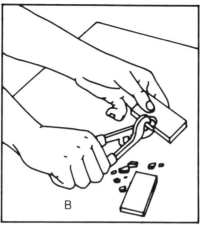
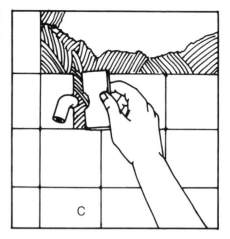

Figure 11-5. Use tile nippers for irregular shapes. A: Draw outline of pipe on tile. B: Nibble out small bites of the tile with nippers. C: Position tile around pipe.

Step Six
Cutting the Tile

Margin of Error: Exact

You may need to cut the border tiles to fit. Many tile dealers will loan you a simple tile cutter; or you can rent a wet saw.

Measure carefully, allowing for the grout line. Transfer your measurements to the tile to be cut.

Some tiles have small parallel ridges on their backs. Whenever possible, make your cuts in the same direction as these ridges.

When cutting with a nonpower tile cutter (Figure 11-4), score the tile only once. Multiple scores will dull the blade and create jagged edges on the tile. Place the breaking wings, located at the bottom of the handle, about ½" from either edge of the tile and slowly but firmly press down on the handle until the wings break the tile along the scored line. Smooth rough edges with a tile sander.

Cut tiles to fit around pipes and faucets after all the field tiles are laid. Remember to allow for the grout lines when taking your measurements.

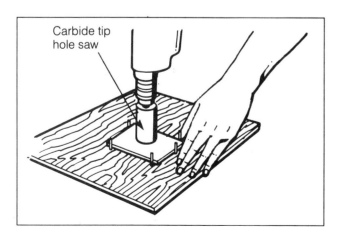

Carbide tip hole saw

Figure 11-6. Use a carbide tip hole saw to cut holes in tiles for plumbing pipes.

There are several ways to make complicated cuts. To fit a tile around a pipe, first cut the tile in two and draw on the tile the outline of the cut you need to make. Score the line with a micro cutter (a small sharp wheel with a handle). Score all over the area to be cut away, then "nibble" the area out with tile nippers (Figure 11-5). Use a tile stone to smooth the edges of the cuts you made with the nippers. To cut small holes for plumbing pipes, you can use a rod saw with a carbide blade, or you can get a tile-cutting attachment for your drill (Figure 11-6).

Step Seven
Trim Tiles and Final Adjustments

Margin of Error: ⅛"

All the edges of the tiled area need to be finished off with edge and corner trim. Simply butter the adhesive on the back of the tiles and stick them onto the wall in the correct position, aligning the grout joints carefully.

Before the adhesive sets up, make any adjustments needed for correct alignment and remove any spacers you have used. Check to see if the tiles are fully set by trying to pull one up. Then clean off any adhesive on the face of the tile. Allow the adhesive to dry for 24 to 48 hours, depending on the type of adhesive.

Step Eight
Grouting

Margin of Error: Not applicable

Grout comes in many colors, to match or contrast with your tile. In addition to highlighting the tile, grout seals the gaps between tiles and keeps out water and dirt. It must be properly applied, or water can get behind the tile and cause rot.

There are many types of grout. Some include sand and some do not. Mix the grout with water, according to the

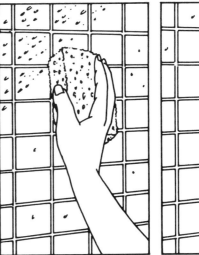

Figure 11-7. Spread grout with a grout trowel or squeegee.

Figure 11-8. Remove excess grout with a grout trowel or squeegee.

Figure 11-9. Remove remaining grout with a damp sponge.

Figure 11-10. Remove film with a soft clean rag.

manufacturer's instructions. A latex additive is often used to make the grout spread easier and bond better and to reduce shrinkage. The grout should be thick enough to be easily troweled into the joints, not soupy.

For standard 4 ⅛" tiles, where the joints are very narrow, use a nonsanded grout. For wider grout joints, with other types of tiles, use a sanded grout, which holds up better.

The easiest way to apply grout is with a grout trowel (a rubber-faced float) or a squeegee, although you can do it with your fingers and a large sponge. Spread the grout on the surface of the tile, forcing it into the joints (Figure 11-7). It is crucial that the joints be completely filled, so that there are no bubbles or gaps. Grout only a small area at a time, because the grout begins to harden as soon as it is spread.

Scrape off the excess grout by wiping diagonally across the tiles with your grout trowel or squeegee (Figure 11-8). With a clean, damp sponge, wipe away any remaining grout (Figure 11-9). Wipe the grout away as you go. Grout a small area, then wipe it down; don't wait until you have grouted the entire area. Rinse the sponge frequently in clean water. Continue to rinse and wring out the sponge until the joints are smooth and level with the tiles.

Allow the grout to dry about 30 minutes, until a hazy film appears. Wipe the film away with a soft clean cloth (Figure 11-10). Use a margin trowel or the end of a toothbrush handle to tool the joints and clean the intersections. To tool the joints, run the trowel or toothbrush handle down the grout line to push the grout in and to remove excess grout.

Step Nine
Installing the Soap Dish and Other Accessories

Margin of Error: ⅛"

After you have laid your field tiles and the adhesive has dried, it's time to install the soap dish (Figure 11-11) and towel rack. Simply butter the back of the accessory and place it in the space you left open for that purpose. To hold the accessories in place until the adhesive dries, tape them tightly to the wall with duct tape. Just be sure that the adhesive behind the field tiles is completely dry, or the tape may pull the tiles off the wall.

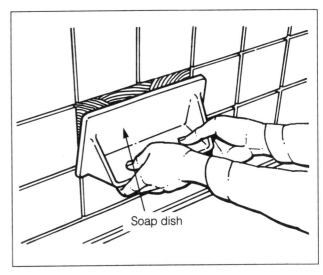

Soap dish

Figure 11-11. Remember to leave untiled spaces for the soap dish and other accessories.

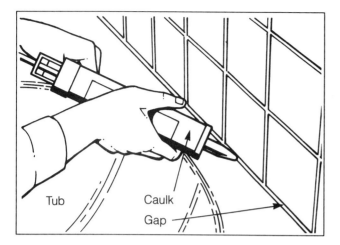

Figure 11-12. Caulk the gap between the tub and the wall.

Step Ten
Caulking and Sealing

Margin of Error: Not applicable

The final step in tiling your bathroom wall is to run a bead of silicone caulk around the edge of the tub, along the top where it meets the wall, and at the base where it meets the floor (Figure 11-12). Also caulk around the plumbing pipes and around the rims of the soap dish and towel rods. Wait until the grout is completely dry before caulking. It's also a good idea to fill the tub with water before caulking to maximize the size of the gap to be filled. Leave the water in the tub until the caulking has cured.

After the grout has completely dried and cured, it needs to be sealed with a silicone-based sealer, painted on according to the manufacturer's instructions. Drying time ranges from three days to two weeks, so be sure to read the label before you begin.

TILE FLOORS

Although the general rules for installing tile on walls apply to any tile job, there are specific techniques that apply only to floors.

Before you begin, cover any drains with tape so that debris won't fall in and clog them. Line the sink with cardboard to avoid scratching it. You should tile your floors before the toilet and cabinets are installed; never try to lay tile around a cabinet or toilet.

Before you order your materials, determine what you will need to finish the exposed edges properly. For example, if your finished floor level is going to be higher than the adjoining room or hallway, you should get bullnose tile to create a smooth transition. If the floor will meet a carpeted edge at the same level, then a regular square-edged tile will work.

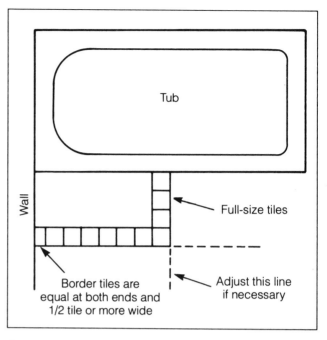

Figure 11-13. Layout for tile floor.

Step One
The Underlayment

Margin of Error: 1/4"

Proper backing for the tile floor is very.important. The backing may consist of exterior-grade heel-proof plywood, luan underlayment panel, or mortar-based backer board. Don't use particle board, flakeboard, or masonite as underlayment for ceramic tile. The total floor sheathing should be at least 1 1/4" thick over the floor joists on 16" centers. Otherwise, flex may cause tiles to pop out of place. If your subfloor has been seriously dented and roughed up during construction, or if you need added stiffness, you may want to add a layer of underlayment over the original subfloor. (For information on installing an underlayment, see Chapter 9, Vinyl Floors.) If you do install a new underlayment over the subfloor, staple polyethylene plastic on top of the subfloor to protect it from water.

When applying a plywood underlayment, leave a slight gap between panels and about 1/4" along the edges to allow for expansion and contraction. Fill low areas and seams with a quick-drying latex patching compound, using a wide application blade to create as flat a surface as possible.

Stagger the joints of the underlayment in a brick-joint fashion, and be sure that underlayment seams do not fall directly over existing subfloor seams. With whatever underlayment you use, except for backer board, countersink all nails slightly.

If you use a mortar-based backer board, cover the joints and seams with joint compound and then seal them with a moisture-resistant sealant.

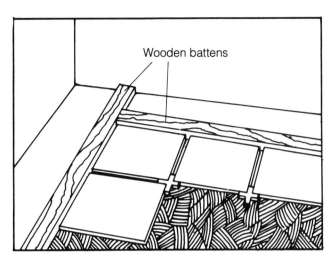

Figure 11-14. Nail battens at a right angle so you will have a square corner from which to start laying your floor tiles.

You can lay tile over concrete, but it must cure for at least 28 days first.

Step Two
Establishing the Layout Lines

Margin of Error: 1/8"

It is crucial that you establish accurate working lines. You want a layout in which there are full-size tiles in areas of high visibility, and in which all edge (border) tiles are at least half a tile wide.

Many floor tiles have matching baseboard pieces, and some baseboard tiles are coved. If you use coved baseboard, be sure to take it into account when making the layout lines.

Begin by making a dry run. Lay the dry tiles out from the two most visible walls to see which layout will work best in your room.

If the tiles at the end of each row (border tiles) are less than half a tile wide, adjust the row by moving it the width of half a tile to either the left or the right. This will ensure that all the cut tiles are more than half a tile in width (Figure 11-13). (See Step Three under Tile Walls.)

Next, check to be sure that the corner that you are working off of is square. If it isn't, tack battens to the floor a tile's width away from the wall along both walls. Then snap another line parallel to the first and outside of it by the thickness of two grout lines (to allow for the grout lines along the wall between the first and second rows of tiles).

Nail battens made from 1x2s or 1x3s along the second (working) line so that you have a firm position to start running your tiles (Figure 11-14). Make sure your battens are straight and form a right angle, and use your spacers.

Once the tiles are laid in the field, remove the battens and lay your tile along the edges, cutting the tiles as necessary. When you are cutting tiles, always remember to leave room for the grout line.

Step Three
Spreading the Adhesive, Laying and Cutting Tiles, and Grouting

Margin of Error: 1/8"

These steps are basically the same as those outlined earlier for installing tiles on walls.

Instead of laying the tiles in a pyramid fashion, as for walls, lay them row by row, cutting them as needed when you get to the end of a row. Use spacers at each joint. Then, with your eye close to the floor, sight down the edges and adjust the tiles if necessary.

TILE COUNTER TOPS

For counter top backing, you can use either backer board or exterior grade plywood. Be sure to cut out all openings for your sink and fixtures before applying the tile, and then cut the tile to fit around these openings. When you are ordering your tile, be sure to take into account all the trim and backsplash tile you'll need, and allow about 10 percent extra.

Step One
Laying Out the Tile

Margin of Error: 1/8"

If you are tiling around a sink, mark the center point of the sink. If there is no sink, mark the center point of the counter top. Make a dry run of the tiles along the edge to see whether the center line needs to be moved half a tile's width so that you don't end up with less than half a tile on the ends. (See Step Three under Tile Walls.)

It's a good idea to make a dry run of how all the tiles will go around the sink and then label them so that you will know where each tile goes after you have cut them. Use a full tile along the front of the counter top, even if that means the tile along the backsplash will be less than half a tile wide.

Step Two
Setting the Tile

Margin of Error: 1/8"

Set the front edge pieces first, before spreading the adhesive for the field tiles. Butter the back of the edge tiles with adhesive and place them along the front edge of the counter top (Figure 11-15). Use spacer legs to maintain your grout line layout.

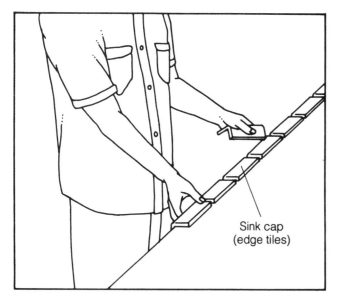

Figure 11-15. Lay the edge tiles before laying the field tiles.

Figure 11-16. Use a framing square to install the first row of field tiles.

Unless the sink is mounted to be on top of the tile, lay the sink trim next. Be sure to caulk between the sink and the plywood backing before setting the sink trim.

If you are using quarter-round tiles, you can either miter the corners or use specially molded quarter-round corners.

Now you are ready to set the field tiles (Figure 11-16). Use a framing square to make sure your lines are straight. Lay each row of tile from front to back, starting either at the center or from the sides of the sink. Cut the tile as needed.

Install the backsplash last (Figure 11-17). The backsplash is usually made of bullnose tiles that can be applied directly above the back field tiles. Use spacer legs to maintain your grout line space at the bottom of the backsplash. Butter the back of the bullnose tiles and stick them to the wall. You can also apply adhesive on the wall; just be sure to stay below the top of the tile so that adhesive does not show above the tile. After grouting, you will need to caulk this joint.

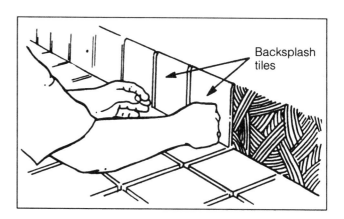

Figure 11-17. The backsplash is made of bullnose tiles.

Step Three
Grouting and Finishing

Margin of Error: Not applicable

These steps are the same as those described earlier for applying tiles to walls.

12 Painting and Wallpapering

PAINTING AND WALLPAPERING are simple projects, and with the information in this chapter they can be enjoyable and effective as well. You will learn how to prepare a wall, estimate materials, and use your tools correctly.

Neither painting nor wallpapering demands a lot of you physically. The toughest part is painting or papering the ceiling. The key thing is proper preparation. You may be tempted to take shortcuts here. Don't.

Before You Begin

SAFETY

It's important to understand, develop, and adhere to proper safety practices for any project you undertake. For both painting and wallpapering, safety precautions include the following.

☐ Always use the appropriate tool for the job.

☐ Keep blades sharp. A dull blade requires excessive force, and can easily slip.

☐ Always wear the proper respirator or face mask when sanding or working with chemicals.

☐ Wipe up spills immediately.

☐ Don't smoke or allow open flames, such as pilot lights, around solvents or solvent-based paints.

☐ Store and dispose of rags properly to avoid spontaneous combustion.

USEFUL TERMS

Booking the wallpaper means to loosely fold presoaked wallpaper, pasted side to pasted side, to allow a few minutes for curing (expansion of the adhesive) before applying it to the wall.

Cutting in means to use a 3" or 4" brush to paint corners and edges that cannot be covered by a roller, such as where wall meets wall, where wall meets ceiling, and next to the trim.

Feathering is a series of light strokes with brush or roller, lifting the applicator lightly at the end of each stroke to blend in the paint.

A **sash brush** is a 1 ½" angled brush made for detail painting of windows and narrow trim pieces.

Sizing is a liquid that is painted on the wall before papering. Sizing dries to a tacky feel, ensuring proper adhesion of

wallpaper. It also makes it easy to remove the wallpaper in the future.

A **trim brush** is a 2" brush for painting door trim and other wide moldings.

TSP (trisodium phosphate) is an industrial cleaner.

PERMITS AND CODES

Some areas require permits whenever you are spending more than a certain amount of money on construction, repairs, or remodeling. Check with your local building inspector to see if you need a permit. Usually only a small fee is required, and this ordinance is often not enforced. Other than this, no permits or inspections apply in either painting or wallpapering projects.

PAINTING

WHAT YOU WILL NEED

Time. Preparation time will depend on the extent of the problems encountered. Painting (including cutting in) takes approximately 30 to 60 minutes per 100 square feet.

Tools for Painting. Figure 12-1 shows some of the tools you will need for your painting project.

For the most part, it is best to use paint rollers for the large areas, brushes for smaller areas, and specialized tools for corners and trim. For tight areas like small bathrooms and closets, where there may be lots of trim and little room to maneuver a roller, consider using a paint pad instead. Always purchase high-quality tools, or you will regret it later. Cheap rollers, pads, and brushes can cause uneven application of the paint and an unprofessional looking result.

Rollers come in many different types, with naps that differ according to their use. A good synthetic roller cover will work as well as a lambswool roller. Read the package and be sure to buy the proper types for your application. The roller you purchase should have a handle with nylon bearings, a comfortable grip, a threaded hole for an extension, and a beveled end. A fairly new option on the market, the electric roller, supplies the roller with a continuous flow of paint. An air compressor with a painting attachment can also save you time and effort (see Chapter 1, Tools).

You will also need to purchase a pan to hold the paint and a metal or plastic grid, or screen, to go in the pan. The screen ensures that the roller is properly loaded with paint.

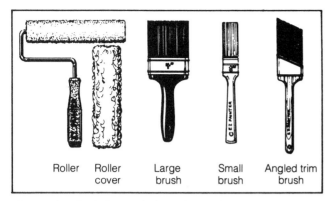

Figure 12-1. Tools for a typical painting project.

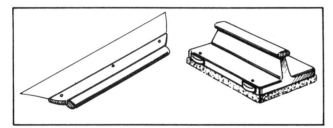

Figure 12-2. The paint guide and paint edger are used for cutting in—painting the corners where ceilings meet walls and where walls intersect.

Brushes come in two types, with synthetic or natural bristles. Natural bristles are recommended for oil-based paints but not for water-based paints. Synthetic filament brushes (nylon or polyester) must be used for water-thinned paint, although they work with solvent coatings as well. Polyester brushes should not be used with shellacs and lacquers. Always purchase high-quality brushes. Most jobs require a 4" brush for cutting in, a 2" brush for baseboards and trim, and a 1 ½" to 2" angled sash brush for windows and smaller trim. Paint pads should have beveled edges and a curled rear edge.

You may also need some specialty tools, such as a paint guide and edger, shown in Figure 12-2.

Materials for Painting. There are two types of paint: alkyd (oil-based) paint and latex (water-based) paint. Latex paints are generally used in areas that don't require frequent washing. Cleanup with these paints requires only water and is much easier than with oil-based paints, which require a solvent for cleaning up.

Oil-based paints are applied in areas that require frequent washing. They are resistant to damage, and are often applied over metal or wood.

Paints also come in several finishes: primers, gloss, and flat. Primers (undercoats and sealers) are used as bases or undercoats beneath the finish coat. Flat paints with no gloss are often used on walls and ceilings. Gloss finishes (from low to high gloss) are used on woodwork and in bathrooms and kitchens because of their water-repellent nature.

Tools for Prepping Walls

Safety glasses or goggles

Respirator or face mask

Ear protectors

Rubber gloves

Paint scraper

Fan

Hand sanding block

Orbital sander

Screwdriver

Putty knife

Sponge

Cap or scarf

Old clothes

Materials for Prepping Walls

Primer

Sizing (for wallpapering)

Tools for Painting

Drop cloths

Ladders

Buckets

Paint edger

Brushes, 4", 3", and 1 ½"

Angled sash brushes, 1 ½" and 2"

Roller pan with screen

Roller covers with appropriate naps

Roller handle

Roller extender

Paint guide

Materials for Painting

Masking tape, 2" wide

Newspaper

Adhesive pad or primer

Paint

Paint thinner (for oil-based paints)

Aluminum foil

Rags

Estimating Paint

To estimate the amount of paint you will need, first calculate the area of the walls and ceiling in square feet. Multiply the perimeter of the room by the height of the walls. For the ceiling, multiply the length by the width. Add these two numbers together to determine the number of square feet you will need to cover. After you figure the square footage, add a little more for touch-up. A gallon of base coat will cover 350 to 450 square feet. Purchase all the paint you will need in one order; it's difficult to match colors exactly in a second batch.

Most Common Mistakes

The single most common mistake in any project is failure to read and follow the manufacturer's instructions for tools and materials. The most common mistakes in painting include the following.

☐ Not preparing a clean, sanded, and primed (if necessary) surface.

☐ Failing to mix the paints properly.

☐ Applying too much paint, or not enough paint, to the applicator.

☐ Using a water-logged applicator.

☐ Not solving dampness problems in the walls or ceilings before painting.

Step One
Prepping the Walls

Margin of Error: Not applicable

Most of us believe that we know everything we need to know about painting. This assumption often leads to poor quality work, both in preparation and in the actual paint application. You need to understand—and use—the proper procedures to ensure a high-quality painting or papering job.

If a job is worth doing, it's worth doing right the first time; and proper preparation is the key. Because it seems to lead to more work, preparation is a step that is too often left out, and the final result reflects this omission. It's too easy just to start painting or papering, without going through the necessary preparatory steps. The paint job may even look pretty good for a while. But sooner or later, the poor quality will show up.

Prepping the wall for a new covering is much the same whether you are papering or painting, although prepping the wall for papering involves a few more steps.

1. Turn the electricity off and remove everything from the walls and ceilings, including electrical wall and ceiling light fixtures, switch plates, and outlet plates, if these have already been installed. After you have safely wrapped all disconnected light fixture wires, you can turn the electricity back on.

2. Vacuum and/or mop the floors and all ledges to remove dust and debris. Cover the floor with a drop cloth.

3. A primer coat is recommended in many cases. For example, drywall must be well sealed with a primer before it is painted, or it will absorb the paint. Or you can apply an adhesive pad to the wall. This is a liquid just like a primer, but it dries to a tacky consistency. Ask your paint supplier for recommendations for your particular job.

4. Always apply a coat of liquid sizing to the surface before hanging wallpaper. The sizing gives a better adhesive to the wallpaper and also makes removal easier in the future.

5. Mask off windows (and the woodwork and trim, if they have been installed) with newspaper and masking tape that is at least 2" wide.

6. Before you begin, assemble all the ladders, buckets, brushes, paints, and everything else you are going to need.

Prepping the Trim and Woodwork

If your woodwork and trim are already installed, you will paint them last. But you need to prep them before beginning to paint, or else the debris from prepping will settle on the new paint. First, fill all dents and gouges with wood putty or patching compound. Don't use fast-drying compound, which is hard to sand. If the gouge is over ⅛" deep, use two layers. Woodwork and trim are usually painted with an enamel or glossy paint, which is both durable and easy to clean.

Getting Ready to Apply the Paint

Properly applying the paint is the final step in a professional-looking paint job. You have prepped the surfaces and chosen the right paint and applicators; now the fun starts.

Before you begin applying the paint, be sure that it is properly mixed. Professionals use a system called "boxing," which ensures that there are no mismatches among different cans of paint. Mix all your paint into one large container until the color and consistency are uniform. It is important to mix enough paint to cover all surfaces; matching can be difficult if you run out.

Air often causes a scum on the surface of oil-based paints. If this happens, you will need to strain the paint through a nylon stocking to remove this "skin." When thinning paint with either a thinner for oil-based paints or water for water-based, thin slowly to avoid overthinning.

After you open each can of paint, use a nail and hammer to punch holes in the rim of the can so that excess paint will drip back in (Figure 12-3).

Plan to paint your room in the following sequence.

1. Ceilings

2. Walls

3. Trim (if installed), windows, door, then baseboard

Paint the ceilings first so that any drips that fall on the walls will be covered later. When painting the walls, always paint from the top down, and do the trim last, again so that any paint that drips down can be covered.

Needless to say, you should wear old clothes. Unless you want to try some unusual hair coloring combinations, wear a hat and scarf or hooded sweatshirt while doing the ceilings.

Step Two
Cutting In

Margin of Error: Not applicable

Cutting in is the process of applying paint to all the places shown in Figure 12-4—all corners where ceilings meet walls and where walls intersect, and next to all molding,

Figure 12-3. Use a nail to punch holes in the rim of the paint can to let excess paint drip back in.

Figure 12-4. Cut in these areas before painting the large surfaces.

trim, and baseboards. Because rollers and sprayers cannot neatly reach these areas, use a 3" or 4" brush to paint all these edges before doing the large surfaces. You can also use a paint edger, a sponge-type brush with a small set of wheels on the side that enable it to make an even close cut. Use a paint edger or straightedge next to trim or baseboard to be sure that no paint gets on the wood. Do all necessary cutting in before painting the large surfaces.

Step Three
Painting the Large Surfaces

Margin of Error: Not applicable

After finishing the cutting in, you are ready to paint the ceiling and then the walls. Be sure the area is well lighted so you can see any ridges or drips. Painting the ceilings is physically difficult. Painting overhead can cause back and neck strain and an occasional eye full of paint. Safety goggles (and yoga) are a must when painting ceilings. Use a high-quality roller with an extension so that you can easily reach all areas of the ceiling. With such a roller, you won't need a ladder, except for touch-ups and cutting in. Use the same roller with extension for the high parts of the walls. If your ceilings are especially high, you may want to erect a low scaffold, using sawhorses or ladders with a 2x12 board between them.

To use a roller, pour the paint into the roller tray or paint pan so that the paint in the reservoir is 1/2" deep. This amount of paint will enable you to load the roller fully without underloading or overloading it. You can save on cleanup time by lining your tray with heavy-duty aluminum foil before pouring in the paint.

Roll or dip your roller into the paint reservoir and roll it around until the paint saturates the roller. Then run the roller a couple of times over the washboard area of the tray to remove excess paint so that it won't drip. You may be surprised at how much paint the roller holds, so be careful to saturate it thoroughly.

To avoid splattering, apply the paint slowly on the ceiling and walls. In the beginning, use overlapping V-shaped strokes, as shown in Figure 12-5. Begin at a corner and work across the wall or ceiling, covering about 3 square feet at a time. After you have made your V-shaped zigzag patterns, fill in the unpainted areas with parallel strokes without lifting the roller from the surface. Increase the pressure on the roller as you work to deliver the paint smoothly.

When you are rolling into unpainted areas, feather the paint in with a series of light strokes, and lift the roller at the end of each stroke. When you need to remove the roller to reload, begin the next section, rolling in a zigzag into the outer border of the area you just completed. Then lightly roll the area between the two sections. Paint the entire surface. Don't allow the paint to dry on part of the wall or ceiling.

Figure 12-5. Spread paint in unpainted areas with zigzag V-shaped strokes.

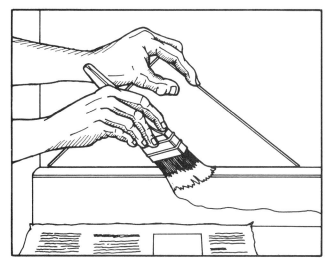

Figure 12-6. When painting baseboards, use a paint guide or masking tape on the walls and cover the floor with newspaper.

Step Four
Painting the Trim and Woodwork

Margin of Error: Exact

After you have finished all the large surfaces, you are ready to paint the trim and woodwork, if they have been installed. To do this successfully, you need to change your mental set about painting. Until now, you have been working on large surfaces, and detailing has not been important. Now you are changing from rough work to finish work. Attention to detail and care at this stage means the difference between a job that is professional-looking and one that is sloppy.

You will also be using different tools. During this stage you will use the smaller angled brushes and a metal paint guide. Use a 1½" angled sash brush on narrow molding, and a 2" trim brush on wider trim. As you apply the paint to the trim and woodwork, keep a supply of clean rags nearby to immediately wipe off any paint that gets on the previously painted surfaces. With oil-based paints, use a little mineral spirits or paint thinner. With water-based paints, use a mild detergent and water. Oil-based paints are most often used on trim because they give a surface that is both durable and easy to clean.

Always paint horizontal surfaces first and then vertical surfaces. Begin with the trim closest to the ceiling and work down. Do the baseboards last. Paint the top edge of the baseboard first, then the floor edge, and finally the center, using a larger brush. Be sure to cover the edge of the wall with a paint guide or masking tape (Figure 12-6).

Paint the inner sections of doors and windows before the outer portions. Windows especially require great care and patience. Apply the paint right down to the glass, so that the paint creates a seal between the wood and the glass. You can either tape the glass or remove the excess paint later with a razor-blade knife. If you are applying masking tape to the panes, leave a hairline gap of glass exposed between the tape and the wood to be sure you have a good paint seal between the wood and glass. Remove the tape as soon as the paint is completely dry.

When painting double-paned windows, you need to follow a certain order. You will need to be able to raise and lower the sashes to reach all areas. Begin by painting the exterior sash. Paint the horizontal side pieces, then the vertical, and then the muntins (the pieces that divide the window into small sections), as shown in Figure 12-7. Paint the lower part of the sash first, then raise the window and do the upper part. Repeat this process with the interior sash. Finally, paint the frame and trim, first the top sides and then the sill.

Raise and lower the sashes a few times while the paint is drying to be sure they don't dry stuck. Don't paint the jambs (the area where the window slides) unless absolutely necessary. To help the window move more smoothly after painting, rub a candle or a bar of soap over the jambs after the paint is completely dry.

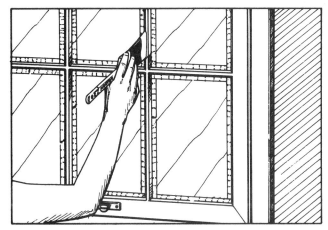

Figure 12-7. Use an angled trim brush to paint the muntins. Paint slightly onto the glass to ensure a good seal between wood and glass.

Doors are best painted when removed from their hinges and set on sawhorses. Flat doors are easily painted with rollers; panel doors require much greater care. First remove all hardware. With panel doors, paint the molding and the inside edges of the panel cavities first, and then the panels. Finally, paint all the horizontal and vertical pieces around the panels. If the door opens into the room, paint the door's latch edge, the jamb, and the door side of the door stop as well. When the door is dry, replace the hardware and rehang the door.

WALLPAPERING

WHAT YOU WILL NEED

Time. Wallpapering averages 10 to 20 minutes per sheet, longer if you must work around trim.

Tools for Wallpapering

Steel tape measure

Wallpaper level

Water tray

Seam roller

Wallpaper brush

Razor knife with lots of blades

Broadknife

Large sponge

Bucket

Pencil

Ladder

Materials for Wallpapering. An enormous number of different types of wall coverings are available. Consult your supplier or a book on wallpapering for complete information. Just be sure to purchase the prepasted type with adhesive on the back.

Wallpaper

Wallpaper paste (if needed)

Paint remover or mineral spirits

Estimating Wallpaper Materials. To calculate the square footage to be covered, multiply the perimeter of the room by the height of the walls and divide the result by 30. (The average roll covers 36 square feet, but there will be some waste for trim and pattern matching.) Subtract half a roll for each normal-size door or window opening. This is the number of rolls of wallpaper you will need. Purchase all rolls in one order to avoid variations in stock.

Most Common Mistakes

Preparation, patience, and an eye for detail are all you need to avoid these common mistakes.

☐ Not sanding, cleaning, and sizing the walls before applying the wallpaper.

☐ Failing to soak the prepasted wallpaper long enough.

☐ Failing to allow the wallcovering to cure the proper amount of time after soaking.

☐ Letting the adhesive dry on the woodwork.

☐ Not positioning the strips of wallpaper level and plumb.

☐ Not getting air pockets out when smoothing the covering on the wall.

☐ Not planning for pattern match-up and extra on top and bottom before cutting each strip.

☐ Underestimating the amount of wallpaper needed for the job.

☐ Not allowing the sizing to dry.

☐ Not overlapping the wallpapers that have a tendency to shrink.

☐ Using a seam roller on embossed wallpaper.

Step One
Planning Your Project

Margin of Error: Not applicable

Proper planning is essential to a professional-looking paper-hanging job. Planning involves understanding where the rolls will be applied and where they will meet at seamlines. Once you start to hang the paper, you must also make adjustments so that the patterns match.

First, you have to decide where to start. This decision will determine the location of the point of mismatch—where odd-shaped pieces will need to be cut. You want these mismatch points to be in the least visible locations. This consideration is more important when you are working with a design with a large pattern. With a neutral or nondirectional pattern, you can begin in an inconspicuous corner or area of the room.

Step Two
Prepping the Walls

Margin of Error: Not applicable

Follow the prepping in Step One for painting. The following additional steps are also recommended for a professional-looking job.

It is important for the walls to be very clean and free of any dust or debris.

When working with an untreated new wall, apply an oil-based primer before papering.

Check to see if your new wallpaper requires a sizing. (With cloth-backed vinyl hung over drywall that has never been sealed, you need to apply a vinyl-to-vinyl primer before hanging the paper.)

Now apply the sizing. This step is sometimes skipped, but it is worth the effort. Sizing is a liquid, applied like paint, that dries tacky. It ensures good adhesion and allows the paper to be easily removed in the future.

Step Three
Marking a Level Line

Margin of Error: Exact

Starting at an inconspicuous corner, measure to a point that is a distance from the corner of 1" less than the width of the wallpaper roll. Make a mark at this point. For example, if your wallpaper rolls are 20" wide, make a mark 19" from the corner.

At this mark, you will make an exactly plumb (vertical) line. There is a good chance that the corner is not plumb, so this process guarantees that you will be working from a plumb line. A common, and drastic, mistake in hanging wallpaper is to hang it out of plumb.

You can use a chalkline, a 4' level, or a wallpaper level (a straightedge with a level bubble) to mark this line on the wall. Be sure that the level bubble is reading true level, and then mark the line from the ceiling to the floor.

Step Four
Cutting the Wallpaper

Margin of Error: Exact

You are now ready to cut your first piece of wallpaper. You need to cut the paper so that there is a 2" overlap at the ceiling and floor. This excess will be trimmed away later. Also, you will want the pattern to break at the ceiling line. This pattern break line can be whatever you find most attractive. Hold each piece up against the wall before you cut it, and mark where it will meet the baseboard and ceiling line.

Use a straightedge and scissors or a utility knife to cut the paper. Change the blade often to avoid ripping the paper.

Step Five
Soaking the Wallpaper

Margin of Error: Exact

Unlike the older types of wallpaper that needed paste spread on the back, most wallpapers today are prepasted—the adhesive is already applied. You simply soak the wallpaper in water and hang it. However, be careful to follow the manufacturer's instructions closely. Not only is there a set period of time you need to let the paper soak in the tray, there is also a set period of time during which it must cure after it is removed from the tray and before it is applied. This time varies, but it is usually several minutes.

Most professionals apply paste even to prepasted wallpaper, for added insurance that it will adhere properly. With high-quality paper, applied on a properly prepared wall, additional paste is not needed. If you are in doubt, consult your supplier.

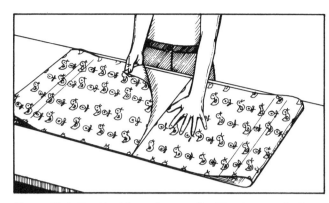

Figure 12-8. "Booking" the wallpaper after it has been soaked. Be careful not to crease the folds.

The paper is placed in a tray of lukewarm water, rolled up, with the pattern side in. Place the tray next to the wall directly below the area to be hung. Upon removing the wallpaper from the tray, fold it, pasted side to pasted side, so that it comes out flat, as shown in Figure 12-8. Be sure to fold the paper loosely and not to crease it at the folds. This is called "booking." Allow the paper to cure to its maximum width before hanging it. Be sure that no dust or debris settles on the paper while it is curing.

Step Six
Hanging the First Sheet

Margin of Error: Exact

You are now ready to hang your first piece of wallpaper. Apply it so that one edge is exactly vertical and aligned with your plumb mark. Leave the bottom fold folded and work at first only with the upper part of the sheet. Be sure the mark for the ceiling is aligned so that there is a pattern break at the ceiling line. If you are working at an inside corner, as is often the case, wrap the 1" overlap into the corner. (See the section on inside corners in Step Eight.)

Use a wallpaper brush to work out any bubbles, stroking the brush from the inside to the outside to push the air out. Start at the top and work your way down the paper. Keep working with the brush until all the bubbles are out and the paper is perfectly smooth on the wall. Be sure you stay aligned with your plumb mark as you work with the brush. Gently lift the bottom edge of the strip to free the sheet of any wrinkles.

When the upper part is smooth, release the bottom fold and position it, using the palms of your hands. Then use the smoothing brush as you did above (Figure 12-9). Be sure that no debris on the wall is poking through. You will trim the paper after the next sheet is hung.

Figure 12-9. Applying the first sheet. Note the 2" overlap at top and bottom.

Step Seven
Hanging the Second and Subsequent Sheets

Margin of Error: Exact

Before cutting the second sheet, be sure that you have allowed for the pattern matching at the seamlines. If your wallpaper has a large pattern, alternate between two different rolls to avoid waste.

The second sheet butts snugly against the first. Do not overlap seams. Apply this sheet as you did the first and maneuver it against the first with your hands. After this sheet is in place, you can go back and trim the first one with a straightedge or broadknife. To ensure a good trim job, change the razor blade for each strip of wallpaper.

After the sheets are in place, go over them with a large damp sponge to get out all the small bubbles and paste. Wipe up any excess paste at the seams and ends with the sponge before it dries. After 20 to 30 minutes, use a seam roller, as shown in Figure 12-10, to be sure that the seams are well secured. Press the roller lightly to avoid glossy areas. If you are working with raised or flocked wallpaper, omit this step.

Step Eight
Corners and Openings

Margin of Error: Exact

Corners, both inside and outside, are the most demanding part of hanging paper. In and of themselves, they are not too difficult. However, the corners are seldom true plumb (ver-

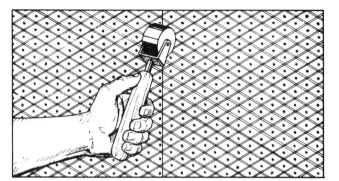

Figure 12-10. Use a seam roller at the seams 20 to 30 minutes after hanging the paper. Do not press too hard or you will gloss the paper.

tical), and this is where the difficulty lies. You need to hang your wallpaper plumb, even if the corners are not. Go slowly here; this is where your skills will be challenged the most.

Inside Corners

At both the top and the bottom of the wall, measure the distances from the strip next to the corner to the corner itself. Add 1" to the greater of the two measurements and cut a sheet lengthwise to this measurement. (If this measurement is within 6" of your full sheet measurement, use a full sheet.) If there is a sheet next to the corner, its edge can be used as a plumb line. (That's why it's so important to install that sheet plumb.)

If there is no sheet at the corner, simply measure out from the corner the dimensions of a wallpaper roll less 1" (for corner overlap) and draw a plumb line. This line will be your guideline in hanging your first sheet.

Now simply hang the sheet as described earlier and wrap the excess into the corner. Since few corners are perfectly plumb, you will need to strike a plumbline on the adjacent wall, again at a distance from the corner of 1" less than the dimension of the roll. Then apply another sheet and wrap the excess so it overlaps the first sheet. Finally, use your broadknife and razor knife to cut the overlapping sheets in the corner and peel away the two excess pieces.

Outside Corners

If the outside corner is exactly plumb (although they seldom are), you can simply wrap the paper around the corner and begin from its edge on the other side of the corner. If the corner is not plumb, a little more attention is needed.

As with inside corners, measure at the top and bottom of the wall the distance from the last sheet to the corner. Add 1" to the longest measurement and cut a sheet that size. If the measurement is within 6" of a full sheet, use a full sheet.

Hang this sheet, cutting a diagonal slit at the corner at the top so it will bend around the corner. Hang it so that it is smooth and fold it smoothly around the corner. Now strike another plumbline on the intersecting wall. The plumbline should be the width of the roll from the corner. Now simply

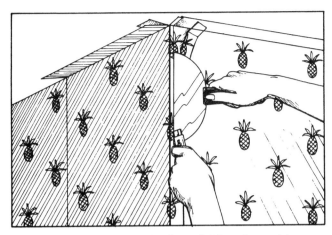

Figure 12-11. Mark a plumb line ¼" to ½" from the outside corner. Cut the paper at this line with a broadknife or razor knife. Remove the excess paper.

hang your intersecting piece to that plumbline. After both pieces are in place, make a new plumbline ¼" to ½" from the corner on the side of the corner. At this line, cut through both pieces of paper and peel away the excess, as shown in Figure 12-11.

Windows and Doors

Openings for windows and doors, fireplaces, and built-in bookshelves and cabinets offer a challenge to the novice. In these areas, don't try to cut the paper first and then apply it. Cut it after it is in place. If you are having trouble with a complex area, take the time to step back and think about it before proceeding.

To work around window openings, simply hang your paper so that it is aligned with the adjacent piece and press it loosely against the window trim. Now cut along the sides of the opening, leaving a 2" excess, which you will trim away later. Using a sharp razor, cut a 45-degree slit at the cor-

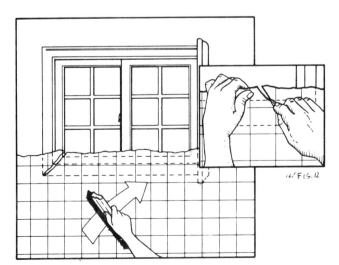

Figure 12-12. Use a razor to cut a 45-degree slit at both the top and bottom corners of the molding. Cut away excess paper.

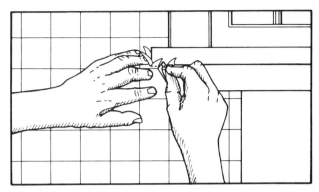

Figure 12-13. Make small cuts to fit the paper around sills.

ners, both top and bottom, to the outside edge of the molding, ending exactly at the molding's edge (Figure 12-12).

Press the paper against the molding with the wallpaper brush and use your broadknife and razor knife to cut away the excess. Leave a hairline gap between the molding and the paper.

As shown in Figure 12-13, you will need to make a series of small diagonal cuts around the sill area to fit all the little corners. Go slow and make small cuts to avoid over-cutting. Press the paper tightly against the molding and trim it where needed.

Finally, you will need to paper the area above and below the window with short strips. Be sure the vertical patterns below the window are aligned with those above, and that the patterns match where they meet the paper hung to the sides

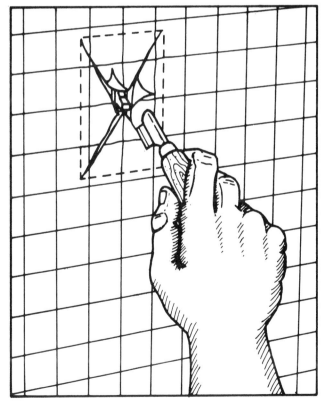

Figure 12-14. Cover over light switches and electrical outlets. Make diagonal cuts in the paper and trim the excess.

of the windows. Wipe any excess glue off the woodwork before it dries.

Light Switches and Outlets

After turning off the electricity and removing the plates from the light switches and outlets (if they have been installed), simply cover over the holes as you hang the wallpaper. Then use your razor knife to make two diagonal cuts across the outlet or switch, about 3" long, to expose the switch or outlet (Figure 12-14). Cut away the excess. You can cover the plates by tucking the wallpaper into the wall and replacing the cover.

13 Kitchen Cabinets and Counter Tops

COMPLETING A KITCHEN is one of the more difficult tasks the owner-builder faces. Aside from installing the cabinets, it involves the electrical, plumbing, gas, and venting systems, as well as many specialized tasks, such as installing flooring and trim. This chapter explains how to install kitchen cabinets and both laminated and preformed counter tops. An overview outlines the basics of installing sinks and appliances; however, these tasks are different in each home, and you will need to gather more information before installing your appliances and plumbing.

You will need to use many of the techniques you learned in other chapters for such tasks as installing finish flooring and drywall and completing the plumbing and electrical systems. Refer to these chapters as necessary.

Before You Begin

SAFETY

☐ Work patiently. If you become frustrated or try to hurry a job like installing cabinets, the chances are great that you will make a mistake.

☐ Unplug tools when making adjustments or changing blades.

☐ Keep work surfaces and traffic areas free of scraps and debris.

☐ If an object such as a cabinet or appliance is too heavy or awkward to lift easily, get help in moving it. Bend from the knees when picking up large or heavy items.

☐ Turn off all utilities before beginning work on them. Remember that pilot lights must be relit.

☐ Use the proper protection, take precautions, and plan ahead. Never bypass safety to save money or to rush a project.

WHAT YOU WILL NEED

Time. The time you will need to install your kitchen depends entirely on the scope of the task; it is not possible to make a meaningful general estimate.

Tools
Hammer
Level
Tape measure
Standard and phillips head screwdrivers
Cordless electric screwdriver
Electric drill
Plane
Crowbar
Dolly
Wrenches
Plumber's wrenches
1" to 1½" pipe wrench
Pry bar
Putty knife
Saw
Nailset
Circular saw
Chalkline
Plumb bob
C-clamps or handscrews
Compass
Router with carbide-tipped laminate trimming bit
Saber saw with fine-tooth blade
Vacuum cleaner
Sanding block and sandpaper
Fine-cut file
Carpenter's square
Combination square
Ladder
Mud tray
Mud knives
Vacuum cleaner

Materials. Aside from such major items as the cabinets and counter top materials, you will need fixtures, flooring materials, paint, and materials to complete the plumbing and electrical hookups.

Materials for Preparation
Newspaper

Wire mesh

Drywall

Spackling/patching plaster

Cement filler

Plywood

Drywall tape

Drywall compound

Primer/sealer

Paint

Rollers and brushes

Materials for Installation

New cabinetry (custom or ready-made)

Counter top material: laminate, end splash, and 1x2 ledger board

Scrap wood (2x4) for jacks

Paper for templates

Shims

Silicone sealant

Wood screws/self-drilling screws

Natural bristle paintbrushes

Brown wrapping paper

Rolling pin

Sanding block and 80 grit sandpaper

Laminate adhesive/contact cement

Masking tape

Plumber's putty

PERMITS AND CODES

It's important to check your local building code to find out which phases of your kitchen installation need to be inspected.

It is assumed that you submitted detailed drawings to indicate all structural, wiring, and plumbing work intended, and that these areas have been inspected during your rough inspections.

DESIGN

In many homes, kitchens are multiuse centers, designed as much for entertainment as for cooking and eating. This book is geared toward construction rather than design. However, many fine books on kitchen design are available to help you plan your new kitchen and to meet your family's needs. Magazines, home centers, and home shows are other sources of inspiration.

One of the major considerations in designing your kitchen is the "work triangle"—the arrangement of the stove, refrigerator, and sink (Figure 13-1). For maximum efficiency, these major workstations should be placed 4' to 7' from each other, for a total of no more than 21'. If more than one person

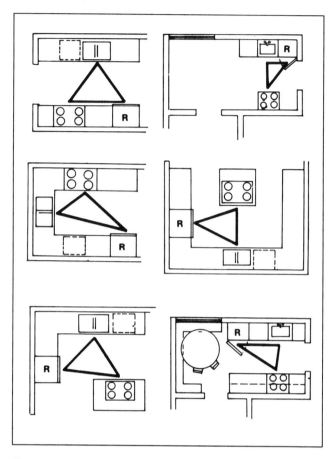

Figure 13-1. Sample kitchen layouts and work triangles.

works in your kitchen at a time, plan your new work area so that you won't get in each other's way.

One of the most important elements of an efficient kitchen, one that is a pleasure to work in, is plenty of counter space for preparing food, cooking, and cleaning up. Your plan should also make efficient use of storage space, with plenty of cabinets, drawers, and shelves.

Step One
Preparation

Margin of Error: ¼"

Most Common Mistakes

☐ Not marking location of studs.

☐ Neglecting to have all electrical and plumbing systems in place and inspected.

Walls

Before you install your new cabinets and appliances, the walls must be painted. Counter tops will be installed after the cabinetry is in place.

You will also need to locate and mark the wall studs on the floor and ceiling as a reference for attaching the new cabinetry. There are several ways to find the studs in a finished wall. Studs are usually located every 16" on center. Tap along the wall with a hammer until you hear a solid sound; look for seams in the drywall; or use a magnetic or density-type stud finder to locate the nails in the studs. Mark the position of the stud on floor and ceiling and snap a chalkline to mark the center of the stud on the wall; or make your marks on small pieces of tape at floor and ceiling if you don't want to put chalk on the wall.

Floors

Ceramic tile floors are usually laid before the cabinets are installed, while vinyl flooring can be laid either before or after the cabinets are in place. Prepare the subfloor as described in Chapter 9, Vinyl Floors. Fill any holes or gouges, and nail down protruding boards. Remove any glue or paint that may have spilled. Countersink nail and screw heads that are sticking up above the surface.

Sequence of Installation

Here is the sequence in which the kitchen is put together. The first step is preparation: primer and paint on the walls, and ceramic tile or hardwood on the floors. (Ceramic tile and hardwood flooring are always laid before cabinets are installed, while vinyl flooring can be laid either before or after the cabinets are in place.) Second, install the cabinets: wall cabinets, then base cabinets and islands. After the cabinets are in place, install the doors, drawers, and hardware. Third, install the plywood base (for tile counter tops), the recessed sink, and the counter top itself. (If you are using a surface-mounted sink, it goes over the counter top.) Faucets and other fittings come next. Fourth, install the appliance—disposal, dishwasher, over the range hood/vent, wall oven, range, etc. Finally, install lighting fixtures and trim.

Step Two
Installing Cabinets

Margin of Error: 1/16" gap between cabinets; level and plumb within 1/8"

Most Common Mistakes
- ☐ Not installing cabinets level and plumb.
- ☐ Not attaching cabinets to studs.
- ☐ Damaging or marring cabinets.
- ☐ Not aligning cabinet doors.
- ☐ Damaging the walls during installation.
- ☐ Not cutting sink opening to proper dimension.
- ☐ Not properly fitting and installing drop-in appliances.
- ☐ Not making exact measurements and cuts.
- ☐ Not installing all needed utilities.

Accurate measurements are crucial to ensure a snug fit when installing stock cabinets. Custom-made cabinetry is usually

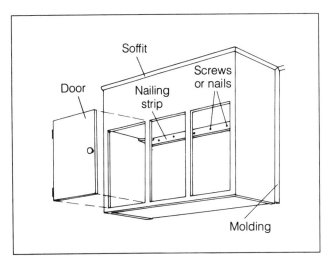

Figure 13-2. The anatomy of a wall cabinet.

sold with a warranty conditional on having the cabinets installed by the dealer. In fact, the dealer will probably send someone to your house to take the measurements.

Both standard and custom cabinets are usually made to uniform sizes. This is a help in designing kitchens, and ensures that sinks, dishwashers, and counter tops in standard sizes will fit.

Inside the cabinets, and sometimes on the top and bottom, are pieces called **nailing strips**. This is where you secure the cabinets to the studs. Rather than attaching the cabinets with nails, however, 3" buglehead screws, installed with a drill or screwgun, are much stronger and easier to install. If you can't get to the nailing strip of the lower cabinets, toenail the sides of the cabinets to the floor, through the shims.

The most important thing to remember about installing cabinets is to begin from the highest level of the floor to set your cabinet heights. It's easy to raise a cabinet with shims, but trimming off all the bottoms is a nightmare. Check the floor level carefully to find the highest point over which a cabinet will go, and determine your level lines from there.

Most ready-made cabinets come with a scribe allowance on their sides to allow you to scribe them to irregular walls. To scribe a line, set the cabinet in position against the wall and plumb the front. Stick a strip of masking tape along the side of the cabinet to be scribed. Set the points of a compass to the width of the widest gap between the side of the cabinet and the wall. Run the compass along the wall and the irregularities will be marked on the tape. Now you can plane or sand down to the line so the cabinet will rest flush against the wall. If the scribe edge is not included on your cabinets, or if your cabinets go all the way to the ceiling, plan on trimming the edges with molding.

Install the upper wall cabinets first so that the lower ones aren't in your way. (Figure 13-2 shows the anatomy of a wall cabinet.) These cabinets have no support except for their attachment to the wall, so they must be securely attached.

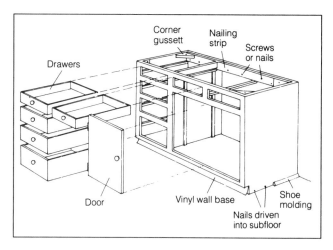

Figure 13-3. The anatomy of a base cabinet.

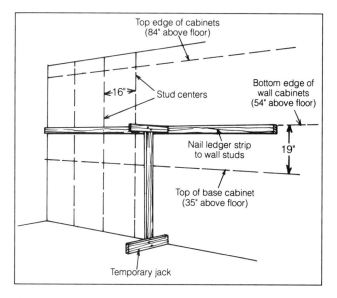

Figure 13-4. Marking reference lines for placement of wall cabinets.

Before installing any of your new cabinets, remove all the drawers, doors, and hardware and label them carefully for easy reassembly. This will make the units much lighter and more manageable.

Installing Wall Cabinets

The first step in installing wall cabinets is to measure the height of the base—usually about 35". (Figure 13-3 shows the anatomy of a base cabinet.) Measure this distance up from the floor and draw a horizontal line across the walls. (If your floor is not level, measure from the highest point.) Now add the thickness of your finished counter top above the first mark and mark this level as well. Work carefully; this line must be true horizontal, because it indicates where the surface of the counter top will be.

Above this second line, measure up to the point where the bottom of the upper cabinet will be (usually 18" to 19"). Use a level to draw another horizontal line across the wall. This

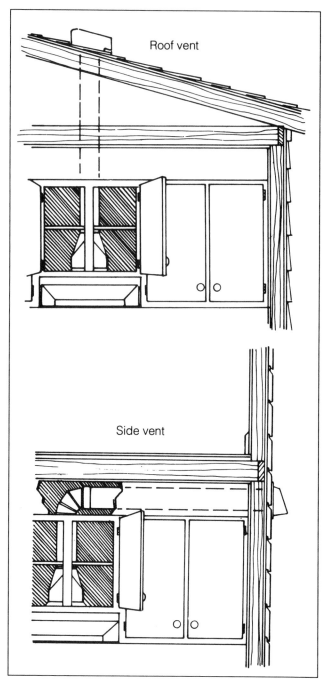

Figure 13-5. Ranges must be vented to the outside. These openings need to be cut before upper cabinets are installed.

line (approximately 54" above the floor) must also be true horizontal and parallel to the line for the counter top.

Screw a temporary 1x2 ledger board to the wall so that the top of the board is even with the line for the bottom of the wall cabinets (Figure 13-4). Be sure you are screwing into the studs. Now mark the cabinet widths along the length of the ledger strip. You want the joints of the upper cabinets to be directly over the joints of the lower cabinets. Mark the position of the base cabinets on the wall and use a level to make sure the upper cabinets are plumb at the joints. Don't forget to add or subtract any scribe or trim differences.

Figure 13-6. Once the upper cabinet is level and plumb, attach it with screws through the nailing strip and into the wall studs.

Figure 13-7. Check base cabinets for level and shim as necessary.

You will need to make some temporary jacks to support the wall cabinets while you attach them to the wall. The jacks should reach exactly from the floor to the bottom of the upper cabinets. Wide blocks of wood nailed to both ends of a 2x4 work well for this purpose. Put the jacks in position near the ledger (see Figure 13-4). Shim to adjust for uneven floors.

Before you begin installing your upper cabinets, you need to cut openings for any hoods, ducts, or vents (Figure 13-5). With the cabinet upside down on the floor, center the hood upside down on the bottom of the cabinet. Trace the outline of the vent hole you plan to cut out along the line for the opening. Also cut out the hole where the duct will leave the cabinet and enter the wall or ceiling. The manufacturer's instructions will guide you in this process.

You are now ready to hang the cabinets. Mark the location of the wall studs on the hanging cleats of each cabinet. Then drill pilot holes for screws at these points; or use self-drilling bugle-head screws. These require no predrilling except when they are used at face frames. Start in a corner if possible, or at one end. Work out or across from this cabinet to install the rest.

Lift the first cabinet into position onto the ledger board and the temporary jacks. Check to see that the cabinet is both level and plumb. If necessary, shim the back of the cabinet to bring it into plumb. Screw 3" No. 8 flat-head wood screws or self-drilling bugle-head screws through the nailing strip at the back of the cabinet and into the wall studs (Figure 13-6). Use two screws at the top and two at the bottom, if possible. Use longer screws if necessary to ensure that they penetrate at least 1¼" into the studs. Each unit should be attached to

at least two studs. If only one stud is located behind a unit, use a toggle bolt as an additional fastener.

Attach all of the upper cabinets in this manner. Then go back and screw the cabinet fronts together. When connecting adjoining cabinets, it may be necessary to loosen the wall screws to allow the faces of the cabinet to be attached flush to each other. Use handscrews or a C-clamp with soft wood scraps between the jaws of the clamp and the face frames of the cabinets to hold the units flush while you screw them together. Many prefabricated cabinets have predrilled holes in the face, top, and bottom.

When all cabinets are attached, check them once again for level and plumb. Then remove the jacks and the ledger board.

Installing Base Cabinets

Position your lower cabinets in place and level them with shims. If your lower cabinets have separate bases, set the bases in position, shimming as necessary to keep them level. Where the cabinet bases meet in a corner, square the intersection with a framing square. Then anchor the bases to the floor and place the cabinet units on top of them. Cabinets that do not have freestanding bases must be leveled and anchored in position individually (Figure 13-7). Again, start in a corner, if possible, or on an end. You will work off this first cabinet, so if it is not level and plumb at the right height, the others won't be either. Be sure to measure and leave room for your appliances, adding ¼" for clearance—more if you will be adding end panels (plywood pieces that will cover the exposed ends of the cabinets). Check the position of your base units against the horizontal line that you drew on the wall, making sure the tops are uniformly on the line you made for the cabinet height.

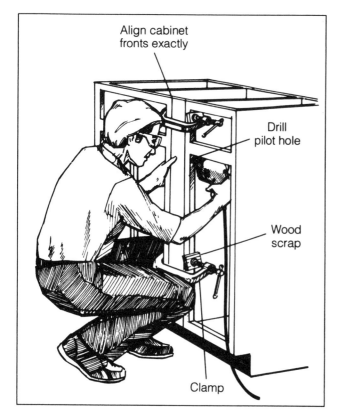

Figure 13-8. Align cabinet fronts exactly and clamp them in place before screwing them to the wall.

Align cabinet fronts exactly

Drill pilot hole

Wood scrap

Clamp

As you join the cabinet units together (Figure 13-8), slide a level down the entire length of the cabinets, adding shims to bring them up to level. Place your level front to back on the top of the cabinet to check for level, again shimming if necessary. As with the upper cabinets, use clamps to hold the units in place while you screw them together. Then screw the cabinet into the wall with 3" screws through the top nailing strip and into the wall studs. If the wall is uneven, there will be gaps between the back of the cabinet and the wall. To avoid pulling your cabinets out of plumb as you secure them to the wall, shim the gaps at the fastening point (where the stud is located) before screwing the cabinet into place. Carefully saw or chisel off the ends of the shims.

Repair any holes you have made in the walls that will show. If your cabinets are not prefinished, now is the time to finish or paint them.

Installing Island Cabinets

Island cabinets must have all of the individual units screwed together into one big piece and placed into position before leveling and plumbing on all four sides. If the base is separate, level and plumb it. Then fasten it to the floor with countersunk screws, toenails, or sheet metal angles. Anchor the cabinets to the base. Units with built-in bases are also screwed together before leveling. Long screws are available to secure the level island to the floor, through the base. When installing an island unit, it is important to square it to the cab-

inets along the wall and to the wall itself, as well as lining the unit up properly with the overhead fixtures (stove vent or lighting). Measure on both ceiling and floor to determine the exact location.

Step Three
Installing Laminate
Counter Tops

Margin of Error: Level within ⅛"

Most Common Mistakes

☐ Not scribing counter tops to the contours of the wall.

☐ Not applying the finish surface accurately.

☐ Not cutting the sink opening to the proper dimensions.

☐ Scratching the counter top while installing it.

☐ Puncturing the counter top surface with screws while fastening it from below.

☐ Not checking the corners and ends to be sure they are square before cutting the counter top.

☐ Not providing adequate ventilation when using contact cement to adhere laminates.

☐ Not spreading enough adhesive when laminating the counter top, causing the laminate to lift up in the corners or along the edges.

The two common types of counter tops are preformed laminate and laminate custom-built at the site. Preformed tops come assembled from the manufacturer, while custom tops are built at the job site using raw materials (laminates and plywood or particle board). Other common counter top materials include tile and butcher block.

Preformed Counter Tops

Preformed counter tops are available only in standard sizes, so you usually have to purchase one that is a little longer than you need and cut it to length. These counter tops are available with mitered corners, and have a built-in backsplash with a ½" scribe lip to accommodate irregularities in your walls. A flush end trim piece with a heat-sensitive adhesive backing is literally ironed in place. The backsplash end is glued and nailed into position.

1. To measure for your counter, add the counter overhang (usually between ¾" and 1" in front and on open ends) and add it to the length of your base cabinets.

2. If an end splash is to be included (Figure 13-9), subtract ¾" from the length of the counter top on that side. Plan your cut carefully for an end splash. Mark the cut on the bottom and cut out from the bottom with a circular saw fitted with a fine-tooth plywood blade to protect against chipping. Cut the excess off with a handsaw. Smooth the edges of the cut with a file or sandpaper.

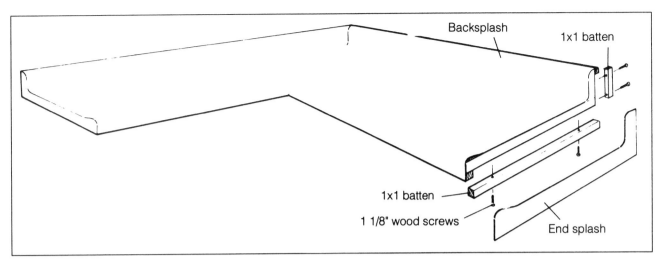

Figure 13-9. Components of a preformed counter top.

3. The end splash is screwed directly onto the edge of the counter top or into wood battens previously attached to the edge. Apply silicone sealant to all surfaces to be joined and hold the end splash in place with C-clamps while driving in the screws. The end splash, a preshaped strip of matching laminate, is glued to the end of the counter top. The end splash usually requires filing or trimming with a router, so install it before you install the top itself.

4. U- and L-shaped counter tops need to be ordered mitered to order; it is difficult to accurately miter these sections at home. These premitered sections have small slots for draw bolts cut into the bottom edges. Coat the edges with silicone sealant before aligning the edges and tightening the bolts.

5. Like cabinets, counter tops rarely fit perfectly against the back or side walls. They come with a scribing strip (Figure 13-10) that can be trimmed to the exact contours of your irregular walls, as described earlier. After scribing, you can plane or sand down to the line so the counter top will rest flush against the wall. The scribing provides a counter top lip that is parallel to the cabinet face, as well as accommodating any unevenness in the wall. This will ensure that your prefabricated miter joints will fit if you installed the base cabinets square. Make all scribe adjustments before trimming the square ends. Once your contours are correct, position the counter top on the cabinet base. Check that all is level, and shim where needed. Check also that drawers and doors open freely.

6. Fasten down the counter top by running screws up from below through the top frame and corners. If there are no corner brackets, install them on the base units. This will allow you to install the counter top easily. Self-drilling bugle head screws work well for this job. Round-head screws work best here. They will not be seen, and so do not have to be counter sunk, and they bear more weight than do other kinds of screws. Use screws that are long enough to penetrate ½" into the backing material. Be careful; one screw that is too

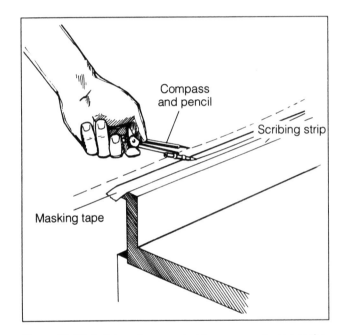

Figure 13-10. Scribing allows you to trim your counter top to the exact contours of your wall.

long can ruin the laminate by penetrating the surface or by making a small lump in the surface.

Custom Laminate Counter Tops

You can also create your own laminate counter top. This option is somewhat more difficult, but it offers greater variety in design and color. Most dealers stock 4'x8' sheets that are 1/16" thick, but they can be special ordered in widths from 2' to 5' and in lengths from 6' to 12'. Store the laminate in your kitchen for at least 48 hours before installing it to allow it to adjust to the temperature and humidity.

The laminate is applied to a base of ¾" particle board, which extends out to form a lip in front of the cabinet. The lip is often made thicker by screwing and gluing on a thin additional strip of particle board under the front edge before laminating.

Roller-nose or flush-cutting router bits are used to trim the laminate edges and joints.

Make certain that the surface upon which you will be applying the laminate is even and smooth.

Applying Laminate to Counter Edges

1. When planning your cuts, try to reserve the factory edges for the places you won't be able to reach with a router, such as where counter top meets backsplash. Cut the laminate with a saber saw fitted with a fine-tooth blade. Leave a margin of ½" on all sides except for the factory edge.

2. Use a soft paintbrush or a vacuum cleaner to completely remove any dust from the counter and the back of the laminate strip being applied. Spread contact cement onto the counter edge with a ¾" natural bristle brush, covering the entire surface. Then brush the contact cement onto the back of the strip of laminate and allow both pieces to set for about 15 minutes. The contact cement is ready for bonding when a piece of brown paper will not stick to it. When working with contact cement, make sure your work area is very well ventilated.

3. Drape strips of brown wrapping paper over the edge of the counter so that the two cement-covered surfaces won't come in contact with each other. Position the strip of laminate close to the counter edge so that the ½" margin extends evenly above and below the edge of the counter. This may be a two-person job. If one end of the strip is to meet an inside corner, start by butting that end into the corner, pulling out the first piece of brown paper and pressing the strip onto the counter's edge. Work your way along the edge of the counter, removing one strip of paper at a time and pressing the laminate into position. Be sure the laminate is exactly where you want it. Once the entire strip is in position, roll over it several times with a rolling pin or a hand roller, using firm, even pressure to ensure a good bond.

4. Trim the laminate with a router fitted with a carbide-tipped flush-cutting bit. Hold the router in position with the lower part of the faceplate flat against the newly laminated strip and the bit held just above the excess. Slowly lower the router until the bit meets the counter top, then move the router along the strip, trimming flush with the counter top. Trim the excess laminate on all sides of the edge, moving in a counterclockwise motion.

5. After laminating all of the counter edges and trimming them with the router, smooth the top edges with a sanding block fitted with 100 grit sandpaper or with a file. Then dust thoroughly. Don't touch the sandpaper to the laminate surface; it will leave permanent scratches. Smears or globs of cement should be allowed to dry and then rubbed off the surface with a rag. Solvent will clean up any spills.

Applying Laminate to Counter Tops

1. Mark all corner miters on the particle board. Scribe a factory edge to fit your wall and transfer these miter marks to the laminate, leaving ½" or so overhanging in front and on the end.

2. Cut the laminate with a straightedge and a circular saw fitted with a fine-tooth plywood blade, a saber saw, a router, or a laminate cutter. (A laminate cutter is a blade made to fit your utility knife.)

3. All scribing, fitting, and adjustments need to be made before you begin gluing. You should dry fit all the pieces together before installing them. If you have a complicated shape, a cardboard template can be helpful. Always begin in a corner and work out from there as you apply the laminate.

4. The easiest way to spread the adhesive for the counter top is with a paint tray and a mohair-covered paint roller. Spread newspapers on the floor. Lay the plastic laminate upside down on top of the newspapers and roll the contact cement on, covering the entire surface. Apply a slightly thicker coat of cement near the edges. Next cover the counter surface with adhesive and allow both pieces to dry for about 15 minutes, or until brown paper will not stick to the adhesive.

5. Dowels or wood strips work better than brown paper to keep the laminate from sticking to the cemented counter surface (Figure 13-11). Place the dowels at intervals of 1'. Then lay the laminate, adhesive side down, on the dowels. Put the factory edge against the backsplash and set the tip of the diagonal cut into the corner. This corner is where you will begin.

6. Pull out the dowel nearest the corner, while pressing the laminate into position. Be sure the laminate is exactly where you want it. Use a sweeping motion so that no air bubbles are trapped beneath the laminate. Work along the counter, pulling out strips of wood and pressing the laminate down. Then immediately roll the surface with your rolling pin or hand roller, applying extra pressure near the edges. When placing the second piece of laminate, make certain the diagonal seam at the corner is very tight and roll the seam thoroughly.

7. If the laminate fails to bond, or if a bubble forms at some point, place a piece of the brown wrapping paper over the spot and place a hot iron (set for cotton) on top of the paper until the laminate feels hot to the touch. The heat will soften the contact cement enough to regain some of its stickiness. Then use the roller again with a firm, steady pressure until the laminate has cooled.

8. Before router-trimming the counter top, put masking tape around the newly laminated edges to avoid marring them. Router off the excess as you did with the edges, moving the router steadily from left to right. Then replace the bit with a 22-degree bevel bit and bevel the seam at the top of the counter, again moving from left to right (Figure 13-12). It's

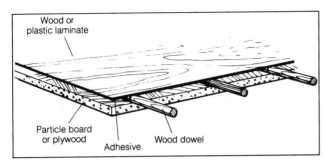

Figure 13-11. Dowels or wood strips keep the laminate from sticking to the particle board base while you position it.

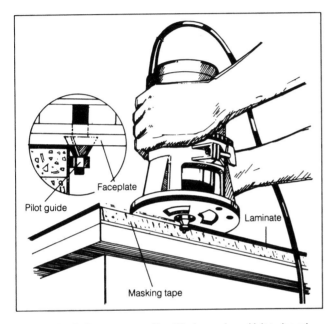

Figure 13-12. Use a router with a 22-degree bevel bit to bevel the seam at the top of the counter.

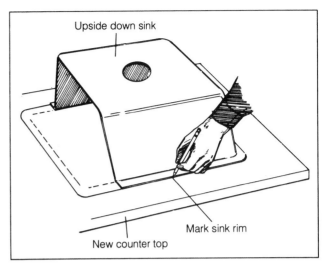

Figure 13-13. Outline the sink and then draw another line ½" inside of the first line to show you where to cut.

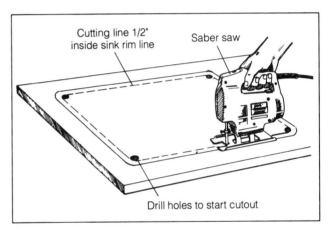

Figure 13-14. Cut along the inside pencil line with a saber saw.

best to practice this technique on scrap before doing the actual project.

9. Finish the bevels with a ¼" single-cut mill file, applying pressure on the downward stroke. Check each angle with your finger to be sure there are no rough edges that could result in cracks later. Inside corners are particularly likely to crack if they are not filed smooth.

Step Four
Installing the Sink and Dishwasher

Most Common Mistakes

☐ Not cutting the sink opening to the proper dimensions.

☐ Not lining up the opening with the sink base cabinet.

No matter what material you have chosen for your counter, some general rules apply to the installation of sinks. Surface-mounted sinks and self-rimming stainless steel sinks are generally installed with caulking and clamps, while self-rimming porcelain sinks need only caulk. Recessed sinks, where the unit is set on the plywood base, are most commonly used with tiled counter tops; porcelain and cast iron sinks are recommended for recessed installation. Stainless steel sinks are not recommended for use with tile because the steel and the tile expand and contract at different rates. However, a double-walled stainless steel sink may work with tile.

1. Position the sink upside down on the newly laminated counter top or the plywood base. Try to center it, but leave at least 1½" and not more than 3" from the front edge of the counter. Draw a pencil line around the edge of the sink, as shown in Figure 13-13; then remove the sink. (If you are using a metal rim for a sink that is not self-rimming, use the rim's outside edge as a template rather than the sink itself.) Now measure inward and draw another line ½" inside of the first line. Be sure of your marks.

2. Drill a hole large enough for a saber saw blade to the inside of each corner of the inside line. Cut along the inside pencil line with your saber saw while another person supports the cut from below (Figure 13-14). Lift the waste piece out.

3. Run a bead of caulk along the entire bottom edge of the upside down self-rimming sink. Position the sink carefully over the opening and press it down firmly until the excess caulk oozes out along the edges. Follow the manufacturer's instructions for any additional hardware that comes with the sink. Typically, a clip or a small metal stud fits in a small channel on the bottom, and a clip secures the rim to the particle board base.

Installing the Hardware and Plumbing the Sink

After the sink is installed, all that remains is to install the hardware and hook up the water supply and drain pipes. (See Chapter 3, Plumbing.) This chapter gives a general outline rather than full instructions for plumbing a sink. Refer to the manufacturer's instructions, or to a book on plumbing.

1. Faucets are mounted through the prepunched holes in the back of the sink, using an adjustable wrench and plumber's putty. Faucets must be installed so that the valves work the right way. The left-hand valve is always connected to the hot water supply and the right-hand side to the cold. Follow the manufacturer's instructions.

2. Supply line connector kits are available that contain everything you need to connect the water supply line—flexible copper pipe, hardware, shutoff valves, and instructions. You will need to specify the diameter and material of the supply line and whether or not you will be adding a dishwasher. The dishwasher requires a special T-shaped shutoff valve.

3. After connecting the water supply line, install the basket strainer. Pack the rim of the hole in the sink with plumber's putty and push the strainer down into place. From beneath the sink, slip on a rubber gasket and metal ring, as shown in Figure 13-15. Then slip on the lock nut and tighten it until the putty oozes out. Clean off all excess putty.

4. Next, install the P-trap. Connect the tailpiece to the basket strainer, using a slip nut over a washer. Then connect the P-trap to the tailpiece with another washer and slip nut. Attach the P-trap to the wall stub-out with a curved drain expansion pipe (see Figure 13-15).

Installing the Dishwasher

The three main connections of a dishwasher are the electrical cord (which will plug into an outlet under or near the dishwasher), the rubber drain hose, and the hot water supply line. An air gap is often needed as well to prevent waste material from backing up into the dishwasher. The waste line for a dishwasher leads up to the vent and drains into a line leading to the disposal.

1. Drill a hole, large enough to accommodate all three connections, through the cabinet wall between the sink and the dishwasher compartment.

2. Thread the black rubber drain hose, the electrical cord, and the water supply tubing through this opening. Leave enough supply tubing in the dishwasher compartment to reach the connections at the front of the dishwasher.

3. You will need to remove a knockout from the disposal collar to attach the drain hose. Fit the hose to the collar with the adjustable hose clamps provided.

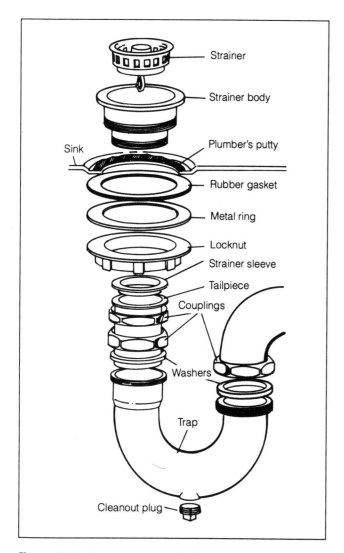

Figure 13-15. Components of a sink drain.

4. If you are adding an air gap, attach the dishwasher hose to the longest of the two hoses in the air gap, using hose clamps. Hook the shorter hose in the air gap to the collar of the disposal unit, again removing a knockout and attaching with hose clamps.

5. Attach the hot water supply line. The tubing must fit snugly to the fitting on the front of the dishwasher and to the shutoff valve. Secure it with the hose clamps provided.

Step Five
Installing Other Appliances

Hood/Vent

Install the hood/vent before the range or cooktop so that you can reach all connections easily (Figure 13-16).

1. Connect a section of 6" or 7" metal ducting to the duct pipe entering the cabinet and to the hood. If these holes are too close for a turn, you may need to have a sheet metal box with nailing flanges made to your specifications. Connect the hood collar or transition fitting to the duct by tightening the metal collar provided, or by wrapping duct tape around both collar and duct.

2. Lift the hood into position and trace the holes for attaching it to the underside of the cabinet, if you did not do this earlier. Drill pilot holes and attach the unit with the screws provided.

3. Connect the wiring in accordance with the manufacturer's instructions.

Ranges and Ovens

Drop-in ranges and ovens are connected in much the same way as their freestanding counterparts. However, they do not simply slide into place, but must be installed on top of a low base cabinet unit for support (Figure 13-17). These can be hung from flanges from the counter level or screwed to the base cabinet below and/or to adjacent units.

1. The electric range/oven has its own pigtail cord that plugs into a separate 220-volt circuit. Attach the unit according to the manufacturer's instructions.

2. Hook up the gas supply line with flexible tubing, adding dope putty and nuts at either end. A range with a pilotless ignition system must also be plugged into a 110-volt circuit. You must call the utility company to check your range hookup before you position it permanently.

Built-In Ovens

Gas and electrical connections for built-ins are the same as for drop-in types. These are most commonly located in an adjacent cabinet. Duct attachments are the same as for hood/vents. The oven will be attached to the surrounding cabinetry.

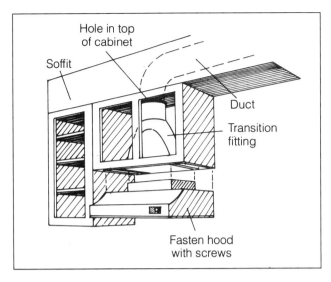

Figure 13-16. The hood/vent is installed before the range or cooktop.

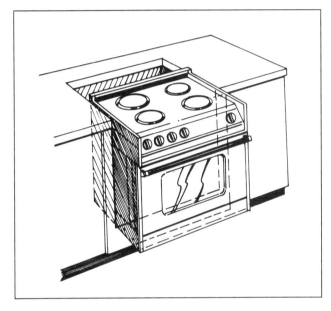

Figure 13-17. A drop-in range is installed on top of a low base cabinet unit.

1. Lift the unit into place and tighten the screws into the sides, top, or bottom of the cabinets. The model you have purchased will dictate whether the face flanges will overlap the cutout edges of the cabinet or whether you will need to install special trim pieces.

2. A microwave oven simply plugs into any grounded 110-volt circuit; it may not even need to be trimmed or attached.

14 Weatherizing

WEATHERIZING A HOME is a two-part process. The first part, which must be completed before the walls are enclosed, includes insulating, ventilating, and installing vapor barriers, as described in Chapter 4, Insulation and Ventilation. The second part, which is done after the house is completed, includes caulking, installing weatherstripping, and other forms of energy conservation. This second step is the subject of this chapter.

Some manufacturers these days install weatherstripping on their doors and windows. If you feel that the manufacturer's weatherstripping is inadequate, you can add more of your own. With today's high energy costs, every leak that can be sealed will save you money.

However, it may be possible to seal your home too tightly. Recent studies have called into question the quality of the air inside tightly sealed homes. You may want to allow in a certain amount of fresh air just to make sure that your indoor air-pollution level is not too high. And you should check the radon level in your basement or crawlspace. Inexpensive radon testing kits are available at many home supply centers and hardware stores in areas where this is a concern.

Before You Begin

SAFETY

Although weatherstripping your new home doesn't sound like a dangerous way to spend a weekend, simple carelessness can lead to potentially harmful situations.

☐ Wear gloves when working with fiberglass.

☐ Wear safety glasses or goggles when hammering.

☐ When sealing your house weather tight, provide adequate ventilation. Ventilation prevents the unhealthy build-up of moisture and fumes from materials used in building the home.

☐ Do not extend the insulation of the gas water heater to the floor. If you do, you will cut off the air supply to the pilot light and the burner.

☐ Do not tape or insulate too close to the vent stack on top of a gas water heater.

USEFUL TERMS

Bulb vinyl weatherstripping is composed of a rounded piece of vinyl (the bulb) attached to a piece of stiff aluminum that is nailed to the inside of the door or window frame.

Caulk is a flexible seal applied with a pressurized applicator to seal cracks and gaps.

A **door threshold** is a piece of wood or metal placed beneath the door to shed water and seal out drafts.

A **downspout** is the portion of a rain gutter that drains from the roof to the ground.

Glazing compound is a type of putty used to seal glass into window frames.

A **jalousie window** is a window formed of horizontal glass slats.

Oakum is a loose, stringy hemp fiber made from old ropes that is used as a caulking material for deep cracks.

A **soffit** is the underside of an enclosed eave on a roof.

Weatherstripping is material added to the movable parts or the jambs of doors and windows to seal them against air infiltration.

Window film is a 6 mil plastic covering that is stapled or adhered to the window in lieu of storm closures.

A **window sash** is the framework that holds glass panes in a window.

WHAT YOU WILL NEED

Time. The amount of time needed to weatherstrip your home depends on the extent of work needed. Most door and window caulking and weatherstripping can be accomplished by one person in a weekend.

Tools. Weatherizing your home requires only the most basic tools found in any home toolbox.

Hacksaw with fine-tooth blade

Pry bar

Circular saw

Drill

Hammer

Tack hammer

Nailset

Caulking gun

Screwdrivers

Steel tape measure

Scissors

Tinsnips

Plane

Putty knife

Utility knife

Hair drier

Because caulking is a major part of weatherstripping, an air compressor with a caulking gun attachment will make this job go more smoothly and easily.

Materials. The materials you need will of course depend on the kind of weatherizing you plan to do. The following list should cover most of your needs.

Thresholds

Caulk

Finishing nails

Galvanized shingle nails

Wood dough

Sandpaper, grades 100 through 180

Stain

Sealer

Weatherstripping materials

Door-bottom seals

Window glazing compound

Fiberglass or oakum

Window film

Programmable thermostat

Masking tape

Insulating jacket for hot water heater

Duct tape

Pipe insulating tape

Pipe insulating sleeves

Insulating fireplace baffle

Glass fireplace doors

Once you have decided which materials best suit your situation, you'll want to make a complete list before you go shopping. Take measurements of all the doors and windows that need weatherstripping and estimate the amount of caulking you will need. Note the number and sizes of openings you want to cover with window film. Also write down any specialty items, such as thresholds, a programmable thermostat, or a water heater blanket.

PERMITS AND CODES

Most weatherizing tasks are simple enough that they don't require permits.

Most Common Mistakes

☐ Neglecting to do a thorough weatherizing audit of your home.

☐ Neglecting to insulate properly in conjunction with weatherizing.

☐ Applying caulk, sealants, and weatherstripping in temperatures too cold for the material being used.

☐ Not providing for adequate ventilation and air circulation when making your home weather tight.

☐ Sealing off the room that houses the thermostat control.

☐ Neglecting to save receipts for materials and labor used in making your home more energy efficient, which may entitle you to state and federal tax credits. Check with your local energy supplier.

Step One
The Weatherizing Audit

Margin of Error: Not applicable

Although insulation reduces heat loss, heated or cooled air can still leak out through seemingly inconsequential cracks. In fact, these "inconsequential" cracks can add up to the equivalent of a 2-square-foot hole!

There are a number of ways to go about tracking down and sealing these leaks. You can contract for an energy audit (a scientific, thermographic audit, conducted by a professional, that produces a comprehensive report on the leaks in your home). Usually, however, your local utility company will conduct an audit at your home free of charge. You may not get the detail in the free audit that you would in one you paid for, but it will provide you with useful information and a place to begin. The utility company may also suggest improvements that would make you eligible for energy tax credits. A third option is to do your energy audit yourself, following the guidelines in this chapter.

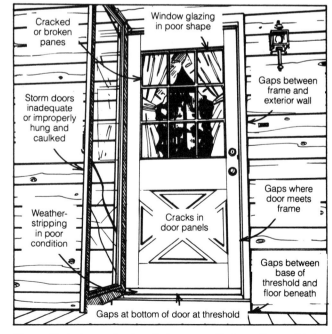

Figure 14-1. Areas of a door that are most likely to allow cold air to enter and heated air to escape.

Doors

The first step in weatherizing your home is to gather information. Begin by checking each exterior door for the following most probable leaks (Figure 14-1).

1. Are there any noticeable cracks at areas where the door meets the frame?

2. Is there any space at the joint between the frame and the interior and exterior walls of the house?

3. Is there a gap between the threshold and the bottom of the door?

4. Is there space between the base of the threshold and the floor underneath?

5. If your door has glass panes, are they properly glazed?

6. Are any of the panes cracked or broken?

7. Do the doors have weatherstripping?

8. If you live in a cold region, does your home have adequate storm doors? Are they properly hung and caulked?

Take the same approach to interior doors. Keeping them properly sealed helps reduce room-to-room infiltration and makes it possible to control more precisely the heating and cooling of individual rooms. However, if your home has a forced air heating system, leave a 3/8" gap at the bottom of interior doors to allow the air to return to the furnace.

Windows

Next take stock of all the windows.

1. Do any of the moving parts allow air to leak?

2. Are there any gaps or flaws in the construction around the frame?

3. Are the seams around the window trim caulked?

4. Do any windows that are not covered with draperies, shades, or blinds have any insulative value?

5. Do the windows have weatherstripping?

6. Are storm windows installed? If so, are they properly fitted and caulked, to eliminate gaps where the window meets the framing?

Other Openings

Once you have covered all the conventional openings to your home, begin looking for other, not so obvious ones (Figure 14-2). Are there any air leaks in the following places?

1. Foundation cracks or cracks in basement walls?

2. Separations between any two elements of the house, such as chimney and roof?

3. Utility pipes?

4. Phone and electric cable lines?

5. Mail slot?

6. Clothes dryer vent?

Energy-Saving Maintenance Tips

Energy can leak from your home in ways other than through openings. How do you and your home rate on energy conservation?

1. Keep your furnace clean and well-tuned for maximum efficiency.

2. Check the furnace, air-conditioning, and range filters frequently and clean or replace as necessary.

3. Keep your heating and cooling registers clean and unblocked.

4. Set your thermostat a little lower in the winter and a little higher in the summer.

5. Turn your thermostat down when you are out of the house.

6. Make sure your thermostat is clean and in good repair.

7. Consider installing a programmable thermostat.

8. Make sure your thermostat is properly located. A thermostat that is located too near a heat or cold source or on an outside wall can waste a lot of energy.

9. Don't routinely heat or cool seldom-used rooms.

10. Check annually for leaks in heating and cooling ducts.

11. Make sure your water heater is functioning well, and set it at an efficient temperature.

12. Insulate your hot and cold water pipes and heating and air-conditioning ducts.

13. Turn the water heater off and extinguish other pilot lights when you go away on vacation.

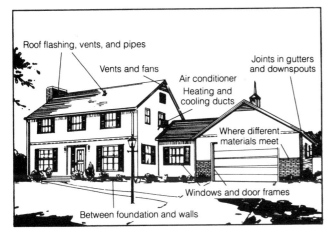

Figure 14-2. Areas of the house that are most likely to allow energy leakage.

7. Outside light connections?

8. TV antenna entry?

9. Electrical outlets?

10. Cracks and splits in siding?

11. Gaps or loose mortar between any blocks, bricks, or stone facing?

12. Air ducts for heating and cooling systems?

13. Window air conditioning units?

14. Leaking basement windows?

15. Roof flashing?

16. Split or loose roof shingles?

17. Poor drainage around the house?

18. Damaged, blocked, or poorly connected downspouts?

Don't try to make your garage air tight. Because of the possibility of gasoline leaks, codes require one 6x14 vent per car in the garage, within 18" of the floor. Your garage may be cold in the winter, but it will be safe all year round.

Step Two
Caulking

Margin of Error: Not applicable

Caulk is a material that forms a flexible seal to stop air and moisture infiltration. The better types of caulk will last up to 20 years. Although these may cost more, they will save you the time and expense of recaulking frequently.

There are no practical alternatives to caulking. If you weatherstrip and caulk in addition to insulating your home, you will reduce your energy bill considerably. Also, caulk-

ing is your first line of defense against insect invasions and fungus.

Caulk is used on window and door frames, siding, corner joints, foundations, and almost any area where you may find a seam or crack.

Take another look at your weatherization audit. It's likely that many of the places you noticed are prime targets for caulking.

There are many types of caulk for various uses, so check with your home center to find out what type will work best for your particular needs. Generally, you apply caulk from a tube with a caulking gun or from a pressurized can, as shown in Figure 14-3. A standard cartridge of caulk will give you approximately 25′ of ¼" bead. You can also purchase rope caulk, which comes in a coil and is simply unwound and stuffed into cracks and crevices.

The trick to a good caulking job is learning to draw an even bead. This may take a little practice. Be sure to hold the caulking gun at a consistent angle of about 45 degrees and draw the bead consistently, rather than in a stop-and-start fashion. Always release the trigger before pulling the gun away to prevent excess caulk from oozing out.

Since caulking is a major element of weatherizing, access to an air compressor will come in very handy. Fitted with a caulking gun attachment, the air compressor allows a smooth, almost effortless application in all areas needing caulk (Figure 14-4).

Be sure the seam or crack you are filling is free from any debris. Use a putty knife or a large screwdriver to scrape the

Figure 14-3. Use a pressurized can to caulk between interior window or door frame and wall.

Figure 14-4. If you have a lot of caulking to do, an air compressor fitted with a caulking gun will make your task easier.

opening clean. Also, the crack should be dry before you apply the caulk. Any moisture in the crack will be trapped inside once the caulk sets up.

If you are covering both sides of a crack or seam with a single wide bead of caulk, make sure it adheres to both sides. The caulk needs to be inside the crack, so if the bead you draw is on the surface of the material, use a putty knife to force it into the crack or seam and smooth it out. With a little practice you will learn the proper angle at which to hold the can or gun so that the caulk is forced immediately into the crack as it comes out of the tube.

The gap between the door frame and the exterior and interior walls can be ⅜" or even more. This space must be carefully and completely caulked to seal it against energy loss in cold regions.

Cracks in the foundation or basement walls can be terrific energy losers. Before caulking them, be certain the crack is very clean. Remove any loose mortar, dirt, and moisture. You may want to paint the area with a primer if the material is porous.

Smaller cracks can be sealed off with a liberal bead of caulk, forced smoothly into the crack. Extra deep or wide cracks should be stuffed with polyethylene foam, fiberglass, or oakum to within 1/2" of the surface. Then caulk over this material to provide a seal.

Where dissimilar materials meet—at wall and foundation, wall and roof, chimney and house, or porch and house—a seal of caulk will reward you with lower energy bills.

Where utility pipes, vent pipes, exterior plumbing, and electrical or phone connections enter the house, caulk the separation. If any of these openings penetrate the ceiling below an unheated attic or the wall to the garage, caulk there as well.

Keep all gutters, downspouts, soffits, and eaves clean and in good repair. Caulk wooden gutters to prevent decay. Tie the downspouts to storm drains, or otherwise divert runoff water away from the foundation, to avoid wet basements, decay problems in crawlspaces, and cracks and settling in the foundation.

Step Three
Window Treatments

Margin of Error: Not applicable

Properly sized and placed, snug-fitting windows can save a significant amount on your heating and cooling bills. Casement windows offer almost 100 percent of their sash opening for ventilation, while double-hung and sliding types offer only 50 percent. A well-placed sliding door will give up to three times the ventilation of an average window.

Energy-efficient sash and frame windows with double-pane insulating glass have low infiltration rates and will easily pay for themselves in energy saved.

Figure 14-5. Use a blow drier to shrink the window insulation film to fit.

Storm Glazing

If you live in a cold area, storm doors and windows, specially fitted with heavy-duty weatherstripping and sealed glass inserts, go a long way in sealing up a house. There are many do-it-yourself storm window and door kits available. Never caulk the little openings in storm windows. They allow moisture to escape rather than condensing on the glass.

If you don't want to go to the expense of providing storm windows for all of your windows, consider the option of adding window insulator film on the inside. This clear plastic film is used with double-sided tape to shrink around your window, window grouping, or glass door. Application of window film is quite simple.

1. Make sure that your interior window molding is clean and dry.

2. Measure and cut the double-sided tape and apply it to the outside edges or faces of the window molding.

3. Unfold the film and cut it to the size of the window (including trim), allowing 2" extra all around.

4. Starting at the top, press the film securely to the tape. The film will have wrinkles.

5. Now shrink the film with a hair blow drier set on the highest setting, being careful not to touch the film (Figure 14-5). Just aim the hot air evenly over the entire covering. This takes the wrinkles out and leaves you with a clear "pane."

6. Trim the excess film with a sharp utility knife or scissors.

This film reduces air leakage by approximately 97 percent, thus reducing frost build-up on windows. The film and tape can be removed at the end of the season.

Window Coverings

Window treatments—draperies, roller shades, venetian blinds, roman shades, louvres, quilts, sunscreens, and exterior awnings—are another prime area for home energy management. Window treatments can offer greater energy savings at less cost than storm windows or double-paned glass. The combinations are almost endless, ranging from

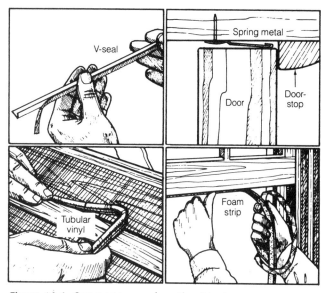

Figure 14-6. Common types of weatherstripping.

Figure 14-7. Apply caulk before installing the threshold.

simple sew-it-yourself, install-it-yourself window quilt systems through professionally installed, passive solar gain controls that can turn your house into an efficient heat pump. Many of these good-looking and energy-efficient offerings are inexpensive as well.

Step Four
Weatherstripping

Margin of Error: Not applicable

Weatherstripping products are numerous, and some are quite specialized. All are designed to seal a gap or space where heated or cooled air is leaking away, most often around doors and windows (Figure 14-6). Modern materials and designs have resulted in a great number of effective, durable, inexpensive, and easy-to-install products. Whatever type of weatherstripping you choose to apply to your doors and windows, it must be applied to a clean, dry surface and in temperatures above 50°.

Weatherstripping Door Bottoms

The first step in weatherstripping your doors is to assess each doorway carefully and plan accordingly. Don't overlook doors that lead to unheated rooms, the basement, or the garage. They can drain off just as much energy as doors to the outside. If you heat your garage, even occasionally, it will be worthwhile to seal that door as well, to minimize cross drafts from required ventilation. Special sweeps and shoes are made to keep rain as well as cold air out at the door bottom. If the weatherstripping that you buy does not come with its own fasteners, use appropriate galvanized fasteners placed 3" apart.

Installing Thresholds

The threshold seals the door at the bottom and is most effective when used in conjunction with a door shoe or door sweep.

Wooden thresholds add warmth in appearance as well as energy efficiency, but they are a bit more costly than aluminum. Bronze and stainless steel are considerably more expensive. All types come in standard door widths. Choose your threshold to match your floor.

Wooden Thresholds. To install a wooden threshold, follow these steps.

1. Trim the threshold to the length between the door jambs. Use a template to mark the door stop notches and cut them out. Be sure that water running off the door will drip on the sloping portion of the threshold, to drain outside.

2. Thoroughly clean the door sill and spread a generous amount of caulk on it (Figure 14-7) to ensure an airtight seal between the sill and the threshold.

3. Use your hammer to tap the new threshold gently into place.

4. Drill pilot holes slightly smaller than the finishing rails you will use to secure the threshold. Then nail the threshold into its permanent position.

5. Countersink the nails with a nailset and fill the holes with wood dough. Sand lightly when dry.

6. Apply a stain if you like. Then apply a water-repellant finish or two coats of a penetrating sealer.

Aluminium and Other Metal Thresholds. Metal thresholds are installed in the same way as wooden ones, except that you will need a hacksaw if it is necessary to cut the threshold to fit, and a metal file to smooth the resulting rough edge. Predrill pilot holes in the floor to make it easier to install the screws.

Figure 14-8 shows a **thermal threshold,** an aluminum threshold with a vinyl gasket that presses against the bottom of the door for a tight seal against drafts.

Interlocking Thresholds. The interlocking threshold shown in Figure 14-9 is recommended in cold climates. Although

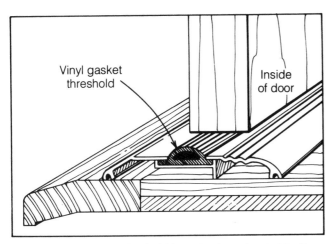

Figure 14-8. A thermal threshold is an aluminum threshold with a vinyl gasket.

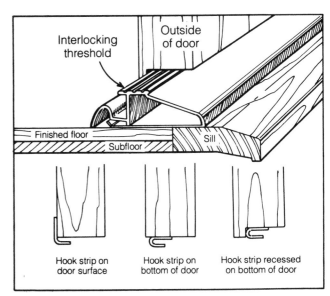

Figure 14-9. Interlocking thresholds are very efficient, but are difficult to install.

it is very efficient when in good condition, the interlocking elements are difficult to install and easily damaged. Special tools are required, and complicated adjustments are often necessary.

Half Thresholds. Half thresholds are used where two floors of different heights come together. They are available in metal, or they can be made by adapting a wooden threshold (Figure 14-10).

Door Bottom Seals

These seals, or sweeps, are either self-adhesive or fastened with screws. They are used to close the gap between the threshold and the bottom of the door (Figure 14-11). It is not usually necessary to remove the door to install them. They are easy to apply, inexpensive, and are sized for standard doors. Look for products that have clear instructions and, if they are attached with screws, slotted screw holes for easy adjustment.

Self-adhesive door seals are applied to the bottom of the interior side of inward-swinging doors.

Storm/entry door seals, which can be used on either the inside or the outside of the door, require a drill and a screwdriver for application. First cut the seal to fit. Then remove the paper backing from the adhesive side and press the seal against the door, positioning it for maximum contact with the floor. Drill pilot holes and secure the seal with screws (see Figure 14-11).

Door Shoes

Installing door shoes (Figure 14-12) and automatic door bottoms requires that the door be removed. For effectiveness, durability, and appearance, door shoes rate about the same as door bottom sweeps. If the clearance between door and threshold is between ½" and ³⁄₁₆", application should be no problem. If the clearance is less than that, you'll need to trim

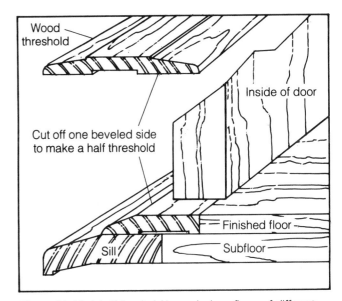

Figure 14-10. A half threshold is used where floors of different heights come together.

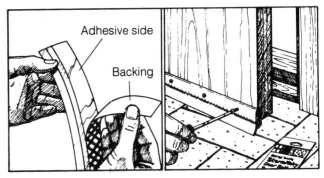

Figure 14-11. Door bottom seals may be either self-adhesive (left) or fastened with screws (right).

some off the bottom of the door. If the clearance is more than ½", you can install a thicker threshold.

To install door shoes, follow these instructions.

1. Measure the length of the door to determine if it will need to be trimmed. The door shoe should come with instructions specifying proper clearance. Once the shoe is installed, the door must be able to open and close easily, yet still fit tightly against all the frame elements.

2. Remove the door and lay it across a couple of sawhorses or on a sturdy work table.

3. Clamp a 2x4 or other straightedge to the door bottom to act as a cutting guide. Then saw or plane the door, if necessary.

4. Measure and cut the door shoe to the width of the door and notch the outside drip edge so that it clears the doorstops. For this you will need a hacksaw or a jigsaw with a metal-cutting blade.

5. Attach the shoe to the bottom of the door with the screws provided.

6. Slip the vinyl ridge into place in the shoe and trim it to length with a utility knife.

7. Rehang the door and check the fit.

Weatherstripping Around Doors

The best products for door treatments are of the plastic V-seal variety or of spring metal. Unlike windows, doors are opened and closed frequently all year round. They require something sturdy enough to take constant use.

V-Seal. This self-adhesive, sticky-back type of weatherstripping is made of durable plastic. It is easy to install and inexpensive. V-seal should be installed in the seams around the door or window frame so that a tight seal is achieved when the door or window is closed.

1. Before installing the weatherstripping, be sure the door does not fit too tightly on the top or along the hinge side. You may need to plane or sand the surface a bit for a smoother fit.

2. Use a steel tape measure to measure the length of the frame on both sides of the door and the width across the top.

3. Cut the plastic V-seal to the correct length with scissors and bend it to make the V.

4. Remove a couple of inches of the paper liner and position the V-seal in place at the top of the door frame so that the bottom of the V-shape points inward toward the house.

5. As you bring the plastic down, simultaneously pull off the paper backing and press it into place with your finger or a screwdriver (Figure 14-13). If you get it slightly out of line, simply pull it back up. The V-seal will adhere again when you press it back into position.

6. Run the V-seal all the way down to the bottom of the door frame. Do the same on the opposite side of the frame and at the top.

Spring Metal. Although it is one of the more expensive weatherstripping materials, spring metal is very durable and is virtually invisible when the door is closed. This type of weatherstripping is a bit more difficult to install than V-seal, and requires a tack hammer and awl or screwdriver.

1. Spring metal weatherstripping is installed on the door jamb next to the door stop. Measure along the sides and the top of the door jamb and cut the spring metal to the correct length with tinsnips.

2. Begin with the hinge side of the doorway, then do the latch side above and below the latch plate. Install the top piece last, miter-cutting the corners.

3. Position the spring metal so that the edge does not quite contact the stop.

4. Tack each strip at the ends to align it and stretch it flat against the jamb before nailing along its length.

5. Some manufacturers provide a strip to fit behind the strike plate. You can also trim a piece to fit.

6. When all the strips have been installed, use an awl or a screwdriver to score along the outside edge to spring the metal into position (Figure 14-14).

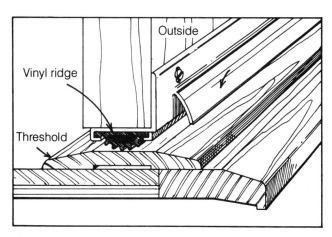

Figure 14-12. Installation of a door shoe requires that the door be removed.

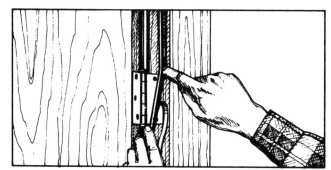

Figure 14-13. Use your finger or a screwdriver to press the V-seal into place.

Figure 14-14. Run an awl or screwdriver along the outside edge to spring the metal into place.

Weatherstripping Around Windows

Double-Hung Windows. V-seal and spring metal weatherstripping are most effective and durable on double-hung windows. Different manufacturers require different fastening techniques; some types are pressure sensitive, others require tacking. For durability, spring metal is the best, although plastic V-seal runs a very close second.

A double-hung window is the most complicated kind of window to weatherstrip; but once you've done one, you can handle other types of windows easily. Figure 14-15 shows where to apply weatherstripping on a double-hung window.

1. As with any weatherstripping job, begin by cleaning the surfaces thoroughly.

2. Measure from the base of the inner channel to 2" above the top rail of the upper sash. Cut four strips to that length for the inner and outer vertical sash channels. You can easily cut the plastic with scissors, but you will need tinsnips for the metal.

3. Next, bend the plastic V. Lower the top sash and slip the weatherstripping materials down into both channels with the point of the V facing toward the house, or the springy part of the metal toward the outside. You will have to secure the top weather stripping, move the top sash back up, and then secure the bottom.

4. Be careful not to cover the pulleys, and be sure that the sash cords or chains can run free. This may require a bit of custom trimming.

5. Place the weatherstripping full length along the top of the upper sash rail and along the bottom of the lower sash rail.

6. The midsection of the window calls for extra care. The sashes can travel past each other far enough to catch one leg of the V and jam or crumple it. Thin, self-adhesive dacron fuzz weatherstripping eliminates this problem. It is applied to the top rail of the bottom sash, so it can't be seen.

Casement Windows and Awning Windows. Casement and awning windows cannot normally be sealed by placing weatherstripping outside. Adhesive-backed foam weatherstripping works best on the interior frame where the sash makes contact. Self-adhesive foam rubber weatherstripping

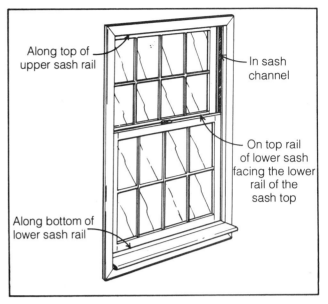

Figure 14-15. Location of weatherstripping on a double-hung window.

is easy to install, widely available in various sizes, and quite inexpensive. Its major drawback is that it wears quickly and cannot be used where friction occurs, such as in the sash channel of a double-hung window.

Sliding Windows. Weatherstripping for sliding metal windows is best installed by a professional glazier or weatherstripping contractor; in fact, it is difficult to find anything on the market for home installation. Sliding wood windows are best insulated with V-seal or spring metal, but adhesive-backed foam or bulb vinyl can also be used effectively if you are closing windows down for the winter and do not plan to open them until warm weather returns. Tubular or bulb vinyl weatherstripping is reusable season after season. This type is especially good for sliding glass doors. Simply cut it to the proper length and press the flanged protrusion into the gap to be sealed, either inside or out. Some types must be secured with nails or staples spaced 4" to 6" apart. You should never paint tubular or bulb vinyl weatherstripping; paint stiffens the vinyl and diminishes its sealing ability.

If both sashes move, weatherstrip them as you would for a double-hung window. If only one sash is movable, use spring metal in the channel where the sash closes against the frame and bulb vinyl on the top, bottom, and where the sashes join.

Jalousie Windows. The design of jalousie windows makes them nearly impossible to weatherstrip. A clear vinyl strip installed across the bottom of each pane is a partial solution, but these make the side gaps larger. You might consider replacing these windows. If that is not an option, storm windows are recommended.

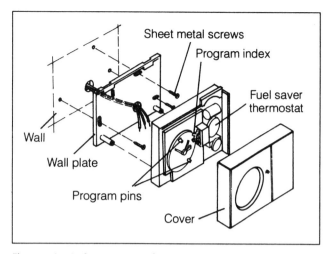

Figure 14-16. Components of a programmable thermostat.

Step Five
Energy Conservation

Margin of Error: Not applicable

The following conservation measures can help you reduce your energy costs significantly.

The Thermostat

The location of your thermostat does make a difference. If it is too near a heat or cold source, or on an outside wall, it can waste energy and cause unnecessary problems.

When weatherizing the interior doors of your home, you don't want to seal off the room with the thermostat. The purpose of tightening interior doors is to avoid heating or cooling unused rooms.

Turn the thermostat down at night when you are sleeping and during the day when the house is unoccupied and use the heat or air conditioner only in the morning and evening. An easily installed programmable thermostat (Figure 14-16) can also help you save energy. To install a programmable thermostat, follow these steps. (Also see Chapter 3, Electricity, for information on working with electrical fixtures.)

1. Turn the power off at the furnace or by removing the fuse or pulling the circuit breaker that runs the furnace fan.

2. Lift the cover plate off the old thermostat. Carefully write down which numbered or lettered wire is connected to which wire on the existing thermostat, and tag each wire with masking tape.

3. Remove the thermostat and inspect the wiring. If it is discolored, or if the insulation is cracked, cut back the insulation to a solid material and rewrap the wire with electrical tape to within 1" of the end.

4. Pull the wires through the opening of the new programmable thermostat wall plate and fasten the wires into place with the color-coded screws provided.

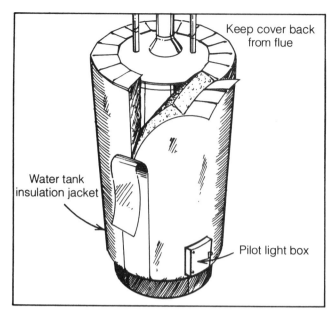

Figure 14-17. Insulating jacket for gas water heater.

5. Screw the thermostat into the wall and install the cover plate.

6. Restore the power to the furnace and then relight the pilot light. Program the thermostat according to the manufacturer's directions.

The Water Heater

Turning down the water heater to 110° will save you energy and prevent accidental scalding. If a hotter temperature is needed, as for a dishwasher, you can get a small water heating booster, which is installed in the plumbing system just before the appliance it serves.

Wrap your water heater in an insulating jacket to improve its heat retention (Figure 14-17). This insulation will also keep the heating element from working so hard to keep the water hot. If you have a gas hot water heater, be careful not to install the jacket too close to the floor. The pilot light and the burner require combustion air.

When you leave home for a vacation, turn off all pilot lights, including the one in the water heater. Check with your local gas company first about turning off and relighting gas pilot lights.

Tune-Ups and Maintenance

It pays to have your furnace and central air conditioner tuned up every couple of years before the heating or cooling season. Although these are jobs for a professional, you can take measures to add to the efficiency of these appliances.

Carefully examine all ducts and flues and seal any leaks with duct tape. Leaks usually occur at a bend in the duct. Replace filters regularly on furnaces and air-conditioning units. Keep heating/cooling registers clean and free from blockage. Window air conditioners should be removed and

Figure 14-18. Wrap the air conditioner in plastic and seal it with duct tape if it will not be used for several months.

cleaned in winter, and the space between the air conditioner and the wall should be sealed with insulating material. During the cooler months you can wrap the air conditioner in plastic and seal it with duct tape (Figure 14-18).

Fireplace chimneys allow warm air to drift right up and out of your home. Reduce this loss by installing a fireplace damper, if you don't have one already. This must be custom-fitted in your chimney by a professional, and it can be expensive. Remote cable-controlled dampers are available that mount atop the chimney. Be sure to mount the cable locking gate where you can adjust it while a fire is burning. Glass fireplace doors work well for the fireplace that is used only occasionally. These doors, which seal against the opening, provide good protection against infiltration.

There are kits available to help you seal off infiltration through electrical outlets. These nonflammable foam pad inserts fit right over the outlet under the plate. Cap the sockets when not in use.

Don't overwork your utilities. Insulating hot and cold water pipes as well as heating and cooling ducts can prevent sweating on hot days and freezing on cold days. Condensation can reduce the effectiveness of your subfloor insulation and cause many other moisture-related problems.

Pipe insulating sleeves are available for most size pipes, as well as one-size-fits-all foam tape. The sleeves are slit on one side to pop over the pipe like a long hot dog bun.

If you have air conditioning, you will find two lines running from the central unit to the condenser outside. The larger line is called a vapor line; this is the coldest portion of the whole system. If the vapor line is not insulated, the efficiency of the system may be reduced by 20 percent or more. The smaller line is the liquid return line; it carries heat from the central unit, and should not be insulated.

15 Home Security

WHETHER YOU LIVE in the city, the suburbs, or the country, home security is probably an important issue for you. There are no guarantees against crime. But with a relatively small investment of time, energy, and money, you can implement home security measures that will decrease the chance that your home will be broken into and increase the chance that stolen items will be recovered. Unfortunately, it's easy to overlook these measures and feel secure—until something happens.

This chapter will introduce some effective security measures you can implement around your home. You will learn how to install deadbolts and entrance locks in exterior doors. Other issues of home security addressed in this chapter include windows, safes and hiding places, alarm systems, and fire security.

Before You Begin

SAFETY

Safety is not directly addressed in this chapter because home security is, in itself, safety. Keep in mind, however, the general safe use practices for any tool you use in making your home secure.

USEFUL TERMS

A **deadbolt**, or **deadlocking latch**, is a 1" steel bar that, when engaged, extends all the way through the jamb.

Door lights are windows in or immediately adjacent to a door.

A **double-cylinder lock** is a lock for which a key is needed to unlock both inside and outside. These locks may violate the fire code in residential buildings.

A **hole saw** is a drill accessory used to create the big hole in which to insert the lockset or deadbolt cylinder into a door. Common sizes are from 1" to 2 ½".

A **mortise** is a notch or square hole cut out specifically to fit a full mortise lock or a hinge.

Polycarbonate is a shatterproof plastic that is up to 250 times stronger than regular glass.

Safety glass is two-ply glass with an adhesive plastic core. It is almost as strong as tempered glass, and unlike tempered glass, it can be cut to size.

A **spade bit** is a drill accessory used to create a hole in the door edge in which to insert the latch set.

A **spring latch** is a throw that has no hardened insert and so is easily jimmied with a credit card.

Tempered glass is four to five times stronger than regular glass. It is made by heating glass almost to the melting point and then chilling it rapidly, causing a tough skin to form around the glass. It cannot be cut without shattering.

WHAT YOU WILL NEED

Time. Making your home secure is not something you can set a time limit on; the process involves different amounts of time for every home. One person can install a deadbolt or a lock set in 2 to 4 hours.

Tools and Materials. The tools and materials you need will depend on what you plan to do to make your home more secure. The following lists should cover most of your needs.

Tools

Electric engraver
⅜" drill
Hole saws
Spade bits
Pencil
Utility knife
Chisel
Hammer
Wood rasp
Pry bar
Screwdrivers
Wrenches

Materials

Sheet metal
Glazing
Polycarbonate (shatterproof plastic)
Solid wood door
Metal door
3" flat head wood screws
Set screws
Sheet metal screws
Dead bolt
Key-in-knob lock
Full-mortise lock

Rim-mounted lock

10-penny nails

Keyed window locks

Bolt-action window locks

Lever-type sash locks

Hasp

Window bars

False book, soft drink can, etc., for hiding valuables

Push-button combination deadbolt

Plywood

Battery-operated alarm

Self-contained alarm

Smoke detector

PERMITS AND CODES

Permits are not usually required for improving security around the home. However, some areas require a permit if more than a certain dollar amount of work is being done. Replacing windows and doors may fall into this category. Also, electrical work for exterior lighting and alarm systems will require permits and inspections.

Codes do not allow the use of double-cylinder deadbolts in most areas, because they can prevent a quick exit in case of fire if you can't find the key.

The Home Security Survey

The best start for a home security program is to have a security survey done by a professional. In many areas, the police department will send a security expert out to your home to walk the house and grounds with you. If this service is not available in your community, the police department may provide pamphlets and other materials that will help you to make your own survey. Several excellent books are available on this subject. If you plan to have a professional install an alarm system, the company will often perform the survey. Be careful, though, because the company may have a hidden agenda—to sell you a more expensive security system than you really need.

To do the survey correctly, you need to think like a burglar. Imagine that you want to break into your own home without getting caught. Where would you enter? When? How? What do you think would give you away? A little imagination will lead you to the following conclusions. Almost all home security is focused on these two facts.

1. Your points of entry—doors and windows—are where you are most vulnerable.

2. Burglars fear two things: being heard and being seen.

A walk around your home, preferably with a professional, will quickly highlight your weaknesses in home security (Figure 15-1). If you, like most of us, lock yourself out from time to time, you already know these points of vulnerability.

You probably also know just how easy it is to break into most homes.

Windows are the most vulnerable points of entry. The standard types of screens and window latches do little to deter a burglar. The common window screen can be opened from the outside with a butter knife in just 5 seconds; the traditional window latch requires a penknife and about 5 seconds more.

Many doors are equipped with simple spring locks; these can be opened with a credit card or a knife. Even if the door is equipped with a deadbolt, a forceful shove can often break the wooden jamb and trim around the door.

As you walk through and around your home, make a list of all vulnerable areas, especially those easily accessible from the ground, including the basement windows and doors. Also check out second-floor entry points that are easily accessible with a ladder.

Ask yourself, What would make my house a house that a burglar would not want to hit? (Park the Jaguar in the back, the Volkswagen in the front.) Are there high bushes for burglars to hide in? Are some areas of the house hidden? Is the lighting adequate? Are valuable items visible through first-floor windows?

If burglars believe they would be in full view of the neighbors, you can be sure they will avoid your home. You may need to cut back some shrubs. If you live where there are no neighbors close by, a noisy alarm system is your best deterrent. At night, motion-sensitive lighting is recommended out of doors.

Are your exterior doors hollow core, or the more durable solid wood or metal doors? Do any trees offer easy access to second-story windows? Are sliding-glass or garage doors vulnerable? What type of windows do you have? Are they vulnerable?

After you complete your security survey, you are ready to make some decisions about what security measures you want to implement. This chapter discusses ways of installing these measures yourself.

Doors

Because windows are usually easier to penetrate, doors are not always the preferred entrance for a burglar. However, doors are usually their preferred exit.

Sixty percent of all home burglars enter through doors. The common key-in-knob lock can be opened in a matter of seconds by a professional with a credit card or a screwdriver. However, there are simple things you can do to make this technique unusable.

First, let's look at the door itself (Figure 15-2). There are many different types of entrance doors, and some are more secure than others. Many include a window or a glass panel directly adjoining the door (a door light). Unless your door locks with a key from the inside, a burglar can break the

Tree leading
to open window

Valuables visible
through window

Broken light
fixture

Screen can be
easily cut

Glass near
door locks

Uncollected
mail

Open basement window

High bushes
offer hiding places

Hinges on
outside of door

Uncollected newspapers

Figure 15-1. A security survey will reveal any vulnerable areas around your home.

glass, reach around to the inside, and open the door. If the panel is large enough, he or she can simply crawl in.

If you have such a door, and don't want to replace it, you can replace the glass panel with tempered glass. This glass is many times stronger than common plate glass, and should deter any intruder. Many doors already have tempered glass, since local codes often require it. If yours does not, take the exact measurements to a glass company and order a replacement pane. It may be fairly expensive. This glass has to be special ordered because it must be heat-treated (tempered) by the manufacturer and shipped to the glass company.

If your door has standard glass in it or near it, and you don't want to replace it with tempered glass just yet, you might consider replacing it with frosted glass, or setting a mortised bolt in the door bottom instead.

Study the door itself. Is it sturdy and well built? Some newer homes have doors that have a hollow interior (hollow-core doors). These are very easy to break. You may want to

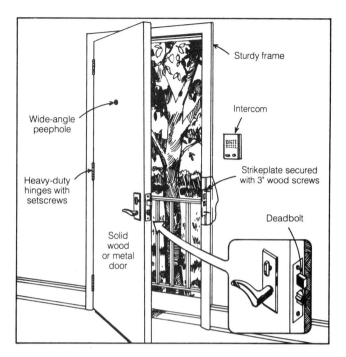

Figure 15-2. Features of a secure entry door.

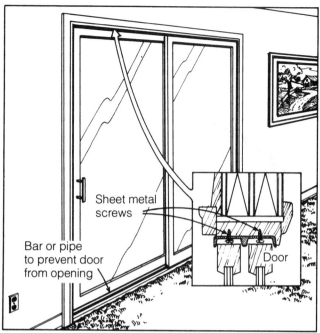

Figure 15-4. Insert screws into the top track to prevent sliding doors from being lifted off the track.

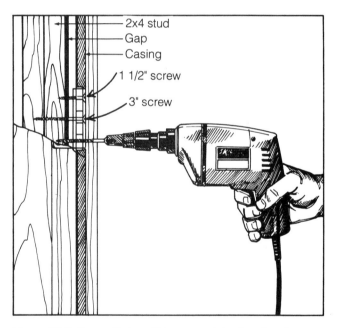

Figure 15-3. Screws that are 3" long penetrate deep into the studs.

replace the door with a new metal or solid-core door. Although replacing a door may seem intimidating, it is actually quite simple. You can buy doors in all standard sizes with the hinges and hardware preinstalled. Simply remove the pins from the hinges of the old door, remove the door, slip the new door on, and replace the hinges. Just be sure that the hinges, handles, locks, and strikeplates match up. For more information on hanging doors, see Chapter 8, Doors and Trim.

Hinges are another area of possible vulnerability. If the hinges are exposed to the exterior, you have a problem. This

is why residential exterior doors always open toward the interior. Burglars can simply pop out the hinges and remove the door, even when it is locked tight! This situation is a burglar's dream, and it was some carpenter's mistake. In this case, you need to reinstall the hinges so they are exposed to the inside, or to replace them with hinges that use fixed pins that cannot be removed. You can also buy kits that allow you to retrofit existing hinges with these setscrews to make them nonremovable.

Even a high-quality metal or solid wood door may not be secure if the frame around the door is not sturdy. In that case, the door can be penetrated with a pry bar, or even with a good shove. The jamb (the exposed frame around the door) is made of 1"-thick wood, which is easy to break through. However, the jamb is attached to a 2x4 framing member. If all hinge screws, strikeplate screws, and bolts penetrate into this thicker member, the door is much more secure. Replace all the screws in the hinges and strikeplates with high-quality 3" screws, which will penetrate deep into this frame, as shown in Figure 15-3. Also be sure to use a long deadbolt.

You may want to replace not only the door but the surrounding frame (jamb, trim, and molding) as well. Test to see if the frame is movable (if there is a gap between the frame and the wall stud). If it is, a burglar can simply push on the frame with a crowbar and pop the door open. Metal stripping can be installed around existing doors to make the door more secure. Although replacing both the frame and the door is somewhat complex, it is easier than you may think. You can buy doors "prehung." This means that the door is hung in its new frame with all hardware and hinges in place. You simply take out the old door and frame and replace it with a new,

Figure 15-5. The four most common types of locks.

more secure one. Depending on the type of siding and interior finish, this can be anywhere from a two-hour to a two-day job. Remember, you are not concerned just with the quality of the door itself, but also with the way it is attached to its frame.

There are a few other things to consider about doors. For one thing, you should install a door viewer or peephole that has a 180-degree viewing area—and use it. If you have sliding glass doors, check to see if they can be lifted off their tracks from the outside. If so, screw some screws in at the top that protrude into the track but do not obstruct the door from sliding (Figure 15-4). You can also fit a bar or pipe into the inside track so that the door can't be slid open when it is in place. These bars are available in any hardware or home center; or you can cut dowels to length yourself.

Small pin locks are available for sliding glass doors in keyed and spring-loaded styles. Be sure to install the spring type far enough in from the edge that a burglar can't get at it.

Deadbolts and Entrance Locks

Regardless of the strength of your doors, an intruder can always gain entry if your locks are not adequate. There are several standard measurements for locking mechanisms, so the bored holes on your door and frame may not match the new lock to be installed. Try to match the holes, or enlarge the existing holes. You can buy hardware that aids in adapting to existing holes. If all this fails, you will need to install a new door.

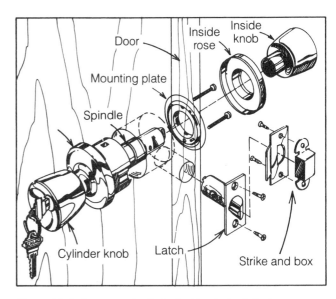

Figure 15-6. Components of a typical key-in-knob lock.

Four commonly used lock systems are described below and illustrated in Figure 15-5. The key-in-knob and the deadbolt are the most often used and the easiest to install.

1. Key-in-knob locks (Figure 15-6) are the most common exterior door locks, but they can be easily jimmied. The better ones have a hardened steel pin with a beveled latch.

2. Deadbolt locks are an excellent way to make your entry door secure. Look for a bolt at least 1" long; a rotating steel pin within the bolt to resist hacksaws; and a free-spinning brass cover over the outside cylinder to resist wrenches.

3. Full mortise locks almost always have to be installed by a professional locksmith. They offer double-lock protection, including a deadbolt.

4. Rim-mounted locks are sometimes called "mortised-in" bolts. They mount to the interior surface of the door and make a good second lock because of their ability to resist prying. They're always mounted on the lower half of the door so that smoke will not obscure them.

Installing a Deadbolt or Entrance Lock

Installing a deadbolt or entrance lock is a simple project that requires only a few special tools and should take no more than 2 or 3 hours. The main thing is to be sure of the exact location of the holes before you start to drill. If the hole is improperly drilled, it can ruin the door. Also be sure to use long (3") screws for the strikeplate so that they will penetrate deeply into the frame behind the jamb.

Step One
Choosing the Proper Lockset

Margin of Error: Not applicable

Most Common Mistake

☐ Purchasing an inadequate lockset.

The first step in installing a new lock is to choose your hardware. It pays to buy a nationally known brand, one whose quality and reliability you can trust. This is not an area where you want to cut corners with a bargain brand.

There are several different styles and designs to choose from. If the design is a decorative feature, it is a matter of taste. However, you need to decide whether you want to use

a double-cylinder, double-keyed lock or not. The advantage to these locks is that you can lock the door on the way out and no one can open it from the outside *or* the inside without a key. This stops burglars from carrying things out the door. The disadvantage is that they can hinder your escape in case of fire. You may want to use an entrance lock that only key locks from the outside, combined with a mortise bolt in the threshold. In case of fire, you will be able to get out quickly without fumbling for the key to open a keyed deadbolt.

Latch bolts come in varying lengths. It's best to choose the longest available. If it's ever needed, it will be worth the extra money.

Step Two
Using the Template

Margin of Error: Exact

Most Common Mistakes

☐ Marking the location of the hole incorrectly.

☐ Locating the new latch at the same level as the previous strikeplate.

In the package with your new entrance lock or deadbolt you will find a small paper template that shows you exactly where to drill the two holes you need install the system. You need to drill a hole through the door face for the lock or deadbolt and one through the edge of the door for the latch. These two holes must be perfectly aligned so that the mechanism will fit together properly. The template is your guide to this alignment.

First, decide how high on the door you want to install the lock. Entrance locks are usually 36" above the floor. The main thing to watch for here is the location of the old strikeplate. Even if you are not reusing this strikeplate (and usually you will not), be sure to center the new strikeplate directly on top of the old one.

Instructions on the use of the template are provided in the package. Generally you simply bend the template and wrap

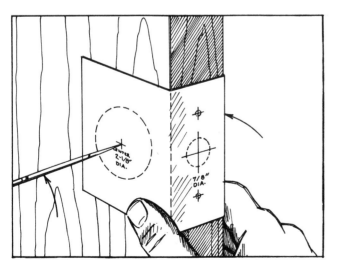

Figure 15-7. Use a paper template to mark the location of the holes you will drill on the edge and face of the door.

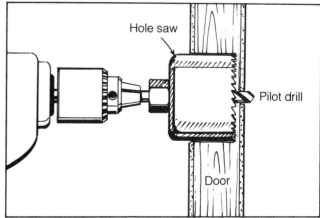

Figure 15-8. Use a hole saw to drill a large hole for the lock.

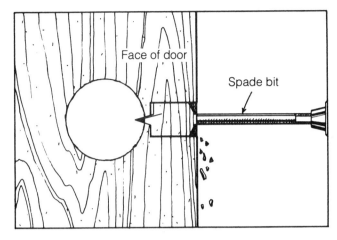

Figure 15-9. Use a spade bit to drill a hole for the spring latch.

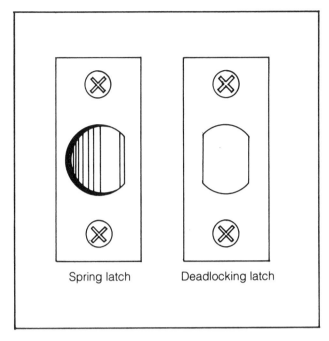

Figure 15-10. Typical latch plates.

it around the door so that the specified side is on the face of the door and the other side is on the edge. The hole center should be directly across from the center of the strike plate opening (Figure 15-7). Two holes in the template show you where to make your marks on the edge and on the face. Use a sharp instrument or a nail to mark the door at these two points. Double-check your work; you don't want to have to scrap a good door because a hole was drilled in the wrong place.

Step Three
Drilling the Holes

Margin of Error: Exact

Most Common Mistakes

☐ Drilling in the wrong place.

☐ Not drilling straight.

☐ Using the wrong size drill bits.

☐ Drilling all the way through both faces of the door in one pass.

☐ Using a dull drill bit.

☐ Using a small (¼") drill.

You are now ready to start drilling. Use a ⅜" drill to provide adequate power to drill cleanly through the solid door. Metal doors usually come predrilled. Use the bit size specified by the manufacturer (often 2 ⅛"). You will need a long spade bit to drill through the edge of the door for the hole for the latch, and you will need a hole saw for the hole through the face for the lock or deadbolt itself.

When drilling the larger lock hole through the door face, it is important to get a clean cut with no unsightly splinters. To do this, you need to use a hole saw, which has a small pilot drill in the center. This pilot drill has two functions. First, it allows you to exactly line up the center of the hole saw when you start to drill. Second, it will poke out the other side of the door before the hole saw itself penetrates all the way

through. As soon as you see the pilot bit pop out the other side, stop drilling immediately (Figure 15-8). Remove the hole saw and start drilling again from the other side of the door. This will create a clean, splinter-free hole. After the hole is drilled in the face, you are ready to drill the latch hole in the edge. Don't drill this hole first.

To drill the latch hole on the edge of the door, use the spade bit specified by the manufacturer. Drill this latch hole until it meets the larger hole you just drilled, as shown in Figure 15-9. To ensure a straight hole, always be careful to hold the drill level as you work.

Step Four
Installing the Spring Latch Plate

Margin of Error: Exact

Most Common Mistakes

☐ Chiseling the hole sloppily or too deep.

☐ Not drilling pilot holes before inserting screws.

After both holes are drilled, you are ready to install the spring latch (Figure 15-10). This is a simple task and should come out perfectly if you take your time and do it accurately. You will need to mortise (inset) the plate into the edge of the door so that it is flush with the surface of the edge.

Place the latch so that the hidden part goes in the hole you drilled in the edge of the door. It is a good idea to actually install the spring latch first, to be sure that the plate is properly located.

When you are sure the plate is properly placed, hold it snug or screw it against the edge of the door. With a sharp

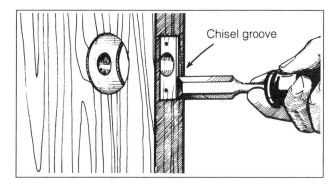

Figure 15-11. Chisel out the area where the spring latch plate will be installed.

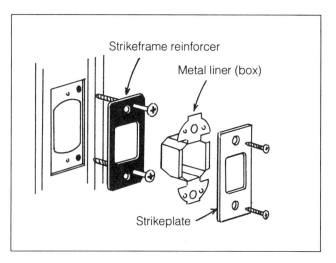

Figure 15-12. Components of the strike plate.

pencil or a utility knife, outline the plate on the door. Now chisel out this area to a depth equal to the thickness of the latch plate (Figure 15-11). This will allow the latch plate to sit flush with the surface of the edge of the door. Use a sharp chisel the same width as the plate. Use light taps of the hammer so you don't gouge out too much wood and sink the plate too deeply. Work slowly and carefully to create a tight fit.

When you are satisfied with the fit, install the latch with the two screws provided. It's best to drill pilot holes first with a bit that is slightly thinner than the shank of the screw. The pilot holes ensure that the screws will go in straight and not split the wood. After the pilot holes are drilled, hold the plate firmly against the door and install the screws.

With the spring latch in place, you can install the lock mechanism. This comes in two pieces, inserted from either side of the door. Needless to say, be sure the keyed knob is pointing out. Slip in the keyed knob, threading it through the latch piece, and then the interior sections. These two are usually joined together by two screws that insert through the interior section and screw into the exterior section, tightly sandwiching the door and holding the lock firmly in place. Now place the cover plate over the interior section. Finally, slip on the knob. There is usually a spring clip that you must press down to slip on the knob. When you release this clip, it fits into a groove and secures the knob.

Step Five
Installing the Strikeplate

Margin of Error: Exact

Most Common Mistakes

☐ Not aligning the strikeplate with the spring latch.

☐ Installing the strikeplate in the old holes if they are in poor condition.

☐ Not using long enough screws.

When your entrance lock is in and functioning properly, you can install the strikeplate in the frame of the door. Don't install this in the holes created by the previous strikeplate screws, or the screws will not have good wood to bite into. Of course, it's a little late to be thinking about this now, since

the plate must align with the spring latch of the lock you just installed. Just be sure that this alignment is exact so that the latch can be easily inserted into the plate and metal liner.

The strikeplate is usually a three-part assembly made up of a strikeframe reinforcer, metal liner (box), and finished strikeplate (Figure 15-12). The metal liner often requires a fairly deep hole. To create this hole, locate the exact center of the strikeplate on the frame and drill two holes with a $7/8$" bit, one $3/8$" above the center point and one $3/8$" below (refer to the manufacturer's instructions). Then chisel out the wood between the two holes and in the corners to create one hole that the metal liner will fit into tightly. Before installing the metal liner, install the strikeframe reinforcer, using two long (3") screws to penetrate into the wall frame. (Be sure to drill pilot holes first.) Now insert the metal liner and screw in the finished strikeplate. Test your lock. If it works smoothly, pat yourself on the back. If it doesn't, you may need to chisel out more wood or slightly adjust the location of the strikeplate.

Windows

Windows are almost as popular as doors as entry points for burglars. Unfortunately, they are much harder to secure than doors, because the glass can always be broken. There are several things you can do to reduce the threat of entry through your windows. Attaching them to a central alarm system is one of the best ways. Also, the glass can be replaced with tempered glass or polycarbonate, although this is a costly alternative. Metal bars can be installed over the windows, but they don't enhance your view, and they can be a real threat to life in case of fire, unless they are fitted with the required emergency releases. And of course, be sure to keep foliage in front of windows well trimmed so that burglars can't hide there. Aside from these techniques, the best way of making your windows secure is to install security latches and other such devices. They are inexpensive, easy

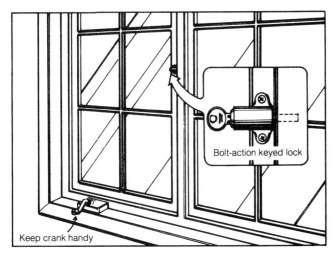

Figure 15-13. Be sure your casement windows are well secured by removing the crank and installing a lock.

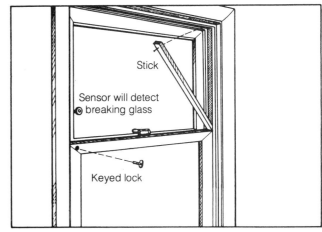

Figure 15-14. Three ways of securing windows.

to install, and will usually deter burglars, unless they are willing to break the glass despite the noise.

You may think that your window locks are effective, but this is often not the case. Clasp locks can be opened with a butter knife inserted between the two windows. There are several different types of window locks on the market, all of which are relatively easy to install. Some locks even allow you to leave the window locked in a partially opened position. Battery-operated alarms are also available for individual windows.

Keyed locks for windows are available, but they can be hard to open in an emergency, such as a fire. Casement windows, which have cranks, can be secured by simply removing the handle (Figure 15-13). Leave the handle in an accessible place in case of fire. You may need to secure only the windows on the first floor, if you feel that your second- and third-floor windows are not vulnerable (Figure 15-14).

Air conditioners or fans installed in windows can also be vulnerable points. They can often be removed by a burglar. Be sure the units are bolted to the house in such a way that they can't be removed from the outside. Also make sure that the window can't be raised.

Safes and Hiding Places

After you have taken all possible precautions to keep burglars from breaking into your home, you may want to take further measures to keep them from locating your valuables if they do break in. This might mean either hiding valuables or placing them in safes so that they can't be removed even if they are located. Some very simple, inexpensive, and clever hiding devices are available. For example, you can buy a small container that looks like a head of lettuce. You put your valuables in it and place it in the refrigerator. Unless the burglar is aware of the gimmick, or decides to make a salad, your jewels are safe. Similar containers are available

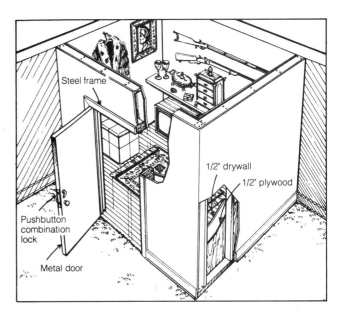

Figure 15-15. A closet can be converted into a small vault.

as soft drink cans, hollow books, or false electrical boxes. If you have wide (6" to 12") baseboards, you can remove a section, hollow out the areas between the studs, put small valuables in, and tack the baseboard back in place.

An imaginative tour of your home will reveal many other effective hiding places. Remember, burglars don't usually have much time to search, so they grab whatever is handy.

If you feel it's worth the effort, you can strengthen a closet to serve as a small vault (Figure 15-15). Simply install a metal or solid wood door and a couple of deadbolts or a pushbutton combination lock. It is a good idea to line the interior with plywood, since drywall can be easily broken through.

Many models and sizes of residential safes are available. Although somewhat more expensive than the other devices described here, they can be effective if properly installed. Smaller onces cemented into place are an excellent deterrent. But remember, a safe will also alert a burglar to what's inside, so be sure it is well secured.

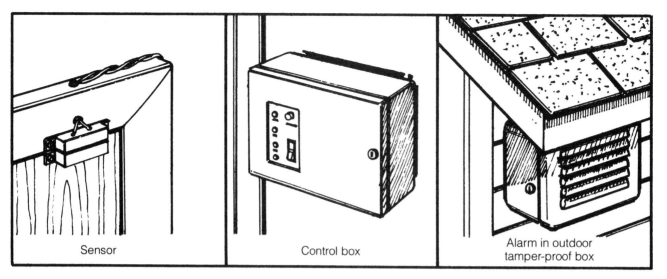

Sensor Control box Alarm in outdoor tamper-proof box

Figure 15-16. The three basic components of an alarm system.

Alarm Systems

Electronic alarms have become very popular in the last decade. Their popularity is well earned; they are both effective and affordable. However, you should not put too much faith in them. When combined with the other safety measures discussed in this chapter, they will make your house relatively secure. But if you rely totally on alarm systems, you will still be vulnerable.

Professional burglars know how to silence or incapacitate even the most complicated alarm systems. Remote alarms—alarms that ring only at the police station or at a private security office—often allow burglars time to get away before help arrives. Also, false alarms are becoming so common that alarms are often ignored.

There are several different types of alarm systems on the market. Some of them are easy to install—you simply screw the sending units into doors and windows. If the door or window is opened or broken, an alarm sounds and a signal is sent to a main receiving unit. Other types are best installed by a professional alarm company. In choosing an alarm system, you need to consider your family's lifestyle. Motion-sensing detectors will be self-defeating if you have pets. If you have several children or frequent overnight guests, alarm systems that demand that you enter a code when entering and leaving the house may not work. I remember arriving at my sister's house alone late one night. I let myself in, and immediately heard the soft buzzer on the alarm system, indicating that I had 10 seconds to enter the code before the alarm would blast. I had forgotten the code and stood there helplessly in the hallway as the alarm woke my sister, her family, and all the neighbors.

Once you've decided which type of alarm system will work best for you, be sure to buy a high-quality system. With luck, and with all the other precautions you've taken, you may never need it. But if you do, you will want it to work properly.

All systems have three basic components, as shown in Figure 15-16: the sensor that senses the intrusion, the control that sounds the alarm once the intrusion has occurred, and the alarm itself. These systems may operate off a battery, off the home's electrical current, or both. Although they are easy to install, the battery-operated units are often not sophisticated enough to satisfy all your needs.

Self-Contained Alarm Systems

Self-contained alarm systems have the alarm, control, and sensor all in one unit (Figure 15-17). They are most commonly used in small houses, offices, and apartments that have a limited number of doors and windows. They can be as simple as a cigarette-box-sized alarm that hangs on the door or chainguard. Others can be plugged into any wall outlet and have a simple motion detector. The more sophisticated models are activated by a change in air pressure, as when a door or window is opened, or work off of high-frequency sound waves. These units are less expensive and easier to install than some others. Their drawbacks are that burglars can quickly locate and disable them because the alarm is with the control. Also, the ones that work off of air pressure or sound waves often give false alarms in response to noise upstairs or high winds.

Alarm Systems with Separate Components

The best alarm systems separate the sensor from the control and from the alarm; these work well if you want to guard several rooms at once. Individual sensors, such as a magnetic contact, are placed on the windows and doors, and the wires

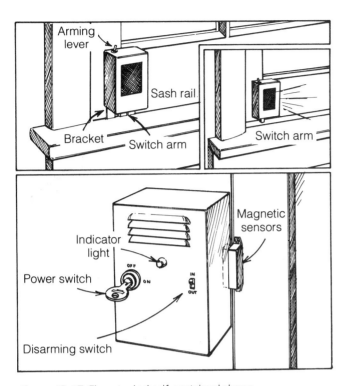

Figure 15-17. Three typical self-contained alarms.

to the alarm control are hidden in the framing. This makes it much more difficult for the burglar to dismantle the system.

These units often have several control stations around the house so that you can activate or deactivate all or part of your sensors. Some units even tell you which doors and windows are open or closed. They also have panic buttons that can be used when you hear someone prowling around outside. These systems are usually activated and deactivated by a code and can sound an alarm at the house, at the police station or security office, or both. Also, some alarms can be wired to dial a number automatically and give a recorded message. You can change the number so that the system will reach you wherever you are, in case the house is burning or you are being robbed.

When you install an alarm, put decals on your doors to let burglars know that your house is equipped with an alarm. They will probably go away.

Neighborhood Watch Programs

As mentioned earlier, burglars fear being observed. Many communities have neighborhood watch programs. Watching out for your neighbors is one of the most effective deterrants to burglars. Your police department will probably give you advice on organizing your block or neighborhood, and may even send an officer out to address your group. They may also provide "Crime Watch" stickers for your windows and doors.

In some areas the police department will loan you an electric engraver with which you can permanently mark your driver's license number on your television, stereo, and other valuables. This will make it possible for the police to return your possessions to you in case they are stolen and retrieved.

Fire Security

Most of this chapter has been devoted to securing your valuables against theft. It is even more important to secure your home itself, and its occupants, against the threat of fire. This section of the chapter is shorter than the sections on home security, not because fire security is less important (in fact, it is much more important), but because fire alarm systems do not require detailed installation instructions. All you have to do is screw the smoke detectors to the walls at the right places.

Fire security is extremely important. To lose your valuables is one thing; to lose a family member is something else altogether. In the United States, almost five thousand people die in fires each year.

Fire safety begins with a thorough investigation of every room in the home to locate possible trouble spots. It is a rare home that does not have at least one of the following fire safety problems.

☐ Overloaded or undersized electrical circuits
☐ Too many electrical appliances plugged into one outlet
☐ Electrical cords running under carpet or rug
☐ Frayed wires
☐ Defective appliances
☐ Oversized fuses
☐ Flammable liquids stored
☐ Gas leaks
☐ Accumulation of grease in stove
☐ Accumulation of grease in range hood
☐ Bare light bulbs, especially in closets
☐ Built-up soot in chimney or wood stove flue
☐ Improperly installed fireplace or wood stove
☐ Inadequate screen in front of open fireplace or wood stove
☐ Outdoor barbecue grease build-up
☐ Children playing with matches
☐ Sloppy workshop area
☐ Smoking in bed
☐ Double cylinder deadbolts
☐ Security bars without an emergency release

The Christmas season poses the following special dangers.

☐ Dry tree
☐ Plastic tree that is not fire retardant
☐ Tree placed near heat source
☐ Children playing with candles

Visit your local fire department and pick up their pamphlets on fire safety. A fire department representative may even visit your home for a security tour.

After you have evaluated all the possible problems and corrected them, you are ready to take some offensive, as opposed to defensive, tactics.

Clean your chimneys and flues as needed to prevent soot, or hire a chimney sweep to do it. As soot accumulates, oily tars leach to the surface. These flammable tars are the source of many residential tars.

Discuss the location and operation of the gas main valve and electrical main breaker with your family. Keep a wrench of appropriate size near the gas main valve.

Plan fire escape routes. Be sure that each bedroom has at least two possible ways out. Teach the entire family, especially children, what to do in case of fire. Make sure they know that most deaths are caused by smoke inhalation, not by the fire itself. Learn how to protect yourself against smoke inhalation. Decide on a meeting place outside your house or at a neighbor's home so you can quickly determine if anyone is missing. Tape the fire department number to all phones. Remind your family to forget about their possessions, just escape. Determine if any special fire escapes or rope ladders are needed for quick escape from the second floor (Figure 15-18). Put a few fire extinguishers, the multipurpose (ABC) type, around the house, especially in the kitchen (Figure 15-19). And finally, conduct a fire drill.

Above all, install smoke detectors. These simple devices are inexpensive and can be installed in just a few minutes. They run off either household electricity or batteries. Experts estimate that half the lives lost in home fires could be saved with these detectors.

Install the detectors between the sleeping areas and the rest of the house. In multilevel homes, codes require at least one on each floor. In a two-story house, install one on the ceiling at the bottom of the staircase. Put one at the bottom of the basement staircase as well. Don't install them near an air supply, open duct, or vent, which can pull the smoke away from the detector. And don't place them near safe sources of heat or smoke, such as a fireplace, which will activate them unnecessarily.

Figure 15-18. A rope ladder is a good idea for second-story bedrooms, especially if there is only one stairway.

Dry chemical extinguisher   Water extinguisher   Carbon dioxide extinguisher

Figure 15-19. Three types of fire extinguishers. The multipurpose ABC type is recommended.

16 Decks

BUILDING A DECK is a fairly complex project that can be demanding both physically and mentally. However, decks don't have to be leakproof or perfectly built, the work goes quickly, and the reward is an enjoyable living area for a small price.

There are two key things to remember in building a deck. First, because the deck is completely exposed to the weather, you must use the proper materials and construction techniques to avoid decay. And second, be sure that your deck is level, plumb, and solidly constructed.

Before You Begin

SAFETY

☐ Always use the appropriate tool for the job.

☐ Keep blades and bits sharp. A dull tool requires excessive force, and can easily slip.

☐ Wear safety goggles and glasses when using power tools, especially if you wear contact lenses.

☐ Always unplug your power tools when making adjustments or changing attachments.

☐ Be sure your electric tools are properly grounded.

☐ Watch power cord placement so it does not interfere with the operation of the tool.

☐ Wear ear protection when operating power tools; some operate at a noise level that is high enough to damage hearing.

☐ Be careful that loose hair and clothing do not get caught in power tools.

☐ Be careful when carrying long boards at the work site.

☐ Wear heavy-soled, sturdy work boots.

☐ To avoid back strain, bend from the knees when lifting large or heavy objects, and be careful when digging post holes.

USEFUL TERMS

All-heart grades are grades of wood that contain no knots or blemishes.

Band joists are 2"-thick top-quality boards that are nailed around the outer joists of the deck and across the end of the joists to form an attractive border.

The **bow** is the deviation from straight and true seen when looking down the edge of a board.

Checking is the tendency of wood to split across the grain as it dries.

Construction common is a grade of redwood that contains sapwood.

The **crown** is the highest point of a warped board seen when looking down the side of the the board; the convex side.

The **cup** is the warp of a board seen from the end of the board; the concave side.

Decking is the boards that make up the surface of the deck.

The **footing** is the concrete base on which the foundation wall or pier rests.

A **girder** is a horizontal support member of a deck framing system that rests on the piers. It is usually 4" wide and the same depth as the joists. The girder is most often parallel to the ledger and supports the opposite end of the joists.

Hot dipped galvanized (HDG) is a rustproof coated metal that is less expensive than aluminum or stainless steel. It is used for nails, bolts, screws, and other metal fasteners.

Joists are a system of floor framing that commonly uses 2x6 lumber or larger, depending on the span.

A **lag screw** is a long screw with a hexagonal head.

A **ledger** is a board of the same size as the joists, attached to a wall, to which the joists are attached perpendicularly.

A **pad** is a cast concrete base designed to spread the load of a pier block over a larger area.

Pier blocks are concrete blocks that support the posts several inches above ground to avoid decay.

Piers are holes dug to below the frost line and filled with concrete to provide a firm footing for the foundation pier blocks.

A **plumb bob** is a heavy object suspended on the end of a string for the purpose of establishing a true vertical line.

Posts are the upright members, usually 4x4, that support the deck.

Pressure-treated lumber has been infused with copper salts that greatly reduce the ability of insects and fungi to grow in the wood. Always wear a mask when cutting this material.

Sapwood is the portion of the tree between the heartwood and the bark. This material is much weaker than heartwood, and should not be used for decking.

Toenailing is nailing at an angle that reduces the chance of nails loosening under stress.

A **torpedo level** is a level 8" or 9" in length with vials to read level, plumb, and 45 degrees.

WHAT YOU WILL NEED

Time. A 12′x12′ deck with a simple foundation and railing can be completed by two people in a few days. Allow 64 to 85 person-hours to completion, longer for complex railing and foundation designs.

Tools. Most deck projects require only common framing tools; no specialized tools are necessary, although you may want to consider renting a pneumatic tool for nailing on the decking. Be sure to use high-quality tools that are capable of doing the job without strain. This is especially important in your choice of power tools.

Safety goggles

Ear plugs

Framing hammer

Torpedo level

4′ level

Plumb bob

Pencil

Nail pouch

Framing square

Combination square

Tape measure

Sawhorses

Shovel

Wheelbarrow or pan for mixing concrete

Cement hoe

Trowel

Pry bar

Caulking gun

Extension cords

Socket or crescent wrench

Power saw

Handsaw

Materials. Your choice of decking and fastener materials is very important. All materials must be chosen for their ability to resist decay and rust. Use only hot dipped galvanized (HDG) nails, bolts, screws, and metal fasteners. Double hot dipped is even better. These galvanized fasteners won't rust quickly. Aluminium and stainless steel nails won't rust either, but HDG is more commonly used and less expensive. Don't use electroplated galvanized (EG); galvanized plating often chips.

Many types of galvanized sheet metal fasteners are commonly used in deck construction. These fasteners simplify your project and strengthen your construction. They can be used in many areas of deck construction, including the following attachments.

☐ Joist to ledger (joist hanger)

☐ Girder to post (post cap)

☐ Joist to girder (right-angle bracket)

☐ Post to foundation (post anchor)

All wood used on the deck must be decay resistant. These woods include redwood, cedar, cypress, and pressure-treated wood. Redwood is perhaps the most attractive and durable wood to use, especially in exposed areas such as the decking and railing. It is decay resistant, dimensionally stable, and relatively straight. It is easy to saw and nail and has little or no pitch. It resists warping, checking, and cupping, and is strong for its light weight. It comes in several grades. Construction common, which contains some sapwood, is acceptable for deck boards. The more expensive all-heart grade is better.

Pressure-treated wood is also commonly used. It is treated to resist decay, and is often green in color. For the best decay resistance, use a pressure-treated frame with redwood decking and railings.

Deck joists are usually 2x6, 2x8, or 2x10 stock, depending on the span. The decking boards are usually 2x4 or 2x6; 2x8 stock is too wide and will cup if used as a decking board. Girders are 4x4, 4x6, or 4x8, or built up with 2x material. Ledgers are 2x6, 2x8, or 2x10, like joists. Posts are usually 4x4 or 4x6.

Your materials list is likely to include the following items.

Pier blocks

Decking

Girder stock

Joist stock

Post stock

Ledger board

Band joists

HDG nails

HDG lag bolts

Joist hangers

Joist hanger nails

Right-angle brackets

Post caps

Post anchors

Railing stock

Flashing

Butyl caulk

Water-repellent sealer

Nylon string

Ready-mix concrete

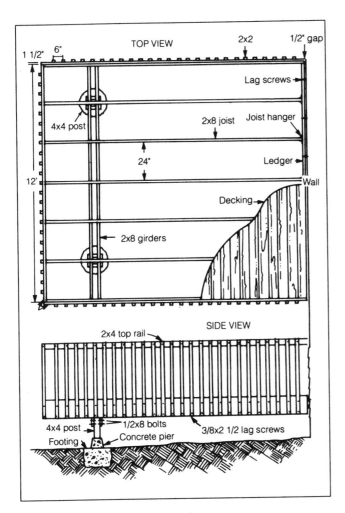

Figure 16-1. A typical set of deck plans.

ESTIMATING MATERIALS

Estimating materials for your deck is fairly simple and straightforward. Use your plans, as shown in Figure 16-1, to determine the following information.

☐ Number of square feet of decking

☐ Size and length of joists, bands, decking, posts, girders, and railings

☐ Number of joist hangers

☐ Number of piers

☐ Amount of concrete for footings

☐ Number and kind of nails, washers, nuts, and bolts

☐ Amount of water-repellent (The California Redwood Association recommends that you treat all the wood with water repellent before the deck is built.)

PERMITS AND CODES

In most areas a permit is required before you begin to build a deck. If the deck is not attached to the house, or if you live in a rural area, you may not need a permit. Check this out before you begin to build.

The permit office will probably require a set of plans. These need not be elaborate, and you can draw them yourself, buy a set, or have a set prepared by a draftsperson or architect. The plans need to specify the following information, as shown in Figure 16-1.

1. Location of the deck. The deck must be far enough away from neighboring property lines, utility easements, and gas, water, and sewer lines.

2. Space between railing pickets. Usually the pickets must be spaced no more than 6" apart.

3. Railing height. If the deck is under 18" high, no railing may be required. Over that, a railing is usually required, and a minimum height of 36" to 42" is usually specified.

4. Foundation piers. The local code will regulate the size, spacing, number, and method of construction of your piers. It will also regulate the depth of your pier holes, depending on the frostline in the area and on the pad width in more temperate climates.

5. The girder. The code will specify the size and location of the girder.

6. Joists. The code will specify the size and spacing of the joists, depending on the type of wood you are using. Joist charts are available at the code office for each type of wood.

7. Fasteners. Many codes detail nailing and fastener requirements for decks.

8. Decking. Your plans must specify the size of deck boards and the type of wood to be used.

9. Posts. Your plans must specify the size of the posts.

10. Bracing. Unless the girder is directly supported by its piers, you will need some sort of diagonal bracing. If applied to the posts, these braces will restrict access under the deck. Some codes allow you to "let in" braces diagonally, in the top or bottom of the joists, to avoid obstructing the access.

11. Earthquakes. In earthquake and hurricane areas, there may be further requirements on how the piers are fastened to the footings, the girders to the posts, and the joists to the girders. The code may also specify the type of metal fasteners that must be used. Additional (or larger) diagonal bracing is often required in areas where hurricanes or earthquakes are common.

DESIGN

There is much to say about the design of a deck, and many books are available on the subject. This chapter provides you not with a primer on deck design but rather with a few design parameters to consider. This is not to understate the importance of design. How well you are pleased with your handiwork and how much you will enjoy using it depend on design more than on construction. Your answers to the following questions will help you to think through your design requirements.

1. At what time of day during each season of the year does the deck location get sun and shade?

2. How much privacy will the deck have from the neighbors? Will this change when the trees lose their leaves or grow taller?

3. What kind of access will there be from the deck to the house? To the yard?

4. How large should the deck be? How much yard must be sacrificed?

5. How should the railings be designed? With planters? With seats?

6. Where should the stairs be placed?

7. Are there any utility lines overhead or below?

8. Should the deck be covered?

9. What will be placed on the deck? A barbecue? A table and chairs?

10. Will the deck block the light or the view from any windows of the house?

11. How should the deck be positioned to take advantage of the views?

12. How much money do you plan to spend?

Step One
Determining the Level and Length of the Ledger

Margin of Error: ¼"

Most Common Mistakes

☐ Choosing the wrong height or location for the deck.

☐ Using badly bowed boards.

☐ Not using redwood, cedar, cypress, or pressure-treated stock.

☐ Not following code.

If you are attaching the deck to your house, as shown in Figure 16-2, you will probably use a ledger board. This board is bolted to the house and the deck is hung on it. If you plan a freestanding deck (not attached to the house), you will not use a ledger. In some areas a deck attached to a house will be taxed, but if it is separated from the house by even an inch or two it will not. This chapter considers the more common approach of attaching the deck to the house. If your deck is freestanding, everything discussed here will still apply, except that the ledger is not bolted to the house. Added bracing may be needed to stabilize a freestanding deck.

The Level of the Ledger

The level, or height, of the ledger determines the height of the deck. The top of the ledger will be 1 ½" below the final top surface of the deck, because 1 ½" decking boards will be nailed on top. The deck joists, which support the decking, or surface of the deck, are sometimes installed so that they sit on top of the ledger, as shown in Figure 16-3, rather than hanging from it. In that case, the top of the ledger will be considerably lower than the final level of the deck (by the width of the joist plus 1 ½"). Before installing the ledger, you need to decide whether to hang the joists from the ledger or to rest them on top. It is stronger to hang them from the ledger with joist hangers, and that is the method described in this chapter and illustrated in Figures 16-16 and 16-17.

No matter what approach you use, you need to be sure that the ledger is low enough so that the level of the finished deck will be at least 1" below the level of the finished floor inside the house (Figure 16-4). This is necessary because you want to step down from the house to the deck. Also, if the deck is higher than the floor, water can run from the deck to the house. If you don't plan to have a door from the house to the deck, the height of the deck is not so important.

To determine the height of the deck, you must first determine the level of the interior finished floor and transfer this level to the outside of the wall where the deck will be built. To do this, measure on the inside and outside from a common reference point, such as a windowsill. (If the sill slopes toward the outside, be sure to adjust your measurements.) Another method is to measure up from the foundation wall, accounting for the floor joist, subfloor, and finished flooring material. Once you have determined the level of the interior finished floor, mark it on the exterior wall. Make a second mark a minimum of 2 ½" below this first mark. This will allow you to install 1 ½" decking and still place the level of the deck 1" below the level of the interior floor. You should place the top of the ledger on this second mark.

The Length of the Ledger

As shown in Figure 16-5, the length of the ledger equals the total width of the deck less 3" (6" if a band joist is used on each side). This 3" is so that the joists at either end of the deck will overlap the ledger and be nailed to it. The joists are 1½" thick, hence the 3" reduction in length so that the finished deck will be the designated length. If possible, the ledger should be one long piece of stock. However, if the length is over 16', it may be hard to find a straight piece that long. In this case, two pieces will work. Always use pieces that are at least 6' long, and preferably longer.

Location on the Wall

It is important that the ledger be securely attached. The lag screws that attach the ledger to the wall need to penetrate something solid, such as wall studs or floor joists. The ledger is often placed at the same level as the floor joists of the first floor, which automatically solves the problem. In this case, the lag screws will penetrate the existing rim joist, behind siding.

If you are installing the ledger at some other level, secure the ledger with lag screws to the wall studs.

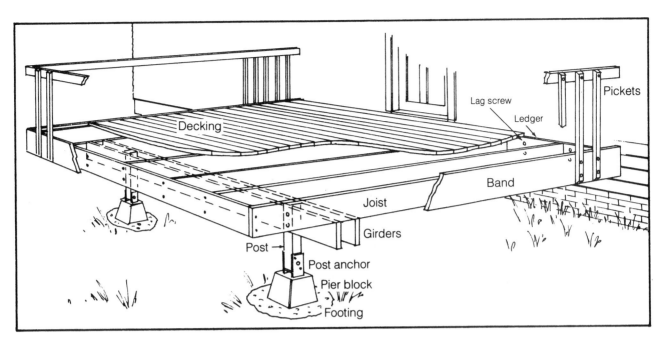

Figure 16-2. Components of a typical deck.

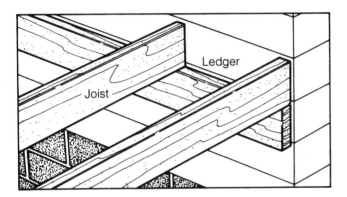

Figure 16-3. An alternate method of installing joists, in which the joists are set on top of the ledger rather than hanging from it.

Choosing the Ledger

Select a straight board for the ledger, one that is the same size as the joists and has little bowing. If the ledger is bowed, the deck will have a corresponding curve upward or downward. Some minor bowing can be forced out as you install the board, but anything greater than about ½" of curve over 12' of board will throw the deck off. The ledger, like all decking materials, should be of redwood, cedar, cypress, or pressure-treated lumber.

At this point you have selected your ledger board and decided on its final placement in regard to both its height and its location on the wall. Now check to see if there are any obstructions, such as hose faucets, dryer vents, gas or water pipes, or electrical wires. Such obstructions may have to be relocated; or perhaps you will have to break the ledger, leaving a gap where the obstruction occurs. Remember that everything below the ledger will be underneath the deck and not easily accessible. You may need to call an electrician or

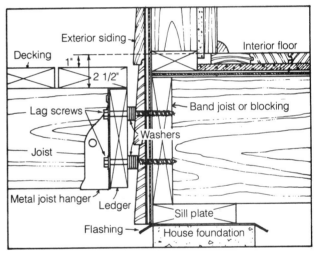

Figure 16-4. The ledger is installed 2½" below the level of the interior finished floor. Be sure it is bolted or screwed into something solid, such as a band joist or blocking.

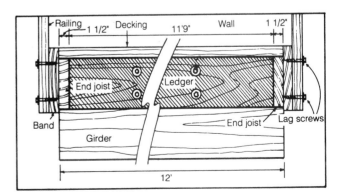

Figure 16-5. The length of the ledger equals the width of the deck less 3" if there is no band, or 6" if a band is used.

a plumber to rearrange wires or pipes. Don't try to work with the electrical or gas system unless you know exactly what you are doing. Mark on the ground the location of any underground pipes or wires before you begin digging the foundation holes so that you can be sure of not disturbing them. If you cannot locate all of the pipes or wires, dig slowly and be prepared to change the location of the foundation piers if necessary.

Step Two
Drilling the Holes for the Ledger

Margin of Error: 1/4"

Most Common Mistakes

☐ Using a badly bowed board.

☐ Not using rustproof lag screws or bolts.

☐ Not installing the ledger at the proper height or location.

☐ Not caulking the bolt holes in the wall before installing the ledger.

☐ Drilling holes where joists will be attached. (If you are bolting the ledger to an existing rim joist, lag screw placement is flexible. If you are bolting the ledger to the wall studs, accurate placement is much more critical. In either case, the lag screw holes and the joist locations should be marked to avoid placing the joists over the lag screws. The joists can be offset if necessary, as long as they remain properly centered, so that the deck planks always break on the center of a joist.)

Now you are ready to install the ledger. First, cut your ledger board to the proper length—the total length of the deck less 3" (6" if band joists are to be used). Next, you will drill holes in the ledger for the lag screws or bolts that will hold the ledger to the house. Drill these holes with a bit that is 1/8" larger than the actual screws so that you will have a little play for adjustments. With the ledger resting on sawhorses, begin at one end and mark the lag screw locations. These holes are usually drilled in pairs, one above the other, every 32", or staggered singly every 16". Drill a pair of holes 12" in from either end, or on the end studs, then drill the rest of the holes. Drill all holes at least 2" in from the edge of the board. (You should check the local code to see if there are any regulations concerning the size and location of these lag screws.)

You are now ready to attach the ledger temporarily to the wall, mark the corresponding holes on the siding, remove the ledger, and drill the holes in the siding. If there is a slight bow, or crown, in the board, remember to turn it up toward the sky.

Place the top edge of the ledger at the mark on the wall that represents the top of the ledger, and nail one end temporarily in place. Then place a 4' or 8' level on the board, get it exactly level, and nail in the other end. The ledger is now temporarily nailed in its proper place. Check once again to be sure it is exactly level before marking the holes. Then, with a felt-tipped

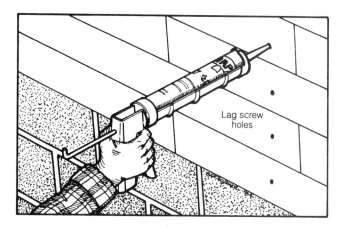

Figure 16-6. Squirt butyl caulk into lag screw holes before attaching the ledger.

pen or sharp pencil, mark on the wall the location of the lag screw holes that you drilled in the ledger. Now remove the ledger.

You are now ready to drill the lag screw holes in the wall. Do not use the same size bit that you used on the ledger. Use a bit that is one size *smaller* than the shank of the lag screw. This will ensure that the lag screw gets a good bite into the wall. Drill all the holes, being sure to hold the drill straight so the lag screws will go in straight. Be sure that you are drilling into solid wood.

Before you attach the ledger permanently to the wall with lag screws, squirt some butyl caulk into the holes in the wall (Figure 16-6). This will keep rainwater from flowing into the structure through the lag screw holes.

Step Three
Attaching the Ledger
Permanently to the Wall

Margin of Error: 1/4"

Most Common Mistakes

☐ Not leaving a space between the wall and the ledger.

☐ Not screwing the ledger into something solid.

☐ Not using rustproof lag screws.

☐ Using galvanized washers with aluminum siding.

☐ Not caulking the holes in the wall before installing the ledger.

There is still one crucial detail you must attend to before you attach the ledger to the wall. If the ledger were attached directly against the wall, with the back surface of the ledger tight against the wall, rainwater running down the wall would get trapped between the wall and the back surface of the ledger and promote decay. To avoid this, you need to leave a small space (1/2" to 3/4") so that water can run down to the ground.

The easiest way to provide this space is to install washers on the lag screws between the ledger and the wall (see Figure

16-4). These washers should be rust-proof HDG. If your wall is aluminum siding, use aluminum washers. Galvanized metal touching aluminum causes corrosion. Stack these washers to leave an adequate gap. If your siding is not flat but has different surface levels (beveled siding, aluminum siding, or shingles), more or fewer washers can be installed on the top screws than on the lower ones, to compensate for any unevenness. You want the ledger to be installed plumb.

After you have squirted caulk into the holes, threaded the lag screws into the ledger, and installed the proper number of washers on each screw, lift the ledger into place, tap the screws into the wall, and tighten the screws with a socket or crescent wrench. Be sure that the screws are biting solidly into the wall, especially the last 2". You are now ready to install the two end joists and to locate your pier holes.

If you are using pressure-treated lumber or a decay-resistant lumber that is not all heart, you should paint the cut ends of the ledger with a water-repellent at this point.

Step Four
Attaching the Two Outer Joists

Margin of Error: 1/4"

Most Common Mistakes

☐ Joists not set at right angle to the wall.

☐ Joists not placed level.

☐ Joists not nailed to cover exposed ends of the ledger.

☐ Crown not pointing skyward.

Once the ledger is permanently attached to the wall, you can install the two outermost joists at either end of the ledger. These joists will be used as reference points for locating the foundation pier holes. The outer joists are set in place, and lines are drawn on them at a prescribed distance from the wall. A string stretched between these lines determines the location of the foundation pier holes. The distance from the house to the supporting girder will depend on the size of your deck, the type of wood being used for the joists, and the size and spacing of the joists.

There are several techniques for locating your pier holes, including one that involves setting up batter boards and layout lines. However, unless your lot slopes steeply, the following method—using the two outer joists as reference points—is the simplest and easiest.

Choose two of your straightest joists to install at either end of the ledger. As you look down the joists you will notice that they have a small bow, or crown. Almost no piece of wood is ever perfectly straight. When the board is installed, this crown should always point up toward the sky (Figure 16-7) to allow gravity to straighten it out. This is called "crowning" the joist. Don't worry about cutting the joists to the proper length at this time; you will cut them after you apply the decking.

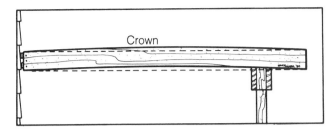

Figure 16-7. The crown of the board should always point up.

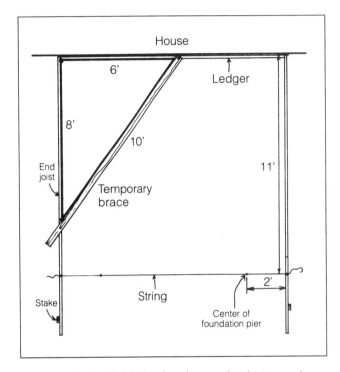

Figure 16-8. Use a 3-4-5 triangle to be sure that the two end joists are exactly at right angles to the ledger.

Nail these outer joists to the ledger with the joists covering up the exposed ends of the ledger. One person nails the joist to the ledger, using three or four 16-penny HDG nails, while another person supports the other end of the joist in a level position. After the end is nailed to the ledger, drive a temporary 2x4 stake into the ground to hold the "floating" end level and at a right angle to the ledger. To position this stake, after nailing the joist to the ledger and while another person is supporting the other end, place a framing square at the intersection of the joist and the ledger. Adjust this intersection to approximately 90 degrees, drive a stake in the ground to support the joist there, place a 4' or 6' level on the joist to be sure it is level, and nail the joist to the stake. At this point none of these measurements need to be too accurate. The next step, using a 3-4-5 triangle, will ensure that these two outside joists are at a true right angle to the ledger (Figure 16-8).

Measure 8' along the end joist on the outside edge and mark that point. Measure along the back of the ledger 6' and mark there. Now measure between these two marks. If that line is exactly 10' long, then you have a true right angle. If it is not, then adjust your stake and joist until the line between the marks is exactly 10′ long. Then nail some temporary braces from the ledger to the joists at an angle to keep the joists in place. Your two outer joists are now level, at true right angles to the ledger, and temporarily supported by stakes.

Step Five
Locating and Digging the Foundation Pier Holes

Margin of Error: 1"

Most Common Mistakes

☐ Locating a pier hole on top of an underground pipe or wire.

☐ Not checking the plans for the exact locations of the holes.

☐ Not digging deep enough for local codes and frost lines.

☐ Not digging until you find stable, undisturbed soil.

The location of the pier block holes is unique to each deck project. Local building codes will have some say here, and design is also a factor. Piers are often inset from the sides of the deck to hide them from view. The girder that is supported by these piers is often inset from the end of the deck a foot or two, because of the visual appeal of a cantilevered or overhanging deck. Since the cantilevered portion is not directly supported by posts, codes specify how far it may extend be-

yond the nearest post. A maximum overhang of one quarter of the total span is common.

At this point, with the two outer joists in place, you have a clear outline of the edges of the deck. From this outline you can locate your foundation pier holes. Check your plans to determine their exact locations. For example, say that your plans call for two holes, the centers of which are exactly 11′ from the wall and 2′ in from the outer edges of the sides of the deck. Measure along each of the two outer joists 11′ from the wall, mark the joists, drive nails at those points, and tie a string from joist to joist between the two nails. Now measure 2′ from the outside edges of the joists along the string and mark the string at those points by tying on a short length of string (Figure 16-9). Make the knot tight so the marker string won't slip. If decorative band joists are to be added over the two side joists just installed (see Figure 16-2), be sure to allow for them in your measurements. The string markers locate the centers of the two pier holes. Now use a plumb bob to transfer these marks to the ground. Drive in two small stakes (Figure 16-10), then mark out for the radius of the pier holes and begin digging.

How to dig a hole may seem obvious, but there are a few things to keep in mind. The size of the hole is important. The diameter of a pier hole is usually about 16", but check your local code. The depth of the hole is often regulated by code as well. Required depths range from 12" to 60", depending on the frost level in your area. (The colder the climate, the deeper the hole needs to be.)

Once you have determined the necessary diameter and depth, simply dig your holes, being sure to dig good, straight

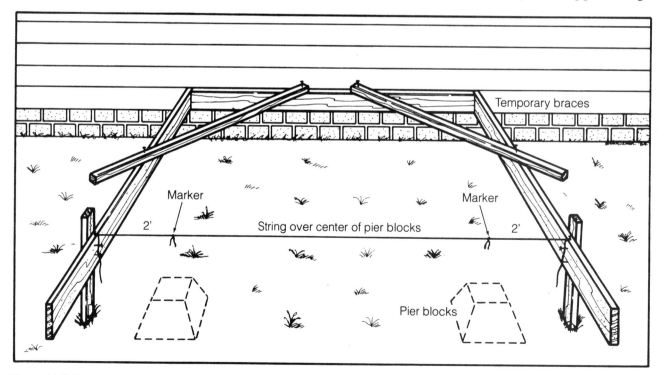

Figure 16-9. The two end joists are installed level and square to the ledger. The string is stretched exactly over the centers of the pier blocks.

(not sloping) walls. Dig until you hit stable, undisturbed soil. Never backfill a hole with loose dirt before pouring concrete. Loose dirt will compact and cause settling.

In many warm, temperate areas, pads are often used instead of piers to support the pier blocks. Piers are used in these regions when the ground slopes, or near a drop-off.

Step Six
Pouring the Footings
and Setting the Pier Blocks

Margin of Error: ½"

Most Common Mistakes

☐ Tops of pier blocks not level.

☐ Pier blocks not properly aligned.

☐ Not using metal connectors where required.

☐ Failing to reinforce piers with steel, if required by code.

☐ Neglecting to have reinforcement inspected before concrete is poured, if required by code.

Now you are ready to mix up some concrete, pour it in the hole, drop a pier block in the fresh concrete, and level and align it.

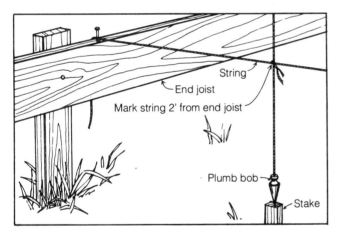

Figure 16-10. Use a plumb bob to locate the center of the pier block holes.

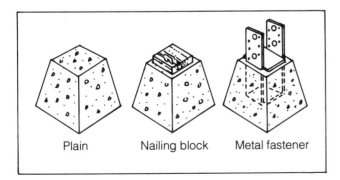

Figure 16-11. Three types of pier blocks. The one with the metal fastener is used for extra strength in earthquake areas.

A few words on mixing concrete are in order here. For small deck jobs, you can buy ready-mixed concrete with all the ingredients in one bag. Simply mix in some water and pour the concrete in the hole. Be sure to buy enough bags to fill all your holes; you don't want to run short and have to rush back to the store to finish off a hole. You will be surprised how many bags it can take to fill a hole. As a rule of thumb, a 60-pound bag of ready-mixed concrete will fill about ½ square foot, or a hole 6"x12"x12".

Mix the concrete with water in a wheelbarrow with a cement hoe, following the instructions on the bag. Pour the concrete in the hole to within an inch or so of the top and smooth it out with a piece of 2x4 or a trowel until it is fairly level. Now you are ready to place the pier blocks.

Pier blocks serve as a transition from the concrete foundation footings to the posts supporting the girder. They can be built at the site, and are also available at building centers and hardware stores in a range of styles and sizes (Figure 16-11). It really isn't worth the hassle of pouring them yourself.

The most common type of pier is a truncated concrete pyramid on top of which the wooden post sits. The weight of the deck keeps the post in contact with the pier block. Another version has a small piece of redwood or pressure-treated lumber embedded in the top so that the post can be toenailed to the block. Metal fasteners are often used in areas where there are earthquakes. They are embedded in the pier blocks and the posts are bolted to them to prevent the posts from shaking off the pier blocks in an earthquake. Other metal anchors are intended to be placed in wet concrete piers. They have corrugated metal tangs that are gripped by the concrete as it sets, and are often required where earthquakes or hurricanes are likely. Be sure that the metal anchors are the correct size for the posts you plan to use.

After the hole has been filled with concrete, and the concrete leveled and smoothed, drop the pier block into the fresh concrete and work it down until at least 3" or 4" of the base

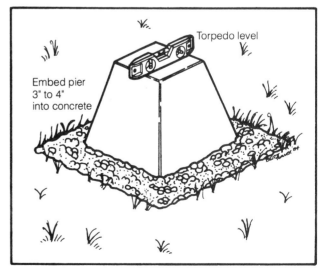

Figure 16-12. Use a small torpedo level to level the pier blocks.

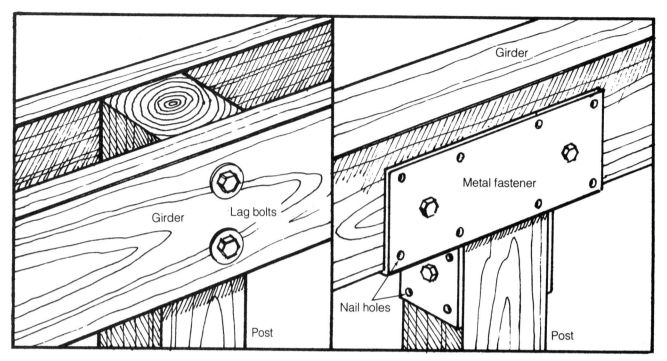

Figure 16-13. Two types of girder/post attachments. Left, 2x6 girders are bolted to either side of the post. Right, 4x6 girder rests on top of the post and is attached with metal fasteners.

of the block is embedded. As you set the blocks, be sure of two things: that they are properly aligned and that they are level. Some regions require up to 12" clearance from the post bottoms to grade, to avoid pest and decay problems. In this case, heavy paper tubes are placed on the top of the pier or pad. These tubes are filled with concrete, and the metal anchors worked down into the top while the concrete is fresh.

To check alignment, simply drop the plumb bob from the mark on your string to be sure the tip of the plumb bob is in the center of the pier block. To check level, place a small torpedo level in both directions as well as diagonally across the top of the pier blocks (Figure 16-12). Tap and move the blocks around until the tops are level. After your pier blocks or metal anchors are properly set, allow the concrete to harden—this can take anywhere from 8 to 24 hours. For a strong footing, keep the concrete damp while it is hardening. You can do this by sprinkling water on the pour as it dries, or by laying wet cloths across the top.

Step Seven
Installing the Girders

Margin of Error: 1/4"

Most Common Mistakes

☐ Using badly bowed girder stock.

☐ Not installing the girder level.

Now you are ready to install your girders. You can do this in one of two ways. One way is to prebuild the girder/post sys-

tem, bolt or nail the entire thing together, and then move it into place as one large piece. The other way is to build it in place piece by piece. If the posts and girders are not too long and heavy, the first procedure is recommended.

There are several different girder/post variations. As shown in Figure 16-13, two girders can sandwich the posts and be bolted or nailed to them; or the girder can rest on top of the posts, attached with metal fasteners. As long as it passes your local code, either system will work. The former method, which uses two girders, one on either side of the post, is the method described in our example.

To build your girder/post system, first cut the wooden posts to the proper height. In our example, the height of the posts is the same as the level of the bottom of the joists. Note that the bottoms of the joists rest on top of the girders and that the tops of the girders are at the same height as the tops of the posts.

To determine this post height, go back to your two outer joists and check to be sure they are still exactly level. Now move the string that is on top of the joists so that it is connected on the bottoms of the joists. The level of the bottoms of the joists is the same level as the top of the posts you are about to cut. Now measure from the top of each pier block (or metal fastener) to the string, as shown in Figure 16-14, and cut the posts to correspond to each of these measurements. It's that simple; just be sure you are accurate and that you make good straight cuts so the posts will sit smoothly on the pier blocks. To make these cuts, mark around the circumference of the post, cut one side, then rotate the post and cut the side opposite the first cut.

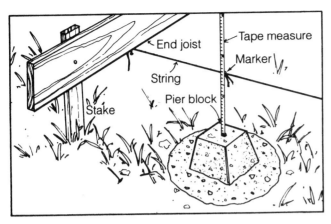

Figure 16-14. To determine the height of the post, measure from top of pier to top of post, as indicated by the strings attached to the bottoms of the end joists.

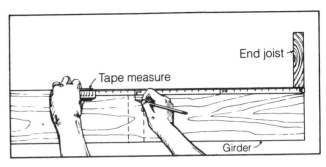

Figure 16-15. Hook tape to nail at center of first joist and mark ledger every 24".

After the posts are cut, cut your pieces of girder stock to the specified length—the length of the ledger plus 3". If possible, use one continuous piece of girder stock for each piece; but if the deck is too long, and you have to use two pieces, be sure that the pieces meet at a post so they can both be attached to it. If the girder is made of pieces, the proper cantilever must be maintained. At least three posts must be used to piece in a girder. Your girder stock must be completely straight. Select these pieces yourself at the lumber yard. The deck joists sit on top of the girders, so if they are bowed, the entire deck will rise or fall. If there is a slight bow, or crown (no greater than ½" in 12'), point it up toward the sky and it will settle down in time.

Once all your posts and girders are cut, you can assemble them into one unit, if the girder is to be bolted or nailed to either side of the posts rather than resting on top of the posts. This is usually done with bolts, nails, or nails and metal fasteners. For most decks, bolts are the recommended method of assembly. If you are resting a single girder on top of the posts, use metal fasteners as shown in Figure 16-13. Remember to use HDG fasteners, bolts, or nails so they won't rust. If you are bolting the deck together, drill your holes good and straight so the bolts will go through straight. Use a bit that is ⅛" larger than the bolt, to allow for making final adjustments.

Once the entire unit is assembled, move it into place under your two outer joists. If you are assembling the pieces in place, the application is pretty much the same. Before toenailing the outer joists to the top of the girder, measure out from the wall to be sure the girder is the proper distance from the wall, and check the posts for plumb. Once you have ascertained this, check once again to be sure that the joists are still at right angles to the ledger and still level. Then nail the joists into the girders with 16-penny galvanized nails; or use L-shaped metal fasteners, called framing clips, and nails.

Step Eight
Laying Out the Joists
on the Girder

Margin of Error: ¼"

Most Common Mistakes

☐ Nailing the joist hanger on the wrong side of the mark.

☐ Improper layout.

Now that your girder is in place, you are ready to install your remaining joists. To do this you need to make marks, called layout marks, on the girder to show where each joist will be located. These marks reflect the ledger joist layout marks, and are easily transferred.

Joists are usually located so that their centers occur exactly every 16" or 24". This is called their "on center" distance. The distance between joists depends on several important factors:

1. Size of the joists

2. Length of span from ledger to girder

3. Type of wood used for the joists

4. Any heavy loading, such as planter boxes or snow

The size and spacing of the joists are among the few crucial dimensions specified in deck building. If you undersize or overspace the joists, the deck could collapse, and the building inspector will stop the project. You should talk to a building code officer about the proper sizing for your joists and span. Joists placed 24" on center are usually adequate.

After you have determined the joist spacing layout on the girder, make a second mark ¾" to one side of the center line, so that you will know on which side of this line the joists will sit (Figure 16-15).

After you have made all your marks, begin nailing the metal joist hangers on the ledger, using the special stubby joist hanger nails that are provided with the hangers. Nail only on one side of the hanger, allowing the other side to float free until the joist is inserted. Nail the hangers so that the *interior* edge is nailed along the line you drew to mark the side of the joist, as shown in Figure 16-16. As you go along, be

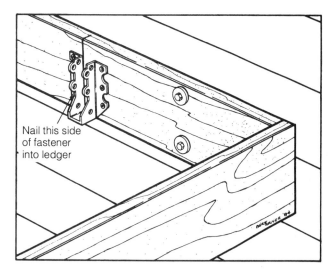

Figure 16-16. Nail one side of hanger to the ledger. Align the edge of the joist hanger with the line that marks the side of the joist.

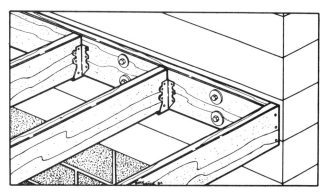

Figure 16-17. Joists are installed and nailed to joist hangers.

sure that you are clear about which side of the mark the joist hanger and joist should be nailed to, to ensure proper spacing. Even a seasoned professional can accidentally put the joist hanger on the wrong side of the mark.

Step Nine
Installing the Remaining Joists

Margin of Error: ¼"

Most Common Mistakes

☐ Not crowning the joists.

☐ Placing a joist on the wrong side of the layout mark.

Installing the remaining joists is fairly simple. Some builders cut the joists to length before installing them. However, it is generally better to cut the joists after almost all of the decking boards are in place. The reason for this will become clear later on. Now, simply insert the joists into the joist hangers, as shown in Figure 16-17. As you do this, be sure that you crown each joist, pointing the bow skyward. Insert the joist in the hanger, then nail the loose side of the hanger

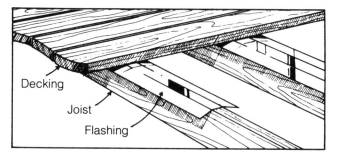

Figure 16-18. Galvanized metal flashing is sometimes installed over the joists.

tightly into the joist and into the ledger. Then use two 16-penny HDG nails to toenail the other end of the joist to the girder. Special HDG sheet metal fasteners can be used here instead of just nails. These fasteners are especially recommended in earthquake and hurricane areas.

Step Ten
Applying the Deck Boards

Margin of Error: ¼"

Most Common Mistakes

☐ Not leaving a gap between the deck boards.

☐ Not using enough nails.

☐ Not forcing the bow out of crooked boards.

You are now ready to start installing your deck boards. This part of the job is quite satisfying because it goes so quickly and easily and you really start to see your deck coming together. You need to consider individually each board you apply, since they will all be seen. Examine each piece and put the most attractive pieces in the high-visibility areas. Also look at each side of each board to see which side you want exposed. Check to see how badly the boards are bowed. Reject any very bad pieces, because they will look crooked once the decking is down.

Some builders, especially in areas of heavy rain or snow, place aluminum or galvanized metal flashing on top of each joist before the deck boards are applied, as shown in Figure 16-18. This flashing helps keep water from getting trapped between the decking and the joists, causing rot. You will need to find out whether this is a good idea in your area.

Begin applying your decking from the wall and work toward the yard. Be sure that the first course you apply next to the wall is made of good, straight pieces, because this course is used as a guide, and if it is crooked it affects all the other courses. Also be sure to leave a gap between the first course and the wall, so that water can drain down the wall.

If possible, purchase boards that are long enough to span the entire width of the deck. If the deck is too wide, this may not be possible. Lengths over 14' are often very crooked. If two pieces are needed, the pieces must always join directly over the center of a joist, to provide a bearing surface for each

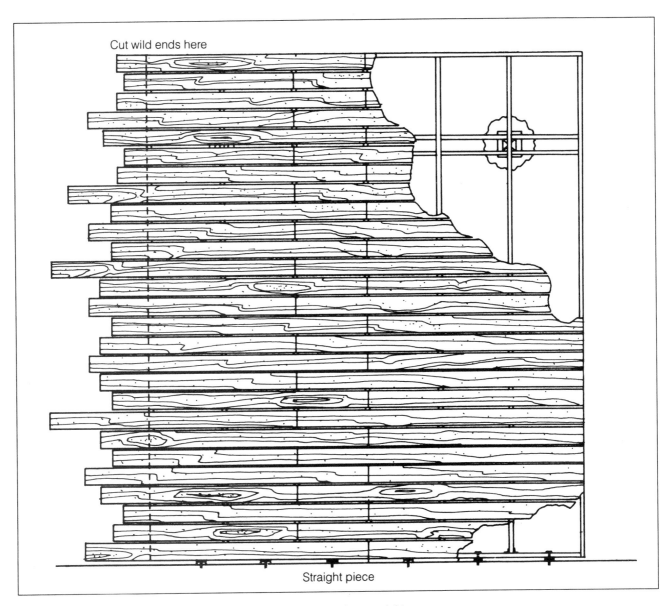

Cut wild ends here

Straight piece

Figure 16-19. Stagger the joists so that every other course joins over the same joists.

end. Don't join all the courses over the same joist; the joints will look like a big suture running down the deck. Stagger the joints so that every other course joins over the same joists, as shown in Figure 16-19, or place them randomly.

Use double-dipped HDG 16-penny nails rather than finishing nails in deck construction. Stainless steel or aluminum nails work even better because they do not stain the wood, but they may be hard to find. Do not use electroplated nails, which often rust. Except with redwood, it is best to use two nails in 2x4 boards and three nails in 2x6 boards, at each point where the board crosses a joist. For 2x4 and 2x6 redwood boards, two nails per joist are enough. All nails should penetrate 1 ½" into the joists.

You might also consider using reduced-corrosion deck screws. Although they are expensive compared to nails, they greatly reduce cupping as the deck ages.

Use a combination square and mark a true perpendicular line across each joist so that you can place all your nails in a straight line. This takes a little more time, but your finished deck will have a much nicer appearance.

The process of nailing on the deck boards is straightforward, but there are a few things to remember. First, you need to leave a gap of ¼" between each course of deck boards to allow water to drain off the deck. To maintain this gap, stick a flat carpenter's pencil between the courses as they are applied (Figure 16-20).

No board is perfect, and many will have bows that need to be pulled out. This is done by forcing the bow out as you nail from one end to the other. This is why you should never nail from both ends to the middle—you may trap the bow. Put in your spacer to create the gap and then force the board into place. It will usually straighten out if it is not too badly bowed. Use a smooth-headed hammer and try not to mar the wood (although the first few rains will probably draw out most of the dents). You can also use a pry bar, as shown in Figure 16-21, to force the board straight.

It is easiest to place the deck boards flush with the outside edge of one of the outer joists, let them "run wild" at the other end, and then cut this end all at once. The only other trick to applying the decking is to drill pilot holes when nailing near the end of a board, especially where two deck boards join together in the center of a joist and both ends must be nailed to the joist. If you try to nail that close to the end of the board, you will often split the wood. To prevent this splitting, drill a pilot hole that is slightly smaller than the shank of the nail, and then drive the nail into the pilot hole (Figure 16-22).

Begin to lay the deck boards from the house end of the deck. Measure out from the wall every few courses to be sure that all boards are equidistant from the wall as you progress. Make adjustments gradually and continue until you are one course away from the end of the joists. (See Step Eleven for what to do then.) Stand up and look down on the deck to be sure you are not trapping any bows in the boards and that the deck looks good in general.

Step Eleven
Cutting the Ends of the
Joists and Deck Boards

Margin of Error: 1/4"

Most Common Mistake

☐ Not cutting square ends.

As you apply the last deck boards, there is one final adjustment to be made. You are aiming for the outer edge of the last course of decking boards to be exactly flush with the ends of

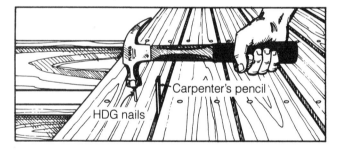

Figure 16-20. Leave a gap between deck boards for drainage.

Figure 16-21. Use a pry bar to remove bow in deck boards.

the joists. Some builders don't bother with this detail, but it gives a nice finished look. The only way to make this work out is to cut the joists after the next-to-last course but before the last course is nailed on. This way, you can tell exactly how things are going to work out. After the next-to-the-last course is in place, measure to the ends of the joists for their final cut (Figure 16-23). The measurement is made from the outer edge of the next-to-the-last course to a point on the joists that equals the width of a deck board plus the gap between courses. For example, a 5 1/2" board with a 1/4" gap would equal 5 3/4". To place the band joists under the last piece of decking, deduct 1 1/2".

Mark the ends of the joists at this point and draw a line across them with a combination square. When you cut the joists at this point, be sure to make good, straight cuts so the band will fit on properly. After you have cut the ends of all the joists, install your last course of deck boards. Be sure this is a good, straight course, because it will be more visible than the others. After this last course is in place you will install the band joist (the piece that goes across the ends of all the joists). This band joist is often one size wider than the joist, to act as a curb for the decking.

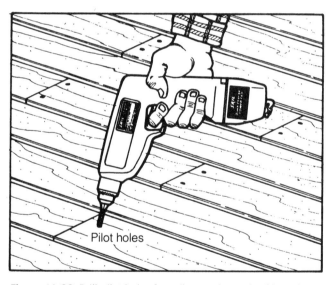

Figure 16-22. Drill pilot holes for nails near the ends of boards to prevent splitting the boards.

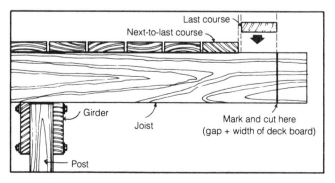

Figure 16-23. Cut joists to align the last deck board with the end of the joists. This method assumes that you are using a band joist.

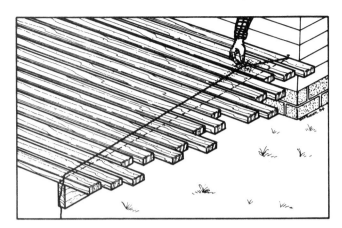

Figure 16-24. Snap a chalkline to show where to cut the wild ends of the decking boards.

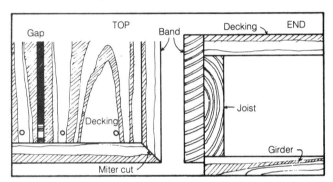

Figure 16-25. Facing the deck with band joists.

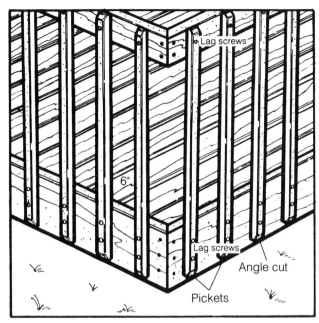

Figure 16-26. A simple, common railing design.

Now you are ready to cut the "wild ends" of the deck boards. To do this, make a mark at either end of the deck and snap a chalkline to get a perfectly straight cut. Cut off the ends of the decking boards to this line (Figure 16-24). Be sure your marks are where you want them to be, allowing for any cantilever over the joists. Set your saw to cut through the deck boards plus ¼". Stop cutting every now and then and check to be sure you are on the line and not veering into or away from the joists.

It's a good idea to paint the exposed ends of the joists and deck boards with a water-repellent material for added protection.

Step Twelve
Facing the Deck
with Band Joists

Margin of Error: ¼"

Most Common Mistake

☐ Not measuring correctly to allow for mitered ends.

After you have cut the ends of the joists and the decking boards, you are ready to apply the band joists (Figure 16-25). Sometimes the band consists only of a piece at the end of the joists where you just made your final cut. Usually, however,

the band covers all outer edges of the deck and is one size (2") wider than the joists so that it acts as a curb for the decking (refer to Figure 16-2).

The ends of the individual band pieces can be miter cut so that there is no exposed end grain. Mitering is not difficult, but it requires accurate measuring and cutting. When measuring, remember that the measurements taken off the deck refer to the inside cut of the miter, not the outside. You may want to practice on some scrap pieces to be sure you have set the correct angle on the saw. Nail the bands to the joists with HDG 10-penny nails.

Step Thirteen
Installing the Railings

Margin of Error: ¼"

Most Common Mistakes

☐ Pickets not plumb.

☐ Top railing not level.

☐ Pickets not evenly spaced.

☐ Pickets too far apart.

There are many different styles of railings for decks. Choose a style that will fit your budget, time, and energy, as well as the overall look and use of the deck. Different styles do different things. You can incorporate planters, benches, tables, and stairs into your design. A wide, flat board installed at the top of the railing is useful for holding flower pots, paper plates, drinks, and other amenities of outdoor living. A book on deck design or a ride around the neighborhood will inspire your imagination.

This section describes a simple, sturdy, widely used railing design (Figure 16-26). Most of the procedures outlined here will apply to almost any design. If you are copying a design from an existing deck, a close inspection of the deck with sketchbook in hand should enable you to understand how the railing is put together. The main thing you want to be sure of when installing a railing is its stability. There is nothing worse than building the entire deck railing, only to discover that it is weak and unstable. You don't want to be worried about a guest taking an unexpected plunge onto your lawn.

Most areas have codes that apply to deck railings. If the deck is more than a certain distance from the ground, often 18", a railing may be required. If the deck is closer to the ground, the railing may be optional. Also, the code will allow only a certain gap between the pickets, usually 6" to 8", so that a child cannot slip through. The height of the railing is regulated, too—usually 36" to 42".

The simple railing shown here is constructed of 2x2 pickets with a 2x4 top railing. Stability is provided by using two HDG lag screws to bolt the bottom of the pickets to the band joists, and by tying the railing to the house with lag screws. Remember to leave a gap between the railing and the house as you did with the ledger.

First, cut all the pickets. An angle cut on the top and bottom of each picket is an esthetically pleasing touch. After cutting the pickets, cut the 2x4 top rail. If possible, use pieces that are long enough to span the entire length of each section. If the span is too long and two pieces are needed, you must join them over a 2x4 picket. Determine where this break should occur for the most balanced appearance. Miter cut the top rail at corners so that no end grain is exposed.

After all your pieces are cut, it is best to screw or bolt the pickets to the top rail and then install the entire assembly onto the deck. Mark the top rail so that the pickets will be properly spaced. Then secure the pickets to the rail using screws or lag screws that are as long as possible without poking through the back side of the railing (about 2 ½", or 4" if you have band joists). Be sure that your marks are correct and that the pickets are attached to the proper side of the marks. A misplaced picket will show all too clearly. At the bottom of each picket, drill two holes that are one size smaller than the lag screws that will hold the pickets to the joists and band. Drill these holes so that they are as far apart as possible and yet at least 1 ½" from the edge of the band and joists.

You are now ready to install the entire assembly onto the deck. This is done by using two more lag screws at the base where the picket meets the joists and band. Use 2 ½" or longer lag screws, and be sure they are HDG or treated to minimize rust. Align the picket/railing assembly so that it is in its exact location and nail a few pickets temporarily in place to hold it there. Then use a level to locate where each picket will meet the band and joists. By using the level at each picket, you can be sure that they are all plumb and parallel to each other. It is imperative that the assembly be in its exact location before you start plumbing the pickets. If you move the assembly, even slightly, after the pickets are attached at the bottom, the entire assembly will look askew and will need to be redone. Once you have ascertained that the assembly is in its exact location, and have plumbed your pickets, hold the picket in place while you mark the location of the holes on the band and joists. Then drill these holes on the band and joists, using a drill bit one size smaller than the shank of the lag screw.

With deck or lag screws, secure the pickets to the band joist. Do this around the entire deck until all the railings are in place. Where two railings intersect at their miter-cut railing top, drill pilot holes through one top railing into the other, and then nail them together with 16-penny HDG finishing nails, two from one direction and two from the other. Your railings are now complete.

Step Fourteen
Waterproofing the Deck

Margin of Error: Not applicable

Most Common Mistake

☐ Not sealing the deck.

If you live where freeze damage is likely, or if you want to maintain the original color of the decking, it's a good idea to treat the decking with water repellent. In temperate areas, decks are often left untreated so they can "breathe." The water repellent can be applied with a brush, roller, or spray. It goes on quite easily because it is thin. Unlike paint, it penetrates quickly. A water repellent with a mildewcide will help redwood keep its red color and not turn gray with age. (As mentioned earlier, the California Redwood Association recommends applying the sealer to all pieces before construction, including the edges, ends, and bottoms.) The sealer can serve as a base coat for other finishes. Apply another coat of the sealer to a redwood deck after it is completed, and reapply every 12 to 18 months to prevent darkening and to preserve the beautiful redwood color.

Unsealed wood is vulnerable to staining and decay, and may have to be replaced prematurely. Because of its enduring qualities, this is not as much of a problem with redwood as it is with other woods.

Color or a bleached effect can be added to redwood and other woods by using decking stains and bleaches. Use a lightly pigmented, oil-based decking stain to show the wood grain or a heavily pigmented, heavy-bodied stain if you prefer an opaque effect. Whether the wood is smooth or rough sawn, use a brush rather than a spray for application. "Shake and shingle" paints, sprays, varnishes, and lacquers are not recommended for decks.

Bleaching agents can be applied to give the decking a silvery, weathered look. If no finish is applied to redwood, it will initially darken and then weather to a driftwood gray.

17 Stairs

BUILDING STAIRS is among the most difficult and intimidating of all carpentry tasks. People expect stairs to be of certain proportions, and the codes specify many details so that these expectations will be met. Aside from understanding the code requirements and the variables that are unique to your site, some simple math is required to design a staircase.

As with electrical or plumbing work, tasks that at first seem complicated are relatively simple if you take them one step at a time, so to speak. And as with electrical work, an appropriate design with detailed drawings is an important starting point, because the design determines the materials you will need to order.

This chapter describes the design and construction of a simple, straight, exterior deck stair. Interior stairs are also discussed, but all stairs are an extension of the simple basic stair.

All stairs must be constructed for strength, and exterior stairs must use rust resistant-fasteners and be made to drain water. Handrails are required on all stairs over 18" high.

Ramps are an option that should be considered. All ramps must conform to the specifications developed for wheelchair access, as outlined at the end of this chapter.

Before You Begin

SAFETY

☐ Wear safety glasses or goggles whenever you are using power tools, especially if you wear contact lenses.

☐ Wear ear protectors when using power tools. Some tools operate at noise levels that can damage hearing.

☐ Be careful not to let loose hair and clothing get caught in tools. Roll up your sleeves and remove jewelry.

☐ Keep blades sharp. A dull blade requires excessive force and can easily slip.

☐ Always use the right tool for the job.

☐ Don't try to drill, shape, or saw anything that isn't firmly secured.

☐ Don't work with tools when you are tired. That's when most accidents occur.

☐ Read the owner's manual for all tools and understand their proper usage.

☐ Keep tools out of reach of small children.

☐ Unplug all power tools when changing settings or parts.

WHAT YOU WILL NEED

Time. The time it will take you to build a staircase depends on your familiarity with the construction methods employed; the requirements of your particular site; and the complexity of your design.

Tools

Framing square

Hammer

Handsaw

Level

Pencils

Power saw

Tape measure

Saw horses

Stair nuts

Screwdrivers

Wrenches

Dust masks

Goggles

Materials. The example illustrated in this chapter, a staircase off a redwood-covered deck, uses 2x6 redwood decking for the treads and 4x4 posts to support the handrail. The diagonal pieces that support the treads, called *stringers*, should be pressure-treated 2x12s or other structurally rated 2x12 material.

Many choices are available for the handrail and spindles; this chapter outlines some code requirements.

You can use either nails or screws; just be sure that they are galvanized or otherwise rust resistant. Joist hangers, framing clips, and other metal framing aids can all simplify the construction and make it stronger.

The amount and kind of materials required—wood, metal, and concrete for the pad—all depend on the design of the stairs.

USEFUL TERMS

Figure 17-1 shows the anatomy of a typical exterior staircase.

A **cleat** is a board attached to another board, flat side to flat side. Once used to support the treads, cleats tend to work their fasteners loose and are no longer used for this purpose. Two kinds of cleats are still used. One type is attached to

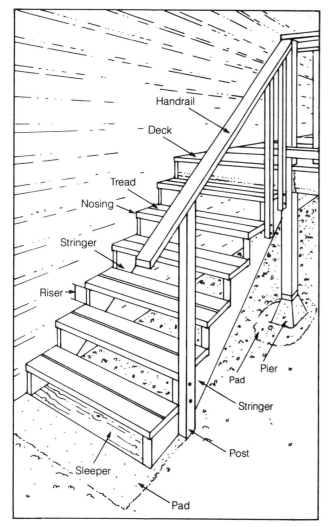

Figure 17-1. Anatomy of a typical exterior staircase attached to a deck.

the stringer to stiffen it, and the other type is attached to the upstairs floor framing to support the top end of the stringer.

A **handrail** is a narrow rail beside the stairs to hold onto as you go up and down. If your stairs are not enclosed on one side, two handrails are necessary.

A **landing** is a flat spot in the staircase where a turn is made.

The **nosing** is the portion of a stair tread that extends beyond the face of the riser.

The **riser** is the height from step to step; or the board itself, installed vertically, that fills this space.

The **run** is the width of a tread, or the cut made on the stringer to support the tread.

A **sleeper** is much like a cleat, but it is used at the bottom of the stringer instead of the top. The sleeper is most often

partly embedded in concrete, so it must be of decay-resistant material.

A **spindle** or **baluster** is an upright support for the handrail.

Stair nuts are small clamps that are attached to the outside edge of a framing square to define the rise and run on a stringer.

Stringers are the diagonal beams that support both treads and risers. Builders call stringers by different names in different parts of the country. A carpenter who mentions a "carriage," "rough string," or "rough horse" is probably talking about a stringer.

A **tread** is the flat portion of the stair that you step on.

PERMITS AND CODES

The building codes address stairs with an eye to uniformity, so you don't have to "relearn" each stair you use. Most building codes require a minimum clear width of 36", wide enough so that two people can pass one another easily. However, the fire code requires a clear width of 44" for interior or exterior stairs used as an escape route from the building.

With this one exception, the building codes prevail in determining minimum standards. Here are some basic rules that most communities use.

1. If the total height from grade to the decking surface is 12", a stair is required.

2. The rise from stair to stair must be no less than 4" and no more than 8". This dimension may not vary more than 3/8" between any two steps in one staircase.

3. Tread depth must be between 9" and 12". Some communities allow deeper treads in some exterior applications.

4. Stairs over 40" wide require a third stringer, centered between the two outboard stringers.

5. The bottom step in an exterior location is often required to be made of cast concrete, to resist pest and fungus attack.

6. Stairways over 44" wide must have two handrails.

7. The handrail must qualify as "graspable."

8. Some communities require nonskid surfaces on exterior stairs, especially where icing is common. Some areas require supports even for short stringers if additional snow loads are expected.

In addition, fire codes address the fire resistance of exterior stairs that may be used as an escape route, particularly in high-density urban areas.

All of these requirements affect design to some extent. You need to be familiar with the code in your community to make sure that your design complies.

BUILDING AN EXTERIOR STAIRCASE

Step One
Designing the Staircase

Margin of Error: ⅛"

The staircase discussed here is a straight exterior staircase from the ground to a deck, using the same materials that were used to frame and cover the deck itself.

The first thing you need to determine is the total height the stairs will climb. Outdoors, this is often affected by the pitch of the ground; the total height from the deck to the lowest portion of the bottom step must be determined.

Figure 17-2 shows how the basic elements of the staircase are put together. Before you begin construction, you should make a similar sketch of your own project. If the deck is 44" above grade, but the soil drops about 4" where the bottom step will be, the approximate total rise is 44" + 4", or 48". The rise of each step should be about 7". Divide 48" by 7", for a rise of about 6⅞"—close enough. This tells you that you need seven risers, each 6⅞" high. The decking surface itself is the top tread, and therefore you will have one fewer treads than risers, or six treads. On stairs that protrude from the top surface, there are an equal number of risers and treads.

In this example, it would also be possible to divide the total rise by 6" and design a staircase with eight 6" risers and seven treads. If your local code is flexible, and you have the space, keep in mind that deeper treads and lower risers are more negotiable and easier to use, and therefore safer. If your space is limited, just remember that the smaller the riser height, the more total run length the treads will need.

In this example, the treads are 2x6 redwood boards, just like the decking. As shown in Figure 17-3, swo flat 2x6s, plus a gap of ¼" for drainage, give a full tread width of 11¼" (5½" + 5½" + ¼"). Code requires the tread to extend 1" beyond the face of the riser, or in this case the riser cut. Because this is a deck staircase, risers won't be used to fill in the vertical spaces. Therefore the tread width cut needs to be 11¼" minus 1", or 10¼".

Step Two
Calculating the Materials List

Margin of Error: Not applicable

Once you have determined the number of treads in your staircase, and the desired stair width, it is easy to calculate the amount of 2x6 you will need. Making a sketch like the one in Figure 17-2 will help you make up your materials list. In this example, the stairs are 48" wide, and each tread uses

two 2x6's. Therefore each step requires 8' of 2x6 from stringer to stringer.

Stringers are ordered at least 2' longer than the diagonally spanned distance, so that you will have enough material to cut the piece on the ends to accommodate the top vertical face and the bottom base. Stringers are always made from structurally rated 2x12 or larger boards. It's a good idea to hand pick them for straightness. If pressure-treated material is available, it is a good choice for exterior stringers.

The handrails and spindles should reflect the overall design of your deck. The railing posts are 4x4, just like the deck rail supports. The railing height is typically 36", measured from the front of the tread.

The bottoms of the stringers are usually attached to a 2x4 decay-resistant board, called a *sleeper*, which is set flush into a reinforced concrete pad slightly above grade level. (As

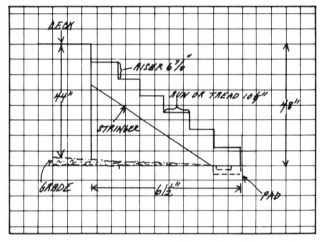

Figure 17-2. The basic elements and dimensions of an exterior staircase.

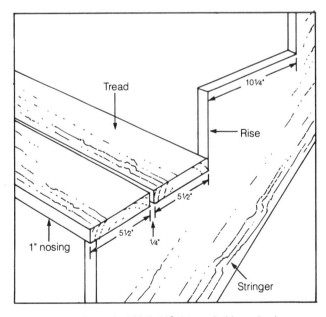

Figure 17-3. Full tread width is 11¼" (two 2x6 boards plus a gap of ¼" so that water can drain off the step).

mentioned earlier, some communities require the whole of the bottom step to be cast concrete.) The pad or step should be the same width as the stair treads, or slightly wider. The sleeper should be positioned to support the heel of the stringers. Be sure to order enough steel and concrete, and the necessary form boards and stakes.

"Welded wire mesh," small steel wires welded together in a grid pattern, is often used to reinforce the pad. The mesh is cut 4" or 5" smaller than the form, in both dimensions, so that the edges are covered by a minimum of 2" of concrete. The mesh is held off the bottom of the form by small precast concrete blocks about 2" high, called "dobes" (pronounced "dough-bee," a contraction of "adobe").

Full-cast concrete bottom steps usually require two or more pieces of mesh. Larger projects usually also need some small rebar for durability.

Step Three
Building the Support Pad

Margin of Error: 1/4"

Use two long boards, cantilevered off the decking and squared to the deck edge, to mark the outside edges of the stringers (Figure 17-4). Now mark off the distance out to the bottom riser face cut. In our example, six treads, each 10 1/4" deep, means a total run of 61 1/2". Measure out 61 1/2" and make a mark. Use a plumb bob or level to drop this line to grade. This

is the front edge of the support pad. Remember that the tread will overhang the cut by 1".

Measure back from the front edge the width of one tread plus whatever is needed to support the heels of the stringers. Usually another 8" of pad width is sufficient, but this depends on the angle at which the stringer is installed. Marking the stringer itself is the easiest way to determine the actual pad width, and to keep its sizes and visual impact to a minimum.

Once you have all the dimensions of the pad, excavate or form this area as necessary to provide a concrete pad at least 4" thick. Install welded wire mesh reinforcement and pour the concrete. Then add the sleeper.

The sleeper is placed 2" or 3" in from the back edge and is usually the full length of the pad. When your form is filled within 1" of the top, place the sleeper into position and then finish filling the form with the concrete. The top of the sleeper should stick up about 1/8" to 1/4", so that the stringer does not rest directly on the concrete. You may have to pound the sleeper down or backfill beneath it to adjust its height. Besides holding the stringer slightly off the pad, all the sleeper does is provide a place to secure the stringers to the pad, so exact placement is not critical.

Step Four
Cutting the First Stringer

Margin of Error: 1/8"

Place the first stringer board, in our example a pressure-treated 2x12, flat across some sawhorses, crown facing up. Clamp a board onto your framing square, as shown in Figure 17-5, or use stair nuts, as shown in Figure 17-6, to set the rise and run. (In our example, that means a 6 7/8" rise and a 10 1/4" run.) Starting at the top, mark the rise and run of the first step on the stringer in pencil. Now move the square down so the first run and second riser marks intersect on the stringer's edge, as shown in Figure 17-7.

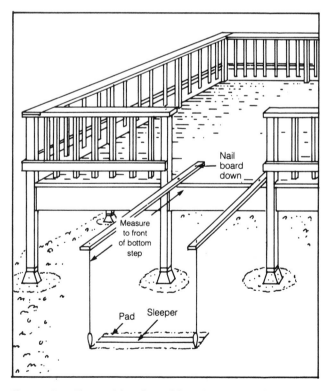

Figure 17-4. The outside edges of the stringer are marked with two long boards cantilevered off the deck.

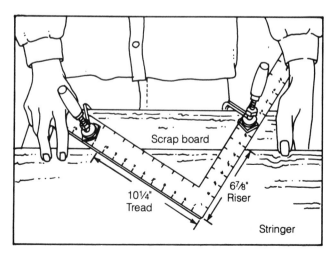

Figure 17-5. Use a square clamped onto a scrap board to mark the rise and run on the stringer.

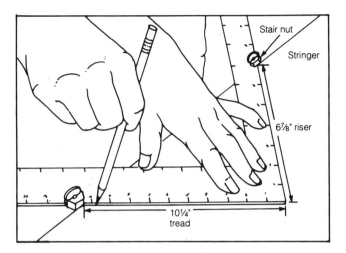

Figure 17-6. You can also use stair nuts to set the rise and run on the stringer.

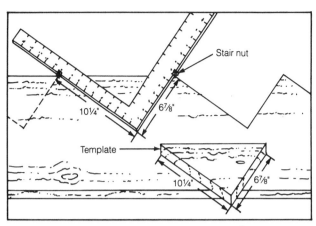

Figure 17-7. After marking the first rise and run, move the square down the stringer and mark the next rise and run.

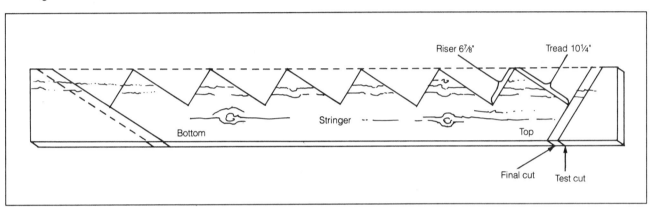

Figure 17-8. After you have marked all the steps on the stringer, mark lines at the top and bottom for test trim cuts.

Repeat this process until you have the correct number of tread cuts and a riser mark on the bottom of the stringer. Using the same tape or nut location, put the square on the other edge of the stringer and mark a continuation of the top riser cut line and an inverted tread cut at the bottom of the last riser. Once all treads and risers are marked on the stringer, make another mark 1" outside of the top riser cut line and 1" below the line at the bottom, as shown in Figure 17-8. Use a power saw to cut the stringer on the outside line.

Now place this trimmed but uncut stringer in position temporarily. The tread and riser marks should be level and plumb respectively, but displaced 1" up and 1" out. If these marks are not level and plumb, you have made a mistake that must be corrected. Double check your calculations and review your drawings for conceptual mistakes. Also check the marks themselves; a ⅛" error multiplied by seven risers makes a ⅞" error over all.

Assuming that the riser and tread marks are OK, and that the top plumb mark and bottom horizontal marks will meet their supports closely, it's time to cut the riser and tread notches and to make the final top and bottom cuts. Never overcut the notches with a power saw. Instead, you should cut to the intersecting corner of the marks and finish the cut

with a handsaw, as shown in Figure 17-9, to retain as much strength as possible in the stringer.

After you have cut the first stringer, you need to check its fit before cutting the others. To do this, mark one riser height down from the top of the decking, on the fascia or framing. Hold the top tread cut of the stringer even with this line; the bottom cut should rest securely on the pad. Mark the location of each stringer and test fit this first stringer at each spot. You may have to adjust the bottom cuts to accommodate any unevenness in the pad or stringer.

Step Five
Dropping the Stringer

Margin of Error: ⅛"

So far, we have been "pretending" that we will walk on the tread cuts themselves—but the thickness of the tread will be added to the bottom riser height, as shown in Figure 17-10, and the riser height at the top will be reduced by the same amount (in our case, about 1½").

In practice, however, instead of reducing each tread cut by this amount, the whole stringer is lowered by the thickness

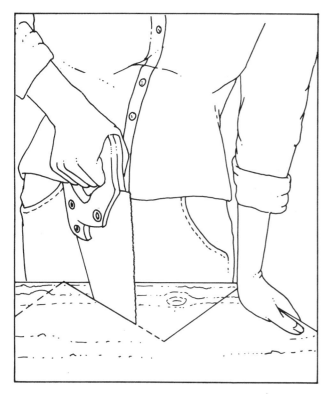

Figure 17-9. When cutting the riser and tread notches, use a handsaw rather than a power saw to finish the cut.

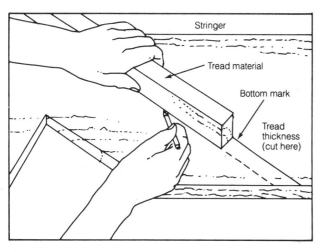

Figure 17-10. For greatest accuracy, use a scrap of the tread material to mark the bottom of the stringer.

of the tread—in our example, 1½". (This is called "dropping the stringer.") After 1½" is trimmed off the bottom of the stringer, the bottom riser height is 5⅜" (6⅞" - 1½" = 5⅜"). You may simply mark the bottom of the stringer with a tape measure, but using a scrap of the actual tread material is more reliable (see Figure 17-10). Since the tread thickness will vary with different materials, using a scrap gives the appropriate distance in each specific case.

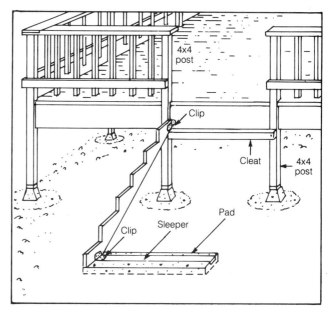

Figure 17-11. Install outside stringers first, attaching them to both the deck and the sleeper, and then install intermediate stringers.

Step Six
Installing the Stringers

Margin of Error: ⅛"

Using the first stringer as a template, cut the other two stringers. Then remove the cantilevered boards you used to locate the support pad.

The top of the stringer is usually attached to the deck with joist hangers or framing clips, and the bottom is clipped or toenailed into the sleeper in the support pad. Always install the outside stringers first, as shown in Figure 17-11, checking for level tread cuts between them, then place intermediate stringers in line with the outside ones.

Particularly where heavy snow loads are expected, long stringers need additional posts to support snow load on the stairs. If you do use additional posts, it's a good idea to use the same post to support both the stringer and the handrail, if possible. These posts are bolted into position on the stringer, leaving 6" or more of free space at the bottom to accommodate a pier block or a cast pier support. For information on installing pier blocks, see Chapter 16, Decks.

Some communities allow you to use a cleat to stiffen the stringer. Typically, a structurally rated 2x4 is bolted or screwed to the side of the stringer below the rise and run notches. The cleats are applied on the inside of the stringer, so they are not visible. Cleats look cleaner than posts, but if the staircase is long or snow is expected, local codes will probably require posts. If you do use a cleat, remember that it will be exposed to the weather. Use a decay-resistant material, caulked to reduce pest and water access to the joint between the cleat and the stringer.

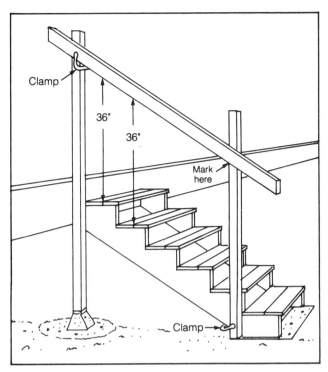

Figure 17-12. Use clamps to hold the handrail and its 4x4 support before bolting them in place.

Step Seven
Installing the Treads

Margin of Error: 1/4"

Installing treads is the easy part of building stairs. Starting at the bottom, hold the first 2x6 flush with the riser cut and nail or screw it into the stringers, using two galvanized #16 nails or treated screws per joint. Then nail on the second 2x6, again using two nails per joint. Remember the 1" nosing, and place the front nails back accordingly. Repeat this process step by step until all the treads are secured.

Step Eight
Installing Handrail Supports

Margin of Error: 1/4"

Handrail supports can be installed either before or after the treads. Clamp a redwood or pressure-treated 4x4 post to the stringers to hold it in place, as shown in Figure 17-12. Then attach it with galvanized carriage bolts, washers, and nuts.

Whatever style you choose for your handrail, remember that the top cuts must be angled to meet the railing. The bottoms are also often cut at the same angle for a neater appearance. Bolt the posts to the outside of the stringer so that you can use the full width of the stairs.

Too many 4x4 posts on your staircase look massive and can detract from the visual appeal of the lighter spindles. For this reason, our example uses one 4x4 on each outside

Figure 17-13. Plumb each spindle and mark its final location before attaching it to the stringer.

stringer, with the deck rail 4x4s supporting the tops of the handrail. To provide a place to attach the spindles and support a flat 2x4 rail on the top, another 2x4 is notched into the 4x4s vertically. Notching is not required, however, and you may want to simply cleat this support piece between the 4x4s so that the inside edge of the support is positioned plumb above the side of the stringer. In this way, flat spindles can be secured at the top and bottom without the careful notching that is common to interior stairs.

Because most deck railings are 42" high and the code requires 36" rail height measured from the front of each step, the stair rail almost always hits the deck rail 4x4 below the top of this support. When a deck rail is only 36" high, or if the top step protrudes at the deck level, the handrail will be above the top of the 4x4 in order to meet the 36" height stipulation for the stairs. In this case, you must add more 4x4 posts to support the tops of the stair rails. Temporarily clamp the rail support 2x4 in position and with a level make a mark on the support piece directly across from the bottom of the deck top rail. Mark back 1 3/4" (half the width of the 4x4), and with your level draw a plumb line here. The 4x4 will be aligned on one side to this line, so the post will support both the stair rail and a new short top rail and support piece between the stair rail and the deck rail.

Step Nine
Installing the Handrails and Spindles

Margin of Error: 1/4"

Handrail height requirements vary from community to community, and the "graspable" rule means that in some areas you can't use a flat 2x6 for a handrail.

Spindle requirements for stairs also vary, as they do for the deck. The maximum allowable space between them is 6" in some areas and 8" in others. Like the stairs themselves, these pieces need to be sturdy and secure for safety reasons.

Use a level, as shown in Figure 17-13, to be sure that your spindles are plumb before nailing or bolting them to the stringer. Then drive finish nails down through the handrail into the beveled top of the spindle.

STAIRS WITH LANDINGS

Code limits the total height of stair rise without a landing to 12'; any higher and a landing is required.

While it may at first appear to be twice as complicated to build stairs with a landing as it is to build straight stairs, it isn't really. The landing should be considered as just another

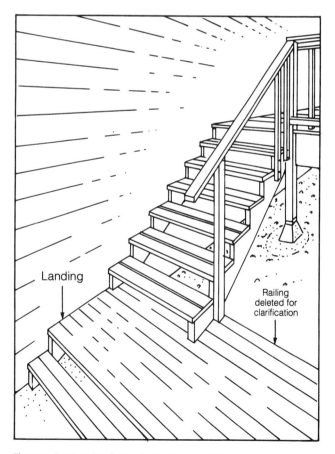

Landing

Railing deleted for clarification

Figure 17-14. A landing, which allows a stairway to make a right-angle turn, can be considered to be simply a large stair.

step, and it is calculated, designed, and framed into position as a large intermediate step, as shown in Figure 17-14. On interior stairs, your calculations should include the subfloor and finish flooring thicknesses. Stringer dimensions are then calculated just as with a straight stair.

RAMPS

Ramps are often a good idea even if you don't have an elderly or handicapped member of your household. They are easier than stairs for children to use, and create a more formal impression when used as access to a deck. They also make it easier to get sacks of soil and other heavy supplies up to a deck container garden.

All ramps must conform to standard wheelchair ramp requirements. In most communities those minimum standards are straightforward. The slope of the ramp must not exceed 1 in 12. Lower pitches are fine, but the overall ramp length must not exceed 30'. These two stipulations limit total rise, on flat terrain, to 2½' ($1/12 \times 30' = 2½'$).

The ramp width must be at least 36" clear, with the handrail height 32" above the ramp surface. Design of the ramp also requires careful consideration of such factors as landings, turns, curbs, and the possibility of icing. Like the stairs, ramps must be sturdy and braced as necessary for durability.

INTERIOR STAIRS

The basic design of interior stairs differs only slightly from the design of exterior stairs. However, because interior stairs must meet numerous code requirements and fit the available space, interior stair calculations are usually done by an architect or designer, during the planning phase.

Besides getting you from one floor to another, interior stairs must control dust, air flow, and noise, and wood-framed stairs must be protected from fire. For these reasons, risers are almost always used to fill in the vertical spaces on interior stairs.

Risers are usually installed before the treads on interior stairs, and the joints are often caulked to control dust. Riser boards are typically 1" thick, and their thickness is added to the tread cut dimensions when the stringer is marked and cut. It's often a good idea to carpet stairs to control noise, but the carpet's thickness must be considered to keep the riser heights within the 3/8" required by code once the stringer is cut.

Interior treads are made of hardwood, particle board, or plywood. These materials vary in thickness, and their thickness determines how much the stringer is dropped.

The riser boards should be secured with screws to the tread from behind, every 6" or so, to minimize the chance of squeaks developing.

The areas below the stairs are drywalled for fire protection, and little triangular "closets" often result. Full-width fireblocks are installed in these spaces from the tread-riser intersection down to the bottom of the stringer, every 4' or so, to protect the stairs and provide nailing for the drywall. In multistory buildings the stairs are often stacked in the same well to avoid these odd spaces.

Building code requirements limit the riser height to 8" inside, and often allow as little rise as 5". A minimum tread width of 11" is required for stairs that would be considered an escape route in case of fire.

Different Floor Finish Thicknesses

The final thickness of your finish floor affects the stringer design. If all your floor surfaces are the same, dropping the stringer is the same as for the deck stairs. If the surfaces are of different heights, the simplest way to accommodate these differences is to use scraps to build each level to its finish height. The total riser height is then determined off these surfaces and the stringer is designed as described earlier. Calculating these differences with your tape is not as reliable. Just be sure to include the finished surface height when you mark the top riser drop on the framing, and test fit the stringer carefully.

Headroom

It's important to provide adequate headroom for interior stairs. The code stipulates a minimum clearance of 6'8", but on a long staircase this minimum can be claustrophobic. A full 7' clearance, or even more, is advisable for most stairs. This distance is always measured from the front of a tread vertically to the ceiling.

Other Differences Between Interior and Exterior Stairs

Interior stairs are often easier to construct than exterior stairs, because they are often placed between two interior walls. In this case, all the studs in these walls can be used to support the stringers vertically, by nailing or screwing through the stringer into the studs.

However, the drywall next to these studs is vulnerable to damage. A kick could knock a hole right through the drywall. Even if you plan to install a baseboard, code generally requires blocking behind the drywall, above the tread, to prevent such damage. Blocking is also used to support

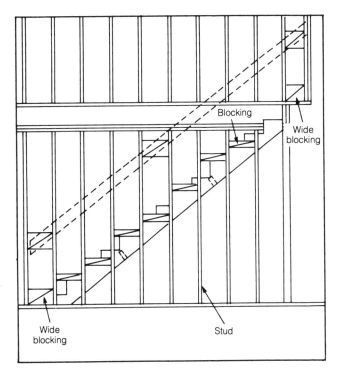

Figure 17-15. Blocking for an interior staircase prevents damage to drywall.

handrails and to provide a place to secure a curved handrail return. The blocking adjacent to the tread is also useful to secure any baseboards or trim around the treads, as shown in Figure 7-15.

Stairs will always flex to some degree, and this flexing often results in squeaks. Outside, these noises are hardly noticeable, but inside they can be annoying. To prevent squeaks, make close, careful cuts, and use both screws and adhesive to secure the treads and risers. Bugle-head drywall screws and flexible panel adhesive, available in caulking gun cartridges, are recommended.

Handrails and spindles are generally much fancier on interior stairs. Hardwood rails with 90° elbows, shaped to match the straight sections, are used to return the rail neatly to the wall or to turn a corner. Spindles may be plain or fancy, but they must be spaced as recommended by local code. If you choose to use a diagonal board above the steps themselves as a base for the spindles, the 6" to 8" maximum gap rule also applies to the little triangle below the board. Just like the stairs, the rails need to be stout and well built to pass the final inspection, as well as for your safety.

Index